MECHANICS OF CURVED COMPOSITES

SOLID MECHANICS AND ITS APPLICATIONS
Volume 78

Aims and Scope of the Series

The fundamental questions arising in mechanics are: *Why?*, *How?*, and *How much?* The aim of this series is to provide lucid accounts written bij authoritative researchers giving vision and insight in answering these questions on the subject of mechanics as it relates to solids.

The scope of the series covers the entire spectrum of solid mechanics. Thus it includes the foundation of mechanics; variational formulations; computational mechanics; statics, kinematics and dynamics of rigid and elastic bodies: vibrations of solids and structures; dynamical systems and chaos; the theories of elasticity, plasticity and viscoelasticity; composite materials; rods, beams, shells and membranes; structural control and stability; soils, rocks and geomechanics; fracture; tribology; experimental mechanics; biomechanics and machine design.

The median level of presentation is the first year graduate student. Some texts are monographs defining the current state of the field; others are accessible to final year undergraduates; but essentially the emphasis is on readability and clarity.

For a list of related mechanics titles, see final pages.

A C.I.P. Catalogue record for this book is available from the Library of Congress.

ISBN 0-7923-6477-5

Published by Kluwer Academic Publishers,
P.O. Box 17, 3300 AA Dordrecht, The Netherlands.

Sold and distributed in North, Central and South America
by Kluwer Academic Publishers,
101 Philip Drive, Norwell, MA 02061, U.S.A.

In all other countries, sold and distributed
by Kluwer Academic Publishers,
P.O. Box 322, 3300 AH Dordrecht, The Netherlands.

Printed on acid-free paper

Printed in the Netherlands.

Mechanics of Curved Composites

by

S.D. Akbarov
Yildiz Technical University,
Istanbul, Turkey and
Institute of Mathematics and Mechanics of
Academy of Science of Azerbaijan,
Baku, Azerbaijan

and

A.N. Guz
Institute of Mechanics of National Academy of
Science of Ukraine,
Kiev, Ukraine

KLUWER ACADEMIC PUBLISHERS
DORDRECHT / BOSTON / LONDON

TABLE OF CONTENTS

PREFACE

This book is the first to focus on mechanical aspects of fibrous and layered composite material with curved structure. By *mechanical aspects* we mean statics, vibration, stability loss, elastic and fracture problems. By *curved structures* we mean that the reinforcing layers or fibres are not straight: they have some initial curvature, bending or distortion. This curvature may occur as a result of design, or as a consequence of some technological process.

During the last two decades, we and our students have investigated problems relating to curved composites intensively. These investigations have allowed us to study stresses and strains in regions of a composite which are small compared to the curvature wavelength. These new, accurate, techniques were developed in the framework of continuum theories for piecewise homogeneous bodies. We use the exact equations of elasticity or viscoelasticity for anisotropic bodies, and consider linear and non-linear problems in the framework of this continuum theory as well as in the framework of the piecewise homogeneous model. For the latter the method of solution of related problems is proposed. We have focussed our attention on self-balanced stresses which arise from the curvature, but have provided sufficient information for the study of other effects.

We assume that the reader is familiar with the theory of elasticity for anisotropic bodies, with partial differential equations and integral transformations, and also with the Finite Element Method.

We have designed the book for graduate researchers, for mechanical engineers designing composite materials for automobiles, trucks, flywheels etc; for civil engineers contemplating the use of composites in infrastructure; for aerospace engineers studying advanced airframe design; and for biomedical engineers developing lightweight composites for bone replacement and repair.

We provide bibliographical notes at the end of each chapter. For the papers which originally appeared in Russian, we have tried to cite only ones which have appeared in English translation; there are, however, some which have not been translated.

ACKNOWLEDGMENTS

We must mention several individuals and organisations that were of enormous help in writing this book. First we wish to thank the editor, Prof. Graham M. L. Gladwell for his invaluable help in editing.

We also wish to thank the scientific collaborators at the Institute of Mathematics and Mechanics of Academy of Science of Azerbaijan (Baku, Azerbaijan), the Institute of Mechanics of National Academy of Science of Ukraine (Kiev, Ukraine), and the Yildiz Technical University (Istanbul, Turkey), for their assistance in the investigations described in this book.

We also thank our colleagues Ercument Akat, Nazmiye Yahnioglu, Zafer Kutug and Arzu Cilli for help in producing the text and figures.

S.D. Akbarov and A.N. Guz

ABOUT THE CONTENTS

This book consists of an introduction, ten chapters and two supplements. In the introduction some necessary information on the classification of composite materials and the classification of the curving in their structure are given together with the mechanical properties of some promising fiber and matrix materials. Various approaches in mechanics of composite materials with curved structures are analysed, and the scope of this book is determined.

In Chapter 1 a simple new version of continuum theory for composite materials with periodic plane-curved structure is presented. It is supposed that the curves are small-scale, in other words, the period of the curvature of the reinforcing elements in the structure of the composite materials is significantly less than the size of the elements of objects fabricated from these composite materials. First, some preliminary remarks on classical continuum approaches are given. Then the three-dimensional exact equations of motion, geometrical relations and boundary conditions in both geometrically linear and non-linear statements, for deformable solid body mechanics are presented. After these preparatory procedures the basic assumptions and relations of the simple new version of continuum theory for composite materials with plane-curved structures are detailed. A solution procedure for the continuum problems is given, then some dynamic and static problems are investigated. In these cases the influence of the periodic plane curving in the structure of the constituent materials on the mechanical behaviour of the composite is studied.

In Chapter 2 a continuum theory for composite materials with general periodic or locally curved structures is discussed, generalising the analysis given in Chapter 1.

In Chapter 3 the theory developed in Chapter 2 is combined with the finite element method (FEM) to solve some two- and three-dimensional static and dynamic problems for strips and rectangular plates. The influence of the parameters defining the curved composite on the stress distribution, and on the natural frequencies of the plates is discussed.

It is evident that any continuum approach is an approximate one and the most accurate information on the stress distribution in the curved composites can be obtained only in the framework of the piecewise-homogeneous body model with the use of the exact equations of deformable solid body mechanics. Therefore in chapters 4-10 we study the problems of curved composites in the framework of the piecewise homogeneous body model. In Chapter 4 we consider plane strain problems of periodically plane-curved composites. First, the method is developed for cases in which the materials of the layers are:1) visco- elastic; 2) rectilinearly anisotropic; 3) curvilinearly anisotropic. Many numerical results and their analyses are given for stress distribution in such composites.

In Chapter 5 the problem formulation and solution method presented in Chapter 4 are developed for three-dimensional problems, namely, for composites with spatially curved layers. The solution method is detailed for periodic curving. Many numerical results are given for the stress distribution in such composites.

In Chapter 6 the stress state in composite material with locally plane-curved layers is investigated. All investigations are carried out for plane-strain state with the use of the relations and assumptions given in Chapter 4. For various local curving form the distribution of the self-balanced stresses on the inter-layer surfaces are studied in detail.

In Chapter 7 the problem formulation and solution methods presented in Chapters 4-6 are developed for fibrous composites with curved structures, and various concrete problems are investigated.

In some combinations of geometric and curvature parameters of the filler layers (or fibers) and of the values of the external force intensities it is necessary to investigate the problems using the geometrically non-linear statement considered in Chapter 8. Using the results of these investigations, we can determine the limit of the intensity of the external forces for which the results obtained in the linear statement are acceptable. Furthermore, we can determine the character of the influence of the geometrical non-linearity on the mechanical behaviour of composites. In this chapter we derive which are very important for investigations of the fracture of unidirectional composites with curved structure in compression.

In Chapter 9 the approach for determination of the normalized mechanical properties of composites with periodically curved layers is developed. For this purpose the results obtained in chapters 4-5 are used. Some numerical results related to the normalized moduli of elasticity are also discussed.

In Chapter 10 fracture problems of curved composites are investigated. First, the *fiber separation* effect in fracture mechanics of composite materials is explained. The crack problems typical to the composites with curved structures is also investigated. Moreover, the approach is suggested for investigation of the fracture of these composites under compression along the reinforcing elements.

In Supplement 1 we propose an approach for investigation of the fracture of the viscoelastic unidirectional composites in compression. Composites with initially insignificantly curved layers are taken as a model for the study of these fracture problems, and the method discussed in the latter section of Chapter 10 is developed for viscoelastic composites.

In Supplement 2 we use the continuum theory developed in Chapter 2 and with employing FEM to investigate some two-dimensional geometrically non-linear and stability problems for strips. We discuss the influence of the geometrical non-linearity on the stress distribution and the influence of the curving parameters of strip material on its stability.

INTRODUCTION

I.1. Types of Composite Materials

According to [70, 73, 74] the classification of composite materials is based on the following principles: *structural-design*, i.e., by the type of the reinforcing elements and their disposition in the matrix; *material*, i.e., by the type of the fiber and matrix material and their properties; *technology*, i.e., by the production process. Here we consider the first classification only.

From the structural-design principle, composite materials are divided into *layered, fibrous* and *particular* ones. By *layered* composite material we understand that the geometrical size of the reinforcing elements in one direction is significantly (some orders) less than the geometrical sizes in the other two directions. By *fibrous* composite material we mean that material for which the geometrical size in two mutually perpendicular directions of the reinforcing elements has the same order and is significantly (some orders) less than the geometrical size in the third mutually perpendicular direction. Finally, by a *particular* composite we mean that material for which the geometrical sizes in the three mutually perpendicular directions of the reinforcing elements have the same order.

Fibrous composites are widely used; fibrous reinforcement is so effective that many materials are much stronger and stiffer in fiber form than they are in bulk form. Moreover fibers allow us to obtain the maximum tensile strength and stiffness of a material. Nevertheless there are obvious disadvantages in using a material in a fiber form, for example, fibers alone cannot support longitudinal compressive loads, and their transverse mechanical properties are generally poorer than the corresponding longitudinal properties. These disadvantages must be taken into account in the production of fibrous composite materials.

Some of the very important fibrous composites are the plastics reinforced by glass, graphite or carbon, boron and other fibers; various polymers are used as matrix.

Plastics are divided into the following three groups with respect to the orientation of the continuous fibers: *unidirected* or *fibrous*; *layered*; *spatially woven*. When the reinforcing fibers are located parallel to each other in a matrix, the material has a *unidirected* structure; when the reinforcing fibers are oriented in various directions, the plastic has a *layered* structure; when the various layers are bound with each other through the transverse connections, the plastic has a *spatially woven* structure.

Although continuous fiber laminated composites are used extensively, the potential for delamination, or separation of the laminae, is still a major problem because the inter-laminar strength is matrix-dominated. Spatially fiber woven composites do not have such disadvantages, but strength and stiffness decrease due to the fiber *curvature*.

Tables I.1 and I.2 show the mechanical properties of some fibers and matrix materials selected from references [73, 74, 142].

These tables show that in modern and prospective composite materials the following situation takes place: the ratio of the modulus of elasticity of fiber material to that of the matrix material ranges from 10 to 500; the strength of matrix materials is two or more orders less than that of fiber materials; Poisson's ratios for both matrix and fiber materials range from 0.2 to 0.4. Note that many composites which are used in various modern techniques have a layered structure; there are many layers located periodically across the thickness. A layered composite is a very suitable model for theoretical investigations since the results obtained for this model can be used to approximate fibrous composites.

Table I.1. The mechanical properties of some selected fibers.

Material	Tensile Strength, Gpa	Tensile Modulus, Gpa	Density g/cm^3	Poisson's Ratio
Glass fibers				
E - glass fibers	3.50	73.5	2.54	0.2
S - glass fibers	4.65	86.5	2.49	---
D - glass fibers	2.45	52.5	2.16	---
C - glass fibers	2.80	70.0	2.49	---
L - glass fibers	1.68	51.0	4.30	---
E - glass fibers	3.50	111.0	2.89	---
Carbon fibers (pitch precursor)				
P-55 (Amoco)	1.72	379.0	1.99	---
P-75 (Amoco)	2.07	517.0	1.99	---
P-100 (Amoco)	2.24	690.0	2.16	---
Boron fibers				
0.004" diameter (Texton)	3.516	400.0	2.57	0.21
0.0056" diameter (Texton)	3.516	400.0	2.49	---
Steel wire	3.43	196.2	7.87	0.3

Table I.2. The mechanical properties of some selected matrix-polymer materials.

Material	Tensile Strength, Mpa	Tensile Modulus, Gpa	Density g/cm^3	Poison's Ratio
Polyester	40 - 90	1.2 - 4.5	1.1 - 1.5	0.35 - 0.42
Vinyl ester	65 - 90	3 - 4	1.15	---
Epoxy	35 - 130	2 - 6	1.1 - 1.4	0.34 - 0.40
Bismaleimide	48 - 78	3.6	1.32	---
Polimide	70 - 120	3.1 - 4.9	1.43 - 1.89	---

I.2. Specific Curving of Reinforcing Elements

In the mechanics of composite materials, problems associated with their structural features occupy a central position. One of the basic structural features of composite materials is curvature of the reinforcing elements. These curvatures may be due to the design features (see [72, 82-85, 118, 119, 138, 142, 143, 145, 150, 153], etc.)

or to technological processes resulting from the action of various factors (see [75, 142, 108, 110, 142],etc.). As an example, Figs.I.1.2.1-I.1.2.3 show typical curvatures of the fibers in the structure of plastics.

The successful practical use of artificially created composite materials is associated, to a considerable extent, with the determination of the stress-strain state in these materials, taking account of their basic features, particularly the distortion of the reinforcing elements. This distortion significantly influences the strength and strain properties of these materials. Besides these, the initial insignificant curving of the reinforcing layers or fibers is taken as a model for the investigation of the various fracture or stability loss problems of unidirected composite materials. Consequently, *establishing mechanics of composite materials with curved structures is urgent both from the viewpoint of fundamental developments in the mechanics of a solid deformable body and from the viewpoint of applications to specific composite-material components used in modern engineering.*

Fig.I.2.1. The view (×28) of the typical curvature of the glass fibers in the structure of glass-fiber reinforced plastic.

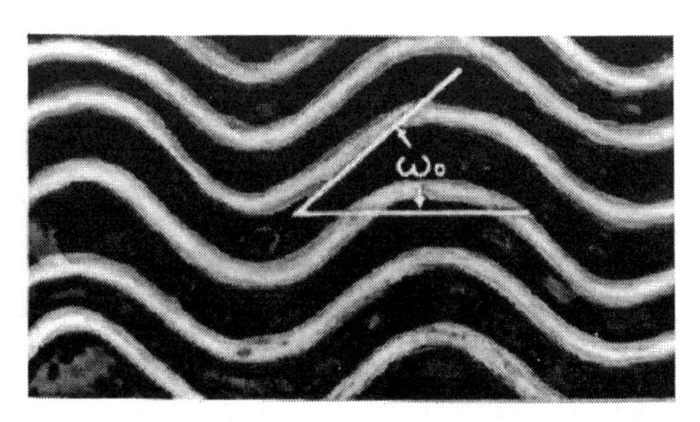

Fig.I.2.2. The typical periodical - sinusoidal curving of fibers in woven glass-fiber reinforced plastics [143].

In almost all investigations related to the mechanics of composites with curved structures, the curving in the structure of laminar or fibrous composite materials is assumed to be *small-scale*, which is usually taken to mean that the dimensions of the curvatures are considerably less than those of the given structural element of composite material. However, there are some investigations (for example, [104]) where an approach to the investigations of mechanical problems of composite materials with *large-scale* structural curvatures has been proposed. Curvatures whose dimensions are of the same (or greater) order of magnitude as the minimal size of the structural element are regarded as large-scale ones. In the present book, *curvature* will mean *small-scale* curvature, unless we state otherwise.

Fig.I.2.3. The local curving of the layers in a multi-layered cylindrical pipe.

Moreover we will distinguish *periodic* (as shown in Fig.I.2.2) and *local* (as shown Fig.I.2.3) curvature of the reinforcing elements; all curving form of reinforcing elements in the cross sections of various composites can be related to these two cases. Further, we distinguish between *co-phase* curvature, shown in Fig.I.2.2, and *anti-phase* curvature shown in Fig.I.2.1. Of course, not all reinforcing elements are curved and bent; some may be considered *ideal,* not curved.

I.3. Background and Brief Review.

There are two basic approaches to the study of the mechanics of composite materials with curved structure.

Continuum approaches may be used to calculate the components of the stress-strain state for areas considerably greater in size than the curving; the influence of distortion in the structure is taken into account by means of quantitative variation in the normalized mechanical characteristics.

The *local* approaches, developed considerably later than the first, enable the influence of reinforcing-element curving to be taken into account in calculating the components of the stress-strain state in areas comparable with, or smaller than the curving. These approaches were developed both in the framework of *continuum theories* and in the framework of a *piecewise homogeneous body model*. These methods are essential for estimating the influence of curving on the *local* distribution of the stress-strain state, which may determine local failure. For regular (periodical) curving, this local failure can transform to global failure of the whole material. Local changes must also be taken into account in investigating high-frequency loading material when the wavelength is comparable with (or less than) the dimensions of the curving. Analogous problems may be encountered in investigating local stability loss of short-wave form.

I.3.1. CONTINUUM APPROACHES

The first attempt was the investigation [62], in which a theory of reinforced media described in [64] was developed for the solution of various problems in which the reinforcing layers in the laminar medium are slightly curved. The basic equations were non-linear. Taking into account that the initial and additional flexure of the reinforcing layers was small, a small parameter was introduced, and the unknowns were expanded as series in terms of this small parameter; the solution to the non-linear boundary problem was reduced to the solution of a recurrent sequence of linear problems. When the functions describing the initial distortion of the reinforcing layers form a random field, expressions characterizing the influence of the reinforcing-layer distortion on the normalized mechanical characteristics were obtained in the first approximation. These expressions were used to find, the influence of reinforcing-elements curving on the modulus of elasticity; it was shown that, in modern laminar composite materials, a decrease of 20% or more may be found in the modulus on account of reinforcing-element curving. This conclusion is confirmed by the experimental data in [127, 142]. The theory [62] was extended in [63] to the case where the binder and reinforcing material are of linearly viscoelastic. Many of results in [62, 63, 127] and others were generalized and developed in [64, 142].

In [56], the reduced-cross section method was used to find the influence of reinforcing-fiber curving on the modulus of elasticity of a composite material by determining the change in amplitude of the waves of curving forms by the action of radial forces in the fiber arcs. Three ideal cases of fiber distribution in one direction and two cases of fiber distribution in two directions were considered.

In [130], it was proposed an approximate method for determining the reduced elastic constants of glass fiber-reinforced plastic, taking account of the reinforcing-fiber curving. All the conclusions obtained here were for a plane stress state, and were confirmed by experiments on model samples.

The influence of the twisted-fiber geometry on the reduced modulus of elasticity and strength of a unidirectional fiber composites was studied in [149]. It was found that there is a limit on the number of turns per unit length below which the twisting of the fibers has little influence on the normalized modulus and the strength. However, above this limit, increase in the number of turns per unit length leads to considerable decrease in the normalized elastic modulus and strength. The theoretical results were tested for unidirectional composites reinforced by graphite fibers.

Besides these investigations there are many others such as [82-85, 112, 128, 140, 145, 150, 153] etc., in which similar results are obtained. The approach developed in these investigations offers the possibility of a rigorous determination of the rigidity of the whole structural element, and the components of the stress-strain state averaged over areas of dimensions significantly greater than those of the curving.

I.3.2. LOCAL APPROACHES

The investigations into local approaches were reviewed in the papers [39-41,44], from which it follows that almost all related investigations had been made by the authors and their students. There are some others by other authors, such as [110, 115], but we do not consider them here.

CHAPTER 1

PLANE-CURVED COMPOSITES

In this chapter we present a simple new version of continuum theory for composite materials with periodic plane-curved structure. We suppose that the curves are small-scale, in other words, the period of the curvature of the reinforcing elements in the structure of the composite materials is significantly less than the size of the elements of objects fabricated from these composite materials. First we give some preliminary remarks on classical continuum approaches. Then we present the three-dimensional exact equations of motion, geometrical relations and boundary conditions in both geometrically linear and non-linear statements, for deformable solid body mechanics. After these preparatory procedures we detail the basic assumptions and relations of the simple new version of continuum theory for composite materials with periodic plane-curved structures. We give a solution procedure for the continuum problems, and then investigate some dynamic and static problems where we study the influence of the periodic plane curving in the structure of the constituent materials on the mechanical behaviour of the composite.

Throughout the investigations we use tensor notation, and sum repeated indices over their ranges; however, underlined repeated indices are not summed.

1.1. Classical Theories

In the classical approaches [56, 62-64, 127, 128, 130, 140, 142] and others, the piecewise homogeneous composite is replaced by a homogeneous anisotropic material with elastic constants which depend on the elastic constants of the matrix and of the layers, and on the curvature of the layers. The layers are assumed to be equally spaced, to have unidirectional fibres, and to have curvature which is unidirectional and periodic. Applications of this classical theory may be found in [64, 127, 130, 142].

We now outline some basic elements of this theory: In any composite there are two components: the reinforcing layers, and the matrix. The reinforcing layers are assumed to have the following properties:

i) their material is homogeneous, anisotropic (usually orthotropic), linearly elastic or viscoelastic

ii) if the layers are curved, then their strain-displacement relations are obtained as the non-linear relations corresponding to a plate with initial imperfections i.e. the curvature is considered as slight imperfection

iii) the strain components are derived by using Kirchhoff's hypothesis, that normal to the undeformed surface is deformed into normal to the deformed surface

iv) the reinforcing layers are significantly stiffer than the matrix layers.

The matrix is treated as follows:

i) it is homogeneous and isotropic

ii) the displacements of each matrix layer are defined in terms of the displacements of the neighbouring reinforcing layers

iii) its strain-displacement relations are linear.

These classical theories are inadequate, as we now show. Fig.1.1.1a shows an ideal composite with straight (uncurved) reinforcing layers; Fig.1.1.1b shows the replacement homogeneous orthotropic medium. If either medium is loaded at infinity by a uniformly distributed stress $\sigma_{11} = p$ then, for plane strain, the equilibrium equations yield $\sigma_{11} = p$, $\sigma_{12} = \sigma_{22} = 0$ everywhere. This adequately reflects the experimentally observed behaviour of the composite.

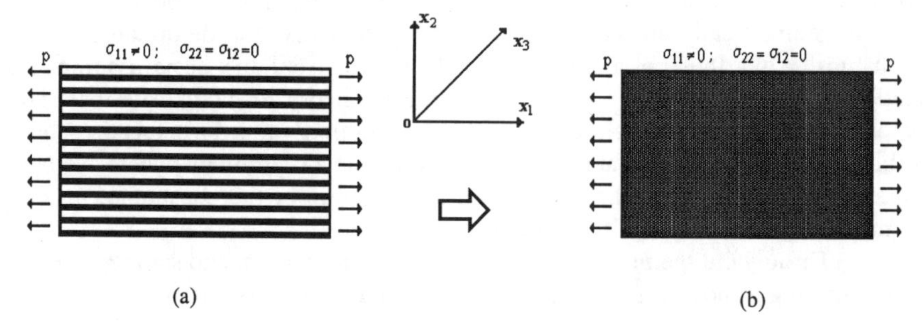

Fig.1.1.1. Schematic presentation of the composite material with ideal (uncurved) structure: (a) piecewise-homogeneous body model; (b) continuum model

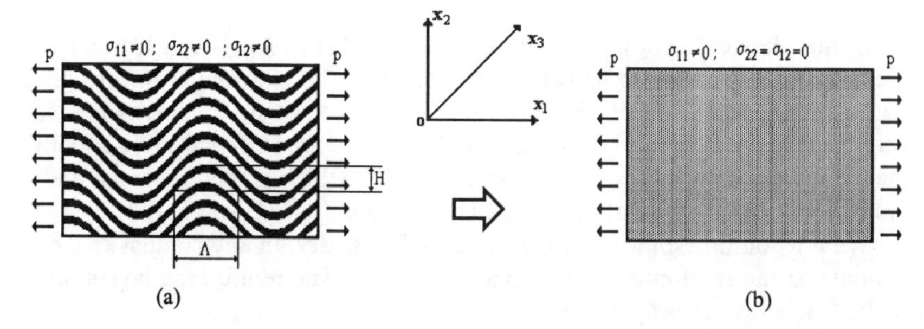

Fig.1.1.2. Schematic presentation of the composite material with plane periodic curved structure: (a) piecewise-homogeneous body model; (b) continuum model

In classical theories, the composite with curved reinforcing layers, shown in Fig.1.1.2a is also replaced by an homogeneous orthotropic medium, shown in Fig.1.1.2b. If E_1 is the Young's modulus in the direction Ox_1, and G_{12} is the shear modulus in the plane Ox_1x_2 for the orthotropic medium in Fig.1.1.1b, then, for example, the Young's modulus in the direction Ox_1 for the replacement medium in Fig.1.1.2b is given by

$$\tilde{E}_1 = \frac{E_1}{1+\frac{E_1}{G_{12}}\varepsilon^2} . \qquad (1.1.1)$$

The parameter ε is given by the formula $\varepsilon = H/\Lambda$, where H is the arrow of rise and Λ is the length of the half period of the curving shown in Fig. 1.1.2a. The, important points is that the composite with curved layers is still replaced by a homogeneous orthotropic medium. But when the latter is loaded at infinity by $\sigma_{11} = p$ then, for plane strain, as we noted before, the stresses in the medium will be $\sigma_{11} = p$, $\sigma_{12} = \sigma_{22} = 0$ everywhere, and this contradicts the known experimental behaviour of a composite with curved layers: the stress σ_{11} is not constant, nor are the other stresses zero; they vary along the layers.

This is why new theories are needed, theories which explain why and how the stresses and displacements in a composite with curved layers vary with the amplitude and wave-length of the layer curvature, if the curving is periodic and unidirectional, or with the other parameters for more complicated, local or multidirectional curving.

1.2. Basic Equations and Boundary Conditions

In this section we present the exact equations of motion, geometrical relations and boundary conditions of deformable solid body mechanics which will be used throughout this book. To derive these equations, we will use Lagrangian coordinates x_j which in their natural state coincide with Cartesian coordinates. We use $S=S_1 \cup S_2$ to denote the surface of the region D occupied by the body; S_1 is the part of S on which the external forces act; S_2 is that part of S on which the displacements are given. We will investigate only the small deformation state and therefore will use the ordinary symmetric stress tensor $\boldsymbol{\sigma}$ with components σ_{ij} and symmetric Green strain tensor $\boldsymbol{\varepsilon}$ with components ε_{ij}.

We write the equation of motion, the geometric relations and the boundary conditions for both geometrically linear and non-linear statements separately.

Geometrically linear statement

The equations of motion:

$$\frac{\partial}{\partial x_j}\sigma_{ij} - \rho\frac{\partial^2}{\partial t^2}u_i = 0, \qquad x_1,x_2,x_3 \in D, \qquad i;j=1,2,3; \qquad (1.2.1)$$

the geometrical relations :

$$\varepsilon_{ij}=\frac{1}{2}\left(\frac{\partial u_j}{\partial x_i}+\frac{\partial u_i}{\partial x_j}\right),\qquad i;j=1,2,3; \tag{1.2.2}$$

and the boundary conditions in the surface portion S_1:

$$\sigma_{ij}N_j=P_i,\qquad x_1,x_2,x_3\in S_1,\quad i;j=1,2,3\,. \tag{1.2.3}$$

Geometrically non-linear statement

The equations of motion:

$$\frac{\partial}{\partial x_i}\left[\sigma_{in}\left(\delta_n^j+\frac{\partial u_j}{\partial x_n}\right)\right]-\rho\frac{\partial^2}{\partial\tau^2}u_j=0\,,\quad x_1,x_2,x_3\in D,\ i;j;n=1,2,3; \tag{1.2.4}$$

the geometrical relations:

$$\varepsilon_{ij}=\frac{1}{2}\left(\frac{\partial u_j}{\partial x_i}+\frac{\partial u_i}{\partial x_j}+\frac{\partial u_n}{\partial x_i}\frac{\partial u_n}{\partial x_j}\right),\quad i;j;n=1,2,3\,, \tag{1.2.5}$$

and the boundary conditions in the surface portion S_1:

$$\left[\sigma_{in}\left(\delta_n^j+\frac{\partial u_j}{\partial x_n}\right)\right]N_i=P_j,\qquad x_1,x_2,x_3\in S_1,\qquad i;j;n=1,2,3. \tag{1.2.6}$$

The boundary conditions for the displacements on the surface portion S_2 for both statements are

$$u_j=f_j,\qquad x_1,x_2,x_3\in S_2,\ j=1,2,3. \tag{1.2.7}$$

For unsteady-state dynamical problems, the appropriate initial conditions must be added to these equations. We used the following notation in (1.2.1)-(1.2.7): u_i is a component of the displacement vector, ρ is the material density, N_i a component of the normal unit vector to the surface S, P_i a component of the given force vector, f_i a component of the given displacement vector and τ is the time.

1.3. Constitutive Relations

1.3.1. GEOMETRICAL NOTATION

We suppose that the curvature (i.e. the flexure) of the reinforcing elements (layers) occurs only in the direction of the Ox_1 axis, and the composite consists of many packets, each of which contains N layers. The structure of this composite material at $x_1 = \mathrm{const}$; $x_3 = \mathrm{const}$ is shown in Fig.1.3.1. The structure of the corresponding composite material with uncurved layers at $x_1 = \mathrm{const}$; $x_3 = \mathrm{const}$ is shown in Fig.1.3.2. We denote the thickness of this packet by ΔH, i.e.

$$\Delta H = h_1 + h_2 + \ldots + h_N \tag{1.3.1}$$

In (1.3.1) $h_i (1 \le i \le N)$ is the thickness of the i-th layer in the packet (Fig.1.3.1) which is

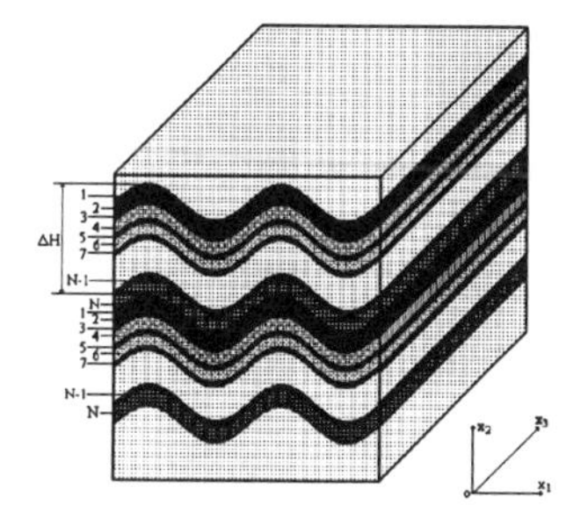

Fig.1.3.1. A part of the composite material with plane periodic curved structure.

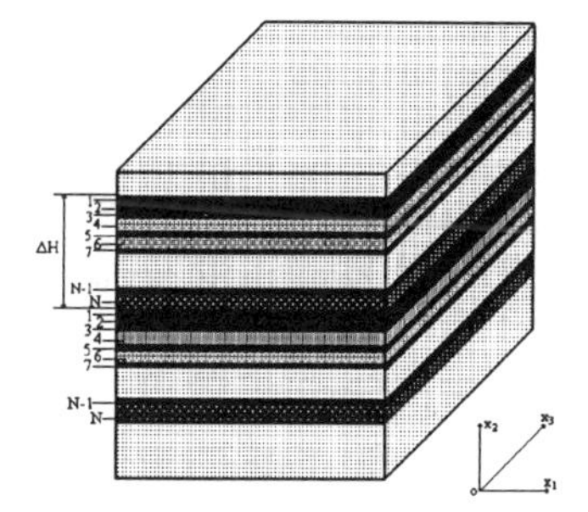

Fig.1.3.2. A part of the composite material with ideal structure.

a representative part of the composite material. We assume that $h_i (1 \le i \le N)$ are constants.

1.3.2. IDEAL COMPOSITES

We denote the normalized mechanical constants of the composite material with ideal structure (Fig.1.3.2) by $\mu^{(0)}_{ij\alpha\beta}$. These enter the following mechanical (constitutive) relations:

$$\langle \sigma_{ij} \rangle = \mu^{(0)}_{ij\alpha\beta} \langle \varepsilon_{\alpha\beta} \rangle, \quad i; j; \alpha; \beta = 1,2,3, \tag{1.3.2}$$

where

$$\left\langle \begin{matrix} \sigma_{ij} \\ \varepsilon_{ij} \end{matrix} \right\rangle = \frac{1}{V} \int_V \begin{matrix} \sigma_{ij} \\ \varepsilon_{ij} \end{matrix} dV, \tag{1.3.3}$$

In general, in (1.3.3) V is the volume of the representative element of the composite material. In Fig.1.3.2, the structure of the representative packet does not depend on x_1

or x_3. Therefore, in (1.3.2) the integrating operation $\frac{1}{V}\int\limits_V (.)dV$ can be replaced by the operation $\frac{1}{\Delta H}\int\limits_{\alpha_1}^{\alpha_2}(.)dx_2$, where $\alpha_2 - \alpha_1 = \Delta H$; α_1 and α_2 being the distances along the Ox_2 axis of the lower and upper planes of the representative packet from the origin of the coordinate system shown in Fig.1.3.2. Thus we can replace (1.3.2) by

$$\left\langle \begin{matrix} \sigma_{ij} \\ \varepsilon_{ij} \end{matrix} \right\rangle = \frac{1}{\Delta H}\int\limits_{\alpha_1}^{\alpha_2} \begin{matrix} \sigma_{ij} \\ \varepsilon_{ij} \end{matrix} dx_2 \,, \quad \alpha_2 - \alpha_1 = \Delta H \tag{1.3.4}$$

Note that below we will omit the brackets $\langle\ \rangle$ and sometimes will use the notation σ_i, ε_j, A^0_{ij} instead of σ_{ij}, ε_{ij}, $\mu^{(0)}_{ij\alpha\beta}$ respectively. In these cases the relations will be written as follows:

$$\sigma_i = A^0_{ij}\varepsilon_j \tag{1.3.5}$$

where

$$\sigma_i = \sigma_{ii} \ (i=1,2,3), \quad \sigma_4 = \sigma_{23}, \quad \sigma_5 = \sigma_{13}, \quad \sigma_6 = \sigma_{12},$$

$$\varepsilon_i = \varepsilon_{ii} \ (i=1,2,3), \ \varepsilon_4 = \varepsilon_{23}, \ \varepsilon_5 = \varepsilon_{13}, \ \varepsilon_6 = \varepsilon_{12}\,. \tag{1.3.6}$$

Thus, the relations (1.3.2) or (1.3.5) are the continuum theory for the composite material with ideal (uncurved) structure (Fig.1.3.2) and the constants $\mu^{(0)}_{ij\alpha\beta}$ or A^0_{ij} are determined through the mechanical and geometrical parameters of the layers of the representative packet. As the methods for calculating these constants are given in [70,73,74] and elsewhere, we do not describe them here.

Below we will consider only the case where the relations (1.3.5) correspond to an orthotropic body with principal axes x_1, x_2, x_3; then the stress-strain equations are

$$\sigma_1 = A^0_{11}\varepsilon_1 + A^0_{12}\varepsilon_2 + A^0_{13}\varepsilon_3, \quad \sigma_4 = 2A^0_{44}\varepsilon_4, \quad A^0_{44} = G^0_{23},$$

$$\sigma_2 = A^0_{12}\varepsilon_1 + A^0_{22}\varepsilon_2 + A^0_{23}\varepsilon_3, \quad \sigma_5 = 2A^0_{55}\varepsilon_5, \quad A^0_{55} = G^0_{13},$$

$$\sigma_3 = A^0_{13}\varepsilon_1 + A^0_{23}\varepsilon_2 + A^0_{33}\varepsilon_3, \quad \sigma_6 = 2A^0_{66}\varepsilon_6, \quad A^0_{66} = G^0_{12}, \tag{1.3.7}$$

and the corresponding strain-stress equations are

$$\varepsilon_1 = a_{11}^0\sigma_1 + a_{12}^0\sigma_2 + a_{13}^0\sigma_3, \quad 2\varepsilon_4 = a_{44}^0\sigma_4, \quad a_{44}^0 = \left(G_{23}^0\right)^{-1}$$

$$\varepsilon_2 = a_{12}^0\sigma_1 + a_{22}^0\sigma_2 + a_{23}^0\sigma_3, \quad 2\varepsilon_5 = a_{55}^0\sigma_5, \quad a_{55}^0 = \left(G_{13}^0\right)^{-1},$$

$$\varepsilon_3 = a_{13}^0\sigma_1 + a_{23}^0\sigma_2 + a_{33}^0\sigma_3, \quad 2\varepsilon_6 = a_{66}^0\sigma_6, \quad a_{66}^0 = \left(G_{12}^0\right)^{-1}, \tag{1.3.8}$$

1.3.3. CURVED COMPOSITES

We now derive the mechanical (constitutive) relations which are analogous to the relations (1.3.2) or (1.3.5) for composite material with periodic plane curved structure (Fig.1.3.1). For this purpose we select the representative packet of the curved layers, and base all discussions on that. The section at $x_3 = \text{const}$ of this packet is shown in Fig.1.3.3 (a); with the centre line of this section we associate the coordinate system Ox_1x_2 (Fig.1.3.3 (a)).

Basic notation and assumptions.
We introduce the following notation:

$$h' = \max\{h_1, h_2, \ldots, h_N\}; \tag{1.3.9}$$

H is the rise of the structural curving; Λ is the half-wavelength of the curvature; r is the radius of curvature of the center line at the point at a distance H from the Ox_1 ; $L = r - H$, and d is the minimum size of the region D occupied by the composite material. The form of the curvature and the parameters are shown in Fig.1.3.3 (a). Note that the parameters d and D, although not shown in Fig. 1.3.3 (a) can easily be defined.

Using the above notations we can formulate the main assumptions of the proposed continuum theory as follows:

1) the composite material has a regular curvature (flexure), periodic along the Ox_1 axis with period 2Λ ; the structure and curvature do not change along the Ox_3 axis which is perpendicular to the plane of Fig.1.3.3 ;

2) the following inequalities are satisfied

$$h' \ll H; \quad \Delta H \ll \Lambda; \quad H \ll \Lambda; \quad \Lambda \ll L; \quad \Lambda \ll d. \tag{1.3.10}$$

The first of these inequalities implies that the thickness of the layers is significantly smaller than the rise of the curve, while the second and third inequalities imply that the thickness of the representative curved packet and the rise of the curve are significantly smaller than the half-period of the curvature. The fourth inequality implies that the curvature is slight, and the last inequality means that the half-period of the curvature is significantly less than the minimum size of the whole body or construction element, i.e., the curvature is small-scale. It can thus be expected that local changes in the stress-deformation state, generated by the presence of curvature must, because of the periodicity, be self-balanced within one period.

To derive the elasticity relations for a representative packet, we use the coordinate system given in Fig. 1.3.3 (a) and carry out the following operations. First, according to the first four inequalities in (1.3.10) we replace the part which is within one half-period of the representative curved packet with the corresponding part of the circular ring with the geometrical parameters shown in Fig.1.3.3 (b). Then we introduce local Cartesian coordinates $x_j^{(m)}$, related to the part of the circular ring corresponding to the m-th half period of the selected representative curved packet and determined as follows:

$$x_2^{(m)} \equiv x_2, \quad x_1^{(m)} = x_1 - (m-1)\Lambda, \quad x_3^{(m)} = x_3, \qquad -\infty \le m \le +\infty . \tag{1.3.11}$$

We note that for odd m the Cartesian coordinates $x_j^{(m)}$ are related to the curvature with convexity "from below", and for even m, to convexity "from above". Now, we introduce a local circular cylindrical coordinate system as shown in (Fig.1.3.3 (b)).

Consider the m-th circular ring part, which according to (1.3.11) is determined by

$$\left|x_1^{(m)}\right| \le \frac{1}{2}\Lambda ;. \tag{1.3.12}$$

The relations between the local Cartesian and polar coordinate systems are as follows:

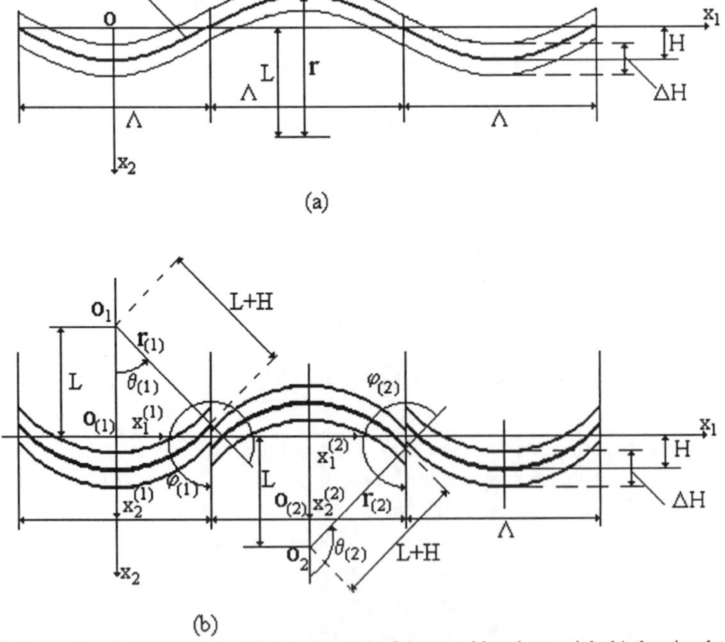

Fig.1.3.3. a) The representative layer (packet) of the considered material; b) the circular arcs which replace the curved segments of a packet.

$$r_{(m)} \cos\theta_{(m)} = x_2^{(m)} - (-1)^m L, \quad r_{(m)} \sin\theta_{(m)} = x_1^{(m)}, \quad \varphi_{(m)} = 2\pi - \theta_{(m)}, \quad (1.3.13)$$

where $\varphi_{(m)}$ denotes the angle measured from the radial direction of the polar coordinate system to the direction of the $Ox_2^{(m)}$ axis.

Using this notation, we can write the stress-strain relation for the part of the circular ring corresponding to the m-th half period of the curved packet in the local cylindrical coordinate system, as follows:

$$\sigma_{rr} = A_{11}^0\varepsilon_{rr} + A_{12}^0\varepsilon_{\theta\theta} + A_{13}^0\varepsilon_{33}, \quad \sigma_{\theta 3} = 2A_{44}^0\varepsilon_{\theta 3}, \quad A_{44}^0 = G_{23}^0,$$

$$\sigma_{\theta\theta} = A_{12}^0\varepsilon_{rr} + A_{22}^0\varepsilon_{\theta\theta} + A_{23}^0\varepsilon_{33}, \quad \sigma_{r3} = 2A_{55}^0\varepsilon_{r3}, \quad A_{55}^0 = G_{13}^0,$$

$$\sigma_{33} = A_{13}^0\varepsilon_{rr} + A_{23}^0\varepsilon_{\theta\theta} + A_{33}^0\varepsilon_{33}, \quad \sigma_{r\theta} = 2A_{66}^0\varepsilon_{r\theta}, \quad A_{66}^0 = G_{12}^0. \quad (1.3.14)$$

These relations may be inverted as follows:

$$\varepsilon_{rr} = a_{11}^0\sigma_{rr} + a_{12}^0\sigma_{\theta\theta} + a_{13}^0\sigma_{33}, \quad 2\varepsilon_{\theta 3} = a_{44}^0\sigma_{\theta 3}, \quad a_{44}^0 = \left(G_{23}^0\right)^{-1},$$

$$\varepsilon_{\theta\theta} = a_{12}^0\sigma_{rr} + a_{22}^0\sigma_{\theta\theta} + a_{23}^0\sigma_{33}, \quad 2\varepsilon_{r3} = a_{55}^0\sigma_{r3}, \quad a_{55}^0 = \left(G_{13}^0\right)^{-1},$$

$$\varepsilon_{33} = a_{13}^0\sigma_{rr} + a_{23}^0\sigma_{\theta\theta} + a_{33}^0\sigma_{33}, \quad 2\varepsilon_{r\theta} = a_{66}^0\sigma_{r\theta}, \quad a_{66}^0 = \left(G_{12}^0\right)^{-1}. \quad (1.3.15)$$

The material constants A_{ij}^0, G_{ij}^0 and a_{ij}^0 which enter the relations (1.3.14), (1.3.15) are the same constants which enter the relations (1.3.7), (1.3.8), respectively.

Stress-strain relations in the local Cartesian coordinate system.
We now consider stress-strain relations (1.3.14) in the local Cartesian system $O_{(m)}x_1^{(m)}x_2^{(m)}x_3^{(m)}$, using the well-known transformation of elastic constants under the rotation of the axes, derived for example, in [125,126] and elsewhere. For the m-th half period (1.3.12), equations (1.3.13) give

$$\cos^2\varphi_{(m)} = \left[x_2^{(m)} - (-1)^m L\right]^2 A_{(m)}^{-1}; \quad \sin^2\varphi_{(m)} = (x_1^{(m)})^2 A_{(m)}^{-1};$$

$$\cos\varphi_{(m)} \sin\varphi_{(m)} = -x_1^{(m)}\left[x_2^{(m)} - (-1)^m L\right] A_{(m)}^{-1};$$

$$A_{(m)} = \left[x_2^{(m)} - (-1)^m L\right]^2 + (x_1^{(m)})^2. \quad (1.3.16)$$

The stress-strain relations for the m-th circular ring part (Fig.1.3.3(b)) in the local Cartesian coordinate system $O_{(m)}x_1^{(m)}x_2^{(m)}x_3^{(m)}$ are as follows:

$$\sigma_i = A_{ij}^{(m)}\varepsilon_j + 2A_{i6}^{(m)}\varepsilon_6,\ i;j=1,2,3, \qquad \sigma_6 = A_{i6}^{(m)}\varepsilon_i + 2A_{66}^{(m)}\varepsilon_6$$

$$\sigma_4 = 2A_{44}^{(m)}\varepsilon_4 + 2A_{45}^{(m)}\varepsilon_5,\ \sigma_5 = 2A_{45}^{(m)}\varepsilon_4 + 2A_{55}^{(m)}\varepsilon_6, \tag{1.3.17}$$

Analogously, we obtain from (1.3.15) the following strain-stress relations which are the inverses of the relations (1.3.17).

$$\varepsilon_i = a_{ij}^{(m)}\sigma_j + 2a_{i6}^{(m)}\sigma_6,\ i;j=1,2,3, \quad \varepsilon_6 = a_{i6}^{(m)}\sigma_i + 2a_{66}^{(m)}\sigma_6$$

$$\varepsilon_4 = 2a_{44}^{(m)}\sigma_4 + 2a_{45}^{(m)}\sigma_5,\ \varepsilon_5 = 2a_{45}^{(m)}\sigma_4 + 2a_{55}^{(m)}\sigma_5. \tag{1.3.18}$$

Note that in (1.3.17) and (1.3.18) use the notation (1.3.6). By the well-known conversion formulas for elastic constants in a rotated coordinate system, we write the expressions for the $A_{ij}^{(m)}$ which enter (1.3.17) in terms of quantities A_{ij}^0 and G_{ij}^0 (1.3.5), (1.3.7) and $\varphi_{(m)}$ (1.3.16). It should be noted that $\varphi_{(m)}$ is denoted by φ in these expressions.

$$A_{11}^{(m)} = A_{11}^0\cos^4\varphi + 2(A_{12}^0 + 2G_{12}^0)\sin^2\varphi\cos^2\varphi + A_{22}^0\sin^4\varphi,$$

$$A_{22}^{(m)} = A_{11}^0\sin^4\varphi + 2(A_{12}^0 + 2G_{12}^0)\sin^2\varphi\cos^2\varphi + A_{22}^0\cos^4\varphi,$$

$$A_{12}^{(m)} = (A_{11}^0 + A_{22}^0 - 2A_{12}^0 - 4G_{12}^0)\sin^2\varphi\cos^2\varphi + A_{12}^0,\ A_{33}^{(m)} = A_{33}^0,$$

$$A_{13}^{(m)} = A_{13}^0\cos^2\varphi + A_{23}^0\sin^2\varphi,\ A_{23}^{(m)} = A_{13}^0\sin^2\varphi + A_{23}^0\cos^2\varphi,$$

$$A_{66}^{(m)} = (A_{11}^0 + A_{22}^0 - 2A_{12}^0 - 4G_{12}^0)\sin^2\varphi\cos^2\varphi + G_{12}^0,$$

$$A_{44}^{(m)} = G_{23}^0\cos^2\varphi + G_{13}^0\sin^2\varphi,\ A_{16}^{(m)} = \Big[A_{22}^0\sin^2\varphi - A_{11}^0\cos^2\varphi +$$

$$\left(A_{12}^0 + 2G_{12}^0\right)\left(\cos^2\varphi - \sin^2\varphi\right)\Big]\sin\varphi\cos\varphi,\ A_{26}^{(m)} = \Big[A_{22}^0\cos^2\varphi - A_{11}^0\sin^2\varphi +$$

$$\left(A_{12}^0 + 2G_{12}^0\right)\left(\sin^2\varphi - \cos^2\varphi\right)\Big]\sin\varphi\cos\varphi,\ A_{36}^{(m)} = (A_{22}^0 - A_{13}^0)\sin\varphi\cos\varphi,$$

$$A_{55}^{(m)} = G_{23}^{0} \sin^{2} \varphi + G_{13}^{0} \cos^{2} \varphi , \quad A_{45}^{(m)} = (G_{23}^{0} - G_{13}^{0}) \sin \varphi \cos \varphi . \qquad (1.3.19)$$

Analogously, for $a_{ij}^{(m)}$ (1.3.18) we can write the following expressions.

$$a_{11}^{(m)} = a_{11}^{0} \cos^{4} \varphi + (a_{66}^{0} + 2a_{12}^{0}) \sin^{2} \varphi \cos^{2} \varphi + a_{22}^{0} \sin^{4} \varphi ,$$

$$a_{22}^{(m)} = a_{11}^{0} \sin^{4} \varphi + (a_{66}^{0} + 2a_{12}^{0}) \sin^{2} \varphi \cos^{2} \varphi + a_{22}^{0} \cos^{4} \varphi$$

$$a_{12}^{(m)} = (a_{11}^{0} + a_{22}^{0} - a_{66}^{0} - 2a_{12}^{0}) \sin^{2} \varphi \cos^{2} \varphi + a_{12}^{0} , \quad a_{33}^{(m)} = a_{33}^{0} ,$$

$$a_{13}^{(m)} = a_{13}^{0} \cos^{2} \varphi + a_{23}^{0} \sin^{2} \varphi , \quad a_{23}^{(m)} = a_{13}^{0} \sin^{2} \varphi + a_{23}^{0} \cos^{2} \varphi ,$$

$$a_{66}^{(m)} = (a_{11}^{0} + a_{22}^{0} - 2a_{12}^{0} - a_{66}^{0}) 4 \sin^{2} \varphi \cos^{2} \varphi + a_{66}^{0} ,$$

$$a_{44}^{(m)} = a_{44}^{0} \cos^{2} \varphi + a_{55}^{0} \sin^{2} \varphi , \; a_{16}^{(m)} = \Big[2a_{22}^{0} \sin^{2} \varphi - 2a_{11}^{0} \cos^{2} \varphi +$$

$$\left(a_{66}^{0} + 2a_{12}^{0} \right) \left(\cos^{2} \varphi - \sin^{2} \varphi \right) \Big] \sin \varphi \cos \varphi , \quad a_{26}^{(m)} = \Big[2a_{22}^{0} \cos^{2} \varphi - 2a_{11}^{0} \sin^{2} \varphi +$$

$$\left(a_{66}^{0} + 2a_{12}^{0} \right) \left(\sin^{2} \varphi - \cos^{2} \varphi \right) \Big] \sin \varphi \cos \varphi , \quad a_{36}^{(m)} = 2 (a_{23}^{0} - a_{13}^{0}) \sin \varphi \cos \varphi ,$$

$$a_{55}^{(m)} = a_{44}^{0} \sin^{2} \varphi + a_{55}^{0} \cos^{2} \varphi , \; a_{45}^{(m)} = (a_{44}^{0} - a_{55}^{0}) \sin \varphi \cos \varphi . \qquad (1.3.20)$$

The expressions (1.3.16)-(1.3.20) completely determine the stress-strain and the strain-stress relations in local Cartesian coordinates $O_{(m)} x_1^{(m)} x_2^{(m)} x_3^{(m)}$ (Fig.1.3.3 (b)) within the m-th circular ring part. However, it is necessary to obtain the elasticity relations for the curved representative packet shown in Fig.1.3.3 (a). For this purpose we continue the transformations of these relations.

Introducing a small parameter, further simplifications.
The fourth inequality in (1.3.8) implies that

$$\varepsilon = \frac{\Lambda}{\pi L} << 1 . \qquad (1.3.21)$$

It follows from this and from Fig.1.3.3, i.e. from the equation $(\Lambda/2)^2 \cong (L + H)^2 - L^2$ that

$$\Lambda^2 \approx 8LH. \tag{1.3.22}$$

So that

$$\varepsilon = \frac{\Lambda}{\pi L} \approx \frac{8}{\pi}\frac{H}{\Lambda} << 1. \tag{1.3.23}$$

Taking these inequalities into account we now simplify the relations (1.3.16), (1.3.19) and (1.3.20). Within the limits of the m-th circular ring part (1.3.12) (i.e., circular ring part which corresponds to the m-th half-period of the representative curved layer (Fig.1.3.3,(a)) the following estimates hold for $x_1^{(m)}$ and $x_2^{(m)}$:

$$\left|x_2^{(m)}\right| \le H + \frac{1}{2}\Delta H, \quad \left|x_1^{(m)}\right| \le \frac{1}{2}\Lambda. \tag{1.3.24}$$

Taking the second, third and fourth inequalities in (1.3.10) and the relations (1.3.22) and (1.3.24), we neglect the quantities $x_2^{(m)}L^{-1}$ in comparison with unity in (1.3.16); i.e. we practically ignore the dependence on the coordinate x_2. As a result we obtain

$$\cos^2\varphi_{(m)} = \left[1 + L_{0m}^2\right]^{-1} + O(L_1),$$

$$\sin^2\varphi_{(m)} = L_{0m}^2\left[1 + L_{0m}^2\right]^{-1} + O(L_1)$$

$$\cos\varphi_{(m)}\sin\varphi_{(m)} = (-1)^m L_{0m}\left[1 + L_{0m}^2\right]^{-1} + O(L_1), \tag{1.3.25}$$

where

$$L_{0m} = x_1^{(m)}L^{-1}, \quad L_1 = L^{-1}\left(H + \frac{1}{2}\Delta H\right). \tag{1.3.26}$$

It follows from (1.3.25) and (1.3.26) that if the terms of order L_1 are neglected in (1.3.16), then we obtain elasticity relations which are independent of the coordinate x_2. From a physical point of view, this simplification is necessary, since it is clear from Fig.1.3.1 that for the composite material with curved structure the basic relations in the framework of the continuum approach cannot depend on x_2.

There is one more aspect of the theory which must be accounted for along with further transformations. Due to the material structure (Fig.1.3.1) the approximated material properties must vary periodically along the Ox_1 axis. This is guaranteed by the expressions (1.3.25): the first two of which have a period Λ, and the last having a period 2Λ. However, since the m-th half period of the periodic curved representative packet has been replaced by the circular ring part shown in Fig.1.3.3, all the expressions (1.3.25) have angular points, i.e. slope discontinuities at $x_1^{(m)} = \pm(1/2)\Lambda$. Since the structure of the composite material varies smoothly with x_1 (Fig.1.3.3), there should be no angular points in the functions describing the normalized properties along the Ox_1 axis.

Thus, it is necessary to approximate expressions (1.3.25) by periodic functions with the same periods, but without discontinuities. First, taking into account the second estimate of (1.3.24), we expand the expression (1.3.25) in a Taylor's series. As a result we obtain the following expressions for the m-th half-period of the structure (1.3.10).

$$\cos^2 \varphi_{(m)} = 1 - L_{0m}^2 + O\left(L_2^4\right) + O\left(L_1\right),$$

$$\sin^2 \varphi_{(m)} = L_{0m}^2 + O\left(L_2^4\right) + O\left(L_1\right),$$

$$\cos \varphi_{(m)} \sin \varphi_{(m)} = (-1)^m L_{0m} + O\left(L_2^3\right) + O\left(L_1\right), \tag{1.3.27}$$

where

$$L_2 = \frac{1}{2} L^{-1} \Lambda. \tag{1.3.28}$$

In expressions (1.3.27) we neglect the terms $O\left(L_2^3\right)$, $O\left(L_2^4\right)$, $O\left(L_1\right)$ and obtain the following approximations:

$$\cos^2 \varphi_{(m)} \approx 1 - L_{0m}^2, \sin^2 \varphi_{(m)} \approx L_{0m}^2, \quad \cos \varphi_{(m)} \sin \varphi_{(m)} \approx (-1)^m L_{0m}. \tag{1.3.29}$$

Note that, because of (1.3.11), the expressions (1.3.29) still have angular points for $x_1^{(m)} = \pm(1/2)\Lambda$. Now we replace (1.3.29) by trigonometric functions, which for small $x_1^{(m)}$ coincide with the series expansions of (1.3.29) with an accuracy not exceeding $\left((1/2)L^{-1}\Lambda\right)^3$. As a result we obtain

$$\cos^2 \varphi_{(m)} \approx 1 - \varepsilon^2 \cos^2 \theta, \qquad \sin^2 \varphi_{(m)} \approx \varepsilon^2 \sin^2 \theta,$$

$$\cos \varphi_{(m)} \sin \varphi_{(m)} \approx -\varepsilon \sin \theta, \tag{1.3.30}$$

where

$$\theta = \frac{\pi x_1}{\Lambda}, \tag{1.3.31}$$

and the small parameter ε is defined in (1.3.23).

Note that in replacing (1.3.29) by relations (1.3.30) we introduce an error at $x_1^{(m)} = \pm(1/2)\Lambda$, however, due to $L_{0m} << 1$ the error will be small; now the angular points have been smoothed out.

The elasticity relations for the representative curved packet (Fig.1.3.3 (a)). Substituting (1.3.30) into (1.3.19), restricting ourselves to an accuracy of ε^2, and performing a number of transformations, we reach expressions for the quantities $A_{ij}^{(m)}$ appearing in (1.3.17) in terms of the A_{ij}^0 and G_{ij}^0 appearing in (1.3.7), and the parameter ε in the form

$$A_{11} = A_{11}^0 + \varepsilon^2\left(-A_{11}^0 + A_{12}^0 + 2G_{12}^0\right)2\sin^2\theta,\ A_{33} = A_{33}^0,$$

$$A_{12} = A_{12}^0 + \varepsilon^2\left(A_{11}^0 + A_{22}^0 - 2A_{12}^0 - 4G_{12}^0\right)\sin^2\theta,$$

$$A_{23} = A_{23}^0 + \varepsilon^2\left(A_{13}^0 - A_{23}^0\right)\sin^2\theta,\ A_{26} = \varepsilon\left(A_{22}^0 - A_{12}^0 - 2G_{12}^0\right)\sin\theta,$$

$$A_{16} = \varepsilon\left(-A_{11}^0 + A_{12}^0 + 2G_{12}^0\right)\sin\theta,\ A_{36} = \varepsilon\left(A_{23}^0 - A_{13}^0\right)\sin\theta\ ,$$

$$A_{22} = A_{22}^0 + \varepsilon^2\left(-A_{22}^0 + A_{12}^0 + 2G_{12}^0\right)2\sin^2\theta,$$

$$A_{13} = A_{13}^0 + \varepsilon^2\left(A_{23}^0 - A_{13}^0\right)\sin^2\theta,\ A_{45} = \varepsilon\left(G_{13}^0 - G_{23}^0\right)\sin\theta,$$

$$A_{44} = G_{23}^0 + \varepsilon^2\left(G_{13}^0 - G_{23}^0\right)\sin^2\theta,\ A_{55} = G_{13}^0 + \varepsilon^2\left(G_{23}^0 - G_{13}^0\right)\sin^2\theta,$$

$$A_{66} = G_{12}^0 + \varepsilon^2\left(A_{11}^0 + A_{22}^0 - 2A_{12}^0 - 4G_{12}^0\right)\sin^2\theta. \qquad (1.3.32)$$

Restricting ourselves in (1.3.32) to accuracy ε^1, we obtain

$$A_{11} = A_{11}^0,\ A_{12} = A_{12}^0,\ A_{13} = A_{13}^0,\ A_{22} = A_{22}^0,\ A_{23} = A_{23}^0,$$

$$A_{33} = A_{33}^0,\ A_{44} = A_{44}^0,\ A_{55} = A_{55}^0,\ A_{66} = A_{66}^0,$$

$$A_{16} = \varepsilon\left(-A_{11}^0 + A_{12}^0 + 2G_{12}^0\right)\sin\theta,\ A_{26} = \varepsilon\left(A_{22}^0 - A_{12}^0 - 2G_{12}^0\right)\sin\theta,$$

$$A_{36} = \varepsilon\left(A_{23}^0 - A_{13}^0\right)\sin\theta,\ A_{45} = \varepsilon\left(G_{13}^0 - G_{23}^0\right)\sin\theta. \qquad (1.3.33)$$

Analogously, substituting (1.3.30) into (1.3.20) and restricting ourselves to an accuracy of ε^2, we obtain the following expressions for $a_{ij}^{(m)}$ appearing in (1.3.18).

$$a_{11} = a_{11}^0 + \varepsilon^2\left(-a_{11}^0 + a_{12}^0 + \frac{1}{2}a_{66}^0\right)2\sin^2\theta,\ a_{33} = a_{33}^0,$$

$$a_{12} = a_{12}^0 + \varepsilon^2\left(a_{11}^0 + a_{22}^0 - 2a_{12}^0 - a_{66}^0\right)\sin^2\theta,$$

$$a_{23} = a_{23}^0 + \varepsilon^2\left(a_{13}^0 - a_{23}^0\right)\sin^2\theta,\ a_{26} = \varepsilon\left(2a_{22}^0 - 2a_{12}^0 - a_{66}^0\right)\sin\theta,$$

$$a_{16} = \varepsilon\left(-2a_{11}^0 + 2a_{12}^0 + a_{66}^0\right)\sin\theta,\ a_{36} = \varepsilon\left(a_{23}^0 - a_{13}^0\right)\sin\theta,$$

$$a_{22} = a_{22}^0 + \varepsilon^2\left(-a_{22}^0 + a_{12}^0 + \frac{1}{2}a_{66}^0\right)2\sin^2\theta,$$

$$a_{13} = a_{13}^0 + \varepsilon^2\left(a_{23}^0 - a_{13}^0\right)\sin^2\theta,\ a_{45} = \varepsilon\left(a_{55}^0 - a_{44}^0\right)\sin\theta,$$

$$a_{44} = a_{44}^0 + \varepsilon^2\left(a_{55}^0 - a_{44}^0\right)\sin^2\theta,\ a_{55} = a_{55}^0 + \varepsilon^2\left(a_{44}^0 - a_{55}^0\right)\sin^2\theta,$$

$$a_{66} = a_{66}^0 + \varepsilon^2\left(a_{11}^0 + a_{22}^0 - 2a_{12}^0 - a_{66}^0\right)4\sin^2\theta. \tag{1.3.34}$$

Restricting ourselves in (1.3.34) to accuracy ε^1 we obtain

$$a_{11} = a_{11}^0,\ a_{12} = a_{12}^0,\ a_{13} = a_{13}^0,\ a_{22} = a_{22}^0,\ a_{23} = a_{23}^0$$

$$a_{33} = a_{33}^0,\ a_{44} = a_{44}^0,\ a_{55} = a_{55}^0,\ a_{66} = a_{66}^0,$$

$$a_{16} = \varepsilon\left(-2a_{11}^0 + 2a_{12}^0 + a_{66}^0\right)\sin\theta,\ a_{26} = \varepsilon\left(2a_{22}^0 - 2a_{12}^0 - a_{66}^0\right)\sin\theta,$$

$$a_{36} = \varepsilon\left(a_{23}^0 - a_{13}^0\right)2\sin\theta,\ a_{45} = \varepsilon\left(a_{55}^0 - a_{44}^0\right)\sin\theta. \tag{1.3.35}$$

Thus, the stress-strain relations for the representative curved packet are obtained as follows

$$\sigma_i = A_{ij}\varepsilon_j + 2A_{i6}\varepsilon_6,\ \sigma_6 = A_{i6}\varepsilon_i + 2A_{66}\varepsilon_6,\ \ i;j=1,2,3,$$

$$\sigma_4 = 2A_{44}\varepsilon_4 + 2A_{45}\varepsilon_5,\ \sigma_5 = 2A_{45}\varepsilon_4 + 2A_{55}\varepsilon_6, \tag{1.3.36}$$

In (1.3.36) $A_{11},...,A_{66}$ are determined by the expressions (1.3.31), (1.3.32). The inverse of the relations (1.3.36) are

$$\varepsilon_i = a_{ij}\sigma_j + 2a_{i6}\sigma_6, \quad \varepsilon_6 = a_{i6}\sigma_i + 2a_{66}\sigma_6, \quad i;j=1,2,3,$$

$$\varepsilon_4 = 2a_{44}\sigma_4 + 2a_{45}\sigma_5, \quad \varepsilon_5 = 2a_{45}\sigma_4 + 2a_{55}\sigma_5, \tag{1.3.37}$$

where $a_{11},...,a_{66}$ are determined by the expressions (1.3.31) and (1.3.34).

It follows from (1.3.33) and (1.3.35) that for small curvature (small ε), within the first-order approximation one must take into account not only the change in normalized (reduced) elastic constants within the orthotropic body model, but also the change in symmetry of the anisotropic body, since within the first approximation $A_{16} \neq 0, A_{26} \neq 0$, $A_{45} \neq 0$ in (1.3.33) and $a_{16} \neq 0$, $a_{26} \neq 0$, $a_{45} \neq 0$ in (1.3.34). It should be noted that this situation is the principal difference of the proposed continuum approach from the classical approaches mentioned in Section 1.1. Thus, we obtain the final basic relations for the medium with the structure shown in Fig.1.3.1; they reduce to expressions (1.3.32), (1.3.36) or (1.3.34), (1.3.37). Note that the elasticity relations (1.3.32) and (1.3.36) or (1.3.34) and (1.3.37) are valid for both geometrically linear (1.2.1)-(1.2.3), (1.2.7) and non-linear problems (1.2.4) -(1.2.7).

1.4. Displacement Equations; Formulation and Solution

1.4.1. THE EQUATIONS OF MOTION

Consider the formulation of continuum problems in terms of displacements. We will investigate only geometrical linear problems (1.2.1)-(1.2.3), (1.2.7).

Taking (1.2.2), (1.3.6), (1.3.31) and (1.3.32) into account, we may write the elasticity relations (1.3.36) of the continuum theory as follows:

$$\sigma_{ij} = \mu_{ij\alpha\beta}\frac{\partial u_\alpha}{\partial x_\beta}, \quad i,j,\alpha,\beta=1,2,3. \tag{1.4.1}$$

In (1.4.1) the following notations are introduced.

$$\mu_{ij\alpha\beta} = \mu^{(0)}_{ij\alpha\beta} + \varepsilon\mu^{(1)}_{ij\alpha\beta}\sin\theta + \varepsilon^2\mu^{(2)}_{ij\alpha\beta}\sin^2\theta, \quad \theta = \frac{\pi x_1}{\Lambda}, \tag{1.4.2}$$

where $\mu^{(0)}_{ij\alpha\beta}, \mu^{(1)}_{ij\alpha\beta}$ and $\mu^{(2)}_{ij\alpha\beta}$ are the components of the tensors $\boldsymbol{\mu}^{(0)}$, $\boldsymbol{\mu}^{(1)}$ and $\boldsymbol{\mu}^{(2)}$, respectively. After some transformations, we obtain the following expressions for the components $\mu^{(0)}_{ij\alpha\beta}, \mu^{(1)}_{ij\alpha\beta}$ and $\mu^{(2)}_{ij\alpha\beta}$:

$$\mu^{(0)}_{ij\alpha\beta} = \delta^{\underline{j}}_{\underline{i}}\delta^{\underline{\beta}}_{\underline{\alpha}}A^{0}_{\underline{i\beta}} + (1-\delta^{\underline{j}}_{\underline{i}})(\delta^{\underline{\alpha}}_{\underline{i}}\delta^{\underline{\beta}}_{\underline{j}} + \delta^{\underline{\beta}}_{\underline{i}}\delta^{\underline{\alpha}}_{\underline{j}})G^{0}_{\underline{ij}},$$

$$\mu^{(1)}_{ij\alpha\beta} = (A^0_{11} - A^0_{12} - 2G^0_{12})\left[\delta^1_i\delta^1_j(\delta^1_\alpha\delta^2_\beta + \delta^2_\alpha\delta^1_\beta) + \delta^1_\alpha\delta^1_\beta(\delta^1_i\delta^2_j + \delta^2_i\delta^1_j)\right]$$

$$+(-A^0_{22} + A^0_{12} + 2G^0_{12})\left[\delta^2_i\delta^2_j(\delta^1_\alpha\delta^2_\beta + \delta^2_\alpha\delta^1_\beta) + \delta^2_\alpha\delta^2_\beta(\delta^1_i\delta^2_j + \delta^2_i\delta^1_j)\right]$$

$$+(A^0_{23} - A^0_{13})\left[\delta^3_i\delta^3_j(\delta^1_\alpha\delta^2_\beta + \delta^2_\alpha\delta^1_\beta) + \delta^3_\alpha\delta^3_\beta(\delta^1_i\delta^2_j + \delta^2_i\delta^1_j)\right] + (G^0_{13} - G^0_{23})$$

$$\times\left[\left(\delta^1_i\delta^3_j + \delta^3_i\delta^1_j\right)\left(\delta^3_\alpha\delta^2_\beta + \delta^2_\alpha\delta^3_\beta\right) + \left(\delta^2_i\delta^3_j + \delta^3_i\delta^2_j\right)\left(\delta^1_\alpha\delta^3_\beta + \delta^3_\alpha\delta^1_\beta\right)\right],$$

$$\mu^{(2)}_{ij\alpha\beta} = 2\left(-A^0_{11} + A^0_{12} + 2G^0_{12}\right)\delta^1_i\delta^1_j\delta^1_\alpha\delta^1_\beta + 2\left(-A^0_{22} + A^0_{12} + 2G^0_{12}\right)$$

$$\times\delta^2_i\delta^2_j\delta^2_\alpha\delta^2_\beta + \left(A^0_{11} + A^0_{22} - 2A^0_{12} - 4G^0_{12}\right)\left[\left(\delta^1_i\delta^1_j\delta^2_\alpha\delta^2_\beta + \delta^2_i\delta^2_j\delta^1_\alpha\delta^1_\beta\right)\right.$$

$$\left. + \left(\delta^1_i\delta^2_j + \delta^2_i\delta^1_j\right)\left(\delta^1_\alpha\delta^2_\beta + \delta^2_\alpha\delta^1_\beta\right)\right] + \left(A^0_{23} - A^0_{13}\right)\left[\delta^3_i\delta^3_j\left(\delta^1_\alpha\delta^1_\beta - \delta^2_\alpha\delta^2_\beta\right)\right.$$

$$\left. + \delta^3_\alpha\delta^3_\beta\left(\delta^1_i\delta^1_j - \delta^2_i\delta^2_j\right)\right] + \left(G^0_{13} - G^0_{23}\right)\left[\left(\delta^2_i\delta^3_j + \delta^3_i\delta^2_j\right)\left(\delta^2_\alpha\delta^3_\beta + \delta^3_\alpha\delta^2_\beta\right)\right.$$

$$\left. - \left(\delta^1_i\delta^3_j + \delta^3_i\delta^1_j\right)\left(\delta^1_\alpha\delta^3_\beta + \delta^3_\alpha\delta^1_\beta\right)\right]. \qquad (1.4.3)$$

Note that in (1.4.3) and henceforward the tensor $\boldsymbol{\mu}^{(0)}$ corresponds to a homogeneous body with rectilinear orthotropy.

Substituting (1.4.1) and (1.4.2) into (1.2.1) we obtain the following equations of motion for the proposed continuum theory, in terms of displacements:

$$L_{j\alpha}u_\alpha + \varepsilon\frac{\partial}{\partial x_i}\left(\mu^{(1)}_{ij\alpha\beta}\sin\theta\frac{\partial}{\partial x_\beta}\right)u_\alpha$$

$$+\varepsilon^2\frac{\partial}{\partial x_i}\left[\mu^{(2)}_{ij\alpha\beta}\sin^2\theta\frac{\partial}{\partial x_\beta}\right]u_\alpha = 0, \quad x_1, x_2, x_3 \in D \qquad (1.4.4)$$

Here $L_{j\alpha}$ are the differential operators of Lame's equation for the rectilinear orthotropic body

$$L_{j\alpha} = \left[A^0_{\underline{j}\underline{\alpha}} + \left(1-\delta^{\underline{\alpha}}_{\underline{j}}\right)G^0_{\underline{j}\underline{\alpha}}\right]\frac{\partial^2}{\partial x_{\underline{j}}\partial x_{\underline{\alpha}}} + \delta^{\underline{\alpha}}_{\underline{j}}\left(1-\delta^{\underline{j}}_{\underline{i}}\right)G^0_{\underline{i}\underline{j}}\frac{\partial^2}{\partial x_{\underline{i}}^2} - \rho\delta^{\underline{\alpha}}_{\underline{j}}\frac{\partial^2}{\partial\tau^2}. \quad (1.4.5)$$

Substituting (1.4.1) into (1.2.3), we obtain boundary conditions on the surface portion S_1, where the stresses are given, in the form

$$N_i\left\{\left[\mu^{(0)}_{ij\alpha\beta} + \varepsilon\mu^{(1)}_{ij\alpha\beta}\sin\theta + \varepsilon^2\mu^{(2)}_{ij\alpha\beta}\sin^2\theta\right]\frac{\partial u_\alpha}{\partial x_\beta}\right\} = P_j,\ x_1, x_2, x_3 \in S_1. \quad (1.4.6)$$

The boundary conditions on the surface portion S_2 remain as (1.2.7).

Thus, the displacement formulation of the continuum problems reduces to equations (1.4.1)-(1.4.6), (1.2.7).

1.4.2. THE SMALL PARAMETER METHOD

We now consider the small parameter method for the solution of the equations. For this purpose we represent all quantities as series in the parameter ε

$$\sigma_{ij} = \sum_{n=0}^{\infty}\varepsilon^n\sigma^{(n)}_{ij};\quad \varepsilon_{ij} = \sum_{n=0}^{\infty}\varepsilon^n\varepsilon^{(n)}_{ij};\quad P_j = \sum_{n=0}^{\infty}\varepsilon^n P^{(n)}_j;\quad f_j = \sum_{n=0}^{\infty}\varepsilon^n f^{(n)}_j. \quad (1.4.7)$$

Substituting (1.4.7) into (1.4.1), (1.4.4), (1.4.6) and (1.2.7) we obtain the following equations for the n-th approximation:

The equation of motion:

$$L_{j\alpha}u^{(n)}_\alpha = -\frac{\partial}{\partial x_i}\left(\mu^{(1)}_{ij\alpha\beta}\sin\theta\frac{\partial}{\partial x_\beta}\right)u^{(n-1)}_\alpha - \frac{\partial}{\partial x_i}\left[\mu^{(2)}_{ij\alpha\beta}\sin^2\theta\frac{\partial}{\partial x_\beta}\right]u^{(n-2)}_\alpha,$$

$$x_1, x_2, x_3 \in D; \quad (1.4.8)$$

the boundary conditions on the surface portion S_1, where the stresses are given:

$$N_i\left[\delta^j_{\underline{i}}\delta^\beta_\alpha A^0_{i\beta} + (1-\delta^j_{\underline{i}})\left(\delta^\alpha_i\delta^\beta_{\underline{j}} + \delta^\beta_i\delta^\alpha_{\underline{j}}\right)G^0_{\underline{i}\underline{j}}\right]\frac{\partial}{\partial x_\beta}u^{(n)}_\alpha$$

$$= -N_i\left[\mu^{(1)}_{\underline{i}j\alpha\beta}\sin\theta\frac{\partial u^{(n-1)}_\alpha}{\partial x_\beta} + \mu^{(2)}_{\underline{i}j\alpha\beta}\sin^2\theta\frac{\partial u^{(n-2)}_\alpha}{\partial x_\beta}\right] + P^{(n)}_{\underline{j}},$$

$$x_1, x_2, x_3 \in S_1; \tag{1.4.9}$$

boundary conditions on the surface portion S_2, where the displacements are given:

$$u_j^{(n)} = f_j^{(n)}, \quad x_1, x_2, x_3 \in S_2; \tag{1.4.10}$$

the expression for the stresses:

$$\sigma_{ij}^{(n)} = \left[\delta_i^j \delta_\alpha^\beta A_{i\beta}^0 + (1-\delta_i^j)\left(\delta_i^\alpha \delta_j^\beta + \delta_i^\beta \delta_j^\alpha\right) G_{ij}^0\right] \frac{\partial}{\partial x_\beta} u_\alpha^{(n)}$$

$$+ \left[\mu_{ij\alpha\beta}^{(1)} \sin\theta \frac{\partial u_\alpha^{(n-1)}}{\partial x_\beta} + \mu_{ij\alpha\beta}^{(2)} \sin^2\theta \frac{\partial u_\alpha^{(n-2)}}{\partial x_\beta}\right]. \tag{1.4.11}$$

We note that the equations (1.4.1)-(1.4.6) were obtained in the general case; consequently, the results (1.4.8)-(1.4.11) refer to an arbitrary approximation. Besides, in the basic relations (1.4.8) the expressions on the left coincide with the corresponding expressions for a linearly elastic uniform orthotropic body with rectilinear orthotropy.

Thus, problems for composite materials with small-scale periodically plane curved structures in an arbitrary n-th approximation have been reduced to problems for a linearly elastic uniform orthotropic body with rectilinear orthotropy. The stresses in the n-th approximation are calculated from the expression (1.4.11), which differs from the corresponding expression for a uniform orthotropic body, since the permutations of n, n-1, and n-2 appear in (1.4.11) approximately.

1.5. Example for Exact Solution

The small parameter method allows us to investigate many continuum problems based on (1.4.1)-(1.4.3). In the general case the problems reduce to a system of partial differential equations with coefficients depending on x_1.

Since the results obtained using the small parameter method are approximate, it is important to obtain the exact solution of some problems of the theory (1.4.1)-(1.4.3). Therefore in this section we consider establishing an example with an exact solution. We will consider a problem relating to a quasi-homogeneous stress-state. Note that problems concerning quasi-homogeneous stress states are quite characteristic, reflecting specific features, and at the same time are comparatively simple (for the mechanics of composite materials with small-scale structural curvatures). By quasi-homogeneous stress states in an infinite body, generated by corresponding loads, we understand stress states which, for a material without structural curving, correspond to a uniform stress state. By this definition quasi-homogeneous stress states in materials with small-scale structural curving are generated by pure shear or pure tension. Note that for quasi-homogeneous states not only are the stresses balanced with the external loads, but also balanced in each period of the curvature.

Consider a problem for a quasi-homogeneous stress state which has an exact solution: the problem of pure shear in the plane Ox_1x_3 (Fig.1.5.1). We investigate the stress-deformation state, when external loads are applied as

$$\sigma_{13} = \tau , \tag{1.5.1}$$

We attempt to satisfy the condition (1.5.1) by the following displacement fields:

$$u_1 \equiv 0, \ u_2 \equiv 0, u_3 = u_3(x_1) \tag{1.5.2}$$

Substituting (1.5.2) into (1.4.1), we derive expressions for the non-vanishing stresses

$$\sigma_{13} = G_{13}^0 \frac{du_3}{dx_1} + \varepsilon^2 \sin^2\theta \left(G_{23}^0 - G_{13}^0\right)\frac{du_3}{dx_1},$$

$$\sigma_{23} = \varepsilon \sin\theta \left(G_{23}^0 - G_{13}^0\right)\frac{du_3}{dx_1}. \tag{1.5.3}$$

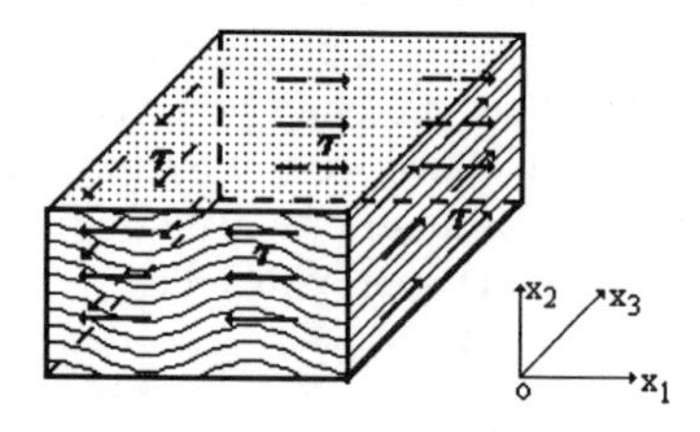

Fig.1.5.1. Pure shear in the Ox_1x_3 plane.

Substituting (1.5.2) into (1.4.4) we find a differential equation for the function u_3:

$$G_{13}^0 \frac{d^2u_3}{dx_1^2} + \frac{d}{dx_1}\left[\varepsilon^2 \sin^2\theta \left(G_{23}^0 - G_{13}^0\right)\frac{du_3}{dx_1}\right] = 0. \tag{1.5.4}$$

This has a first integral

$$\frac{du_3}{dx_1} = \left[G_{13}^0 + \varepsilon^2 \sin^2\theta \left(G_{23}^0 - G_{13}^0\right)\right]^{-1} C, \tag{1.5.5}$$

where C is an arbitrary constant. After a few transformations, and neglecting a second arbitrary constant, we find

$$u_3 = C\frac{\dfrac{\Lambda}{\pi G_{13}^0}}{\sqrt{1+\varepsilon^2 \dfrac{G_{23}^0 - G_{13}^0}{G_{13}^0}}}\arctan\phi(x_1)\,, \qquad (1.5.6)$$

where

$$\tan\phi(x_1) = \sqrt{1+\varepsilon^2 \frac{G_{23}^0 - G_{13}^0}{G_{13}^0}}\tan\theta, \quad \theta = \frac{\pi}{\Lambda}x_1\,. \qquad (1.5.7)$$

Substituting (1.5.5) into (1.5.3) we obtain the following expressions for the stresses.

$$\sigma_{13} = C\,, \qquad (1.5.8)$$

$$\sigma_{23} = \varepsilon\left(G_{23}^0 - G_{13}^0\right)\left[G_{13}^0 + \varepsilon^2 \sin^2\theta\left(G_{23}^0 - G_{13}^0\right)\right]^{-1} C\sin\theta\,. \qquad (1.5.9)$$

From the expressions (1.5.8) and from (1.5.1) we obtain that

$$C = \tau\,. \qquad (1.5.10)$$

Thus, the stresses (1.5.8), (1.5.9) do not vanish and correspond to the external loading (1.5.1). Neglecting terms of order ε^2, we find the following relations for the stresses from (1.5.8), (1.5.9):

$$\sigma_{13} = \tau\,, \quad \sigma_{23} = \varepsilon\tau\left(G_{13}^0\right)^{-1}\left(G_{23}^0 - G_{13}^0\right)\sin\left(\frac{\pi}{\Lambda}x_1\right). \qquad (1.5.11)$$

Note that if

$$\varepsilon = 0 \;\text{ or }\; G_{13}^0 = G_{23}^0 \qquad (1.5.12)$$

we obtain from (1.5.8)-(1.5.10) the classical solution, corresponding to an orthtotropic body (a material without curvature), in the form

$$\sigma_{13} = \tau\,, \quad \sigma_{23} = 0\,, \quad u_3 = \frac{\tau}{G_{13}^0}x_1\,. \qquad (1.5.13)$$

The first condition ($\varepsilon = 0$) corresponds to the absence of curvature, but the second condition of (1.5.12) is not related to this limiting case. Note that the second condition of (1.5.12) cannot be satisfied for composite materials having a predominant orientation of reinforcing elements (fibers), for example, for unidirectional composite materials reinforced along the Ox_1 axis, as for such composites, within the continuum approach, the normalized shear modulus G_{23}^0 is significantly different from the normalized shear modulus G_{13}^0. Consequently, we may conclude that in composite materials having a predominant orientation of periodically plane-curved reinforcing elements, there are two shear stresses: the stress σ_{13} which appears in a material without curvature, and a stress σ_{23} given by (1.5.9), (1.5.10); this latter stress is balanced within each period. Note that this conclusion is based on the exact solution, and not on an approximate one.

1.6. Vibration Problems

1.6.1. A GENERAL SOLUTION PROCEDURE

Along with the small parameter method which has been discussed in section 1.4 and which can be applied to many continuum problems (1.4.1)-(1.4.3), there are many other approximate analytical and numerical methods, one of which is the variational method that can be applied to steady-state dynamical problems. For an example, we consider the application of Hamilton's variational principle to the investigation of the natural frequencies and forms of vibration of a singly connected body with a volume V bounded by the surface $S = S_1 \cup S_2$ and filled by composite material with small-scale periodic curvature in the structure; S_1 (S_2) is the portion of the surface S on which the external forces (displacements) are given.

Theoretically, the investigation of the natural vibration is reduced to the solution of the following problem:

$$\frac{\partial}{\partial x_i}\sigma_{ij} - \rho\frac{\partial^2}{\partial \tau^2}u_j = 0, \quad x_1, x_2, x_3 \in V, \tag{1.6.1}$$

$$N_i\sigma_{ij} = 0, \quad x_1, x_2, x_3 \in S_1, \tag{1.6.2}$$

$$u_j = 0, \quad x_1, x_2, x_3 \in S_2, \tag{1.6.3}$$

We write the elasticity relations for the body material as follows:

$$\mu_{ij\alpha\beta} = \mu_{ij\alpha\beta}^{(0)} + \varepsilon\mu_{ij\alpha\beta}^{(1)}\sin\gamma + \varepsilon^2\mu_{ij\alpha\beta}^{(2)}\sin^2\gamma, \tag{1.6.4}$$

where

$$\gamma = \pi \frac{x_1 + \ell}{\Lambda}. \tag{1.6.5}$$

In (1.6.4) the components $\mu^{(0)}_{ij\alpha\beta}$, $\mu^{(1)}_{ij\alpha\beta}$ and $\mu^{(2)}_{ij\alpha\beta}$ are determined as in (1.4.3). Further, in (1.6.4) the new variable ℓ is introduced which is not in the expressions (1.4.2). Note that for the expressions (1.4.2) the origin of the coordinate system $Ox_1x_2x_3$ is connected to the summit of the curve in the material structure. However, for the body with a concrete form, it is preferable to connect the origin of the coordinate system $Ox_1x_2x_3$ with a certain point of this body instead of the summit. In (1.6.5) the value ℓ is the distance between the origin of the coordinate system $Ox_1x_2x_3$ and the first summit of the curve of the material structure.

We represent the displacement in the following form:

$$u_j = \upsilon_j(x_1, x_2, x_3)\exp(i\omega\tau). \tag{1.6.6}$$

Taking (1.4.2), (1.6.1)-(1.6.3) and (1.6.6) into consideration, after some transformations we obtain the following equation corresponding to Hamilton's variational principle:

$$\int_V \left[\frac{\partial}{\partial x_i} \left(\mu_{ij\alpha\beta} \frac{\partial}{\partial x_\beta} \upsilon_\alpha \right) + \rho\omega^2 \upsilon_j \right] \delta\upsilon_j dV - \int_{S_1} N_i \left(\mu_{ij\alpha\beta} \frac{\partial}{\partial x_\beta} \upsilon_\alpha \right) \delta\upsilon_j dS = 0. \tag{1.6.7}$$

Now we consider the eigenvalue problem formulated as follows:

$$\mu^{(0)}_{ij\alpha\beta} \frac{\partial^2 \upsilon_\alpha}{\partial x_i \partial x_\beta} + \rho\Omega^2 \upsilon_j = 0, \quad x_1, x_2, x_3 \in V, \tag{1.6.8}$$

$$N_i \mu^{(0)}_{ij\alpha\beta} \frac{\partial \upsilon_\alpha}{\partial x_\beta} = 0, \quad x_1, x_2, x_3 \in S_1, \tag{1.6.9}$$

$$\upsilon_j = 0, \quad x_1, x_2, x_3 \in S_2. \tag{1.6.10}$$

This corresponds to a classical body with rectilinear orthotropy which is characterized by the tensor $\boldsymbol{\mu}^{(0)}$ of elasticity constants.

The components of the m-th eigenvector and the eigenvalue are denoted as follows:

$$\upsilon_{\alpha m}, \quad \rho\Omega^2_m, \quad m = 1, 2, \ldots, \infty, \quad \alpha = 1, 2, 3. \tag{1.6.11}$$

After determining the eigenvector and eigenvalue (1.6.11), we represent the displacements υ_α in (1.6.6), (1.6.7) in the series form

$$\upsilon_\alpha = \sum_{m=1}^{\infty} c_m \upsilon_{\alpha m}, \quad c_m = \text{const} \tag{1.6.12}$$

Substituting (1.6.12) and (1.6.4) into (1.6.7), taking (1.6.8)-(1.6.10) and the orthogonality of the eigenvectors $\upsilon_{\alpha m}$ into account and grouping the coefficients δc_n, we obtain from (1.6.6) a set of homogeneous algebraic equations for the constants c_n.

$$\sum_{m=1}^{\infty} c_m \left[\delta_m^n \rho\left(\omega^2 - \Omega_m^2\right) + \varepsilon a_{nm} + \varepsilon^2 b_{nm}\right] = 0, \qquad n = 1,2,...,\infty. \tag{1.6.13}$$

In (1.6.13) the following notation is introduced:

$$a_{nm} = -\int_V (\sin\gamma)\mu_{ij\alpha\beta}^{(1)} \left(\frac{\partial \upsilon_{\alpha m}}{\partial x_\beta}\right)\left(\frac{\partial \upsilon_{jn}}{\partial x_i}\right) dV, \tag{1.6.14}$$

$$b_{nm} = -\int_V \sin^2\gamma \mu_{ij\alpha\beta}^{(2)} \left(\frac{\partial \upsilon_{\alpha m}}{\partial x_\beta}\right)\left(\frac{\partial \upsilon_{jn}}{\partial x_i}\right) dV. \tag{1.6.15}$$

According to the well-known procedure we obtain from (1.6.13) the frequency-equation

$$\det\left(\delta_m^n \rho\left(\omega^2 - \Omega_m^2\right) + \varepsilon a_{nm} + \varepsilon^2 b_{nm}\right) = 0, \quad n; m = 1,2,...,\infty. \tag{1.6.16}$$

In (1.6.16) Ω_m is the natural frequency of the m-th mode of the body with rectilinear orthotropy.

From (1.4.3) by direct verification we establish that the tensor $\mu^{(1)}$ and $\mu^{(2)}$ have the same symmetry properties as the tensor $\mu^{(0)}$. Taking this situation into account, we find from (1.6.14), (1.6.15) that a_{nm} and b_{nm} and hence also the matrix of coefficients in (1.6.13), are symmetric.

1.6.2. AN EXAMPLE

As an example, we consider the strip $|x_1| \le d$ which is infinite in the directions of the Ox_2 and Ox_3 axes (Fig.1.6.1). Note that the Ox_3 axis is directed perpendicularly to the plane of Fig.1.6.1. We investigate the shear vibration of this strip; the displacement field has the form

$$u_1 \equiv 0, \quad u_2 \equiv 0, \quad u_3 = u_3(x_1, \tau). \tag{1.6.17}$$

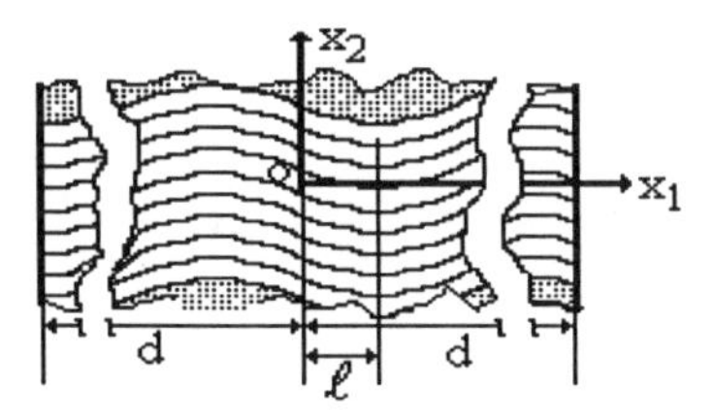

Fig.1.6.1. The geometry of the vibrating strip.

We suppose that the strip is rigidly supported at the ends $x_1 = \pm d$. Thus, from (1.6.17), (1.6.6), (1.6.8)-(1.6.10) we obtain

$$\frac{d^2}{dx_1^2}\upsilon_3 + \rho\frac{\Omega^2}{G_{13}}\upsilon_3 = 0, \quad |x_1| < d; \quad \upsilon_3\big|_{x_1=\pm d} = 0. \tag{1.6.18}$$

According to (1.6.11) the orthonormal eigenfunctions and eigenvalues of the problem (1.6.18) are

$$\upsilon_{3m} = \frac{1}{\sqrt{d}}\sin\left(\pi\frac{m}{d}x_1\right), \quad \rho\Omega_m^2 = \left(\frac{m\pi}{d}\right)^2 G_{13}^0. \tag{1.6.19}$$

Taking the equations (1.6.14), (1.6.15), (1.4.3), (1.6.19) into account, we find after a series of transformations the following expressions for a_{nm} and b_{nm}:

$$a_{nm} = 0, \quad b_{nm} = \frac{2\pi nm}{\Lambda d}\sin\left(\pi\frac{2d}{\Lambda}\right)\cos\left(\pi\frac{\ell}{\Lambda}\right)(-1)^{m+n}\frac{\left(G_{13}^0 - G_{23}^0\right)}{\left[\left(2d\Lambda^{-1}\right)^2 - (m+n)^2\right]}$$

$$\times\frac{(2d\Lambda^{-1})^2 - m^2 - n^2}{(2d\Lambda^{-1})^2 - (m-n)^2}. \tag{1.6.20}$$

For this problem, the frequency equation is obtained from (1.6.13) and (1.6.20) in the form.

$$\sum_{m=1}^{\infty} c_m\left[\delta_m^n\rho\left(\omega^2 - \Omega_m^2\right) + \varepsilon^2 b_{nm}\right] = 0, \quad n, m = 1, 2, \ldots, \infty. \tag{1.6.21}$$

We now analyze two particular cases with respect to the values of parameters $2d\Lambda^{-1}$ and $\ell\Lambda^{-1}$. In the first case we suppose that the thickness d of the strip is an integral number M of half-wavelengths of the curvature, i.e. $d = M\Lambda/2$. For this case $\sin(\pi 2d/\Lambda) = 0$ and hence $b_{nm} = 0$ except when $m \pm n = M$. For the latter, a limiting argument yields

$$m + n = M, \quad b_{nm} = \frac{\pi^2(-1)^{n+m}}{2d\Lambda}\left(G_{13}^0 - G_{23}^0\right)\frac{mn}{m+n}\cos\left(\pi\frac{\ell}{\Lambda}\right),$$

$$m - n = M, \quad b_{nm} = \frac{\pi^2(-1)^{n+m}}{2d\Lambda}\left(G_{13}^0 - G_{23}^0\right)\frac{mn}{m-n}\cos\left(\pi\frac{\ell}{\Lambda}\right). \qquad (1.6.22)$$

For a second particular case we suppose that the origin of the coordinate system associated with the middle plane of the strip shifts (along the Ox_1 axis) a quarter of the curvature period from the summit, then

$$\ell = \frac{1}{2}\Lambda \qquad (1.6.23)$$

and equation (1.6.20) shows that $b_{nm} = 0$ for all m,n. Now the values of ω_n (the frequency of the strip fabricated from the composite material with periodically plane-curved structure) coincide with the values of Ω_n (the frequency of the strip fabricated from the composite material with ideal structure). We conclude that, depending on the size of the body and on the arrangement of the composite material, there is a vibration mode for which the presence of the curvature does not change the values of frequency of any mode.

1.7. Quasi-homogeneous Stress States Corresponding to Pure Shears

We will now use the small parameter method described in section 1.4, to investigate the quasi-homogeneous stress states in an infinite composite material with regular plane-curved structure under pure shear. As in section 1.5, by a quasi-homogeneous stress state in an infinite body, generated by corresponding loads, we will understand a stress state which, for the same material without structural curving, corresponds to a uniform stress state. According to this definition the quasi-homogeneous stresses consist of two parts: 1) those that balance the external loads; 2) those that are self-balanced within one period of the structure curvature. The self-balanced parts of the stresses are caused by the curvatures within the structure; the averages of the self-balanced parts over the curvature period are zero.

Let us examine the quasi-homogeneous stress states corresponding to shears in different planes. We note that the problem of a quasi-homogeneous stress state corresponding to pure shear in the plane Ox_1x_3 has been exactly solved in section 1.5. We now examine problems for shears in other planes, i.e., in Ox_2x_3 and Ox_1x_2, which do not have an exact solutions.

1.7.1. PURE SHEAR IN THE Ox_2x_3 PLANE (Fig.1.7.1)

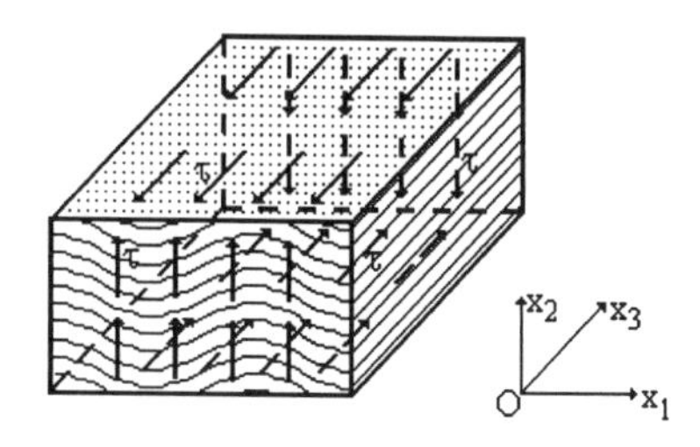

Fig.1.7.1. Pure shear in the Ox_2x_3 plane.

We study the stress-strain state when the applied external load is

$$\sigma_{23} = \tau . \tag{1.7.1}$$

For this case the values of the zeroth approximation are determined as follows:

$$\sigma_{13}^{(0)} = 0; \quad \sigma_{23}^{(0)} = \tau; \quad u_3^{(0)} = \frac{\tau}{G_{23}^0} x_2 . \tag{1.7.2}$$

For the n-th approximation we choose the following displacement field:

$$u_1^{(n)} \equiv 0; \quad u_2^{(n)} \equiv 0; \quad u_3^{(n)} = u_3^{(n)}(x_1, x_2). \tag{1.7.3}$$

From (1.7.1) and (1.4.11), we obtain the non-trivial stress-tensor components in the n-th approximation in the form

$$\sigma_{23}^{(n)} = G_{23}^0 \frac{\partial}{\partial x_2} u_3^{(n)} + \sin\theta \left(G_{13}^0 - G_{23}^0\right) \frac{\partial}{\partial x_1} u_3^{(n-1)}$$

$$+ \sin^2\theta \left(G_{13}^0 - G_{23}^0\right) \frac{\partial}{\partial x_2} u_3^{(n-2)},$$

$$\sigma_{13}^{(n)} = G_{13}^0 \frac{\partial}{\partial x_1} u_3^{(n)} + \sin\theta \left(G_{13}^0 - G_{23}^0\right) \frac{\partial}{\partial x_2} u_3^{(n-1)}$$

$$+ \sin^2\theta \left(G_{23}^0 - G_{13}^0\right) \frac{\partial}{\partial x_1} u_3^{(n-2)}, \quad \theta = \frac{\pi x_1}{\Lambda} . \tag{1.7.4}$$

Equations (1.4.8), (1.4.3) and (1.7.3) yield the following equation for $u_3^{(n)}$

$$\left(G_{13}^{0}\frac{\partial^{2}}{\partial x_{1}^{2}}+G_{23}^{0}\frac{\partial^{2}}{\partial x_{2}^{2}}\right)u_{3}^{(n)}=-\left(G_{13}^{0}-G_{23}^{0}\right)\left\{\left(\frac{\partial}{\partial x_{1}}(\sin\theta)\frac{\partial}{\partial x_{2}}\right.\right.$$

$$\left.\left.+\sin\theta\frac{\partial^{2}}{\partial x_{2}\partial x_{1}}\right)u_{3}^{(n-1)}-\left[\sin^{2}\theta\frac{\partial^{2}}{\partial x_{2}^{2}}-\frac{\partial}{\partial x_{1}}\sin^{2}\theta\frac{\partial}{\partial x_{1}}\right]u_{3}^{(n-2)}\right\} \quad (1.7.5)$$

From (1.7.2), (1.7.4) and (1.7.5) we get the following equations for the first approximation.

$$\sigma_{13}^{(1)}=G_{13}^{0}\frac{\partial}{\partial x_{1}}u_{3}^{(1)}+\tau\frac{G_{13}^{0}-G_{23}^{0}}{G_{23}^{0}}\sin\theta,\ \sigma_{23}^{(1)}=G_{23}^{0}\frac{\partial}{\partial x_{2}}u_{3}^{(1)},$$

$$\left(G_{13}^{0}\frac{\partial^{2}}{\partial x_{1}^{2}}+G_{23}^{0}\frac{\partial^{2}}{\partial x_{2}^{2}}\right)u_{3}^{(1)}=-\tau\frac{G_{13}^{0}-G_{23}^{0}}{G_{23}^{0}}\frac{\pi}{\Lambda}\cos\theta. \quad (1.7.6)$$

In the beginning of this section we noted that in an infinite body the quasi-homogeneous stresses consist of two parts, the first of which is balanced by the external forces. It follows from (1.7.2) that this part is determined only by the zeroth approximation and all boundary conditions are satisfied by this approximation. Therefore for determining the first approximation, unknown constants entering the general solution of the equation (1.7.6) must be taken as zero. In other words, for the infinite body the second (self-balanced) part of the quasi-homogeneous stresses must be determined only by the particular solution to the equation (1.7.6). It should be noted that in the analogous manner we will investigate the equations corresponding to the second approximation. Moreover, in this and following chapters we will investigate other problems related to a quasi-homogeneous stress-state in an infinite body in a similar fashion.

Thus, we choose the particular solution to equation (1.7.6) as follows

$$u_{3}^{(1)}=\tau\frac{G_{13}^{0}-G_{23}^{0}}{G_{13}^{0}G_{23}^{0}}\frac{\pi}{\Lambda}\cos\theta. \quad (1.7.7)$$

From the first two expressions of (1.7.6) we obtain, in the first approximation, the stresses

$$\sigma_{23}^{(1)}=0,\quad \sigma_{13}^{(1)}=0. \quad (1.7.8)$$

From (1.7.2), (1.7.8) and (1.7.4) we derive the second approximation for the stresses:

$$\sigma_{13}^{(2)}=G_{13}^{0}\frac{\partial}{\partial x_{1}}u_{3}^{(2)},\sigma_{23}^{(2)}=G_{23}^{0}\frac{\partial}{\partial x_{2}}u_{3}^{(2)}+\tau\frac{G_{13}^{0}-G_{23}^{0}}{2G_{13}^{0}}2\sin^{2}\theta. \quad (1.7.9)$$

From (1.7.2), (1.7.7), and (1.7.5) we obtain an equation for $u_3^{(2)}$:

$$\left(G_{13}^0 \frac{\partial^2}{\partial x_1^2} + G_{23}^0 \frac{\partial^2}{\partial x_2^2} \right) u_3^{(2)} = 0. \tag{1.7.10}$$

It follows from the expression $\sigma_{23}^{(2)}$ in (1.7.9) that the last term gives the non-self-balanced part. Since $\sigma_{23}^{(2)}$ must be a self-balanced within the curvature period, and $\sigma_{ij}^{(2)} = 0$ $(i \neq 2, j \neq 3)$ we choose the following particular solution for the equation (1.7.10):

$$u_3^{(2)} = -\tau \frac{G_{13}^0 - G_{23}^0}{2G_{13}^0 G_{23}^0} x_2. \tag{1.7.11}$$

Now equation (1.7.9), and (1.7.11) give the self-balanced stresses in the second approximation:

$$\sigma_{23}^{(2)} = \tau \frac{G_{23}^0 - G_{13}^0}{2G_{13}^0} \cos 2\theta, \quad \sigma_{13}^{(2)} = 0. \tag{1.7.12}$$

In accordance with (1.4.7), equations (1.4.7), (1.7.2), (1.7.7), (1.7.8) and (1.7.11) are used together with (1.7.12) to obtain the following expression including the zeroth, first, and second approximations:

$$\sigma_{13} = 0, \ \sigma_{23} = \tau \left(1 + \varepsilon^2 \frac{g_0}{2} \cos 2\theta \right), \ g_0 = \frac{G_{13}^0 - G_{23}^0}{G_{13}^0}$$

$$u_3 = \frac{\tau}{G_{23}^0} \left(x_2 + \varepsilon \frac{\Lambda}{\pi} g_0 \cos\theta + \varepsilon^2 g^0 x_2 \right). \tag{1.7.13}$$

Equation (1.7.13) shows that the additional self-balanced stresses for the quasi-homogeneous stress state are proportional to the relative difference between the shear moduli G_{23}^0 and G_{13}^0.

1.7.2. PURE SHEAR IN THE Ox_1x_2 PLANE (Fig.1.7.2)

The external loads are

$$\sigma_{12} = \tau . \tag{1.7.14}$$

In the zeroth approximation, the stresses are

$$\sigma_{12}^{(0)} = \tau , \; \sigma_{11}^{(0)} = 0, \; \sigma_{22}^{(0)} = 0, \; u_2^{(0)} = \frac{\tau}{G_{12}^0} x_1 . \tag{1.7.15}$$

For the n-th approximation we use the displacement field

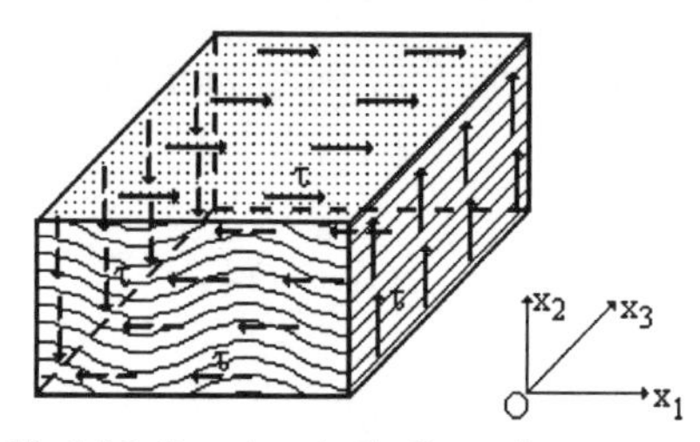

Fig.1.7.2. Pure shear in the Ox_1x_2 plane.

$$u_1^{(n)} = u_1^{(n)}(x_1, x_2), \; u_2^{(n)} = u_2^{(n)}(x_1, x_2), \; u_3^{(n)} = 0, \tag{1.7.16}$$

since the body is infinite in the direction of the Ox_3 axis, and all quantities are independent of x_3.

In the first approximation, from the last expression of (1.7.15), (1.4.11) and (1.4.3) we have

$$\sigma_{11}^{(1)} = A_{11}^0 \frac{\partial u_1^{(1)}}{\partial x_1} + A_{12}^0 \frac{\partial u_2^{(1)}}{\partial x_2} + \tau \left(G_{12}^0 \right)^{-1} \left(A_{11}^0 - A_{12}^0 - 2G_{12}^0 \right) \sin\theta ,$$

$$\sigma_{12}^{(1)} = G_{12}^0 \left(\frac{\partial u_1^{(1)}}{\partial x_2} + \frac{\partial u_2^{(1)}}{\partial x_1} \right), \quad \theta = \frac{\pi x_1}{\Lambda} ,$$

$$\sigma_{22}^{(1)} = A_{12}^0 \frac{\partial u_1^{(1)}}{\partial x_1} + A_{22}^0 \frac{\partial u_2^{(1)}}{\partial x_2} + \tau \left(G_{12}^0 \right)^{-1} \left(- A_{22}^0 + A_{12}^0 + 2G_{12}^0 \right) \sin\theta . \tag{1.7.17}$$

The last expression of (1.7.15), (1.4.8), and (1.4.3) yield the following system of equations for the first approximation:

$$\left(A_{11}^0 \frac{\partial^2}{\partial x_1^2} + G_{12}^0 \frac{\partial^2}{\partial x_2^2} \right) u_1^{(1)} + \left(A_{12}^0 + G_{12}^0 \right) \frac{\partial^2}{\partial x_1 \partial x_2} u_2^{(1)}$$

$$= -\frac{\tau}{G_{12}^{0}}\left(A_{11}^{0} - A_{12}^{0} - 2G_{12}^{0}\right)\frac{\pi}{\Lambda}\cos\theta \ , \ \theta = \frac{\pi x_1}{\Lambda} \ ,$$

$$\left(A_{22}^{0}\frac{\partial^2}{\partial x_1^2} + G_{12}^{0}\frac{\partial^2}{\partial x_2^2}\right)u_2^{(1)} + \left(A_{12}^{0} + G_{12}^{0}\right)\frac{\partial^2}{\partial x_1 \partial x_2}u_1^{(1)} = 0 \ . \qquad (1.7.18)$$

Considering (1.7.17), we choose the solution to the system (1.7.18) in the form

$$u_2^{(1)} = 0 \ , \ u_1^{(1)} = \frac{\tau}{A_{11}^{0}G_{12}^{0}}\left(A_{11}^{0} - A_{12}^{0} - 2G_{12}^{0}\right)\frac{\Lambda}{\pi}\cos\theta \ . \qquad (1.7.19)$$

For the first approximation, (1.7.17) and (1.7.19) give the stresses

$$\sigma_{11}^{(1)} = 0, \ \sigma_{12}^{(1)} = 0 \ , \ \sigma_{22}^{(1)} = \tau\frac{\left(A_{12}^{0}\right)^2 - A_{11}^{0}A_{22}^{0} + 2G_{12}^{0}\left(A_{11}^{0} + A_{12}^{0}\right)}{A_{11}^{0}G_{12}^{0}}\sin\theta \ . \qquad (1.7.20)$$

In the second approximation, equations (1.7.15), (1.7.19), (1.4.11) and (1.4.3) yield

$$\sigma_{11}^{(2)} = A_{11}^{0}\frac{\partial u_1^{(2)}}{\partial x_1} + A_{12}^{0}\frac{\partial u_2^{(2)}}{\partial x_2} \ , \ \sigma_{22}^{(2)} = A_{12}^{0}\frac{\partial u_1^{(2)}}{\partial x_1} + A_{22}^{0}\frac{\partial u_2^{(2)}}{\partial x_2} \ ,$$

$$\sigma_{12}^{(2)} = G_{12}^{0}\left(\frac{\partial u_1^{(2)}}{\partial x_2} + \frac{\partial u_2^{(2)}}{\partial x_1}\right) + \tau C\sin^2\theta \ ,$$

$$C = \left(A_{11}^{0}G_{12}^{0}\right)^{-1}\left[\left(A_{11}^{0} - A_{12}^{0} - 2G_{12}^{0}\right)^2 + \left(A_{11}^{0} - 2A_{12}^{0} + A_{22}^{0} - 4G_{12}^{0}\right)A_{11}^{0}\right]. \qquad (1.7.21)$$

For the second approximation we obtain the following system of differential equations from the (1.4.3), (1.4.8), (1.7.15) and (1.7.19)

$$\left(A_{11}^{0}\frac{\partial^2}{\partial x_1^2} + G_{12}^{0}\frac{\partial^2}{\partial x_2^2}\right)u_1^{(2)} + \left(A_{12}^{0} + G_{12}^{0}\right)\frac{\partial^2}{\partial x_1 \partial x_2}u_2^{(2)} = 0 \ ,$$

$$\left(A_{22}^{0}\frac{\partial^2}{\partial x_2^2} + G_{12}^{0}\frac{\partial^2}{\partial x_1^2}\right)u_2^{(2)} + \left(A_{12}^{0} + G_{12}^{0}\right)\frac{\partial^2}{\partial x_1 \partial x_2}u_1^{(2)} = -\frac{\pi}{\Lambda}\tau C\sin(2\theta). \qquad (1.7.22)$$

Considering (1.7.21) and taking into account the explanations given in the previous subsection, we choose the solution to the system (1.7.22) in the form

$$u_1^{(2)} = 0, \quad u_2^{(2)} = \frac{\Lambda}{4\pi}\frac{\tau C}{G_{12}^0}\sin 2\theta - \frac{\tau C}{2G_{12}^0}x_1. \tag{1.7.23}$$

Equations (1.7.21) and (1.7.23) yield in the second approximation

$$\sigma_{12}^{(2)} = 0; \sigma_{11}^{(2)} = 0; \sigma_{22}^{(2)} = 0. \tag{1.7.24}$$

In accordance with (1.4.7), equations (1.7.15), (1.7.19), (1.7.23), and (1.7.24) give us the following expressions, including the zeroth, first, and second approximations:

$$\sigma_{12} = 0, \ \sigma_{11} = 0, \ \sigma_{22} = \varepsilon\tau\frac{\left(A_{12}^0\right)^2 - A_{11}^0 A_{22}^0 + 2G_{12}^0\left(A_{11}^0 + A_{12}^0\right)}{A_{11}^0 G_{12}^0}\sin\theta.$$

$$u_1 = \varepsilon\frac{\tau}{A_{11}^0 G_{12}^0}\left(A_{11}^0 - A_{12}^0 - 2G_{12}^0\right)\frac{\Lambda}{\pi}\cos\theta,$$

$$u_2 = \frac{\tau}{G_{12}^0}\left[x_1 + \varepsilon^2\left(\frac{\Lambda}{4\pi}\frac{\tau C}{G_{12}^0}\sin 2\theta - \frac{\tau C}{2G_{12}^0}x_1\right)\right]. \tag{1.7.25}$$

Thus, for shear in the Ox_1x_2 plane, self-balanced normal stresses (1.7.25) also occur.

1.7.3. PURE SHEAR IN THE Ox_1x_3 PLANE (Fig.1.5.1)

The exact solution to the present problem was given in section 1.5. Here, for completeness we give the final results including the zeroth, first and second approximations.

$$\sigma_{13} = \tau, \ \sigma_{23} = \tau\varepsilon g_0 \sin\theta, \ g_0 = \frac{G_{23}^0 - G_{13}^0}{G_{13}^0}$$

$$u_3 = \frac{\tau}{G_{13}^0}\left[x_1 - \varepsilon^2\frac{1}{2}g_0\left(x_1 - \frac{\Lambda}{2\pi}\sin(2\theta)\right)\right]. \tag{1.7.26}$$

1.8. Quasi-homogeneous States Corresponding to Tension-Compression

We will examine the quasi-homogeneous stress states corresponding to tension-compression along different coordinate axes. With regard to pure shear, in the preceding section we studied each case separately. For tension-compression, we will examine two cases at the same time. It should be noted that the problems which will be investigated in this section do not have exact solutions.

1.8.1. TRIAXIAL TENSION-COMPRESSION (Fig.1.8.1)

We study the stress-strain state when the external loads are assigned in the form

$$\sigma_{11} = p_1, \quad \sigma_{22} = p_2, \quad \sigma_{33} = p_3. \tag{1.8.1}$$

Zeroth Approximation.
In the zeroth approximation, we choose the displacements in the form

$$u_1^{(0)} = A_1 x_1; \quad u_2^{(0)} = A_2 x_2; \quad u_3^{(0)} = A_3 x_3, \tag{1.8.2}$$

where the constants A_j are determined from the system

$$A_1 A_{11}^0 + A_2 A_{12}^0 + A_3 A_{13}^0 = p_1, \quad A_1 A_{12}^0 + A_2 A_{22}^0 + A_3 A_{23}^0 = p_2,$$

$$A_1 A_{13}^0 + A_2 A_{23}^0 + A_3 A_{33}^0 = p_3. \tag{1.8.3}$$

If, together with A_{ij}^0, the elasticity constants, we introduce a_{ij}^0, the compliance constants constituting the adjoint of the matrix $\left(A_{ij}^0\right)$, then for the constant A_j we obtain

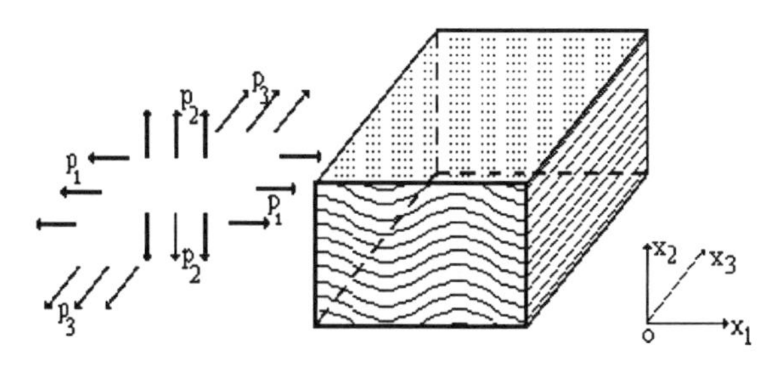

Fig.1.8.1. Triaxial tension-compression.

$$A_1 = a_{11}^0 p_1 + a_{12}^0 p_2 + a_{13}^0 p_3; \quad A_2 = a_{12}^0 p_1 + a_{22}^0 p_2 + a_{23}^0 p_3;$$

$$A_3 = a_{13}^0 p_1 + a_{23}^0 p_2 + a_{33}^0 p_3. \tag{1.8.4}$$

First Approximation.
In the first approximation, (1.8.2), (1.4.11) and (1.4.3) give stresses

$$\sigma_{11}^{(1)} = A_{11}^0 \frac{\partial u_1^{(1)}}{\partial x_1} + A_{12}^0 \frac{\partial u_2^{(1)}}{\partial x_2} + A_{13}^0 \frac{\partial u_3^{(1)}}{\partial x_3},$$

$$\sigma_{12}^{(1)} = G_{12}^0 \left(\frac{\partial u_1^{(1)}}{\partial x_2} + \frac{\partial u_2^{(1)}}{\partial x_1} \right) + C_1 \sin\theta,$$

$$\sigma_{13}^{(1)} = G_{13}^0 \left(\frac{\partial u_1^{(1)}}{\partial x_3} + \frac{\partial u_3^{(1)}}{\partial x_1} \right), \quad \sigma_{23}^{(1)} = G_{23}^0 \left(\frac{\partial u_2^{(1)}}{\partial x_3} + \frac{\partial u_3^{(1)}}{\partial x_2} \right),$$

$$\sigma_{22}^{(1)} = A_{12}^0 \frac{\partial u_1^{(1)}}{\partial x_1} + A_{22}^0 \frac{\partial u_2^{(1)}}{\partial x_2} + A_{23}^0 \frac{\partial u_3^{(1)}}{\partial x_3},$$

$$\sigma_{33}^{(1)} = A_{13}^0 \frac{\partial u_1^{(1)}}{\partial x_1} + A_{23}^0 \frac{\partial u_2^{(1)}}{\partial x_2} + A_{33}^0 \frac{\partial u_3^{(1)}}{\partial x_3}, \tag{1.8.5}$$

where

$$C_1 = A_1\left(A_{11}^0 - A_{12}^0 - 2G_{12}^0\right) + A_2\left(-A_{22}^0 + A_{12}^0 + G_{12}^0\right) + A_3\left(A_{13}^0 - A_{23}^0\right). \tag{1.8.6}$$

Equations (1.8.2), (1.4.8) and (1.4.3) yield the following system in the first approximation

$$\left(A_{11}^0 \frac{\partial^2}{\partial x_1^2} + G_{12}^0 \frac{\partial^2}{\partial x_2^2} + G_{13}^0 \frac{\partial^2}{\partial x_3^2} \right) u_1^{(1)} + \left(A_{12}^0 + G_{12}^0\right) \frac{\partial^2}{\partial x_1 \partial x_2} u_2^{(1)}$$

$$+ \left(A_{13}^0 + G_{13}^0\right) \frac{\partial^2}{\partial x_1 \partial x_3} u_3^{(1)} = 0,$$

$$\left(A_{12}^{0}+G_{12}^{0}\right)\frac{\partial^2}{\partial x_1 \partial x_2}u_1^{(1)}+\left(G_{12}^{0}\frac{\partial^2}{\partial x_1^2}+A_{22}^{0}\frac{\partial^2}{\partial x_2^2}+G_{23}^{0}\frac{\partial^2}{\partial x_3^2}\right)u_2^{(1)}$$

$$+\left(A_{23}^{0}+G_{23}^{0}\right)\frac{\partial^2}{\partial x_1 \partial x_3}u_3^{(1)}=-C_1\frac{\pi}{\Lambda}\cos\theta,$$

$$\left(A_{13}^{0}+G_{13}^{0}\right)\frac{\partial^2}{\partial x_1 \partial x_3}u_1^{(1)}+\left(A_{23}^{0}+G_{23}^{0}\right)\frac{\partial^2}{\partial x_2 \partial x_3}u_2^{(1)}$$

$$+\left(G_{13}^{0}\frac{\partial^2}{\partial x_1^2}+G_{23}^{0}\frac{\partial^2}{\partial x_2^2}+A_{33}^{0}\frac{\partial^2}{\partial x_3^2}\right)u_3^{(1)}=0. \tag{1.8.7}$$

Taking the particular solution as before we choose the solution to the system (1.8.7) in the form

$$u_1^{(1)}=0,\ u_2^{(1)}=\frac{1}{G_{12}^{0}}\frac{\Lambda}{\pi}C_1\cos\theta,\quad u_3^{(1)}=0. \tag{1.8.8}$$

Equations (1.8.5) and (1.8.8) give

$$\sigma_{ij}^{(1)}=0. \tag{1.8.9}$$

Second approximation.

In the second approximation, equations (1.8.2), (1.8.8), (1.4.8) and (1.4.3) yield

$$\sigma_{11}^{(2)}=A_{11}^{0}\frac{\partial u_1^{(2)}}{\partial x_1}+A_{12}^{0}\frac{\partial u_2^{(2)}}{\partial x_2}+A_{13}^{0}\frac{\partial u_3^{(2)}}{\partial x_3}+C_{11}\sin^2\theta,$$

$$\sigma_{22}^{(2)}=A_{12}^{0}\frac{\partial u_1^{(2)}}{\partial x_1}+A_{22}^{0}\frac{\partial u_2^{(2)}}{\partial x_2}+A_{23}^{0}\frac{\partial u_3^{(2)}}{\partial x_3}+C_{22}\sin^2\theta,$$

$$\sigma_{33}^{(2)}=A_{13}^{0}\frac{\partial u_1^{(2)}}{\partial x_1}+A_{23}^{0}\frac{\partial u_2^{(2)}}{\partial x_2}+A_{33}^{0}\frac{\partial u_3^{(3)}}{\partial x_3}+C_{33}\sin^2\theta,$$

$$\sigma_{12}^{(2)}=G_{12}^{0}\left(\frac{\partial u_1^{(2)}}{\partial x_2}+\frac{\partial u_2^{(2)}}{\partial x_1}\right),\quad \sigma_{13}^{(2)}=G_{13}^{0}\left(\frac{\partial u_1^{(2)}}{\partial x_3}+\frac{\partial u_3^{(2)}}{\partial x_1}\right),$$

$$\sigma_{23}^{(2)} = G_{23}^{0}\left(\frac{\partial u_2^{(2)}}{\partial x_3} + \frac{\partial u_3^{(2)}}{\partial x_2}\right) \tag{1.8.10}$$

where the C_{ii}, i=1,2,3 are

$$C_{11} = 2A_1\left(-A_{11}^0 + A_{12}^0 + 2G_{12}^0\right) + A_2\left(A_{11}^0 + A_{22}^0 - 2A_{12}^0 - 4G_{12}^0\right)$$

$$+ A_3\left(A_{23}^0 - A_{13}^0\right) - C_1\left(G_{12}^0\right)^{-1}\left(A_{11}^0 - A_{12}^0 - 2G_{12}^0\right),$$

$$C_{22} = A_1\left(A_{11}^0 + A_{22}^0 - 2A_{12}^0 - 4G_{12}^0\right) + 2A_2\left(-A_{22}^0 + A_{12}^0 + 2G_{12}^0\right)$$

$$+ A_3\left(A_{13}^0 - A_{23}^0\right) - C_1(G_{12}^0)^{-1}\left(-A_{22}^0 + A_{12}^0 + 2G_{12}^0\right),$$

$$C_{33} = \left(A_{23}^0 - A_{13}^0\right)\left(A_1 - A_2 + C_1\left(G_{12}^0\right)^{-1}\right). \tag{1.8.11}$$

Equations (1.8.2), (1.8.8), (1.4.8) and (1.4.3) give the following equations for the second approximation:

$$L_{j\alpha}u_{\alpha}^{(2)} = -\delta_j^2 C_{11}\frac{\pi}{\Lambda}\sin 2\theta;\ j,\alpha = 1,2,3. \tag{1.8.12}$$

The differential Lamé operators $L_{j\alpha}$ have the form (1.4.6) without inertia terms or they correspond to the left hand side of the equations (1.8.7).

Considering (1.8.10), we take the solution to the equation (1.8.12) in the form

$$u_1^{(2)} = B_1x_1 + \frac{C_{11}}{A_{11}^0}\frac{\Lambda}{4\pi}\sin 2\theta,\ u_2^{(2)} = B_2x_2,\ u_3^{(2)} = B_3x_3. \tag{1.8.13}$$

Choosing the constants B_1, B_2, and B_3 in (1.8.3) so that

$$B_1A_{11}^0 + B_2A_{12}^0 + B_3A_{13}^0 = -\frac{1}{2}C_{11},\ B_1A_{12}^0 + B_2A_{22}^0 + B_3A_{23}^0 = -\frac{1}{2}C_{22},$$

$$B_1A_{13}^0 + B_2A_{23}^0 + B_3A_{33}^0 = -\frac{1}{2}C_{33}, \tag{1.8.14}$$

we may express for B_1, B_2 and B_3 in terms of the adjoint coefficients:

$$B_1 = -\frac{1}{2}\left(C_{11}a_{11}^0 + C_{22}a_{12}^0 + C_{33}a_{13}^0\right),\ B_2 = -\frac{1}{2}\left(C_{11}a_{12}^0 + C_{22}a_{22}^0 + C_{33}a_{23}^0\right),$$

$$B_3 = -\frac{1}{2}\left(C_{11}a_{13}^0 + C_{22}a_{23}^0 + C_{33}a_{33}^0\right). \qquad (1.8.15)$$

Equations (1.8.10), (1.8.13), and (1.8.14) yield the second order approximations for the stress-tensor components:

$$\sigma_{22}^{(2)} = -\frac{1}{2}\left(C_{22} - A_{22}^0 a_0\right)\cos 2\theta,\ a_0 = \frac{C_{11}}{A_{11}^0},$$

$$\sigma_{33}^{(2)} = -\frac{1}{2}\left(C_{33} - A_{13}^0 a_0\right)\cos 2\theta. \qquad (1.8.16)$$

From (1.4.7), (1.8.1), (1.8.2), (1.8.4), (1.8.8), (1.8.9), (1.8.13), (1.8.15) and (1.8.16), we obtain the following expressions including again the zeroth, first and second approximations:

$$\sigma_{11} = p_1,\ \sigma_{22} = p_2 - \varepsilon^2\frac{1}{2}\left(C_{22} - A_{12}^0 a_0\right)\cos 2\theta,$$

$$a_0 = \frac{C_{11}}{A_{11}^0},\ \sigma_{33} = p_3 - \varepsilon^2\frac{1}{2}\left(C_{33} - A_{13}^0 a_0\right)\cos 2\theta,$$

$$u_1 = A_1 x_1 + \varepsilon^2\left(B_1 x_1 + a_0\frac{\Lambda}{4\pi}\sin 2\theta\right),$$

$$u_2 = A_2 x_2 + \varepsilon\frac{C_1}{G_{12}^0}\frac{\Lambda}{\pi}\cos\theta + \varepsilon^2 B_2 x_2,\ u_3 = A_3 x_3 + \varepsilon^2 B_3 x_3. \qquad (1.8.17)$$

Note that the constants A_j, C_1, C_{jj} and B_j (j=1,2,3) entering (1.8.17) are determined through the expressions (1.8.4), (1.8.6), (1.8.11) and (1.8.15), respectively.

Note that the results (1.8.17) were obtained for an arbitrary triaxial uniform tension-compression. Let us examine a special case in more detail.

1.8.2. UNIAXIAL TENSION-COMPRESSION ALONG Ox_1 AXIS

We will suppose that all quantities are independent of x_3 (plane strain in the Ox_1x_2 plane). In this case we must set

$$p_2 = 0, \ p_3 = 0, \ A_3 = 0. \tag{1.8.18}$$

Using equations (1.8.18), (1.8.3) and (1.8.4), we find

$$A_1 = p_1 A_{22}^0 D_0, \ A_2 = -p_1 A_{12}^0 D_0, \tag{1.8.19}$$

where

$$D_0 = \left(A_{11}^0 A_{22}^0 - \left(A_{12}^0 \right)^2 \right)^{-1}. \tag{1.8.20}$$

From (1.8.6), (1.8.18) and (1.8.19), we obtain

$$C_1 = p_1 \left[1 - 2G_{12}^0 \left(A_{22}^0 + A_{12}^0 \right) D_0 \right]. \tag{1.8.21}$$

Substituting (1.8.18)-(1.8.21) into (1.8.11) we find out the constants C_{ii} (i=1,2,3), as follows

$$C_{11} = p_1 \left[A_{12}^0 \left(A_{11}^0 - A_{22}^0 \right) D_0 - \left(G_{12}^0 \right)^{-1} \left(A_{11}^0 - A_{12}^0 - 2G_{12}^0 \right) \right],$$

$$C_{22} = p_1 \left[\left(G_{12}^0 \right)^{-1} \left(A_{22}^0 - A_{12}^0 - 2G_{12}^0 \right) + A_{22}^0 \left(A_{11}^0 - A_{22}^0 \right) D_0 \right],$$

$$C_{33} = p_1 \left(A_{23}^0 - A_{13}^0 \right) \left[G_{12}^0 - \left(A_{22}^0 + A_{12}^0 \right) D_0 \right]. \tag{1.8.22}$$

Equations (1.8.17)-(1.8.22) give the following expressions for stresses:

$$\sigma_{11} = p_1, \ \sigma_{22} = -\varepsilon^2 \frac{1}{2} p_1 \left\{ -2 + \frac{A_{22}^0 - A_{12}^0}{G_{12}^0} + A_{22}^0 \left(A_{11}^0 - A_{22}^0 \right) D_0 \right.$$

$$\left. - \frac{A_{12}^0}{A_{11}^0} \left[2 - \frac{A_{11}^0 - A_{12}^0}{G_{12}^0} + A_{12}^0 \left(A_{11}^0 - A_{22}^0 \right) D_0 \right] \right\} \cos 2\theta,$$

$$\sigma_{33} = -\varepsilon^2 \frac{1}{2} p_1 \left\{ \frac{A_{23}^0 - A_{13}^0}{G_{12}^0} - \left(A_{23}^0 - A_{13}^0 \right) \left(A_{22}^0 + A_{13}^0 \right) D_0 \right.$$

$$-\frac{A_{13}^{0}}{A_{11}^{0}}\left[2-\frac{A_{11}^{0}-A_{12}^{0}}{G_{12}^{0}}+A_{12}^{0}\left(A_{11}^{0}-A_{22}^{0}\right)D_{0}\right]\Bigg\}\cos 2\theta . \qquad (1.8.23)$$

Note that the expression for σ_{33} in (1.8.23) corresponds to the additional stress which arises from regular small-scale plane-curving of the composite material structure. As we consider the plane strain in the plane Ox_1x_2, we must add the additional term which corresponds to the plane strain state. According to (1.8.18) and (1.8.2) this additional term becomes

$$\sigma_{33}=A_{13}^{0}\frac{\partial u_{1}^{(0)}}{\partial x_{1}}+A_{23}^{0}\frac{\partial u_{2}^{(0)}}{\partial x_{2}}=A_{13}^{0}A_{1}+A_{23}^{0}A_{2}$$

$$=p_{1}\left(A_{22}^{0}A_{13}^{0}-A_{23}^{0}A_{12}^{0}\right)D_{0}. \qquad (1.8.24)$$

The results obtained in this section explain that under tension-compression along the permanently reinforced direction, the Ox_1 axis, of composite materials with regular small-scale plane-curved structure, there are normal stresses σ_{22} and σ_{33}, which are also self-balanced within each curvature period. These results will later form the basis for the explanation of some effects in the fracture mechanics of unidirectional composite materials.

1.9. Some Detailed Results on Quasi-Homogeneous States

Let us examine a specific application of these results to laminated composites consisting of alternating layers of two materials. We assume that the material of each layer is isotropic and these layers are located on the planes $x_2 = \text{const}$. (Fig.1.3.1); in the ideal (uncurved) case these layers would be normal to the Ox_2 axis (Fig.1.3.2).

1.9.1. VALUES OF THE NORMALIZED ELASTICITY CONSTANTS

In a continuum approximation, the layered material with ideal (uncurved) structure is transversely isotropic, the isotropy axis being directed along the Ox_2 axis (Fig.1.3.2). Therefore, considering the notation (1.3.7) we have

$$A_{23}^{0}=A_{12}^{0},\quad A_{11}^{0}=A_{33}^{0},\quad G_{12}^{0}=G_{23}^{0},\quad A_{11}^{0}-A_{12}^{0}=2G_{13}^{0} \qquad (1.9.1)$$

The values related to the matrix and to the reinforcing material will be labelled 1 and 2 respectively: λ_k, μ_k are Lamé constants; E_k are Young's moduli; ν_k are Poisson's ratios; η_k are the concentrations of the components, where

$$\lambda_k = \frac{\nu_k E_k}{(1-2\nu_k)(1+\nu_k)}, \quad \mu_k = \frac{E_k}{2(1+\nu_k)}, \quad k=1,2. \tag{1.9.2}$$

The values of the normalized elastic constants of the composite material with ideal (uncurved) structure can easily be calculated by using the well-known expressions presented in various references, see [70]. Given the stated location of the axes in the present chapter, we obtain the following expressions:

$$G_{12}^0 = G_{23}^0 = \frac{\mu_1\mu_2}{\mu_1\eta_2 + \mu_2\eta_1}, \quad G_{13}^0 = \eta_1\mu_1 + \eta_2\mu_2,$$

$$\frac{1}{2}\left(A_{11}^0 - A_{13}^0\right) = \eta_1\mu_1 + \eta_2\mu_2, \quad A_{23}^0 = A_{12}^0 = \lambda_1\eta_1$$

$$+\lambda_2\eta_2 - \eta_1\eta_2(\lambda_1 - \lambda_2)\frac{(\lambda_1 + 2\mu_1) - (\lambda_2 + 2\mu_2)}{(\lambda_1 + 2\mu_1)\eta_2 + (\lambda_2 + 2\mu_2)\eta_1},$$

$$\frac{1}{2}\left(A_{11}^0 + A_{13}^0\right) = (\lambda_1 + \mu_1)\eta_1 + (\lambda_2 + \mu_2)\eta_2$$

$$-\eta_1\eta_2 \frac{(\lambda_1 - \lambda_2)^2}{(\lambda_1 + 2\mu_1)\eta_2 + (\lambda_2 + 2\mu_2)\eta_1},$$

$$A_{22}^0 = (\lambda_1 + 2\mu_1)\eta_1 + (\lambda_2 + 2\mu_2)\eta_2$$

$$-\eta_1\eta_2 \frac{((\lambda_1 + 2\mu_1) - (\lambda_1 + 2\mu_2))^2}{(\lambda_1 + 2\mu_1)\eta_2 + (\lambda_2 + 2\mu_2)\eta_1}. \tag{1.9.3}$$

With the values given in (1.9.3), we will explore the quantitative value of the self-balanced stresses for several quasi-homogeneous stress states.

1.9.2. PURE SHEAR IN THE Ox_1x_3 PLANE

Inserting (1.9.3) into (1.7.26) and considering the relation $\eta_1 + \eta_2 = 1$, we obtain the self-balanced stresses

$$\sigma_{23} = -\varepsilon\tau \frac{(\mu_1 - \mu_2)^2 \eta_1\eta_2}{\mu_1\mu_2 + (\mu_1 - \mu_2)^2 \eta_1\eta_2} \sin\theta. \tag{1.9.4}$$

We suppose the materials of the layers are quite different in their elastic properties, and the concentration of the stiffer layers is not very small, i.e., when the following conditions are satisfied

$$\mu_2 >> \mu_1, \quad \eta_2\mu_2 >> \eta_1\mu_1, \quad \mu_2\eta_1\eta_2 >> \mu_1 . \tag{1.9.5}$$

From (1.9.4) and (1.9.5), we find

$$\sigma_{23} \approx -\varepsilon\tau \sin\theta . \tag{1.9.6}$$

The maximum values of the self-balanced stresses are of order ε.

1.9.3. UNIAXIAL TENSION-COMPRESSION ALONG THE Ox_1 AXIS

We use equations (1.8.23) and (1.9.3) to determine the stresses, and assume that conditions (1.9.5) are satisfied. From (1.8.23), (1.9.3) and (1.9.5), we obtain

$$\sigma_{11} = p_1, \sigma_{22} \approx -\frac{1}{2}p_1\varepsilon^2\left[-1+2\eta^{(1)}\left(1+\eta^{(2)}\right)+\frac{2\nu^{(1)}\eta^{(1)}}{1-2\nu^{(1)}}\left(1+\eta^{(2)}\right)\right]\cos 2\theta . \tag{1.9.7}$$

To estimate the results we assume that $\eta^{(1)} \approx \eta^{(2)}$ and $\nu^{(1)} = 0.4$, and find

$$\sigma_{11} = p_1, \quad \sigma_{22} \approx -1.75\varepsilon^2 p_1 \cos 2\theta . \tag{1.9.8}$$

Equation (1.9.8) shows that for tension-compression along the Ox_1 axis, the self-balanced stress is of order $\varepsilon^2 p_1$.

Thus, in this and the previous two sections we have given a complete study of the quasi-homogeneous stress states in composite materials with regular small-scale curvature in the structure under the influence of pure shear and tension-compression loads. The study used a simple version of the continuum theory of composite materials with periodically plane-curved structures which was stated in section 1.3. It was shown that, together with the stresses which arise in materials without curvature, stresses also arise along other axes and these stresses are self-balanced within the periods of the curvatures.

1.10. Composites with Large-scale Curving

1.10.1. BASIC ASSUMPTIONS

So far, it has been supposed that the curving of the reinforcing elements in the structure of the composite materials is small-scale, i.e. the half-period of the periodic curving form is significantly less than the smallest dimension of construction elements.

This situation has been taken into account by the last inequality in (1.3.10). Now we consider the case where the curving in the structure of the composite material is large-scale, i.e.

$$\Lambda >> d, \tag{1.10.1}$$

where Λ is the half-period of the representative packet and d is the minimum dimension of a body constructed from the considered composite material.

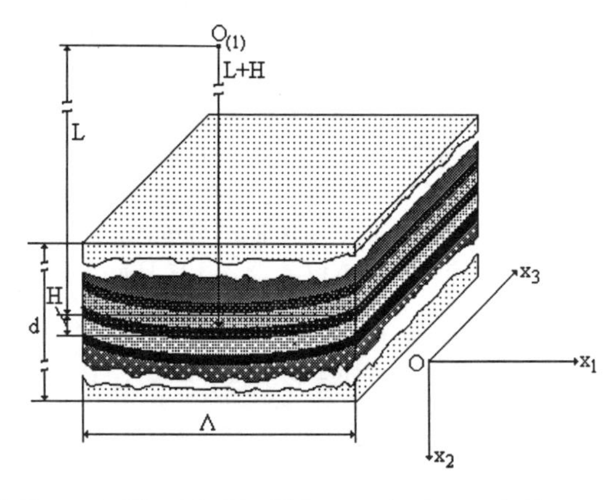

Fig. 1.10.1. The part of the material with large-scale plane periodic curved structure.

We assume that the radius of curvature r=L+H (Fig.1.10.1) of the curved layers at the point corresponding to the maximum rise H (Fig.1.10.1) is significantly larger than the parameter Λ. Moreover, we assume that

$$H << \Lambda << L \tag{1.10.2}$$

With these assumptions, the representative curved packet can be taken as a part of a circular ring with central radius L+H. Hence, in the framework of the continuum approach the elasticity relations can be written as (1.3.16)-(1.3.20), where m=1; we will therefore omit the index (m) (i.e. the index (1)) in these relations.

1.10.2. THE SOLUTION METHOD

We introduce the dimensionless coordinates

$$\bar{x}_i = \frac{x_i}{\Lambda}, \tag{1.10.3}$$

and will omit the over bars. Furthermore, according to (1.10.2) we introduce the small parameter

$$\varepsilon = \frac{\Lambda}{L} << 1 . \qquad (1.10.4)$$

Taking into account (1.10.3) and (1.10.4) we can rewrite the expressions (1.3.16) as follows:

$$\cos\varphi = (1+\varepsilon x_2)\left[(1+\varepsilon x_2)^2 + (\varepsilon x_1)^2\right]^{-\frac{1}{2}}, \sin\varphi = -\varepsilon x_1\left[(1+\varepsilon x_2)^2 + (\varepsilon x_1)^2\right]. \quad (1.10.5)$$

Consider the geometrical linear problem (1.2.1)-(1.2.3), (1.2.7) and, according to (1.10.5), represent all quantities as series (1.4.7) in the parameter ε (1.10.4). Hence, substituting (1.4.7), (1.10.5) into (1.3.17), (1.3.19) we obtain the following elasticity relations for the n-th approximation.

$$\sigma_{ij}^{(n)} = \left[\delta_{ij}\delta_{\alpha\beta}A_{i\beta}^{0} + (1-\delta_{ij})(\delta_{i\alpha}\delta_{j\beta} + \delta_{i\beta}\delta_{j\alpha})G_{ij}^{0}\right]\frac{\partial}{\partial x_\beta}u_\alpha^{(n)} + \sum_{k=1}^{n-1} N_{ij\alpha}^{(n-k)}u_\alpha^{(k)} , \qquad (1.10.6)$$

where $N_{ij\alpha}^{(n-k)}$ are the first order linear differential operators with (n-k) order polynomial coefficients.

Substituting (1.10.6) into (1.2.1) and (1.2.3) we obtain the equation of motion

$$\left\{\left[\delta_{ij}\delta_{\alpha\beta}A_{i\beta}^{0} + (1-\delta_{ij})(\delta_{i\alpha}\delta_{j\beta} + \delta_{i\beta}\delta_{j\alpha})G_{ij}^{0}\right]\frac{\partial^2}{\partial x_i \partial x_\beta}u_\alpha^{(n)} - \rho\frac{\partial^2}{\partial \tau^2}\right\}u_\alpha^{(n)} =$$

$$-\sum_{k=0}^{n-1}\frac{\partial}{\partial x_i}N_{ij\alpha}^{(n-k)}u_\alpha^{(k)} , \qquad (1.10.7)$$

and the boundary conditions in the surface portion S_1

$$N_i\left[\delta_{ij}\delta_{\alpha\beta}A_{i\beta}^{0} + (1-\delta_{ij})(\delta_{i\alpha}\delta_{j\beta} + \delta_{i\beta}\delta_{j\alpha})G_{ij}^{0}\right]\frac{\partial}{\partial x_\beta}u_\alpha^{(n)} = P_j^{(n)} -$$

$$\sum_{k=1}^{n-1} N_i N_{ij\alpha}^{(n-k)}u_\alpha^{(k)} \qquad (1.10.8)$$

for the n-th approximation. The boundary conditions for the displacements on the surface portion S_2 are written as (1.4.10) and for unsteady-state dynamical problems, the appropriate initial conditions must be added to these equations.

Now we present the expressions for the operators $N_{ij\alpha}^{(1)}$ and note that the expressions of these operators for the subsequent approximations can be obtained by employing well-known operations.

$$N_{11\alpha}^{(1)}=x_1\left(A_{11}^0-A_{12}^0-2G_{12}^0\right)\left(\delta_{\alpha 1}\frac{\partial}{\partial x_2}+\delta_{\alpha 2}\frac{\partial}{\partial x_1}\right);$$

$$N_{22\alpha}^{(1)}=x_1\left(-A_{22}^0+A_{12}^0+2G_{12}^0\right)\left(\delta_{\alpha 1}\frac{\partial}{\partial x_2}+\delta_{\alpha 2}\frac{\partial}{\partial x_1}\right);$$

$$N_{33\alpha}^{(1)}=x_1\left(A_{13}^0-A_{23}^0\right)\left(\delta_{\alpha 1}\frac{\partial}{\partial x_2}+\delta_{\alpha 2}\frac{\partial}{\partial x_1}\right);$$

$$N_{23\alpha}^{(1)}=x_1\left(G_{13}^0-G_{23}^0\right)\left(\delta_{\alpha 1}\frac{\partial}{\partial x_3}+\delta_{\alpha 23}\frac{\partial}{\partial x_2}\right);$$

$$N_{13\alpha}^{(1)}=x_1\left(G_{13}^0-G_{23}^0\right)\left(\delta_{\alpha 3}\frac{\partial}{\partial x_2}+\delta_{\alpha 2}\frac{\partial}{\partial x_3}\right);$$

$$N_{12\alpha}^{(1)}=x_1\left[\left(A_{11}^0-A_{12}^0-2G_{12}^0\right)\delta_{\alpha 1}\frac{\partial}{\partial x_1}+\left(-A_{22}^0+A_{12}^0+2G_{12}^0\right)\delta_{\alpha 2}\frac{\partial}{\partial x_2}\right.$$

$$\left.+\left(A_{13}^0-A_{23}^0\right)\delta_{\alpha 3}\frac{\partial}{\partial x_3}\right]; \qquad (1.10.9)$$

We will now use the expressions (1.10.9) for the solution of a particular problem.

1.10.3. AN APPLICATION

As an example, consider the stress-strain state in a plate which is infinite along the Ox_3 axis and has dimensionless sizes 2γ and 2 along the Ox_2 and Ox_1 axes, respectively (Fig.1.10.2).

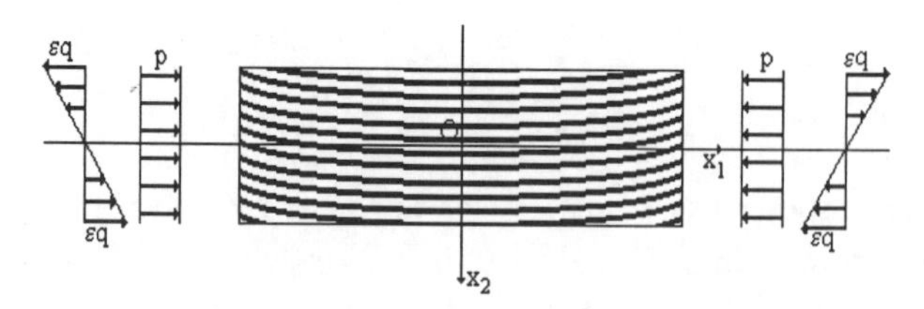

Fig.1.10.2. Distribution of the external forces acting at the ends of the considered strip.

We investigate the case where the upper and lower planes (i.e. $x_2 = \pm\gamma$) of the plate are free, and at the ends (i.e. at $x_1 = \pm 1$), uniformly distributed compressive forces of intensity p act along the Ox_1 axis (Fig.1.10.2). Because of the large scale curving of the reinforcing layers, the compression forces will bend the mid-plane of the plate. To prevent this bending we apply a moment εq (Fig.1.10.2) at the ends, i.e. at $x_1 = \pm 1$. To find the relation between p and q which prevents the bending we investigate the plane strain state in the rectangle $\{-1 \le x_1 \le 1; -\gamma \le x_2 \le \gamma\}$ with the following boundary conditions.

$$\sigma_{22}\big|_{x_2=\pm\gamma} = 0; \quad \sigma_{12}\big|_{x_2=\pm\gamma} = 0; \quad \sigma_{12}\big|_{x_1=\pm1} = 0; \quad \sigma_{11}\big|_{x_1=\pm1} = -p - q\varepsilon x_2 . \qquad (1.10.10)$$

The relation between p and q will be determined from the requirement

$$u_2\big|_{x_2=0} = 0 . \qquad (1.10.11)$$

Consider the solution to this problem in the framework of the zeroth and first approximations. According to (1.4.7), (1.10.6) and (1.10.7) for the plane-strain state in the plane Ox_1x_2 we obtain the following equilibrium equations in the n-th approximation.

$$\left(A_{11}^0 \frac{\partial^2}{\partial x_1^2} + G_{12}^0 \frac{\partial^2}{\partial x_2^2}\right) u_1^{(n)} + \left(A_{12}^0 + G_{12}^0\right) \frac{\partial^2 u_2^{(n)}}{\partial x_1 \partial x_2} = -\sum_{k=0}^{n-1} \left[\frac{\partial}{\partial x_1} \times \right.$$

$$\left. \left(N_{111}^{(n-k)} u_1^{(k)} + N_{112}^{(n-k)} u_2^{(k)}\right) + \frac{\partial}{\partial x_2}\left(N_{211}^{(n-k)} u_1^{(k)} + N_{212}^{(n-k)} u_2^{(k)}\right) \right];$$

$$\left(G_{12}^0 \frac{\partial^2}{\partial x_1^2} + A_{22}^0 \frac{\partial^2}{\partial x_2^2}\right) u_2^{(n)} + \left(A_{12}^0 + G_{12}^0\right) \frac{\partial^2 u_1^{(n)}}{\partial x_1 \partial x_2} = -\sum_{k=0}^{n-1} \left[\frac{\partial}{\partial x_1} \times \right.$$

$$\left. \left(N_{121}^{(n-k)} u_1^{(k)} + N_{121}^{(n-k)} u_2^{(k)}\right) + \frac{\partial}{\partial x_2}\left(N_{221}^{(n-k)} u_1^{(k)} + N_{222}^{(n-k)} u_2^{(k)}\right) \right]. \qquad (1.10.12)$$

From (1.4.7), (1.10.6) and (1.10.10) we obtain the following boundary conditions for the n-th approximation.

$$\left(A_{12}^0 \frac{\partial u_1^{(n)}}{\partial x_1} + A_{22}^0 \frac{\partial u_2^{(n)}}{\partial x_2}\right)\Bigg|_{x_2=\pm\gamma} = -\sum_{k=1}^{n-1} \left(N_{221}^{(n-k)} u_1^{(k)} + N_{222}^{(n-k)} u_2^{(k)}\right)\Big|_{x_2=\pm\gamma} ;$$

$$G_{12}^{0}\left(\frac{\partial u_1^{(n)}}{\partial x_2}+\frac{\partial u_2^{(n)}}{\partial x_1}\right)\Bigg|_{x_2=\pm\gamma}=-\sum_{k=1}^{n-1}\left(N_{121}^{(n-k)}u_1^{(k)}+N_{122}^{(n-k)}u_2^{(k)}\right)\Big|_{x_2=\pm\gamma};$$

$$G_{12}^{0}\left(\frac{\partial u_1^{(n)}}{\partial x_2}+\frac{\partial u_2^{(n)}}{\partial x_1}\right)\Bigg|_{x_1=\pm1}=-\sum_{k=1}^{n-1}\left(N_{121}^{(n-k)}u_1^{(k)}+N_{122}^{(n-k)}u_2^{(k)}\right)\Big|_{x_1=\pm1};$$

$$\left(A_{11}^{0}\frac{\partial u_1^{(n)}}{\partial x_1}+A_{12}^{0}\frac{\partial u_2^{(n)}}{\partial x_2}\right)\Bigg|_{x_1=\pm1}=-p\delta_{n0}-qx_2\delta_{n1}-$$

$$\sum_{k=1}^{n-1}\left(N_{221}^{(n-k)}u_1^{(k)}+N_{222}^{(n-k)}u_2^{(k)}\right)\Big|_{x_1=\pm1}, \tag{1.10.13}$$

where $\delta_{00}=1$, $\delta_{n0}=0$ for $n\neq0$ and $\delta_{11}=1$, $\delta_{n1}=0$ for $n\neq1$.

Consider the determination of the zeroth and first approximations. From (1.10.11) and (1.10.12) for the zeroth approximation, we obtain the equilibrium equations

$$\left(A_{11}^{0}\frac{\partial^2}{\partial x_1^2}+G_{12}^{0}\frac{\partial^2}{\partial x_2^2}\right)u_1^{(0)}+\left(A_{12}^{0}+G_{12}^{0}\right)\frac{\partial^2 u_2^{(0)}}{\partial x_1\partial x_2}=0,$$

$$\left(G_{12}^{0}\frac{\partial^2}{\partial x_1^2}+A_{22}^{0}\frac{\partial^2}{\partial x_2^2}\right)u_2^{(0)}+\left(A_{12}^{0}+G_{12}^{0}\right)\frac{\partial^2 u_1^{(0)}}{\partial x_1\partial x_2}=0, \tag{1.10.14}$$

and boundary conditions

$$\left(A_{12}^{0}\frac{\partial u_1^{(0)}}{\partial x_1}+A_{22}^{0}\frac{\partial u_2^{(0)}}{\partial x_2}\right)\Bigg|_{x_2=\pm\gamma}=0;\quad G_{12}^{0}\left(\frac{\partial u_1^{(0)}}{\partial x_2}+\frac{\partial u_2^{(0)}}{\partial x_1}\right)\Bigg|_{x_2=\pm\gamma}=0;$$

$$G_{12}^{0}\left(\frac{\partial u_1^{(0)}}{\partial x_2}+\frac{\partial u_2^{(0)}}{\partial x_1}\right)\Bigg|_{x_1=\pm1}=0;\quad \left(A_{11}^{0}\frac{\partial u_1^{(0)}}{\partial x_1}+A_{12}^{0}\frac{\partial u_2^{(0)}}{\partial x_2}\right)\Bigg|_{x_1=\pm1}=-p. \tag{1.10.15}$$

It is easy to prove that a particular solution to the problem (1.10.14), (1.10.15) is

$$u_1^{(0)} = pA_{22}^0 D_0 x_1,\ u_2^{(0)} = -pA_{12}^0 D_0 x_2,\ D_0 = \left(A_{11}^0 A_{22}^0 - \left(A_{12}^0\right)^2\right)^{-1}. \quad (1.10.16)$$

From (1.10.16) it follows that in the zeroth approximation the condition (1.10.11) is satisfied. Taking (1.10.16) into account we obtain the following equations and boundary conditions for the first approximation from (1.10.9), (1.10.12) and (1.10.13):

$$\left(A_{11}^0 \frac{\partial^2}{\partial x_1^2} + G_{12}^0 \frac{\partial^2}{\partial x_2^2}\right) u_1^{(1)} + \left(A_{12}^0 + G_{12}^0\right) \frac{\partial^2 u_2^{(1)}}{\partial x_1 \partial x_2} = -A,$$

$$\left(G_{12}^0 \frac{\partial^2}{\partial x_1^2} + A_{22}^0 \frac{\partial^2}{\partial x_2^2}\right) u_2^{(1)} + \left(A_{12}^0 + G_{12}^0\right) \frac{\partial^2 u_1^{(1)}}{\partial x_1 \partial x_2} = 0, \quad (1.10.17)$$

$$\left.\left(A_{12}^0 \frac{\partial u_1^{(1)}}{\partial x_1} + A_{22}^0 \frac{\partial u_2^{(1)}}{\partial x_2}\right)\right|_{x_2=\pm\gamma} = 0; \quad G_{12}^0 \left.\left(\frac{\partial u_1^{(1)}}{\partial x_2} + \frac{\partial u_2^{(1)}}{\partial x_1}\right)\right|_{x_2=\pm\gamma} = -x_1 A;$$

$$G_{12}^0 \left.\left(\frac{\partial u_1^{(1)}}{\partial x_2} + \frac{\partial u_2^{(1)}}{\partial x_1}\right)\right|_{x_1=\pm1} = -A; \left.\left(A_{11}^0 \frac{\partial u_1^{(1)}}{\partial x_1} + A_{12}^0 \frac{\partial u_2^{(1)}}{\partial x_2}\right)\right|_{x_1=\pm1} = -qx_2, \quad (1.10.18)$$

In (1.10.17) and (1.10.18) we use the notation

$$A = p\left(1 - 2G_{12}^0\left(A_{22}^0 + A_{12}^0\right)D_0^{-1}\right). \quad (1.10.19)$$

Note that for the first approximation, equations (1.4.7) and (1.10.11) give

$$\left. u_2^{(1)} \right|_{x_2=0} = 0. \quad (1.10.20)$$

The solution to the problem (1.10.17)-(1.10.20) is expressed as

$$u_1^{(1)} = C_1 x_1 x_2, \quad u_2^{(1)} = C_2 x_2^2. \quad (1.10.21)$$

Substituting (1.10.21) into (1.10.17)-(1.10.19) and performing some transformations we obtain the following expressions for the unknowns C_1, C_2 and q.

$$C_1 = \frac{1}{2} A \left(G_{12}^0\right)^{-1}, \quad C_2 = \frac{1}{2} A_{12}^0 A \left(G_{12}^0 A_{22}^0\right)^{-1},$$

$$q = p\left(G_{12}^{0}A_{22}^{0}\right)^{-1}\left[A_{11}^{0}A_{22}^{0} - \left(A_{12}^{0}\right)^{2} - 2G_{12}^{0}\left(A_{22}^{0} + A_{12}^{0}\right)\right]. \tag{1.10.22}$$

Thus, the required bending moment is

$$M = -\frac{2p\gamma^{3}}{3A_{22}^{0}G_{12}^{0}}\left[A_{11}^{0}A_{22}^{0} - \left(A_{12}^{0}\right)^{2} - 2G_{12}^{0}\left(A_{22}^{0} + A_{12}^{0}\right)\right]\varepsilon \tag{1.10.23}$$

Note that this result can be refined by considering the subsequent approximation, but (1.10.23) has a high accuracy for the sufficiently small amplitude large-scale curving. This effect must be taken into account in stability problems for such composites.

1.11. Bibliographical Notes

The continuum theory presented in section 1.3 was given by A.N. Guz [96, 98]. The application of this theory to the investigations of the quasi-homogeneous stress-state in composite materials with plane-periodically curved structures and the vibration problems of the body fabricated from such composites was also made by A.N.Guz in [99] and in [97], respectively. Note that these investigations have been presented in sections 1.6-1.9.

The continuum theory for the composite materials with the large-scale curving in the structure which has described in the section 1.10 was developed in the paper by A.N. Guz and G.V. Guz [104].

CHAPTER 2

GENERAL CURVED COMPOSITES

In this chapter we discuss a continuum theory for composite materials with general periodic or locally curved structures; this generalises the analysis given in the previous chapter. Again we use tensor notation and sum repeated indices over their ranges, but do not sum underlined repeated indices.

2.1. Some Preliminary Remarks on Geometry

In the previous chapter we considered *periodic plane* curved structures. However, the curving in the structure of unidirectional fibrous and layered composite materials can be in three dimensional space, not just periodic in a plane, – we will call this *spatially* periodic -, or it can be generally (i.e. *spatially*) *local*, or *local in a plane* as shown in Figs. 2.1.1, 2.1.2 and 2.1.3, respectively, at $x_1 = \text{const}$ and $x_3 = \text{const}$. In these figures, it is assumed that the reinforcing layers lie in planes parallel to the plane Ox_1x_3 and these materials consist of alternating packets which contain N different layers, the thickness of which is denoted by ΔH given by formula (1.3.1).

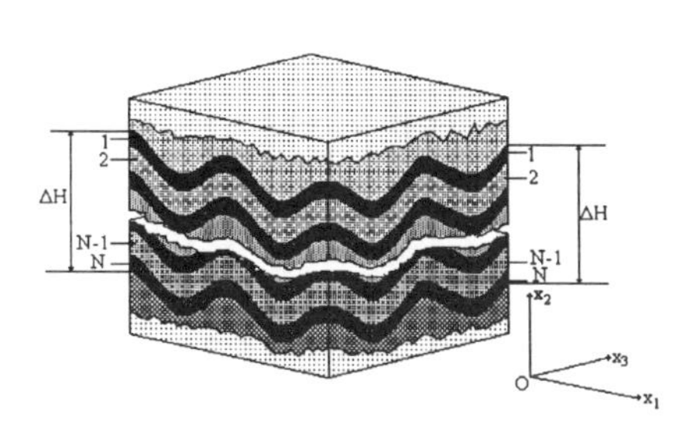

Fig.2.1.1. Part of a composite material with spatially periodic curved structure.

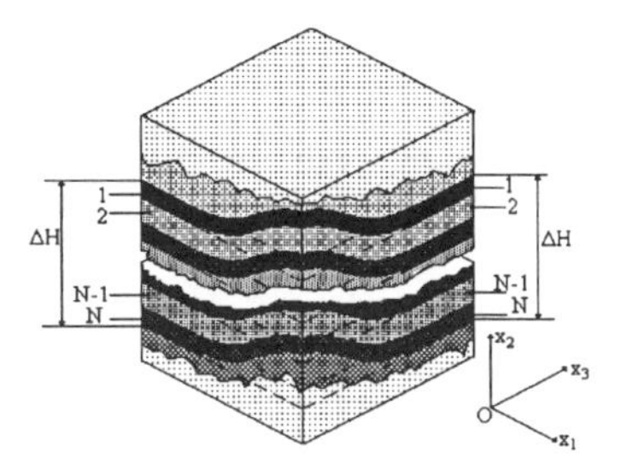

Fig.2.1.2. Part of a composite material with spatially local curved structure.

2.2. Constitutive Relations

As in the previous chapter, we will use the Lagrangian coordinates x_j, which in their natural state coincide with Cartesian coordinates. Moreover, we assume that the equations (1.2.1)-(1.2.7) remain valid. Now we consider the formulation of the elasticity relations for a composite material with arbitrarily curved structure. To do this, we isolate a representative curved packet of the composite material, shown in Figs. 2.2.1. and 2.2.2. In these figures the following notation is introduced: H is the characteristic vertical rise of the structural curve; Λ_1 and Λ_3 are the half-wavelengths of periodic curves in the directions of the Ox_1 and Ox_3 axes, respectively; ΔH is the thickness of the representative packet; h' is the thickness determined by the expression (1.3.9). Note that for local curving in the structure of the composite, the meaning of the parameters Λ_1 and Λ_3 will be different.

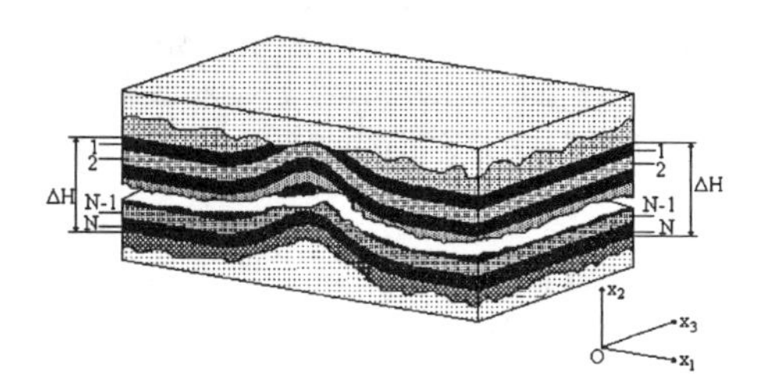

Fig.2.1.3. Part of a composite material with plane-local curved structure.

We introduce a local coordinate system $O'x'_1\, x'_2\, x'_3$ with the origin on the median surface of the isolated representative packet; the axis $O'x'_2$ is directed along the normal vector $\boldsymbol{\tau}'_2$, and the axes $O'x'_1$ and $O'x'_3$ along the tangential vectors $\boldsymbol{\tau}'_1$ and $\boldsymbol{\tau}'_3$ to this surface (Figs. 2.2.1 and 2.2.2). Note that, in the absence of curving, the directions of the vectors $\boldsymbol{\tau}'_1$, $\boldsymbol{\tau}'_2$, and $\boldsymbol{\tau}'_3$ coincide with the directions of the vectors $\boldsymbol{\tau}_1$, $\boldsymbol{\tau}_2$, and $\boldsymbol{\tau}_3$, respectively (Figs. 2.2.1 and 2.2.2). The vectors $\boldsymbol{\tau}'_1$, $\boldsymbol{\tau}'_2$, and $\boldsymbol{\tau}'_3$ are determined below.

The equation of the median surface of the selected representative packet we take as follows:

$$x_2 = F(x_1, x_3) = \varepsilon f(x_1, x_3). \tag{2.2.1}$$

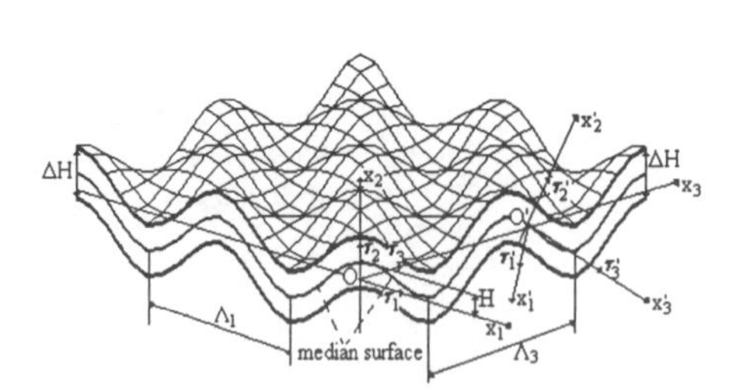

Fig.2.2.1. The geometry of the representative packet for composite material with spatially periodic curved structure.

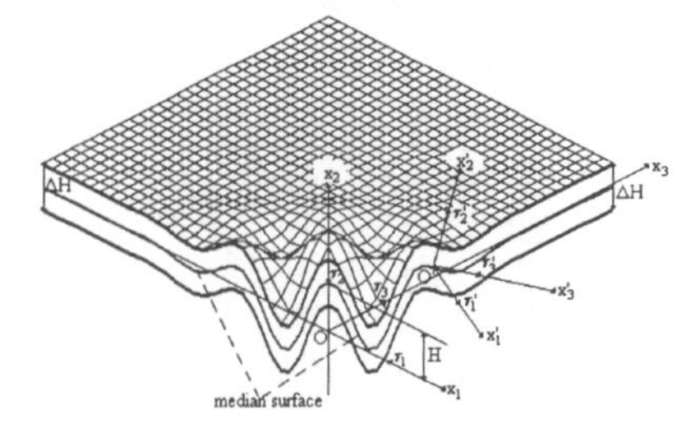

Fig.2.2.2. The geometry of the representative packet for composite material with spatially local curved structure.

where ε is a parameter which is introduced for characterizing the degree of the curving; the actual determination of this parameter will be given for each specified function $F(x_1, x_3)$ in the equation (2.2.1). We suppose that the function F and its first-order partial derivatives are continuous.

We assume that the elasticity relations at the chosen point of the median surface in the local coordinate system $O'x'_1\,x'_2\,x'_3$ are the elasticity relations (1.3.6)-(1.3.8) for the orthotropic body, with the principal axes $O'x'_1$, $O'x'_2$, and $O'x'_3$. This means that in the ideal (uncurved) case the composite material is orthotropic with principal axes Ox_1, Ox_2 and Ox_3 (Figs. 2.1.1-2.1.3). Let us formulate the elasticity relations for the arbitrarily curved representative packet in the global coordinate system $Ox_1x_2x_3$ (Fig.2.2.1). For this purpose we first determine the cosines $\ell_{ij} = \boldsymbol{\tau}'_i . \boldsymbol{\tau}_j$ of the angles between the axes $O'x'_i$ and Ox_j and we rewrite the equation (2.2.1) in the vectorial form

$$\mathbf{r} = x_1\boldsymbol{\tau}'_1 + F(x_1, x_3)\boldsymbol{\tau}'_2 + x_3\boldsymbol{\tau}'_3 \tag{2.2.2}$$

From the equation (2.2.2) and from the relations $\boldsymbol{\tau}'_2 = (\partial\mathbf{r}/\partial x_3 \times \partial\mathbf{r}/\partial x_1)/|(\partial\mathbf{r}/\partial x_3 \times \partial\mathbf{r}/\partial x_1)|$, $\boldsymbol{\tau}''_1 = (\partial\mathbf{r}/\partial x_1)/|\partial\mathbf{r}/\partial x_1|$, $\boldsymbol{\tau}''_3 = \boldsymbol{\tau}''_1 \times \boldsymbol{\tau}'_2$ the triad of vectors $\boldsymbol{\tau}''_1$, $\boldsymbol{\tau}'_2$, $\boldsymbol{\tau}''_3$ are obtained. After rotating this triad about $\boldsymbol{\tau}'_2$ with angle $\psi/2$ where $\sin\psi = F_{,1}F_{,3}V_1V_3$, we obtain the new triad of vectors $\boldsymbol{\tau}'_1$, $\boldsymbol{\tau}'_2$, $\boldsymbol{\tau}'_3$, the expressions of which are

$$\boldsymbol{\tau}'_i = \ell_{ij}\boldsymbol{\tau}_j, \quad i,j=1,2,3, \tag{2.2.3}$$

where

$$\ell_{11} = g_1V_1 + F_{,1}F_{,3}g_2\,V_1V_2, \; \ell_{12} = F_{,1}g_1V_1 - F_{,3}g_2V_1V_2, \; \ell_{13} = -(1+F_{,1}^2)g_2V_1V_2,$$

$$\ell_{21} = -F_{,1}V_2, \; \ell_{22} = V_2; \; \ell_{23} = -V_2F_{,3}, \; \ell_{31} = g_2V_1 - F_{,1}F_{,3}g_1V_1V_2,$$

$$\ell_{32} = F_{,1}g_2V_1 + F_{,3}g_1V_1V_2, \; \ell_{33} = (1+F_{,1}^2)g_1V_1V_2 \; . \tag{2.2.4}$$

In (2.2.4) the following notation is used

$$F_{,1} = \frac{\partial F}{\partial x_1}, \quad F_{,3} = \frac{\partial F}{\partial x_3}, \quad V_1 = \left(1+(F_{,1})^2\right)^{-\frac{1}{2}},$$

$$V_2 = \left(1+(F_{,1})^2+(F_{,3})^2\right)^{-\frac{1}{2}}, \; V_3 = \left(1+F_{,3}^2\right)^{-\frac{1}{2}},$$

$$g_1 = \cos\left(\frac{\psi}{2}\right), \qquad g_2 = \sin\left(\frac{\psi}{2}\right), \quad \psi = \arcsin(F_{,1}F_{,3}V_1V_3)\,. \qquad (2.2.5)$$

Using the well-known transformation formulas for the elastic constants under the rotation of the coordinate system, we obtain the following expressions for the normalized elastic constants of the arbitrarily curved representative packet in the global coordinate system $Ox_1x_2x_3$ (Fig.2.2.1).

$$A_{sp} = A^0_{mn} q_{sm} q_{pn} \ , \qquad (m;n;s;p=1,2,3,4,5,6), \qquad (2.2.6)$$

where A^0_{mn} are the normalized elasticity constants of the considered composite material in the case where its structure is ideal (uncurved), i.e. the constants which enter the relations (1.3.7); q_{sm} denote the following expressions:

$$q_{11} = \ell_{11}^2, \; q_{12} = \ell_{12}^2, \; q_{13} = \ell_{13}^2, \; q_{14} = 2\ell_{12}\ell_{13}, \; q_{15} = 2\ell_{13}\ell_{11}, \; q_{16} = 2\ell_{12}\ell_{11},$$

$$q_{21} = \ell_{21}^2, \; q_{22} = \ell_{22}^2, \; q_{23} = \ell_{23}^2, \; q_{24} = 2\ell_{23}\ell_{22}, \; q_{25} = 2\ell_{23}\ell_{21}, \; q_{26} = 2\ell_{22}\ell_{21},$$

$$q_{31} = \ell_{31}^2, \; q_{32} = \ell_{32}^2, \; q_{33} = \ell_{33}^2, \; q_{34} = 2\ell_{33}\ell_{32}, \; q_{35} = 2\ell_{33}\ell_{31}, \; q_{36} = 2\ell_{32}\ell_{31},$$

$$q_{41} = \ell_{31}\ell_{21}, \; q_{42} = \ell_{32}\ell_{22}, \; q_{43} = \ell_{33}\ell_{23}, \; q_{44} = \ell_{33}\ell_{22} + \ell_{32}\ell_{23},$$

$$q_{45} = \ell_{33}\ell_{21} + \ell_{31}\ell_{23}, \; q_{46} = \ell_{31}\ell_{22} + \ell_{32}\ell_{21}, \; q_{51} = \ell_{31}\ell_{11}, \; q_{52} = \ell_{32}\ell_{12},$$

$$q_{53} = \ell_{33}\ell_{13}, \; q_{54} = \ell_{33}\ell_{12} + \ell_{32}\ell_{13}, \; q_{55} = \ell_{33}\ell_{11} + \ell_{31}\ell_{13},$$

$$q_{56} = \ell_{31}\ell_{12} + \ell_{32}\ell_{11}, \; q_{61} = \ell_{21}\ell_{11}, \; q_{62} = \ell_{12}\ell_{22}, \; q_{63} = \ell_{13}\ell_{23},$$

$$q_{64} = \ell_{13}\ell_{22} + \ell_{12}\ell_{23}, \; q_{65} = \ell_{13}\ell_{21} + \ell_{11}\ell_{23}, \; q_{66} = \ell_{11}\ell_{22} + \ell_{12}\ell_{21}. \qquad (2.2.7)$$

Thus, we obtain the following elasticity relations in the global coordinates x_i.

$$\sigma_i = A_{ij}\varepsilon_j, \quad i;j=1,2,3,4,5,6. \qquad (2.2.8)$$

In (2.2.8) the notation (1.3.6) is used.

Note that the elasticity relations (2.2.6)-(2.2.8) are obtained without any restrictions on the curving parameters. According to (2.2.6)-(2.2.8), in general we can write the following relations for A_{ij} through the constants A^0_{ij} and the function $F(x_1, x_3)$.

$$A_{ij} = A_{ij}(A^0_{nm}, F(x_1, x_3)). \tag{2.2.9}$$

In the special cases, the explicit form of the expressions (2.2.9) will be given in the subsequent sections.

2.3. Explicit Constitutive Relations for Small Curving

By small curving we will understand the curving for which the following conditions for the function F (2.2.1) are satisfied.

$$\varepsilon^2\left((f_{,1})^2 + (f_{,3})^2\right) < 1, \quad 0 \le \varepsilon < 1. \tag{2.3.1}$$

Taking (2.3.1) into account, we can represent the expressions for V_1, V_2 and V_3 in (2.2.4) as power series in the parameter ε. By using (2.2.5)-(2.2.8) we obtain the following explicit form of the expressions (2.2.9).

$$A_{ij} = \begin{cases} A^0_{ij} + \sum_{k=1}^{\infty} \varepsilon^{2k} A_{ijk} & \text{for combinations } ij = 11;12;13;22;23;33;44;55;66 \\ \sum_{k=1}^{\infty} \varepsilon^{2k-1} A_{ijk} & \text{for combinations } ij = 14;16;24;26;34;36;45;56 \\ \sum_{k=1}^{\infty} \varepsilon^{2k} A_{ijk} & \text{for combinations } ij = 15;25;35;46 \end{cases} \tag{2.3.2}$$

Note that the expressions for the coefficients A_{ijk} of the series (2.3.2) can be obtained from the equation (2.2.4)-(2.2.8) after routine transformations; we give only the coefficients A_{ij1} which will be used later for practical problems.

$$A_{111} = a^2 a_{111};\ A_{121} = a^2 a_{121} + b^2 b_{121},\ A_{131} = a^2 a_{131} + b^2 b_{131},\ A_{141} = bb_{141},$$

$$A_{151} = aba_{151},\ A_{161} = aa_{161},\ A_{221} = a^2 a_{221} + b^2 b_{221},\ A_{241} = bb_{241},$$

$$A_{231} = a^2 a_{231} + b^2 b_{231},\ A_{251} = aba_{251},\ A_{261} = aa_{261},\ A_{331} = b^2 b_{331},$$

$$A_{341} = bb_{341},\ A_{351} = aba_{351},\ A_{361} = aa_{361},\ A_{441} = a^2 a_{441} + b^2 b_{441},$$

$$A_{451} = aa_{451},\ A_{461} = aba_{461},\ A_{551} = a^2a_{551} + b^2b_{551},\ A_{561} = bb_{561},$$

$$A_{661} = a^2a_{661} + b^2b_{661}, \tag{2.3.3}$$

where

$$a \equiv a(x_1, x_3) = \frac{\partial f(x_1, x_3)}{\partial x_1} = f_{,1};\ b \equiv b(x_1, x_3) = \frac{\partial f(x_1, x_3)}{\partial x_3} = f_{,3};$$

$$a_{111} = \left(-2A_{11}^0 + 2A_{12}^0 + 4A_{66}^0\right);\ a_{121} = \left(A_{11}^0 - 2A_{12}^0 + A_{22}^0 - 4A_{66}^0\right);$$

$$b_{121} = \left(-A_{12}^0 + A_{13}^0\right);\ a_{131} = \left(-A_{13}^0 + A_{23}^0\right);\ b_{131} = \left(A_{12}^0 - A_{13}^0\right);$$

$$b_{141} = \left(A_{12}^0 - A_{13}^0\right);\ a_{151} = \left(-A_{11}^0 + A_{12}^0 + A_{66}^0\right);\ a_{161} = \left(-A_{11}^0 + A_{12}^0 + 2A_{66}^0\right);$$

$$a_{221} = \left(2A_{12}^0 - 2A_{22}^0 + 4A_{66}^0\right);\ b_{221} = \left(-2A_{22}^0 + 2A_{23}^0 + 4A_{44}^0\right);$$

$$a_{231} = \left(-A_{23}^0 + A_{13}^0\right);\ b_{231} = \left(A_{22}^0 - 2A_{23}^0 + A_{33}^0 - 4A_{44}^0\right);$$

$$b_{241} = \left(A_{22}^0 - A_{23}^0 - 2A_{44}^0\right);\ a_{251} = \left(A_{22}^0 - A_{21}^0 - 2A_{44}^0 + 2A_{55}^0 - 2A_{66}^0\right);$$

$$a_{261} = \left(-A_{12}^0 + A_{22}^0 - 2A_{66}^0\right);\ b_{331} = \left(2A_{23}^0 - 2A_{33}^0 + 4A_{44}^0\right);$$

$$b_{341} = \left(A_{23}^0 - A_{33}^0 + 2A_{44}^0\right);\ a_{351} = \left(-A_{31}^0 - 2A_{55}^0 + A_{23}^0 + 2A_{44}^0\right);$$

$$a_{361} = \left(-A_{13}^0 + A_{23}^0\right);\ a_{441} = \left(-A_{44}^0 + A_{55}^0\right);$$

$$b_{441} = \left(A_{22}^0 - 2A_{23}^0 + A_{33}^0 - 4A_{44}^0\right);\ a_{451} = \left(A_{44}^0 - A_{55}^0\right);$$

$$a_{461} = \left(-A_{12}^0 + A_{13}^0 + A_{22}^0 - A_{23}^0 - A_{44}^0 + A_{55}^0 - 2A_{66}^0\right);$$

$$a_{551} = \left(A_{44}^0 - A_{55}^0\right);\ b_{551} = \left(A_{66}^0 - A_{55}^0\right);$$

$$a_{661} = \left(A_{11}^0 - 2A_{12}^0 + A_{22}^0 - 4A_{66}^0\right);\ b_{661} = \left(-A_{55}^0 + A_{66}^0\right). \tag{2.3.4}$$

For the general plane curving, i.e. for the case where $b = \partial f/\partial x_3 \equiv 0$, the explicit form of the constitutive relations (2.2.9) will be given in the following chapter.

2.4. Displacements Equations for Small Curving; Formulation and Solution

2.4.1. THE EQUATIONS OF MOTION

We now construct a system of equations in terms of displacements for this continuum theory in the case of small curving, i.e. when the constitutive relations are presented as (2.2.8), (2.3.2)-(2.3.4). Here we will consider only the geometrically linear statement and, using the notation (1.3.6), we will represent the constitutive relations (2.2.8), (2.3.2) as follows:

$$\sigma_{ij} = \mu_{ij\alpha\beta} \frac{\partial u_\alpha}{\partial x_\beta}, \quad i, j, \alpha, \beta = 1,2,3 \tag{2.4.1}$$

where

$$\mu_{ij\alpha\beta} = \mu_{ij\alpha\beta}^{(0)} + \sum_{k=1}^{\infty} \varepsilon^{2k-1} \mu_{ij\alpha\beta}^{(2k-1)} + \sum_{k=1}^{\infty} \varepsilon^{2k} \mu_{ij\alpha\beta}^{(2k)}. \tag{2.4.2}$$

In (2.4.2) the following notation is used:

$$\mu_{ij\alpha\beta}^{(0)} = \delta_{\underline{i}}^{j} \delta_{\underline{\alpha}}^{\beta} A_{\underline{i\beta}}^{0} + \left(1 - \delta_{\underline{i}}^{j}\right)\left(\delta_{\underline{i}}^{\alpha} \delta_{\underline{j}}^{\beta} + \delta_{\underline{i}}^{\beta} \delta_{\underline{j}}^{\alpha}\right) G_{\underline{ij}}^{0},$$

$$\mu_{ij\alpha\beta}^{(2k-1)} = \delta_i^1 \delta_j^1 \left(A_{14k} \delta_\alpha^2 \delta_\beta^3 + A_{16k} \delta_\alpha^1 \delta_\beta^2\right) + \delta_i^2 \delta_j^2 \left(A_{24k} \delta_\alpha^2 \delta_\beta^3 + A_{26k} \delta_\alpha^1 \delta_\beta^2\right) +$$

$$\delta_i^3 \delta_j^3 \left(A_{34k} \delta_\alpha^2 \delta_\beta^3 + A_{36k} \delta_\alpha^1 \delta_\beta^2\right) + \delta_j^3 \delta_\alpha^1 \left(A_{45k} \delta_i^2 \delta_\beta^3 + A_{56k} \delta_i^1 \delta_\beta^2\right),$$

$$\mu_{ij\alpha\beta}^{(2k)} = \delta_{\underline{i}}^{j} \delta_{\underline{\alpha}}^{\beta} A_{\underline{i\beta}k} + \left(1 - \delta_i^j\right)\left[\delta_j^3 \delta_\beta^3 \left(A_{44k} \delta_\alpha^2 \delta_i^2 + A_{55k} \delta_\alpha^1 \delta_i^2\right) + \delta_i^1 \delta_j^2 \delta_\alpha^1 \delta_\beta^2 A_{66k}\right]$$

$$+ \delta_\alpha^1 \delta_\beta^3 \left(A_{15k} \delta_i^1 \delta_j^1 + A_{25k} \delta_i^2 \delta_j^2\right) + \delta_\alpha^1 \delta_j^3 \left(A_{35k} \delta_i^3 \delta_\beta^3 + A_{46k} \delta_i^2 \delta_\beta^2\right), \tag{2.4.3}$$

where δ_i^β are the Kronecker symbols.

Thus, considering (2.4.2)-(2.4.3) into account and substituting (2.4.1) into (1.2.1) we get the following equations of motion in terms of displacements:

$$L_{i\alpha} u_\alpha + \sum_{k=1}^{\infty} \varepsilon^{2k} K_{i\alpha k} u_\alpha + \sum_{k=1}^{\infty} \varepsilon^{2k-1} R_{i\alpha k} u_\alpha = 0. \tag{2.4.4}$$

The notation for differential operators is introduced here, in the form

$$L_{i\alpha} = \mu^{(0)}_{ij\alpha\beta}\frac{\partial^2}{\partial x_j \partial x_\beta} - \rho\delta^\alpha_i \frac{\partial^2}{\partial \tau^2}, \quad K_{i\alpha k} = \frac{\partial}{\partial x_j}\left(\mu^{(2k)}_{ij\alpha\beta}\frac{\partial}{\partial x_\beta}\right),$$

$$R_{i\alpha k} = \frac{\partial}{\partial x_j}\left(\mu^{(2k-1)}_{ij\alpha\beta}\frac{\partial}{\partial x_\beta}\right). \tag{2.4.5}$$

Note that the expressions for the operators $L_{i\alpha}$ in (2.4.4) coincide with the expressions (1.4.5).

Substituting (2.4.1) and (2.4.2) into (1.2.3), we obtain the boundary conditions on the surface portion S_1 where stresses are given in the form

$$N_i\left[\mu^{(0)}_{ij\alpha\beta} + \sum_{k=1}^{\infty}\varepsilon^{2k-1}\mu^{(2k-1)}_{ij\alpha\beta} + \sum_{k=1}^{\infty}\varepsilon^{2k}\mu^{(2k)}_{ij\alpha\beta}\right]\frac{\partial u_\alpha}{\partial x_\beta} = P_j. \tag{2.4.6}$$

This gives the complete system of equations in terms of displacements of the continuum theory for composite materials with arbitrarily small curving in the structure. By direct verification it can be proven that in the case where the equation (2.2.1) is taken as

$$x_2 = -H\cos\theta = -\varepsilon\frac{\Lambda}{\pi}\cos\theta, \ \theta = \frac{\pi}{\Lambda}x_1, \ \varepsilon = \frac{\pi H}{\Lambda}, \tag{2.4.7}$$

and instead of the series entering (2.4.2) and (2.4.6), their terms of order up to ε^2 are taken, then the equations (2.4.1)-(2.4.6) coincide with the equations (1.4.1)-(1.4.6) respectively, of the theory presented in the previous chapter. However, it should be noted that the expression for the parameter ε entering the previous chapter by the formula (1.3.23) is different from that entering here by the formula (2.4.7). This difference follows from the approximate character of the theory presented there. The relation

$$\frac{\text{the parameter } \varepsilon \text{ entering (2.4.7)}}{\text{the parameter } \varepsilon \text{ entering (1.3.23)}} = \frac{\pi^2}{8} \approx 1.625. \tag{2.4.8}$$

holds and leads to the conclusion that the results obtained in the previous chapter and related to the small-scale curving can be made more accurate if we use ε determined from (2.4.7) instead of that determined from (1.3.23).

2.4.2. THE SMALL PARAMETER METHOD

First, note that in the general case, the equations (2.4.3)-(2.4.6) correspond to the relations of elasticity theory of an inhomogeneous anisotropic body with properties varying continuously along the axes Ox_1 and Ox_3. It is not possible to obtain the closed form solution to equations (2.4.3)-(2.4.6) in the general case, yet it is possible to derive a general theory based on power series in the small parameter ε. Thus, substituting (1.4.7) into (2.4.1), (2.4.2), (2.4.4) and (2.4.6) we obtain the following equations in terms of the displacements in the q-th approximation.

The equations of motion:

$$L_{i\alpha} u_\alpha^{(q)} + \sum_{k=1}^{[q/2]} K_{i\alpha k} u_\alpha^{(q-2k)} + \sum_{k=1}^{[(q+1)/2]} R_{i\alpha k} u_\alpha^{(q+1-2k)} = 0, \tag{2.4.9}$$

The constitutive relations:

$$\sigma_{ij}^{(q)} = \mu_{ij\alpha\beta}^{(0)} \frac{\partial u_\alpha^{(q)}}{\partial x_\beta} + \sum_{k=1}^{[(q+1)/2]} \mu_{ij\alpha\beta}^{(2k-1)} \frac{\partial u_\alpha^{(q+1-2k)}}{\partial x_\beta} + \sum_{k=1}^{[q/2]} \mu_{ij\alpha\beta}^{(2k)} \frac{\partial u_\alpha^{(q-2k)}}{\partial x_\beta}, \tag{2.4.10}$$

The boundary conditions on the surface portion S_1, *where stresses are given*:

$$\left[\mu_{ij\alpha\beta}^{(0)} \frac{\partial u_\alpha^{(q)}}{\partial x_\beta} + \sum_{k=1}^{[(q+1)/2]} \mu_{ij\alpha\beta}^{(2k-1)} \frac{\partial u_\alpha^{(q+1-2k)}}{\partial x_\beta} + \sum_{k=1}^{[q/2]} \mu_{ij\alpha\beta}^{(2k)} \frac{\partial u_\alpha^{(q-2k)}}{\partial x_\beta}\right] N_i = P_j^{(q)}. \tag{2.4.11}$$

In (2.4.9)-(2.4.11) [q/2] is the integer part of q/2. Note that the boundary conditions on the surface portion S_2, where displacements are given, remain as (1.4.10).

Thus, there is a closed system of equations and corresponding mechanical and geometrical relations for determining the stress and strain in each approximation. The zeroth approximation is analogous to the relations for a homogeneous rectilinearly orthotropic body; the influence of structural distortion in a composite material on the distortion of the stress-strain is characterized by the first, second, and subsequent approximations.

2.5. Example of the Small Parameter Method

As an illustration of the application of the small-parameter method, consider the following simple problem of the stress-strain state in an infinite body with pure shear in the plane Ox_1x_3. It is assumed here that structural curving in the composite occurs only in the direction of the x_1 axis, i.e.

$$F_{,3} \equiv 0. \tag{2.5.1}$$

We investigate the case shown in Fig.2.5.1 in which the external load is

$$\sigma_{13} = \tau \, . \tag{2.5.2}$$

To satisfy the equation (2.5.2) we choose the displacement field

$$u_1^{(q)} \equiv u_2^{(q)} \equiv 0 \, ; \quad u_3^{(q)} = u_3^{(q)}(x_1) \, . \tag{2.5.3}$$

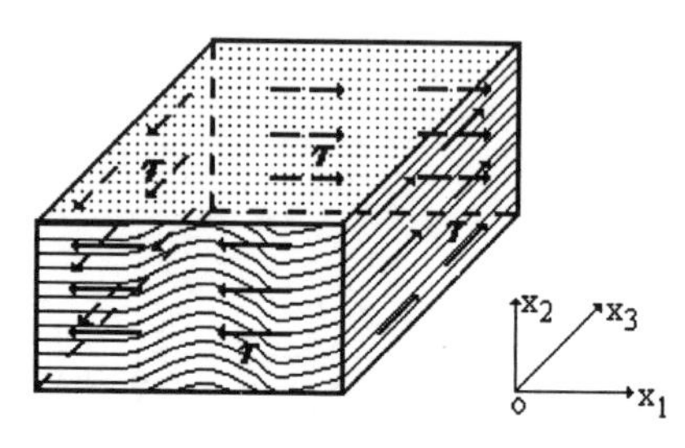

Fig.2.5.1. Pure shear in the plane Ox_1x_3 of the composite material with arbitrarily plane curved structure.

In accordance with equation (1.4.7), the following relations are obtained from equations (2.5.2), (2.4.9), and (2.4.10):

$$\sigma_{13}^{(0)} = \tau \, ; \quad u_3^{(0)} = \frac{\tau}{G_{13}^0} x_1 \, . \tag{2.5.4}$$

It follows from equations (2.5.4), (2.4.9), and (2.4.10) that

$$G_{13}^0 \frac{d^2 u_3^{(1)}}{dx_1^2} = 0 \, ; \quad \sigma_{13}^{(1)} = G_{13}^0 \frac{du_3^{(1)}}{dx_1} \, . \tag{2.5.5}$$

In the first approximation we obtain the following expressions from equations (2.5.5), (2.5.4), (2.4.10) and (2.5.2):

$$u_3^{(1)} \equiv 0 \, , \; \sigma_{23}^{(1)} = \frac{df}{dx_1}\left(G_{23}^0 - G_{13}^0\right)\frac{\tau}{G_{13}} ; \quad \sigma_{ij}^{(1)} \equiv 0 \;\; (ij \neq 23 \text{ or } ij \neq 32) \, . \tag{2.5.6}$$

For the second approximation, equations (2.5.4) - (2.5.6) and (2.4.9) give

$$G_{13}^0 \frac{d^2 u_3^{(2)}}{dx_1^2} + \frac{dA_{551}}{dx_1} \frac{du_3^{(0)}}{dx_1} = 0 \, . \tag{2.5.7}$$

Substituting the expression for A_{551} from equation (2.3.3), (2.3.4) into equation (2.5.7) and solving this equation, we obtain the second approximation

$$u_3^{(2)} = \frac{\tau}{\left(G_{13}^0\right)^2}\left(G_{13}^0 - G_{23}^0\right)\int (f'(x_1))^2 dx_1 \; ; \; f'(x_1) = \frac{df(x_1)}{dx_1}, \; \sigma_{ij}^{(2)} \equiv 0 . \quad (2.5.8)$$

By continuing in this way we may determine any subsequent approximation.

Thus, within the framework of the first two approximations, the solution is

$$u_1 \equiv 0 \; ; \quad u_2 \equiv 0 \; ; \quad u_3 \approx \frac{\tau}{G_{13}^0}\left(x_1 + \varepsilon^2 \left(G_{13}^0 - G_{23}^0\right)\frac{1}{G_{13}^0} \int (f'(x_1))^2 dx_1 \right);$$

$$\sigma_{11} = \sigma_{22} = \sigma_{33} = \sigma_{12} = 0 \; ; \quad \sigma_{13} = \tau \; ; \quad \sigma_{23} \approx \varepsilon f'(x_1)\left(G_{23}^0 - G_{13}^0\right)\frac{\tau}{G_{13}} . \quad (2.5.9)$$

For periodic curving the function $F(x_1)$ is chosen in the form (2.4.7) and from (2.2.2), (2.4.7) we obtain

$$f(x_1) = -\frac{\Lambda}{\pi}\cos\theta \, , \; \theta = \frac{\pi}{\Lambda} x_1 \quad (2.5.10)$$

Allowing the difference given in (2.4.8) for the definitions of ε, equations (2.5.9) and (2.5.10) agree with (1.5.11).

2.6. An Exact Solution

Closed form solution to the problem with the external load in the form in equation (2.5.2) (Fig.2.5.1) is obtained by means of the field

$$u_1 \equiv 0 \; ; \quad u_2 \equiv 0 \; ; \quad u_3 = u_3(x_1) . \quad (2.6.1)$$

Substituting expressions (2.5.1) and (2.6.1) into equations (2.2.8), (1.2.2) and (1.2.1), we find

$$\frac{d}{dx_1}\left(A_{55}(x_1)\frac{du_3}{dx_1} \right) = 0 . \quad (2.6.2)$$

From (2.5.1), (2.2.5) - (2.2.7), (1.3.7) we obtain

$$A_{55}(x_1) = \frac{G_{13}^0 + \varepsilon^2 (f'(x_1))^2 G_{23}^0}{1 + \varepsilon^2 (f'(x_1))^2} , \quad (2.6.3)$$

and from equations (2.6.2) and (2.6.3) that

$$u_3 = c\int \frac{1+\varepsilon^2 (f'(x_1))^2}{G_{13}^0 + \varepsilon^2 (f'(x_1))^2 G_{23}^0} dx_1; \quad c=\text{const.} \tag{2.6.4}$$

Substituting equation (2.6.4) into equation (2.2.8), we find that the non zero stresses are

$$\sigma_{13} = c; \qquad \sigma_{23} = c\varepsilon f'(x_1) \frac{G_{23}^0 - G_{13}^0}{G_{13}^0 + G_{23}^0 \varepsilon^2 (f'(x_1))^2}. \tag{2.6.5}$$

Equations (2.5.2) and (2.6.5) imply that $c = \tau$.

2.7. Pure Shear of Composite Materials

2.7.1. NOTATION AND ASSUMPTIONS

By the term quasi-homogeneous stress state we understand the stress state which was determined in Section 1.5. In the present section we will apply the continuum theory of (2.2.8)- (2.2.9) to study some examples involving the quasi-homogeneous stresses in a composite material with periodically curved structures; we will use the small parameter method. For the examples, we will select pure shear in the Ox_1x_3 plane and quasi-homogeneous stress states corresponding to triaxial tension-compression.

We suppose that the curving form (2.2.1) in the structure is given as follows:

$$x_2 = -H\cos\theta\cos\varphi, \qquad \theta = \frac{\pi}{\Lambda_1}x_1, \; \varphi = \frac{\pi}{\Lambda_3}x_3, \tag{2.7.1}$$

where H is the rise; $\Lambda_1(\Lambda_3)$ is the half-period length of the curving in the direction of the $Ox_1(Ox_3)$ axis (Fig.2.7.1).

Assuming that

$$\pi H \ll \Lambda_1 \quad \text{and} \quad \pi H \ll \Lambda_3, \tag{2.7.2}$$

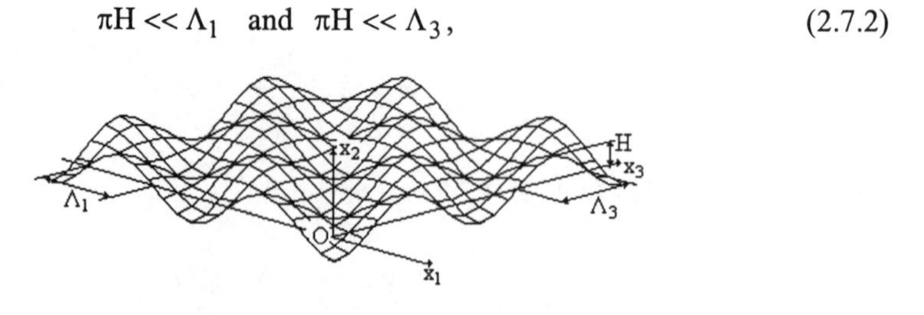

Fig. 2.7.1. The geometry of the spatially periodic curving.

also taking ε as the small parameter

$$\varepsilon = \frac{\pi H}{\Lambda_1}, \tag{2.7.3}$$

and introducing the parameter

$$\gamma_1 = \frac{\Lambda_1}{\Lambda_3}, \tag{2.7.4}$$

to characterize the spatiality of the curving, we find that equations (2.7.1) and (2.3.4) give

$$a(x_1, x_3) = \sin\theta\cos\varphi\,, \quad b(x_1, x_3) = \gamma_1 \cos\theta\sin\varphi\,. \tag{2.7.5}$$

We now discuss the conditions under which the stress distributions obtained in the present and in the following sections are applicable. We will assume that the composite material occupies the infinite three-dimensional region and there are no boundary effects. The stresses that will be obtained will be estimates of stresses in the region which is far from the boundaries. We assume that

$$\Lambda_1 << d \quad \text{and} \quad \Lambda_3 << d\,, \tag{2.7.6}$$

where d is the minimal size of the real region occupied by the composite material; in other words the curving in the structure of the composite material is small-scale. Moreover, for the applicability of the continuum approach (2.2.8), (2.2.9) we must also assume that the following conditions are satisfied

$$\Delta H << \Lambda_1\,, \qquad \Delta H << \Lambda_3\,, \tag{2.7.7}$$

where ΔH is a thickness of the representative curved packet. Thus, the stress distributions obtained below can be applied when the conditions (2.7.6), (2.7.7) are satisfied and these stresses arise in the sections which are sufficiently far from the boundaries of the real region occupied by the composite material. For sufficiently small values of the parameter ε the first non-zero term in the series (1.4.7) besides the zeroth approximation, can give highly accurate information on the stress distribution in the composite material. Note that for some stresses the first non-zero term arises in the first approximation, and for others in the second. Therefore, we will determine only the values of the zeroth , first and second approximations.

2.7.2. PURE SHEAR IN THE Ox_1x_3 PLANE

We explore the stress state in the composite material when $\sigma_{13} = \tau$ (Fig 2.7.2). From now on the values of the zeroth approximation are taken as those of the zeroth approximation of the corresponding problem considered in the previous chapter; (2.5.4) in the present case. Now consider the first and second approximations:

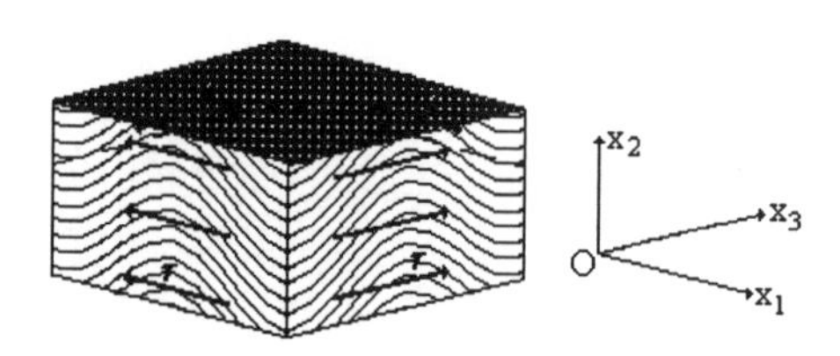

Fig.2.7.2. Pure shear in the Ox_1x_3 plane of the composite material with spatially periodic curved structure.

First approximation.
Equations (2.4.2)-(2.4.5), (2.4.9)-(2.4.11) and (2.5.4) give

$$L_{ij}u_j^{(1)} = -\left(\frac{\partial A_{561}}{\partial x_1} \frac{\partial u_3^{(0)}}{\partial x_1} + \frac{\partial A_{451}}{\partial x_3} \frac{\partial u_3^{(0)}}{\partial x_1} \right) \delta_i^2 , \tag{2.7.8}$$

where L_{ij} are Lamé operators corresponding to the linear static problems of a homogeneous elastic body with rectilinear orthotropy. Note that the expressions for these operators are given in (1.4.5).

We attempt to satisfy the equations (2.7.8) by using the displacement field

$$u_1^{(1)} \equiv 0 , \quad u_3^{(1)} \equiv 0 , \quad u_2^{(1)} = u_2^{(1)}(x_1, x_3) . \tag{2.7.9}$$

Taking (2.7.9), (2.5.4), (2.3.2)-(2.3.4) and (2.7.5) into consideration, we obtain the following equation for $u_2^{(1)}$ from (2.4.9):

$$A_{66}^{0} \frac{\partial^2 u_2^{(1)}}{\partial x_1^2} + A_{44}^{0} \frac{\partial^2 u_2^{(1)}}{\partial x_3^2} = \alpha \gamma_1 \frac{\tau}{G_{13}^{0}} \left(a_{561} + a_{451} \right) \sin(\alpha x_1) \sin(\beta x_3) , \tag{2.7.10}$$

where

$$\alpha = \frac{1}{\Lambda_1} , \quad \beta = \frac{1}{\Lambda_3} . \tag{2.7.11}$$

Taking into account the same assumptions proposed for (1.7.7), we choose the solution to equation (2.7.10) as follows:

$$u_2^{(1)} = B \sin(\alpha x_1) \sin(\beta x_3) , \tag{2.7.12}$$

where

$$B = -\gamma_1 \frac{\tau}{G_{13}^0} \frac{(a_{561} + a_{451})}{\alpha(G_{12}^0 + \gamma_1^2 G_{23}^0)}, \tag{2.7.13}$$

Substituting (2.5.4), (2.7.9) and (2.7.12) into (2.4.10) we obtain the following expression for the non-zero stress contributions:

$$\sigma_{23}^{(1)} = \left[-\frac{\gamma_1^2 G_{23}^0 (G_{12}^0 - 2G_{13}^0 + G_{23}^0)}{G_{12}^0 + \gamma_1^2 G_{23}^0} + G_{23}^0 - G_{13}^0 \right] \frac{\tau}{G_{13}^0} \sin(\alpha x_1) \cos(\beta x_3),$$

$$\sigma_{12}^{(1)} = \gamma_1 \left[-\frac{G_{12}^0 (G_{12}^0 - 2G_{13}^0 + G_{23}^0)}{G_{12}^0 + \gamma_1^2 G_{23}^0} + G_{12}^0 - G_{13}^0 \right] \frac{\tau}{G_{13}^0} \cos(\alpha x_1) \sin(\beta x_3). \tag{2.7.14}$$

Equation (2.7.14) shows that, as $\gamma_1 \to 0$, $\sigma_{23}^{(1)}$ approaches the expression (2.5.6), (2.5.10) obtained for a composite with periodically plane-curved structure; $\sigma_{12}^{(1)} \to 0$.

Second approximation.
Taking into account the expressions (2.7.9) and (2.7.12), we derive from (2.4.9) the following equations for the second approximation:

$$\begin{aligned} L_{ij} u_j^{(2)} = &-\Bigg[\left(\frac{\partial A_{151}}{\partial x_1} + \frac{\partial A_{551}}{\partial x_3} \right) \frac{\tau}{G_{13}^0} + A_{161} \frac{\partial^2 u_2^{(1)}}{\partial x_1^2} + A_{451} \frac{\partial^2 u_2^{(1)}}{\partial x_3^2} \\ &+ (A_{141} + A_{561}) \frac{\partial^2 u_2^{(1)}}{\partial x_1 \partial x_3} + \left(\frac{\partial A_{141}}{\partial x_1} + \frac{\partial A_{451}}{\partial x_3} \right) \frac{\partial u_2^{(1)}}{\partial x_3} + \left(\frac{\partial A_{161}}{\partial x_1} \right. \\ &\left. + \frac{\partial A_{561}}{\partial x_3} \right) \Bigg] \delta_i^1 - \Bigg[\left(\frac{\partial A_{551}}{\partial x_1} + \frac{\partial A_{351}}{\partial x_3} \right) \frac{\tau}{G_{13}^0} + A_{561} \frac{\partial^2 u_2^{(1)}}{\partial x_1^2} + A_{341} \frac{\partial^2 u_2^{(1)}}{\partial x_3^2} \\ &+ A_{341} \frac{\partial^2 u_2^{(1)}}{\partial x_3^2} + (A_{451} + A_{361}) \frac{\partial^2 u_2^{(1)}}{\partial x_1 \partial x_3} + \left(\frac{\partial A_{451}}{\partial x_1} + \frac{\partial A_{341}}{\partial x_3} \right) \frac{\partial u_2^{(1)}}{\partial x_3} \\ &+ \left(\frac{\partial A_{561}}{\partial x_1} + \frac{\partial A_{361}}{\partial x_3} \right) \frac{\partial u_2^{(1)}}{\partial x_1} \Bigg] \delta_i^2 . \end{aligned} \tag{2.7.15}$$

We satisfy the equations (2.7.15) by the following displacement field:

$$u_1^{(2)} = u_1^{(2)}(x_1, x_3); \quad u_2^{(2)} \equiv 0; \quad u_3^{(2)} = u_3^{(2)}(x_1, x_3) \,. \tag{2.7.16}$$

From equation (2.7.9), (2.7.12), (2.7.15) and (2.7.16) we obtain the following for the second order displacements:

$$A_{11}^0 \frac{\partial^2 u_1^{(2)}}{\partial x_1^2} + A_{55}^0 \frac{\partial^2 u_1^{(2)}}{\partial x_3^2} + \left(A_{13}^0 + A_{55}^0\right) \frac{\partial^2 u_3^{(2)}}{\partial x_1 \partial x_3}$$

$$= \frac{\alpha}{2} B_{11} \sin(2\beta x_3) + \frac{\alpha}{2} B_{13} \cos(2\alpha x_1) \sin(2\beta x_3);$$

$$A_{55}^0 \frac{\partial^2 u_3^{(2)}}{\partial x_1^2} + A_{33}^0 \frac{\partial^2 u_3^{(2)}}{\partial x_3^2} + \left(A_{55}^0 + A_{13}^0\right) \frac{\partial^2 u_1^{(2)}}{\partial x_1 \partial x_3}$$

$$= \frac{\alpha}{2} B_{21} \sin(2\alpha x_3) + \frac{\alpha}{2} B_{23} \sin(2\alpha x_1) \cos(2\beta x_3), \tag{2.7.17}$$

where

$$B_{11} = -\left(\gamma_1^2 b_{551} - \gamma_1 a_{551}\right) \frac{\tau}{G_{12}^0} - \alpha B \gamma_1^2 \left(a_{561} - a_{451}\right);$$

$$B_{13} = -\left(a_{151} + \gamma_1 a_{551} + \gamma_1^2 b_{551}\right) \frac{\tau}{G_{13}^0} - \alpha B \left(a_{161} + \gamma_1^2 \left(a_{451} + a_{141} + a_{451}\right)\right);$$

$$B_{21} = \left(-a_{551} + \gamma_1^2 b_{551}\right) \frac{\tau}{G_{13}^0} + \alpha B \left(\gamma_1 a_{561} + \gamma_1 a_{451} - \gamma_1^3 a_{341}\right);$$

$$B_{23} = \left(-a_{551} - \gamma_1^2 \left(b_{551} + a_{351}\right)\right) \frac{\tau}{G_{13}^0} - \alpha B \left(\gamma_1 a_{561} + \gamma_1 \left(a_{451} + a_{361}\right)\right). \tag{2.7.18}$$

The solution to the equation (2.7.17) is chosen to be:

$$u_1^{(2)} = C_1 \sin(2\beta x_3) + C_3 \cos(2\alpha x_1) \sin(2\beta x_3),$$

$$u_3^{(2)} = D_1 \sin(2\alpha x_3) + D_3 \sin(2\alpha x_1) \cos(2\beta x_3), \tag{2.7.19}$$

where

$$C_1 = -\frac{\alpha}{8\beta^2 A_{55}^0} B_{11}, \qquad D_1 = -\frac{1}{8\alpha A_{55}^0} B_{21},$$

$$D_3 = -\frac{B_{23}}{8\alpha\,(A_{55}^0 + \gamma_1^2 A_{33}^0)} - \gamma_1 \frac{A_{13}^0 + A_{55}^0}{A_{55}^0 + \gamma_1^2 A_{33}^0} C_3,$$

$$C_3 = \left[\frac{B_{23}\gamma_1 (A_{13}^0 + A_{55}^0)}{8\alpha\,(A_{55}^0 + \gamma_1^2 A_{33}^0)(A_{11}^0 + \gamma_1^2 A_{55}^0)} - \frac{B_{13}}{8\alpha\,(A_{11}^0 + \gamma_1^2 A_{55}^0)}\right]$$

$$\times \left(1 - \frac{\gamma_1^2 \left(A_{13}^0 + A_{55}^0\right)^2}{(A_{55}^0 + \gamma_1^2 A_{33}^0)(A_{11}^0 + \gamma_1^2 A_{55}^0)}\right)^{-1}. \qquad (2.7.20)$$

Equations (2.7.18)-(2.7.20) and (2.4.10) give the following expressions for the non zero second order stresses:

$$\sigma_{11}^{(2)} = \left[-2\alpha C_3 A_{11}^0 - 2\beta D_3 A_{13}^0 + \gamma_1 a_{151} \frac{\tau}{G_{13}^0} + \frac{\alpha B}{4}\left(\gamma_1^2 a_{141} + a_{161}\right)\right]$$

$$\times \sin(2\alpha x_1) \sin(2\beta x_3) \quad ;$$

$$\sigma_{22}^{(2)} = \left[-2\alpha C_3 A_{12}^0 - 2\beta D_3 A_{23}^0 + \gamma_1 a_{251} \frac{\tau}{G_{13}^0} + \frac{\alpha B}{4}\left(\gamma_1^2 a_{241} + a_{261}\right)\right]$$

$$\times \sin(2\alpha x_1) \sin(2\beta x_3);$$

$$\sigma_{33}^{(2)} = \left[-2\alpha C_3 A_{13}^0 - 2\beta D_3 A_{33}^0 + \gamma_1 a_{351} \frac{\tau}{G_{13}^0} + \frac{\alpha B}{4}\left(\gamma_1^2 a_{341} + a_{361}\right)\right]$$

$$\times \sin(2\alpha x_1) \sin(2\beta x_3);$$

$$\sigma_{13}^{(2)} = \left[2\alpha D_1 - \frac{\tau}{4 G_{13}^0}\left(a_{551} - \gamma_1^2 b_{551}\right) + \frac{\alpha B}{4} \gamma_1 \left(-a_{451} + a_{561}\right)\right] \cos(2\alpha x_1)$$

$$+\left[2\alpha\gamma_1 C_1+\frac{\tau}{4G_{13}^0}\left(a_{551}-\gamma_1^2 b_{551}\right)+\frac{\alpha B}{4}\gamma_1\left(a_{451}-a_{561}\right)\right]\cos(2\beta x_3)$$

$$+\left[2\alpha(D_3+\gamma_1 C_3)-\frac{\tau}{4G_{13}^0}\left(a_{551}+\gamma_1^2 b_{551}\right)-\frac{\alpha\gamma_1}{4}B\left(a_{451}+a_{561}\right)\right]\times$$

$$\cos(2\alpha x_1)\cos(2\beta x_3). \qquad (2.7.21)$$

If we take $\gamma_1=0$ in (2.7.21) and neglect the terms that depend only on x_3, we obtain the expressions (2.5.8). These results give us the expressions

$$u_1\approx\varepsilon^2 u_1^{(2)}(x_1,x_3);\ u_2\approx\varepsilon u_2^{(1)}(x_1,x_3);\ u_3\approx\frac{\tau}{G_{13}}x_1+\varepsilon^2 u_3^{(2)}(x_1,x_3);$$

$$\sigma_{11}\approx\varepsilon^2\sigma_{11}^{(2)}(x_1,x_2);\ \sigma_{22}\approx\varepsilon^2\sigma_{22}^{(2)}(x_1,x_2);\ \sigma_{33}\approx\varepsilon^2\sigma_{33}^{(2)}(x_1,x_2);$$

$$\sigma_{13}\approx\tau+\varepsilon^2\sigma_{13}^{(2)}(x_1,x_2);\ \sigma_{23}\approx\varepsilon\sigma_{23}^{(1)}(x_1,x_2);\ \sigma_{12}\approx\varepsilon\sigma_{12}^{(1)}(x_1,x_2), \qquad (2.7.22)$$

By comparing equation (2.7.22) with (2.6.5) we see that the spatiality of the curving in the structure introduces other non-zero stresses; these are self-balanced within the curving period. Note that, when $\gamma_1\to 0$ and is neglected in (2.7.12)-(2.7.14), and in (2.7.19)-(2.7.21) the terms that depend only on x_3, we obtain the expressions (1.7.26) from (2.7.22). However, in this case the relation (2.4.8) must be taken into account.

2.8. Quasi-homogeneous Stress State Corresponding to Triaxial Tension-compression

We investigate the quasi-homogeneous stress state in the composite material when the external forces are $\sigma_{11}=p_1$, $\sigma_{22}=p_2$ and $\sigma_{33}=p_3$ (Fig.2.8.1).

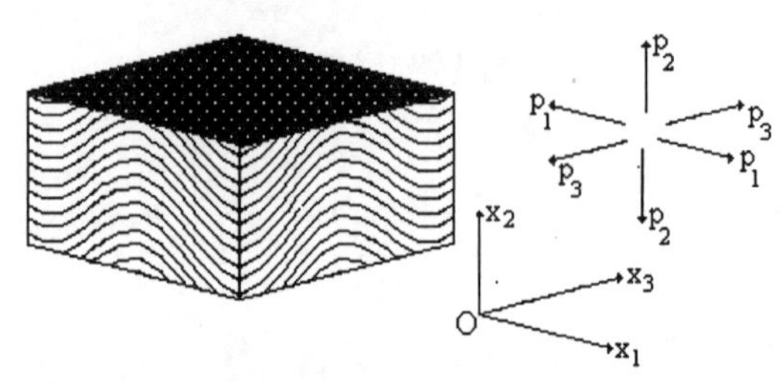

Fig.2.8.1. Triaxial tension-compression of the composite material with spatially periodic curved structure.

Zeroth approximation.
This is given by equations (1.8.2)-(1.8.4).

First approximation.
From equation (1.8.2) and (2.4.9) we find

$$L_{ij}u_j^{(1)} = -\left(A_1 \frac{\partial A_{161}}{\partial x_1} + A_2 \frac{\partial A_{261}}{\partial x_1} + A_3 \frac{\partial A_{361}}{\partial x_1} \right.$$

$$\left. + A_1 \frac{\partial A_{141}}{\partial x_3} + A_2 \frac{\partial A_{241}}{\partial x_3} + A_3 \frac{\partial A_{341}}{\partial x_3} \right) \delta_i^2 . \tag{2.8.1}$$

Taking into account (2.7.4), (2.3.3) and (2.3.4) we choose the solution as follows:

$$u_1^{(1)} \equiv 0 ; \quad u_3^{(1)} \equiv 0 ; \quad u_2^{(1)} = u_2^{(1)}(x_1, x_3) . \tag{2.8.2}$$

Substituting (2.8.2) into (2.8.1) we obtain

$$A_{66}^0 \frac{\partial^2 u_2^{(1)}}{\partial x_1^2} + A_{44}^0 \frac{\partial^2 u_2^{(1)}}{\partial x_3^2} = -\alpha [A_1 a_{161} + A_2 a_{261} + A_3 a_{361}$$

$$+ \gamma_1^2 (A_1 a_{141} + A_2 a_{241} + A_3 a_{341})] \cos(\alpha x_1) \cos(\beta x_3) , \tag{2.8.3}$$

which has the solution

$$u_2^{(1)} = D \cos(\alpha x_1) \cos(\beta x_3) , \tag{2.8.4}$$

where

$$D = \frac{1}{\alpha \left(A_{66}^0 + \gamma_1^2 A_{44}^0 \right)} [A_1 a_{161} + A_2 a_{261} + A_3 a_{361}$$

$$+ \gamma_1^2 (A_1 a_{141} + A_2 a_{241} + A_3 a_{341}) . \tag{2.8.5}$$

From equations (2.8.5), (2.8.2), (2.4.3) and (2.4.10) we obtain the following non-zero stresses in the first approximation:

$$\sigma_{23}^{(1)} = \left[-\alpha D A_{44}^0 + \gamma_1 (A_1 a_{141} + A_2 a_{241} + A_3 a_{341}) \right] \cos(\alpha x_1) \sin(\beta x_3) ,$$

$$\sigma_{12}^{(1)} = \left[-\alpha D A_{66}^{0} + A_1 a_{161} + A_2 a_{261} + A_3 a_{361} \right] \sin(\alpha x_1) \cos(\beta x_3). \quad (2.8.6)$$

when $\gamma_1 = 0$, $\beta \equiv \gamma_1 / \Lambda_1 = 0$, and equations (2.8.5) and (2.8.6) show that $\sigma_{23}^{(1)} = \sigma_{12}^{(1)} = 0$.

Second approximation.
Equations (1.8.2), (2.4.9) and (2.8.2)-(2.8.4) give the equation

$$L_{ij} u_j^{(2)} = -\left[A_1 \left(\frac{\partial A_{111}}{\partial x_1} + \frac{\partial A_{151}}{\partial x_3} \right) + A_2 \left(\frac{\partial A_{121}}{\partial x_1} + \frac{\partial A_{251}}{\partial x_3} \right) \right.$$

$$+ A_3 \left(\frac{\partial A_{131}}{\partial x_1} + \frac{\partial A_{351}}{\partial x_3} \right) + A_{161} \frac{\partial^2 u_2^{(1)}}{\partial x_1^2} + A_{451} \frac{\partial^2 u_2^{(1)}}{\partial x_3^2}$$

$$+ (A_{141} + A_{561}) \frac{\partial^2 u_2^{(1)}}{\partial x_1 \partial x_3} + \frac{\partial u_2^{(1)}}{\partial x_3} \left(\frac{\partial A_{141}}{\partial x_1} + \frac{\partial A_{451}}{\partial x_3} \right)$$

$$\left. + \frac{\partial u_2^{(1)}}{\partial x_1} \left(\frac{\partial A_{161}}{\partial x_1} + \frac{\partial A_{561}}{\partial x_3} \right) \right] \delta_i^1 - \left[A_1 \left(\frac{\partial A_{151}}{\partial x_1} + \frac{\partial A_{131}}{\partial x_3} \right) \right.$$

$$+ A_2 \left(\frac{\partial A_{251}}{\partial x_1} + \frac{\partial A_{231}}{\partial x_3} \right) + A_3 \left(\frac{\partial A_{351}}{\partial x_1} + \frac{\partial A_{331}}{\partial x_3} \right) + A_{561} \frac{\partial^2 u_2^{(1)}}{\partial x_1^2}$$

$$+ A_{341} \frac{\partial^2 u_2^{(1)}}{\partial x_3^2} + (A_{451} + A_{361}) \frac{\partial^2 u_2^{(1)}}{\partial x_1 \partial x_3} + \left(\frac{\partial A_{451}}{\partial x_1} + \frac{\partial A_{341}}{\partial x_3} \right) \frac{\partial u_2^{(1)}}{\partial x_3}$$

$$\left. + \left(\frac{\partial A_{561}}{\partial x_1} + \frac{\partial A_{361}}{\partial x_3} \right) \frac{\partial u_2^{(1)}}{\partial x_1} \right] \delta_i^3 . \quad (2.8.7)$$

We try a displacement field

$$u_1^{(2)} = u_1^{(2)}(x_1, x_3), \quad u_2^{(2)} = C'_2 \, x_2, \quad u_3^{(2)} = u_3^{(2)}(x_1, x_3). \quad (2.8.8)$$

Substituting (2.8.8) into (2.8.7) we obtain

$$A_{11}^{0}\frac{\partial^2 u_1^{(2)}}{\partial x_1^2}+A_{55}^{0}\frac{\partial^2 u_1^{(2)}}{\partial x_3^2}+\left(A_{13}^{0}+A_{55}^{0}\right)\frac{\partial^2 u_3^{(2)}}{\partial x_1 \partial x_3}$$

$$=C_{11}\sin(2\alpha x_1)+C_{21}\sin(2\alpha x_1)\cos(2\beta x_3),$$

$$A_{55}^{0}\frac{\partial^2 u_3^{(2)}}{\partial x_1^2}+A_{33}^{0}\frac{\partial^2 u_3^{(2)}}{\partial x_3^2}+\left(A_{13}^{0}+A_{55}^{0}\right)\frac{\partial^2 u_1^{(2)}}{\partial x_1 \partial x_3}$$

$$=C_{31}\sin(2\beta x_3)+C_{41}\cos(2\alpha x_1)\sin(2\beta x_3), \qquad (2.8.9)$$

where

$$C_{11}=-\frac{\alpha}{2}\left(a_{111}A_1+\left(a_{121}-\gamma_1^2 b_{121}\right)A_2+\left(a_{131}-\gamma_1^2 b_{131}\right)A_3\right.$$

$$\left.+\frac{\alpha}{2}D\left(-a_{161}-\gamma_1^2 a_{451}+\gamma_1^2\left(a_{141}+a_{561}\right)+\gamma_1^2\left(a_{141}+a_{451}\right)-a_{161}-\gamma_1^2 a_{561}\right)\right);$$

$$C_{21}=-\frac{\alpha}{2}\left(\left(a_{111}+\gamma_1^2 a_{151}\right)A_1+\left(a_{121}+\gamma_1^2\left(b_{121}+a_{251}\right)\right)A_2\right.$$

$$\left.+\left(a_{131}+\gamma_1^2\left(b_{131}+a_{351}\right)\right)A_3-D\alpha\left(a_{161}+\gamma_1^2 a_{561}\right)\right);$$

$$C_{31}=-\frac{\alpha}{2}\left(\left(-\gamma_1 a_{131}+\gamma_1^3 b_{131}\right)A_1+\left(-\gamma_1 a_{231}+\gamma_1^3 b_{231}\right)A_2\right.$$

$$\left.+\gamma_1^3 a_{331}A_3+\alpha D\left(\gamma_1 a_{361}-\gamma_1^3 a_{561}\right)\right),$$

$$C_{41}=-\frac{\alpha}{2}\left(\left(\gamma_1 a_{151}+\gamma_1 a_{131}+\gamma_1^3 b_{131}\right)A_1+\left(\gamma_1 a_{251}+\gamma_1 a_{231}+\gamma_1^3 b_{231}\right)A_2\right.$$

$$+\left(\gamma_1 a_{351}+\gamma_1^3 a_{331}\right)A_3+\frac{\alpha}{2}D\left(-\gamma_1 a_{561}-\gamma_1^3 a_{341}-\gamma_1\left(a_{451}+a_{361}\right)\right.$$

$$\left.\left.-\gamma_1\left(a_{451}+\gamma_1^2 a_{341}\right)-\gamma_1\left(a_{561}+a_{361}\right)\right)\right). \qquad (2.8.10)$$

The solution to equations (2.8.9) is found as follows:

$$u_1^{(2)}=B_{13}\sin(2\alpha x_1)+B_{23}\sin(2\alpha x_1)\cos(2\beta x_3)+C'_1\, x_1;$$

$$u_3^{(2)} = D_{13}\sin(2\beta x_3) + D_{23}\cos(2\alpha x_1)\sin(2\beta x_3) + C'_3 x_3 . \qquad (2.8.11)$$

Substituting (2.8.11) into (2.8.9) we find

$$D_{13} = -\frac{C_{31}}{4\alpha^2\gamma_1^2 A_{33}^0},$$

$$B_{23} = -\frac{1}{4\alpha^2}\left[\gamma_1^2\left(A_{13}^0 + A_{55}^0\right)C_{41} + \left(A_{55}^0 + \gamma_1^2 A_{33}^0\right)C_{21}\right.$$

$$\left[\left(A_{11}^0 + \gamma_1^2 A_{55}^0\right)\left(A_{55}^0 + \gamma_1^2 A_{33}^0\right) + \gamma_1^4\left(A_{13}^0 + A_{55}^0\right)^2\right]^{-1},$$

$$D_{23} = \left[\frac{C_{41}}{4\alpha^2} - B_{23}\gamma_1^2\left(A_{13}^0 + A_{55}^0\right)\right]\left(A_{55}^0 + \gamma_1^2 A_{33}^0\right)^{-1}. \qquad (2.8.12)$$

Substituting (2.8.11) into (2.4.10) we obtain for the second order stresses $\sigma_{ii}^{(2)}$ (i =1,2,3), $\sigma_{13}^{(2)}$. In this case the expressions for $\sigma_{ii}^{(2)}$ have a non-balanced part. However, it was noted that the stresses corresponding to the first and subsequent approximations must be self-balanced. To eliminate the non-self-balanced part of the normal stresses in the second approximation, the constants C'_1, C'_2 and C'_3 which enter the expressions for $u_i^{(2)}$ (i=1,2,3) (2.8.8), (2.8.11), must be determined from the following equations:

$$A_{11}^0 C'_1 + A_{12}^0 C'_2 + A_{13}^0 C'_3 + \frac{1}{4}A_1 a_{111} + \frac{1}{4}A_2\left(a_{121} + \gamma_1^2 b_{121}\right)$$

$$+\frac{1}{4}A_3\left(a_{131} + \gamma_1^2 b_{131}\right) - \alpha\gamma_1^2 D a_{141}\frac{1}{4} - \alpha D a_{161}\frac{1}{4} = 0;$$

$$A_{12}^0 C'_1 + A_{22}^0 C'_2 + A_{23}^0 C'_3 + \frac{1}{4}A_1\left(a_{121} + \gamma_1^2 b_{121}\right) + \frac{1}{4}A_2\left(a_{221} + \gamma_1^2 b_{221}\right)$$

$$+\frac{1}{4}A_3\left(a_{231} + \gamma_1^2 b_{231}\right) - \alpha\gamma_1^2 D a_{241}\frac{1}{4} - \alpha D a_{261}\frac{1}{4} = 0,$$

$$A_{13}^0 C'_1 + A_{23}^0 C'_2 + A_{33}^0 C'_3 + \frac{1}{4}A_1\left(a_{131} + \gamma_1^2 b_{131}\right) + \frac{1}{4}A_2\left(a_{231} + \gamma_1^2 b_{231}\right)$$

$$\frac{1}{4}A_3\gamma_1^2 a_{331} - \frac{1}{4}\alpha\gamma_1^2 D - \frac{1}{4}\alpha D a_{361} = 0 . \qquad (2.8.13)$$

Equations (2.8.13), (2.8.11), (2.8.8), (2.8.4), and (2.4.9) yield the expressions for the non-zero second order stresses $\sigma_{ii}^{(2)}$ (i =1,2,3), $\sigma_{13}^{(2)}$. We give just

$$\sigma_{22}^{(2)} = \cos(2\alpha x_1)\Big[2\alpha B_{13}A_{12}^0 - \frac{1}{4}a_{121}A_1 - \frac{1}{4}\left(a_{221} - \gamma_1^2 b_{221}\right)A_2$$

$$- \frac{1}{4}\left(a_{231} - \gamma_1^2 b_{231}\right)A_3 - \alpha D\gamma_1^2 a_{241}\frac{1}{4} + \alpha D a_{261}\frac{1}{4}\Big]$$

$$+ \cos(2\beta x_3)\Big[2\alpha D_{13}A_{23}^0\gamma_1 + \frac{1}{4}A_1\left(a_{121} - \gamma_1^2 b_{121}\right) + \frac{1}{4}A_2\left(a_{221} - \gamma_1^2 b_{221}\right)$$

$$+ \frac{1}{4}A_3\left(a_{231} - \gamma_1^2 b_{231}\right) + \frac{1}{4}\gamma_1^2\alpha D a_{241} - \frac{1}{4}\alpha D a_{261}\Big]$$

$$+ \cos(2\alpha x_1)\cos(2\beta x_3)\Big[2\alpha B_{23}A_{12}^0 + 2\alpha\gamma_1 D_{23}A_{23}^0 - \frac{1}{4}\left(a_{111} + \gamma_1^2 b_{121}\right)A_1$$

$$- \frac{1}{4}\left(a_{221} + \gamma_1^2 b_{221}\right)A_2 - \frac{1}{4}\left(a_{231} + \gamma_1^2 b_{231}\right)A_3 + \alpha D\gamma_1^2 a_{241}\frac{1}{4} + \alpha D a_{261}\frac{1}{4}\Big]. \qquad (2.8.14)$$

As before, substituting $\gamma_1 = 0$ into (2.8.14) and neglecting the terms that depend only on x_3, we obtain the expressions (1.8.23).

Thus, according to (1.4.7), we obtain the following expressions

$$u_1 \approx A_1 x_1 + \varepsilon^2 u_1^{(2)}(x_1, x_3), \quad u_2 \approx A_2 x_2 + \varepsilon \qquad u_2^{(1)}(x_1, x_3) + \varepsilon^2 C'_2 x_2 ,$$

$$u_3 \approx A_3 x_3 + \varepsilon^2 u_3^{(2)}(x_1, x_3), \; \sigma_{ii} \approx p_i + \varepsilon^2 \sigma_{ii}^{(2)}(x_1, x_3), \; i=1, 2, 3,$$

$$\sigma_{12} \approx \varepsilon\sigma_{12}^{(1)}(x_1, x_3), \quad \sigma_{23} \approx \varepsilon\sigma_{23}^{(1)}(x_1, x_3), \quad \sigma_{13} \approx \varepsilon^2\sigma_{13}^{(2)}(x_1, x_3) . \qquad (2.8.15)$$

The right hand sides are determined through equations (1.8.2), (2.8.2), (2.8.6), (2.8.11), and (2.8.14). Based on these results, the solution corresponding to the various particular cases can easily be derived, for example, for uniaxial tension-compression along the Ox_1 axis.

2.9. Approximate Results for Layered Composites

As an example, in this section we examine a specific application of the results obtained in the previous sections for laminated composites consisting of alternating layers of two materials. Moreover, we assume that the material of each layer is isotropic, which means that in the continuum approximation this composite under ideal (uncurved) layout of the layers is transversely isotropic about the Ox_2 axis. We will assume that the normalized mechanical properties of the material are determined from the expressions (1.9.1)-(1.9.3). Based on the earlier investigations, we consider some detailed results on the stress distribution in a composite material with such a spatially periodic curved structure .

2.9.1. PURE SHEAR IN THE Ox_1x_3 PLANE

Consider the stresses σ_{12} and σ_{23} which have order ε, and suppose that the conditions (1.9.5) hold. Substituting (1.9.3) into (2.7.14) and making several transformations we obtain

$$\sigma_{23} \approx -\left(1-\frac{2\gamma_1^2}{1+\gamma_1^2}\right)\tau\varepsilon\sin(\alpha x_1)\cos(\gamma_1\alpha x_3),$$

$$\sigma_{12} \approx -\gamma_1\left(1-\frac{2}{1+\gamma_1^2}\right)\tau\varepsilon\cos(\alpha x_1)\sin(\gamma_1\alpha x_3). \tag{2.9.1}$$

When $\gamma_1 = 0$, the expression obtained for σ_{23} in (2.9.1) coincides with that given by (1.9.6). Also, it follows from the expressions (2.9.1) that the stresses σ_{12} arise as a result of the spatiality of the curving in the structure of the composite material and, when $\gamma_1 > 1$ and γ_1 increases the absolute values of these stresses grow monotonically. Equation (2.9.1) reveals that the amplitude of σ_{23} decreases as γ_1 increases.

2.9.2. UNIAXIAL TENSION-COMPRESSION ALONG THE Ox_1 AXIS

We will use the expressions (2.8.6), (2.8.14), and (2.8.5) and will consider only the stresses σ_{12}, σ_{23}, and σ_{22}. Note that the stresses σ_{12} and σ_{23} are of order ε, but the stress σ_{22} which acts in the direction Ox_2 axis perpendicular to the plane of the layers, is of order ε^2. Supposing that the conditions (1.9.5) and the equations (1.8.18) and (1.8.19) hold, we derive the following expressions from (2.8.6), (2.8.14), and (2.8.15):

$$\sigma_{12} \approx -\varepsilon p_1 \frac{\gamma_1^2}{1+\gamma_1^2}\sin(\alpha x_1)\cos(\gamma_1\alpha x_3),$$

$$\sigma_{23} \approx \varepsilon p_1 \frac{\gamma_1}{1+\gamma_1^2} \cos(\alpha x_1) \sin(\gamma_1 \alpha x_3),$$

$$\sigma_{22} \approx \varepsilon^2 p_1 \{ a_1 \cos(2\alpha x_1) \cos(2\gamma_1 \alpha x_3) + a_2 \cos(2\alpha x_1) + a_3 \cos(2\gamma_1 \alpha x_3) \}, \quad (2.9.2)$$

where

$$a_1 = \eta^{(1)} c_5 (\chi_1 + \eta_2 c_3) \left(2 \left[c_2 c_1 + \gamma_1^4 (\chi_2 + 1)^2 \right] \right)^{-1} \left[\frac{1}{1+\gamma_1^2} \frac{c_2}{2} c_1 + \frac{\gamma_1^3}{1-\nu_2} \left(1 + \gamma_1^2 \chi_2 \right) \right]$$

$$\times \left[1 - \gamma_1^3 \frac{3-\nu_2}{1-\nu_2+2\gamma_1^2} \right] - \frac{\gamma_1^2 \eta_1 c_5}{2c_1(1-\nu_2)} (\chi_1 + \eta_2 c_3)$$

$$+ \frac{1}{4} - \frac{\gamma_1^2}{2} (1 - 3\nu_2) + c_5 \eta_1 \frac{1}{2} \left(1 + \eta_2 \frac{(1-2\nu_2)(1-\nu_1)}{(1-\nu_2)(1-2\nu_2)} \right);$$

$$a_2 = \left(\frac{\eta_1 c_5}{4(1+\gamma_1^2)} [\chi_1 + 2\eta_2 c_3] + \frac{\gamma_1^2 \eta_1 c_5}{4(1+\gamma_1^2)} [\chi_1 (1-\nu_2) - \eta_2 \nu_2 (1-\nu_1)] \right.$$

$$\left. - \frac{1}{4} + \frac{\gamma_1^2 - 1}{2(1+\gamma_1^2)} c_5 + \frac{\eta_1 (1-\gamma_1^2)}{4(1+\gamma_1^2)} c_5 (2c_4 - \eta_2 c_3) \right);$$

$$a_3 = \left(\frac{c_5}{2(1+\gamma_1^2)} \left(\gamma_1^2 - \nu_2 \right) \eta_1 \left(\frac{\chi_1}{2} + \eta_2 c_3 \right) + \frac{1-\gamma_1^2}{2(1+\gamma_1^2)} c_5 - \frac{\eta_1 \left(1 - \gamma_1^2 \right)}{2(1+\gamma_1^2)} c_5 (c_4 - \eta_2 c_3) \right). \quad (2.9.3)$$

In (2.9.3) the following notation is used.

$$c_1 = 1 + \gamma_1^2 \frac{2}{1-\nu_2}; \; c_2 = \gamma_1^2 + \frac{2}{1-\nu_2}; \; c_3 = \frac{\nu_2 (1-\nu_1)}{(1-2\nu_1)(1-\nu_2)};$$

$$c_4 = 1 + \eta_2 \frac{1-\nu_1}{1-2\nu_1}; \; c_5 = -1 + \gamma_1^2 (1-\nu_2); \; \chi_1 = \frac{2\nu_1}{1-2\nu_1}; \; \chi_2 = \frac{2\nu_2}{1-\nu_2}. \quad (2.9.4)$$

If $\gamma_1 = 0$, (2.9.2)-(2.9.4) show that $\sigma_{12} = \sigma_{23} = 0$; neglecting the terms depending only on x_3, from the expression σ_{22} written in (2.9.2)-(2.9.4), we obtain the expression (1.9.7). Note that the expression (1.9.7) has been derived for the composite material with periodically plane-curved structures.

Using the expressions (2.9.2)-(2.9.4) we consider some numerical results and assume that $\eta^{(1)} = \eta^{(2)}$, $\nu^{(2)} = 0.2$, $\nu^{(1)} = 0.4$ and also investigate the distribution of the stresses σ_{12} and σ_{22}. It follows from (2.9.2)-(2.9.4) that the stress σ_{12} has its absolute maxima at the points $\alpha x_1 = \pi/2 + n\pi$, $\beta x_3 = n\pi$, n=0,1,2,..., but the stress σ_{22} has its absolute maxima at points $\alpha x_1 = n\pi/2$, $\beta x_3 = n\pi/2$, at which, the values of the above stresses can be expressed as follows:

$$\sigma_{12} \approx -\varepsilon p_1 s_{12}(\gamma_1), \quad \sigma_{22} \approx -\varepsilon^2 p_1 s_{22}(\gamma_1) \tag{2.9.5}$$

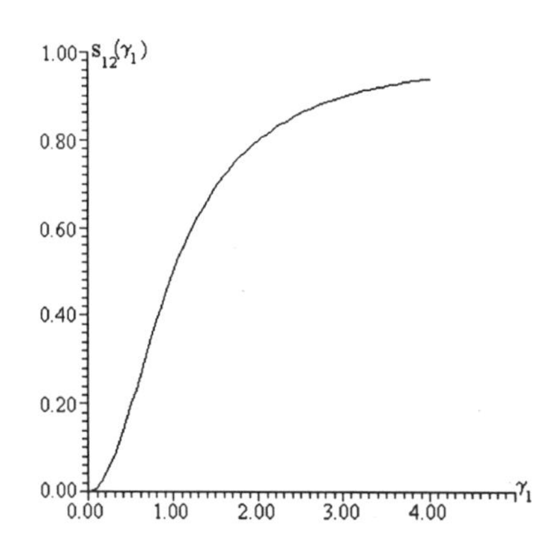

Fig. 2.9.1. The relation between $s_{12}(\gamma_1)$ and γ_1.

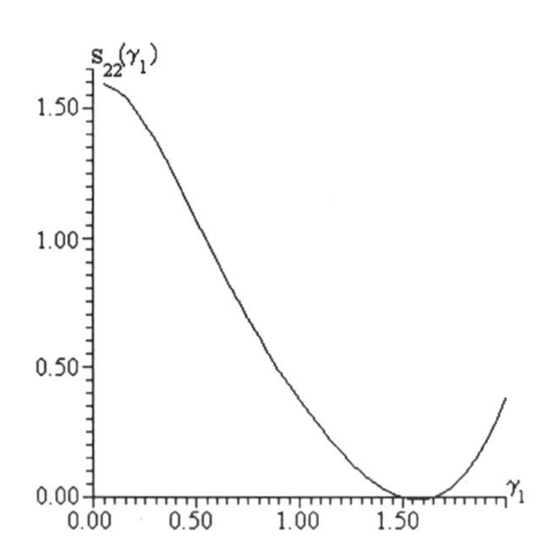

Fig. 2.9.2. The relation between $s_{22}(\gamma_1)$ and γ_1.

Figs. 2.9.1 and 2.9.2 show the relation between $s_{12}(\gamma_1)$ and γ_1, and between $s_{22}(\gamma_1)$ and γ_1 respectively. Fig. 2.9.1 shows that the stress σ_{12}, which arises only as a result of the spatiality of the curving, increases monotonically with γ_1. The first expression of (2.9.2) shows that the values of s_{12} have an upper limit, i.e., $s_{12} \to 1$ or $\sigma_{12} \to -\varepsilon p_1$ as $\gamma_1 \to \infty$. Besides, Fig. 2.9.2 shows that when $0 < \gamma_1 < 1.60$, s_{22} decreases as γ_1 increases; when $1.60 < \gamma_1 < 2.00$ s_{22} increases with γ_1.

Thus, for uniaxial tension-compression of composite material with spatially periodic curving in the structure we can conclude that the spatiality of the curving produces self-balanced stresses of order ε^1; when the composite material has periodic plane-curved structure, the order of the self-balanced stresses is ε^2.

The procedure used in this chapter can be applied to other quasi-homogeneous stress states in composite materials with spatial and periodic curved structures.

2.10. The Applicability of the Proposed Continuum Approach

In the previous section, with concrete problems as an example, in the framework of the restrictions (2.7.2), (2.7.6) and (2.7.7), some applications of the continuum theory (2.2.8), (2.2.9) were given. It should be noted that the applicability ranges of this theory are larger than those of the restrictions (2.7.2), (2.7.6) and (2.7.7). From the well-known mechanical considerations and the construction techniques used under obtaining the relations (2.2.8) and (2.2.9) it follows that the framework of the applicability of this theory can be determined by

$$\Delta H << d \ ; \ \Delta H << \Lambda_1 \text{ and } \Delta H << \Lambda_3 . \tag{2.10.1}$$

The meaning of the notations used in (2.10.1) is given in sections 2.2 and 2.7.

Thus, the continuum theory presented in this chapter can also be applied without the restrictions given by relations (2.7.6), i.e. can also be applied for the cases where

$$d \approx \Lambda_1 ; \Lambda_3 \quad \text{or} \quad d << \Lambda_1 ; \Lambda_3 \tag{2.10.2}$$

As has been noted in Section 1.10, a curvature that satisfies the last conditions in (2.10.2) is called large-scale. Note that in the following chapter, we will investigate some problems of the theory in the framework of the restrictions (2.10.1).

2.11. Bibliographical Notes

The continuum theory described in section 2.2. was proposed by S.D. Akbarov and A.N. Guz [38]. The application of this theory to the investigation of the quasi-homogeneous stress state in the composite materials with spatially periodic curved structures was made by S.D. Akbarov and G.M. Guliev [27].

CHAPTER 3

PROBLEMS FOR CURVED COMPOSITES

In this chapter we combine the theory developed in Chapter 2 with the finite element method (FEM) to solve some two and three dimensional static and dynamic problems for strips and rectangular plates. We discuss the influence of the parameters defining the curved composite on the stress distribution, and on the natural frequencies of the plates.

3.1. Bending of a Strip

In this section, we will consider some plane deformation problems for the continuum theory presented in the previous chapter. It is assumed that the plane deformation takes place in the Ox_1x_2 plane for a rectangular region whose geometry is shown in Fig.3.1.1. Note that the problems correspond to the bending of a strip fabricated from composite material with curved structure, so that there are various edge conditions. The investigations are carried out in the framework of the exact equations of the theory of elasticity and all results are obtained numerically by employing FEM.

3.1.1. BASIC EQUATIONS AND FORMULATION OF THE PROBLEMS

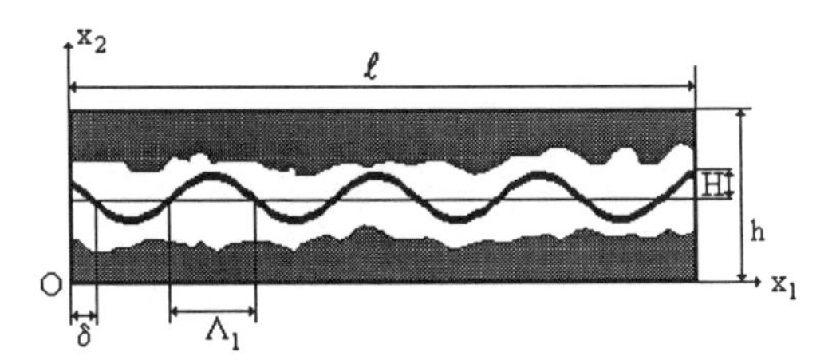

Fig.3.1.1. The geometry of the considered strip.

We assume that the curvature is unidirectional, in the Ox_1 direction, but that the degree of the curving is not very small. Consequently, we will not use the series expansions (2.3.2) for the normalized material properties; instead we will use the relations which are given below.

Explicit form of the constitutive relations (2.2.9) for composite materials with arbitrarily plane curving in the structure.

Taking into account that $\partial F/\partial x_3 \equiv 0$, we obtain from (2.2.4)- (2.2.7) the following explicit form of the constitutive relations (2.2.9):

$$A_{11}(x_1) = A_{11}^0\Phi_1^4(x_1) + 2A_{12}^0\Phi_1^2(x_1)\Phi_2^2(x_1) + A_{22}^0\Phi_2^4(x_1) + 4G_{12}^0\Phi_1^2(x_1)\Phi_2^2(x_1),$$

$$A_{22}(x_1) = A_{11}^0\Phi_2^4(x_1) + 2A_{12}^0\Phi_1^2(x_1)\Phi_2^2(x_1) + A_{22}^0\Phi_1^4(x_1) + 4G_{12}^0\Phi_1^2(x_1)\Phi_2^2(x_1),$$

$$A_{33} = A_{33}^0, \quad A_{44}(x_1) = A_{44}^0\Phi_1^2(x_1) + A_{55}^0\Phi_2^2(x_1),$$

$$A_{55}(x_1) = A_{55}^0\Phi_1^2(x_1) + A_{44}^0\Phi_2^2(x_1),$$

$$A_{66} = \Phi_1^2(x_1)\Phi_2^2(x_1)\left(A_{11}^0 - 2A_{12}^0 + A_{22}^0\right) + A_{22}^0\left(\Phi_1^4(x_1) + \Phi_2^4(x_1) -\right.$$

$$\left. 2\Phi_1^2(x_1)\Phi_2^2(x_1)\right), \quad A_{13}(x_1) = A_{13}^0\Phi_1^2(x_1) + A_{23}^0\Phi_2^2(x_1),$$

$$A_{12} = \Phi_1^2(x_1)\Phi_2^2(x_1)\left(A_{11}^0 - 4A_{66}^0 + A_{22}^0\right) + A_{12}^0\left(\Phi_1^4(x_1) + \Phi_2^4(x_1)\right),$$

$$A_{16} = \Phi_1^3(x_1)\Phi_2(x_1)\left(A_{12}^0 - A_{11}^0 + 2A_{66}^0\right) + \left(A_{22}^0 - A_{12}^0 - 2A_{66}^0\right)\Phi_1(x_1)\Phi_2^3(x_1),$$

$$A_{23}(x_1) = A_{13}^0\Phi_2^2(x_1) + A_{23}^0\Phi_1^2(x_1),$$

$$A_{26} = \Phi_1(x_1)\Phi_2^3(x_1)\left(A_{12}^0 - A_{11}^0 + 2A_{66}^0\right) + \left(A_{22}^0 - A_{12}^0 - 2A_{66}^0\right)\Phi_1^3(x_1)\Phi_2(x_1)$$

$$A_{36} = \Phi_1(x_1)\Phi_2(x_1)\left(A_{23}^0 - A_{13}^0\right), \quad A_{45} = \Phi_1(x_1)\Phi_2(x_1)\left(A_{44}^0 - A_{55}^0\right),$$

$$A_{14} = A_{15} = A_{24} = A_{25} = A_{34} = A_{35} = A_{46} = A_{56} = 0, \qquad (3.1.1)$$

where

$$\Phi_1(x_1) = \frac{1}{\sqrt{1+\left(\varepsilon\frac{df}{dx_1}\right)^2}}, \qquad \Phi_2(x_1) = \varepsilon\frac{df}{dx_1}\Phi_2(x_1). \qquad (3.1.2)$$

In (3.1.2) $\varepsilon f(x_1) = F(x_1)$ is the function which characterizes the curved form in the material structure.

The system equations and boundary conditions.
For deformation in the Ox_1x_2 plane, the mechanical relations (2.2.8), (1.3.6) can be written as follows:

$$\sigma_{11} = A_{11}\varepsilon_{11} + A_{12}\varepsilon_{22} + 2A_{16}\varepsilon_{12}, \quad \sigma_{22} = A_{12}\varepsilon_{11} + A_{22}\varepsilon_{22} + 2A_{26}\varepsilon_{12},$$

$$\sigma_{12} = A_{16}\varepsilon_{11} + A_{26}\varepsilon_{22} + 2A_{66}\varepsilon_{12}. \qquad (3.1.3)$$

We assume that the strip occupies the region $\Omega = \{0 < x_1 < \ell;\ 0 < x_2 < h\}$. With (3.1.3), we have the equilibrium equations

$$\frac{\partial \sigma_{ij}}{\partial x_j} = 0, \quad i; j = 1,2; \qquad (3.1.4)$$

and the geometrical relations

$$\varepsilon_{ij} = \frac{1}{2}\left(\frac{\partial u_i}{\partial x_j} + \frac{\partial u_j}{\partial x_i}\right), \ i; j=1,2. \qquad (3.1.5)$$

These are the basic equations. Now consider the formulation of the boundary conditions which correspond to various concrete problems.

Problem 1.

$$\sigma_{i2}\big|_{x_2=h;x_1\in(0,\ell)} = -p_1(x_1)\delta_2^i;\ \sigma_{i2}\big|_{x_2=0;x_1\in(0,\ell)} = 0;\ \ i=1,2$$

$$u_2\big|_{x_1=0,\ell;x_2\in[o,h]} = 0;\ \sigma_{11}\big|_{x_1=0,\ell;x_2\in[o,h]} = 0. \qquad (3.1.6)$$

The boundary conditions (3.1.6) correspond to the case in which the edges of the strip are simply supported, and normal forces of intensity $p_1(x_1)$ act on the upper plane surface in the direction opposite to that of the Ox_2 axis.

Problem 2.

$$\sigma_{i2}\big|_{x_2=h;x_1\in(0,\ell)} = -p_1(x_1)\delta_2^i;\ \sigma_{i2}\big|_{x_2=0;x_1\in(0,\ell)} = 0;\ \ i=1,2,$$

$$u_1\big|_{x_1=0,\ell;x_2\in[o,h]} = 0;\ u_2\big|_{x_1=0,\ell;x_2\in[o,h]} = 0. \qquad (3.1.7)$$

The conditions (3.1.7) correspond to the case in which the edges of the strip are rigidly supported.

Problem 3.

$$\sigma_{i2}\big|_{x_2=0,h;x_1\in(0,\ell)} = 0;\ \ i=1,2, \qquad \sigma_{11}\big|_{x_1=\ell;x_2\in[0,h]} = 0;$$

$$\sigma_{12}\big|_{x_1=\ell;x_2\in[0,h]} = p\frac{x_2}{h}\left(\frac{x_2}{h}-1\right); \quad u_1\big|_{x_1=0;x_2\in[o,h]} = 0; \quad u_2\big|_{x_1=0;x_2\in[o,h]} = 0 \,. \quad (3.1.8)$$

The equations (3.1.8) correspond to the bending of a cantilever strip (beam) by forces applied at the right hand end, $x_1 = \ell$. The forces are distributed parabolically with absolute maximal value $p/4$.

Since it is difficult to solve these problems analytically, we will use the FEM. First we discuss FEM modelling, and then the solution.

3.1.2. FEM MODELLING

The equations of the theory of elasticity for an anisotropic body are self-adjoint and consequently the problems (3.1.1)-(3.1.8) may be formulated variationally. The variational formulation for the displacement-based FEM starts with the functional

$$\Pi = \frac{1}{2}\iint_{\Omega}\sigma_{ij}\varepsilon_{ij}d\Omega - \int_{S_1}P_i u_i dS, \tag{3.1.9}$$

which is the total potential energy of the elastic system. In (3.1.9) S_1 is the part of the surface of the strip on which the external forces P_i (i=1,2) are given. We assume that the equations (3.1.3), (3.1.5) and boundary conditions for the displacements given in (3.1.6)-(3.1.8) are satisfied for the virtual displacements δu_i. According to the principle of virtual work, the equilibrium state is obtained from

$$\delta\Pi = \iint_{\Omega}\sigma_{ij}\delta\varepsilon_{ij}d\Omega - \int_{S_1}P_i\delta u_i dS = 0, \tag{3.1.10}$$

where

$$\delta\varepsilon_{ij} = \frac{1}{2}\left(\frac{\partial\delta u_i}{\partial x_j} + \frac{\partial\delta u_j}{\partial x_i}\right). \tag{3.1.11}$$

Integration by parts applied to equation (3.1.10) yields the equations (3.1.4) and the boundary conditions (3.1.6)-(3.1.8).

The region $\Omega = \{0 < x_1 < \ell; 0 < x_2 < h\}$ is divided into a finite number of elements Ω_k, i.e.

$$\Omega = \bigcup_{k=1}^{M}\Omega_k \tag{3.1.12}$$

We take the element Ω_k as a rectangle with parameters shown in Fig.3.1.2. The local normalized coordinates for Ω_k are

$$\xi = \frac{x_1 - x_{1o}}{\beta}, \quad \eta = \frac{x_2 - x_{2o}}{\alpha} \tag{3.1.13}$$

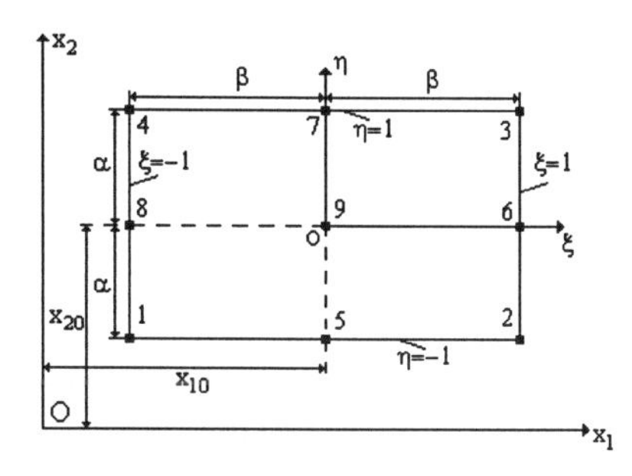

Fig.3.1.2. Selected finite element form and its nodes.

The shape functions are as follows:

$$N_1 = \frac{1}{4}(\xi^2 - \xi)(\eta^2 - \eta), \quad N_2 = \frac{1}{4}(\xi^2 + \xi)(\eta^2 - \eta), \quad N_3 = \frac{1}{4}(\xi^2 + \xi)(\eta^2 + \eta),$$

$$N_4 = \frac{1}{4}(\xi^2 - \xi)(\eta^2 + \eta), \quad N_5 = \frac{1}{4}(\xi^2 - 1)(\eta^2 - \eta), \quad N_6 = -\frac{1}{2}(\xi^2 + \xi)(\eta^2 - 1),$$

$$N_7 = -\frac{1}{2}(\xi^2 - 1)(\eta^2 + \eta), N_8 = -\frac{1}{2}(\xi^2 - \xi)(\eta^2 - 1), N_9 = (\xi^2 - 1)(\eta^2 - 1). \tag{3.1.14}$$

In each Ω_k element the displacements are approximated by

$$\underline{\mathbf{u}}^{k} \approx \underline{\tilde{\mathbf{u}}}^{k} = \underline{\mathbf{N}}^{k}\underline{\mathbf{a}}^{k}, \tag{3.1.15}$$

where

$$\left(\mathbf{a}^k\right)^T = \left\{u_{11}^k, u_{21}^k, u_{12}^k, u_{22}^k, \ldots u_{18}^k, u_{28}^k, u_{19}^k, u_{29}^k\right\},$$

$$\mathbf{N}^k = \begin{Bmatrix} N_1^k & 0 & \ldots & N_9^k & 0 \\ 0 & N_1^k & \ldots & 0 & N_9^k \end{Bmatrix}. \tag{3.1.16}$$

In (3.1.16) the components of the vector $\mathbf{a}^k$ are the values of the displacements at the nodes of the element Ω_k. The second index of these components shows the number of the node.

Thus taking into account (3.1.12), (3.1.15) we obtain the following equation from (3.1.10).

$$\delta\Pi = \frac{\partial\Pi}{\partial\mathbf{a}}\delta\mathbf{a} = 0, \tag{3.1.17}$$

where

$$(\mathbf{a})^T = \left\{\mathbf{a}^1, \mathbf{a}^2, \ldots, \mathbf{a}^M\right\}. \tag{3.1.18}$$

Using the well-known procedure described in [155] and elsewhere we obtain the following equations with respect to $\mathbf{a}$ from equation (3.1.17):

$$\mathbf{Ka} = \mathbf{r}. \tag{3.1.19}$$

In (3.1.19) the components of the $\mathbf{r}$ vector are the nodal forces, and $\mathbf{K}$ is the stiffness matrix. The components of the stiffness matrix $\mathbf{K}$ are determined through the following formulas:

$$\mathbf{K} = \sum_{k=1}^{M}\mathbf{K}^k, \ \mathbf{K}^k = \begin{Bmatrix} \mathbf{K}_{11}^k & \mathbf{K}_{12}^k & \cdots & \mathbf{K}_{19}^k \\ \mathbf{K}_{21}^k & \mathbf{K}_{22}^k & \cdots & \mathbf{K}_{29}^k \\ \cdots & \cdots & \cdots & \cdots \\ \mathbf{K}_{91}^k & \mathbf{K}_{92}^k & \cdots & \mathbf{K}_{99}^k \end{Bmatrix}, \quad \mathbf{K}_{ij}^k = \iint_{\Omega_k} \left(\mathbf{B}_j^k\right)^T \mathbf{D}^k \mathbf{B}_i^k d\Omega_k, \tag{3.1.20}$$

where i;j=1,2,...9 and

$$\mathbf{D}^k = \left.\begin{pmatrix} A_{11}(x_1) & A_{12}(x_1) & A_{13}(x_1) \\ A_{12}(x_1) & A_{22}(x_1) & A_{23}(x_1) \\ A_{13}(x_1) & A_{23}(x_1) & A_{33}(x_1) \end{pmatrix}\right|_{x_1\in\Omega_k}, \quad \mathbf{B}_i^k = \begin{pmatrix} \dfrac{\partial N_i^k}{\partial x_1} & 0 \\ 0 & \dfrac{\partial N_i^k}{\partial x_2} \\ \dfrac{\partial N_i^k}{\partial x_2} & \dfrac{\partial N_i^k}{\partial x_1} \end{pmatrix}. \tag{3.1.21}$$

where the functions $A_{ij}(x_1)$ (i,j =1,2,3) are determined by the expressions given in (3.1.1).

It follows from (3.1.21) that the components of the matrix $\mathbf{D}^k$ are functions of x_1. Therefore it is impossible to use the well-known canned programs (such as LUSAS and ANSYS), which assume that the components of the matrix $\mathbf{D}^k$ are given as constants; we must derive new programs.

There are some difficulties in the calculation of the stresses in the displacement-based FEM; the displacements are approximated in each element through their values at the nodes and the shape functions, and these have only simple, C_0 continuity; the derivatives of the displacements with respect to x_1 or x_2 generally have a discontinuity at the inter-element boundaries. In other words, if we calculate the stresses using the formulas

$$\hat{\boldsymbol{\sigma}} = \mathbf{DBa}, \tag{3.1.22}$$

where

$$\hat{\boldsymbol{\sigma}} = \left\{\hat{\boldsymbol{\sigma}}^1,...,\hat{\boldsymbol{\sigma}}^M\right\}, \quad \mathbf{D} = \left\{\mathbf{D}^1,...,\mathbf{D}^M\right\}, \quad \mathbf{B} = \left\{\mathbf{B}^1,...,\mathbf{B}^M\right\},$$

$$\mathbf{B}^i = \left\{\mathbf{B}_1^i,...,\mathbf{B}_9^M\right\}, \quad \left(\hat{\boldsymbol{\sigma}}^i\right)^T = \left\{\hat{\sigma}_{11}^i, \hat{\sigma}_{22}^i, \hat{\sigma}_{12}^i\right\}, \quad 1 \le i \le M, \tag{3.1.23}$$

and $\hat{\sigma}_{nm}^i$ (n;m=1,2) is the stress σ_{nm} in element Ω_i, then the stresses σ_{nm} (n;m=1,2) will be discontinuous at the inter-element boundaries. This situation follows from the nature of the displacement-based FEM. To obtain the continuous stresses we use the following approximation introduced in [109, 155]:

$$\boldsymbol{\sigma} = \mathbf{N}_\sigma \overline{\boldsymbol{\sigma}} \tag{3.1.24}$$

where

$$\boldsymbol{\sigma} = \begin{pmatrix} \sigma_{11} \\ \sigma_{22} \\ \sigma_{12} \end{pmatrix}; \quad \overline{\boldsymbol{\sigma}} = \left(\overline{\boldsymbol{\sigma}}^1, \overline{\boldsymbol{\sigma}}^2, ..., \overline{\boldsymbol{\sigma}}^M\right); \quad \overline{\boldsymbol{\sigma}}^i = \begin{pmatrix} \overline{\boldsymbol{\sigma}}_{11}^i \\ \overline{\boldsymbol{\sigma}}_{22}^i \\ \overline{\boldsymbol{\sigma}}_{12}^i \end{pmatrix}; \quad (1 \le i \le M);$$

$$\overline{\boldsymbol{\sigma}}_{nm}^i = \left(\overline{\sigma}_{nm1}^i, \overline{\sigma}_{nm2}^i, ..., \overline{\sigma}_{nm9}^i\right), \qquad (n;m=1,2). \tag{3.1.25}$$

Here $\overline{\sigma}_{nmk}^i$ (k=1,2,...,9) is the value of stress σ_{nm} at the k th node of the element Ω_i and $\mathbf{N}_\sigma$ is the matrix of shape functions (3.1.14). Consequently, for matrix $\mathbf{N}_\sigma$, we can write the following expression:

$$\mathbf{N}_\sigma = \left(\mathbf{N}^1, \mathbf{N}^2, ..., \mathbf{N}^9\right),$$

$$\mathbf{N}^i = \begin{pmatrix} N_1^i & 0 & 0 & N_2^i & 0 & 0 & \dots & N_9^i & 0 & 0 \\ 0 & N_1^i & 0 & 0 & N_2^i & 0 & 0 & \dots & N_9^i & 0 \\ 0 & 0 & N_1^i & 0 & 0 & N_2^i & 0 & 0 & \dots & N_9^i \end{pmatrix}. \quad (3.1.26)$$

The values of $\overline{\sigma}^i_{nmk}$ are determined by minimizing the functional

$$Q = \iint_{\Omega} (\boldsymbol{\sigma} - \hat{\boldsymbol{\sigma}})^2 d\Omega \quad (3.1.27)$$

with respect to the $\overline{\sigma}^i_{nmk}$, i.e.

$$\frac{\partial Q}{\partial \overline{\sigma}^i_{nmk}} = 0. \quad (3.1.28)$$

In this way we obtain a continuous stress distribution in the strip. The number of elements Ω_i is determined from the required accuracy.

3.1.3. PERIODIC CURVING: NUMERICAL RESULTS

Preliminary remarks.

In this sub-section we analyse the numerical results for the stress and displacement distribution in the strip shown in Fig.3.1.1; in particular we analyse the effect of the curving in the strip material structure on the stresses and displacements. Suppose that the composite is fabricated from alternating layers of two materials, and these layers are perpendicular to the Ox_2 axis in the absence of the curving. We assume that the material of each layer is isotropic. For these layers we introduce the following notation: E_1, E_2 are the Young's moduli; ν_1, ν_2 are the Poisson's ratios; η_1 and η_2 are the concentrations of the constituents. We will use the expression for the normalized elasticity coefficients A^0_{11}, A^0_{12}, A^0_{22}, A^0_{33}, $A^0_{55} = G^0_{12}$, $A^0_{44} = G^0_{23}$ and $A^0_{66} = G^0_{12}$ which are given by the formulas (1.9.1), (1.9.3).

First, suppose that the curving in the strip material structure is periodic. Therefore, the function $\varepsilon f(x_1)$ entering (3.1.2) and characterizing the form of the curvature is chosen as

$$\varepsilon f(x_1) = H \sin\left(\pi \frac{x_1}{\Lambda_1} + \delta\right) = \frac{H}{\Lambda_1} \Lambda_1 \sin\left(\pi \frac{\ell}{\Lambda_1} \frac{x_1}{\ell} + \delta\right) =$$

$$\varepsilon\Lambda_1 \sin\left(\pi\frac{\ell}{\Lambda_1}\frac{x_1}{\ell}+\delta\right), \ \varepsilon=\frac{H}{\Lambda_1} \tag{3.1.29}$$

Here H is the rise of the curve; Λ_1 is the half wavelength of curvature; δ characterizes the distance by which the origin of the system of coordinates Ox_1x_2 (Fig.3.1.1) shifts (along the Ox_1 axis) relative to the first "nodal" point of the curving form. We assume that $H<\Lambda_1$; the parameter $\varepsilon = H/\Lambda_1$ estimates the degree of curving; if $\varepsilon=0$, then there is no curving. The parameters in (3.1.29) are shown in Fig.3.1.1.

In applying the theory of Chapter 2, we will assume that conditions (2.10.1) are satisfied, i.e. $\Delta H << d$, $\Delta H << \Lambda_1$, $\Delta H << \Lambda_3$. Since we are considering plane strain, and the curving exists only in the direction of the Ox_1 axis we can ignore the third inequality of (2.10.1). We will not use the series expansions (2.3.2) for the normalized mechanical properties A_{ij} which enter (2.2.8) and (3.1.3); therefore we can ignore the inequalities (2.3.1) and consider the case where the inequality $H<\Lambda_1$ is satisfied. Note that the thickness of the strip h which satisfies the inequality $h<\ell$, can be taken as the parameter d which enters the first inequality of (2.10.1) and characterizes the minimal size of the region occupied by the composite material. As the starting point, we assume that

$$h \geq 10\Delta H; \ \Lambda_1 \geq 10\Delta H. \tag{3.1.30}$$

Further, we introduce the following notations: $\ell/\Lambda_1 = N_\Lambda$; $\ell/h=\beta_h$; $\Lambda_1 = N_{\Delta H}\Delta H$. The number of the representative packets in the strip is determined by the following relation

$$N_\ell = \left[\frac{h}{\Delta H}\right] = \left[\frac{N_{\Delta H}N_\Lambda}{\beta_h}\right]. \tag{3.1.31}$$

In (3.1.31) N_ℓ is a number of the representative packets; [x] represent the integer part of x. It should be noted that the values of the parameters entering (3.1.31) must satisfy the inequalities (3.1.30). For example, if $h/\ell = 0.1 \Rightarrow \beta_h = 10$, $\ell/\Lambda_1 = N_\Lambda = 16$, $N_{\Delta H}=10$, then equation (3.1.31) shows that $N_\ell = 16$. If we take $N_\Lambda = 1$, then from equation (3.1.31) and from the first inequality (3.1.30) it follows that $N_{\Delta H} \geq 100$. Note that we can similarly determine the application field of the theory (3.1.1)-(3.1.3) in the other cases, the starting points for which are different from (3.1.30).

Now we analyse the numerical results obtained for the case where the function $p_1(x_1)$ entering the boundary conditions (3.1.6) and (3.1.7) and describing the external forces, is constant, i.e. $p_1(x_1)=p$. Moreover, we assume that $\nu_1=\nu_2=0.3$,

$\eta_1 = \eta_2 = 0.5$, $\delta = \pi/2$ and the curving form in the strip material structure is symmetric about the mid point $x_1 = \ell/2$. Taking the symmetry into account we will use only the region $\{0 \le x_1 \le \ell/2; 0 \le x_2 \le h\}$ for FEM modelling of problems 1 and 2 in the previous sub-section. The region $\{0 \le x_1 \le \ell/2;\ 0 \le x_2 \le h\}$ is divided into 80 elements with 369 nodes and with 720 (for the Problem 1) and 711 (for the Problem 2) number of degrees of freedom (NDOF).

Numerical results obtained for the Problem 1.
Figs. 3.1.3-3.1.4 drawn for $\ell/\Lambda_1 = 16$, $h/\ell = 0.1$ and various ε, show the distributions of normalized displacements $u = E_1 u_1/(\ell p)$, $v = E_1 u_2/(\ell p)$ and stresses σ_{11}/p, σ_{22}/p with respect to x_1/ℓ; the corresponding values of x_2/ℓ and E_2/E_1 are shown in the figure captions. These graphs show that by increasing the parameter ε which characterizes the degree of the curving rise in the structure of the strip material, the absolute values of the normalized displacements grow monotonically. These influences of the parameter ε on the displacement distributions are explained by the decrease of the stiffness of the strip as a result of the curving of the reinforcing elements in the strip material structure.

Note that many other numerical results, which are not given here, show that the influences of the other problem parameters, such as h/ℓ, E_2/E_1, to the displacement distributions in the strip are the same as those in the case where $\varepsilon = 0$. Therefore in almost all the numerical investigations, we will analyse mainly the influence of the problem parameters (i.e. of the strip or of the plate) on the stress distribution in the strip or plate.

Consider Figs. 3.1.4; they show that quantitative and qualitative characteristics

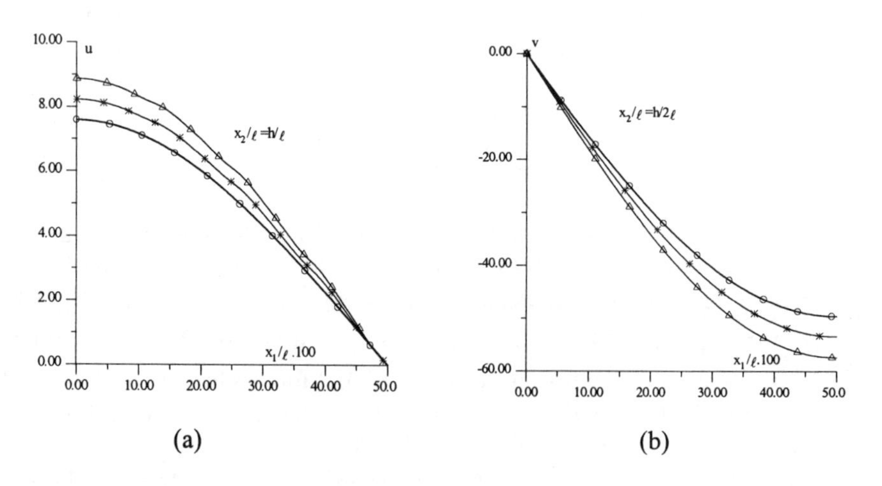

Fig.3.1.3. The influence of the parameter ε on the distribution of displacement $u = E_1 u_1/(\ell p)$ (a), $v = E_1 u_2/(\ell p)$ (b) for $h/\ell = 0.10$, $\ell/\Lambda_1 = 16$, $E_2/E_1 = 5$:
○ -- $\varepsilon = 0.00$; ✳ -- $\varepsilon = 0.10$; △ -- $\varepsilon = 0.15$.

of the distributions of the stresses σ_{11} and σ_{22} with respect to x_1 depend significantly on the parameter ε. In this case the influence of the parameter ε on these distributions grows monotonically with increasing ε, and the form of the distributions of the stresses corresponds to the form of the curving in the strip material structure. These graphs also show that for Problem 1 the maximal influence of the curving on the distributions of the stresses σ_{11} and σ_{22} occurs at $x_1/\ell = 1/2$. Because of the symmetry of the problem about $x_1/\ell = 1/2$ the stress σ_{12} is zero at $x_1/\ell = 1/2$. Note that many other numerical results, which are not given here, indicate that the existence of the curving in the strip material structure does not significantly change the distribution of σ_{12} with respect to x_1. Analogous numerical results are obtained for the distributions of σ_{11} and σ_{12} across the thickness (i.e. along the Ox_2 axis (Fig.3.1.1.)) of the strip. Taking these situations into account, we do not show the dependence of σ_{12} on x_1 or x_2, or σ_{11} on

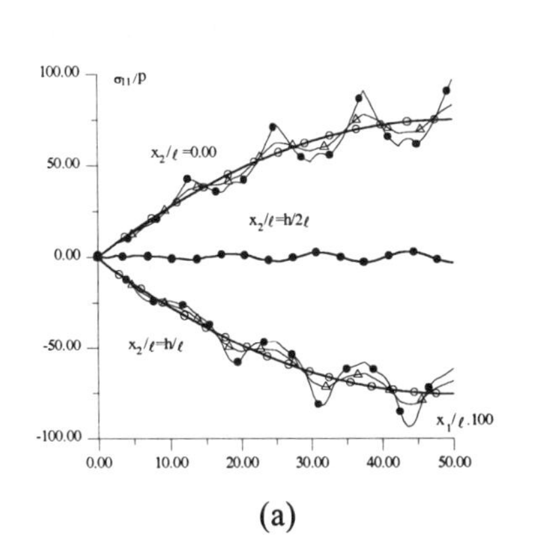

(a)

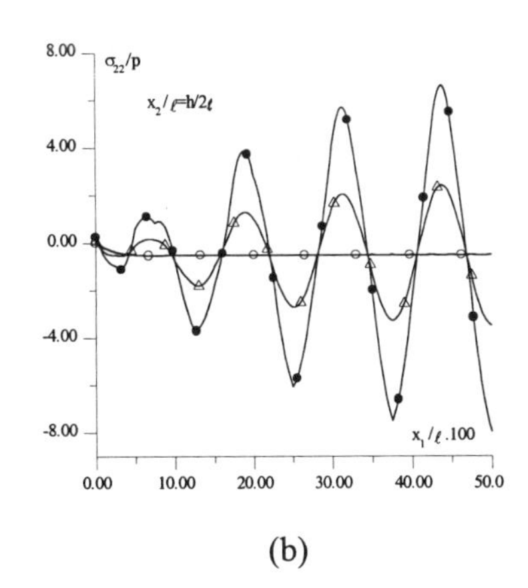

(b)

Fig.3.1.4. The influence of the parameter ε on the distribution of stress σ_{11} (a) and σ_{22} (b) for $h/\ell = 0.10$, $\ell/\Lambda_1 = 16$, $E_2/E_1 = 5$: ○ -- $\varepsilon = 0.00$; △ -- $\varepsilon = 0.05$; • -- $\varepsilon = 0.15$.

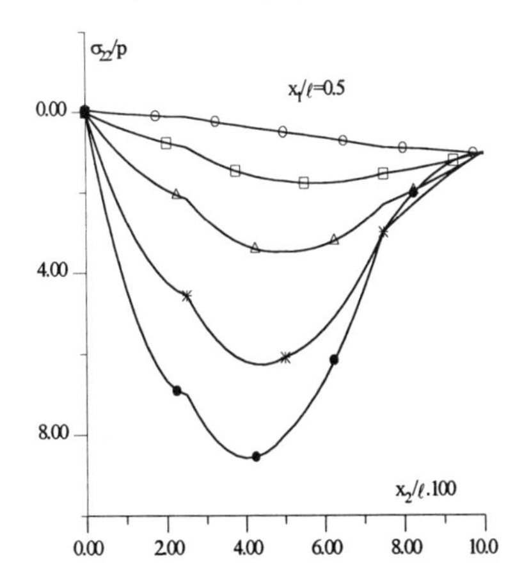

Fig.3.1.5. The influence of the parameter ε on the distribution of stress σ_{22} with respect to x_2, for $h/\ell = 0.10$, $\ell/\Lambda_1 = 16$, $E_2/E_1 = 5$: ○ -- $\varepsilon = 0.00$; □ -- $\varepsilon = 0.02$; △ -- $\varepsilon = 0.05$ ✳ -- $\varepsilon = 0.10$; • -- $\varepsilon = 0.15$.

x_2 . We consider only how σ_{22} varies with x_2 at $x_1/\ell = 1/2$, i.e. at $x_1 = 8\Lambda_1$. These distributions are given in Fig.3.1.5; they show that by increasing ε the stress σ_{22} has its absolute maximal values (with respect to x_2) in the vicinity of $x_2 = h/2$. However, in the case where $\varepsilon=0$ the absolute maximal value of the stress σ_{22} arises at the boundary $x_2 = h$. As is observed from the graphs given in Fig.3.1.5, these graphs have a convexity "from below" and are beneath the curve for $\varepsilon=0$. Note that the curves constructed for $x_1 = 6\Lambda_1;4\Lambda_1;2\Lambda_1$ have the same character as those given in Fig.3.1.5. However, the analogous curves for $x_1 = 7\Lambda_1;5\Lambda_1;3\Lambda_1$, for which the extremal values of σ_{22} are positive, have convexity "from above" and are above the curve for $\varepsilon = 0$; they are not shown here.

Now consider the influence of E_2/E_1 on the stress distributions. We analyse the dependence of σ_{22}/p on x_1/ℓ, and σ_{11}/p on x_1/ℓ constructed for various E_2/E_1 (Fig.3.1.6) and selected values of x_2/ℓ. The graphs show that by increasing E_2/E_1 the influence of the curving in the strip material structure on the stress distributions grows monotonically. The same results hold for the graphs given in Fig.3.1.7, which show the dependence of σ_{22}/p on x_2/ℓ at $x_1/\ell=1/2$.

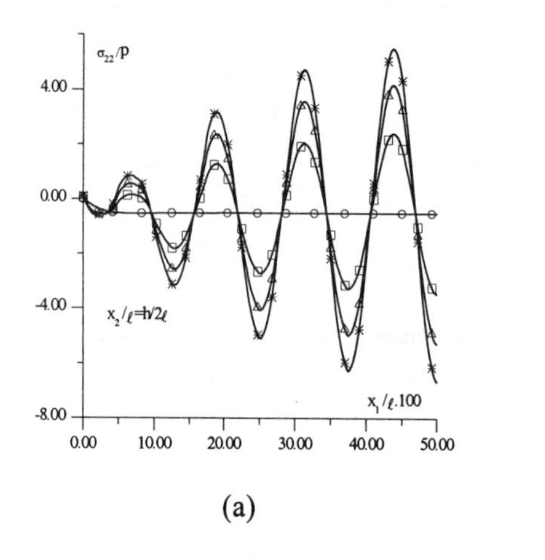

(a)

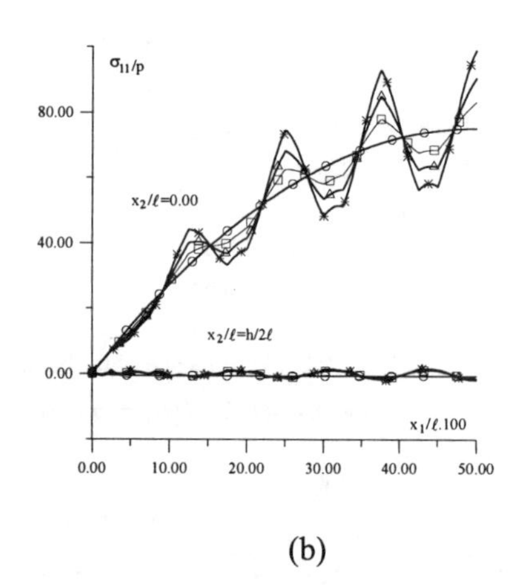

(b)

Fig.3.1.6. The influence of E_2/E_1 on the distribution of stress σ_{22} (a); σ_{11} (b) with respect to x_1. for $h/\ell = 0.10$, $\ell/\Lambda_1 = 16$: ○ -- $E_2/E_1 = 5$; $\varepsilon = 0.00$; and $\varepsilon = 0.05$
□ -- $E_2/E_1 = 5$; △ -- $E_2/E_1 = 10$ ✳ -- $E_2/E_1 = 15$.

The influence of the change of the thickness of the strip, i.e. of h/ℓ, on the stress distributions is shown in Figs. 3.1.8 and 3.1.9; these curves show that for fixed values of the other parameters, the stress perturbations caused by curving in the strip material structure, decrease as h/ℓ increases.

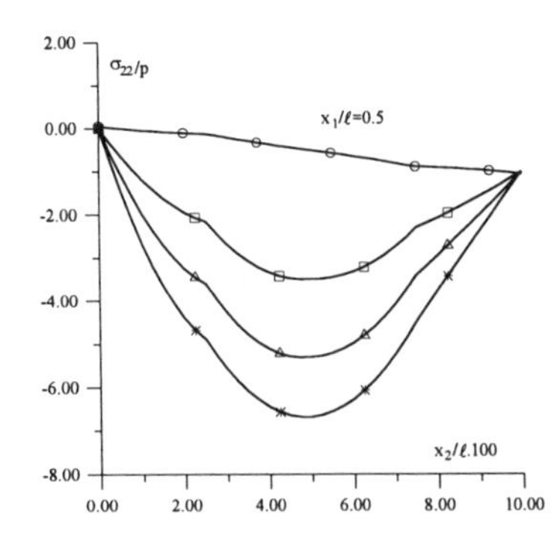

Fig.3.1.7. The influence of the parameter E_2 / E_1 on the distribution of stress σ_{22} with respect to x_2 in the case where $h/\ell = 0.10$, $\ell/\Lambda_1 = 16$: ○ -- $E_2/E_1 = 5$ $\varepsilon = 0.00$;while $\varepsilon = 0.05$; □ -- $E_2/E_1 = 5$; △ -- $E_2/E_1 = 10$; ✳ -- $E_2/E_1 = 15$.

Consider the influence of ℓ/Λ_1, which gives the number of the "half-waves" of the curving along the length of the strip, on the distributions of the stresses σ_{22} and σ_{11} with respect to x_1. The graphs of the distributions are given in Fig. 3.1.10; they show that dependence on ℓ/Λ_1 is non monotonic.

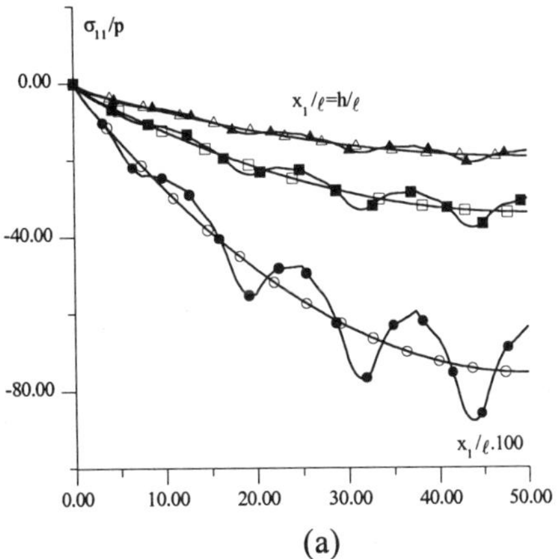

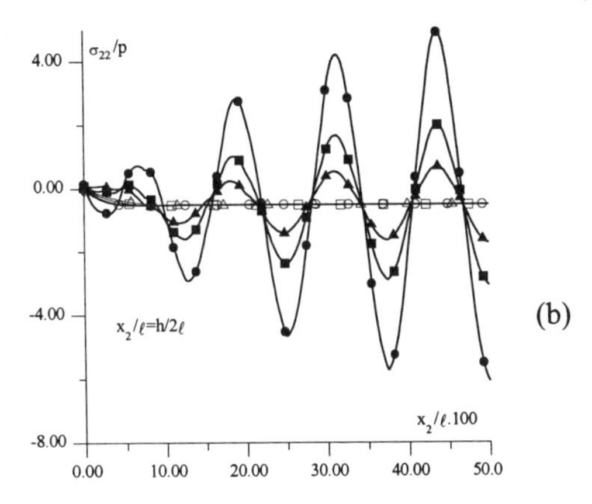

Fig.3.1.8. The influence of h/ℓ on the distribution of stress σ_{11} (a); σ_{22} (b) with respect to x_1 in the case where $E_2/E_1 = 5$ $\ell/\Lambda_1 = 16$: ○ -- $h/\ell = 0.10$, $\varepsilon = 0.00$; ● -- $h/\ell = 0.10$, $\varepsilon = 0.05$; □ -- $h/\ell = 0.15$, $\varepsilon = 0.00$; ■ -- $h/\ell = 0.15$, $\varepsilon = 0.05$; △ -- $h/\ell = 0.20$, $\varepsilon = 0.00$; ▲ -- $h/\ell = 0.20$, $\varepsilon = 0.05$.

Fig.3.1.9. The influence of the parameter h/ℓ to the distribution of stress σ_{22} with respect to x_2 in the case where $E_2/E_1 = 5$ $\ell/\Lambda_1 = 16$: ○ -- $h/\ell = 0.10$, $\varepsilon = 0.00$; ● -- $h/\ell = 0.10$, $\varepsilon = 0.05$
□ -- $h/\ell = 0.15$, $\varepsilon = 0.00$;
■ -- $h/\ell = 0.15$, $\varepsilon = 0.05$;
△ -- $h/\ell = 0.20$, $\varepsilon = 0.00$;
▲ -- $h/\ell = 0.20$, $\varepsilon = 0.05$.

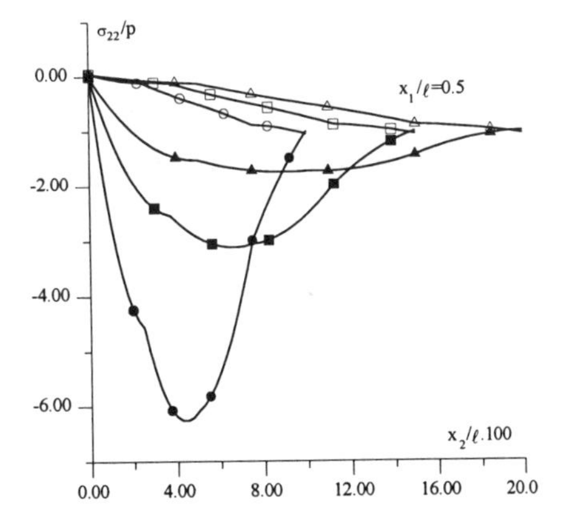

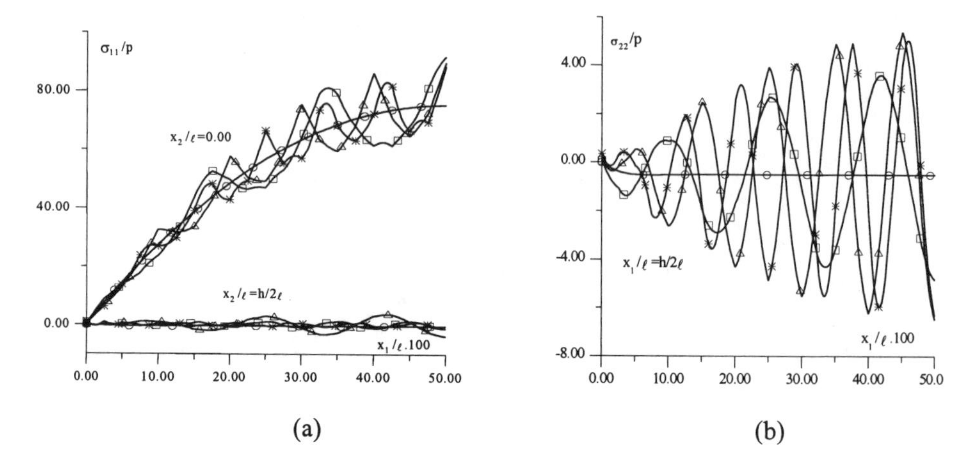

Fig.3.1.10. The influence of ℓ/Λ_1 to the distribution of stress σ_{11} (a); σ_{22} (b) with respect to x_1, in the case where $h/\ell = 0.10$, $E_2/E_1 = 5$: ○ -- $\ell/\Lambda_1 = 00$; $\varepsilon = 0.00$; while $\varepsilon = 0.10$ □ -- $\ell/\Lambda_1 = 12$; △ -- $\ell/\Lambda_1 = 20$; ✳ -- $\ell/\Lambda_1 = 28$.

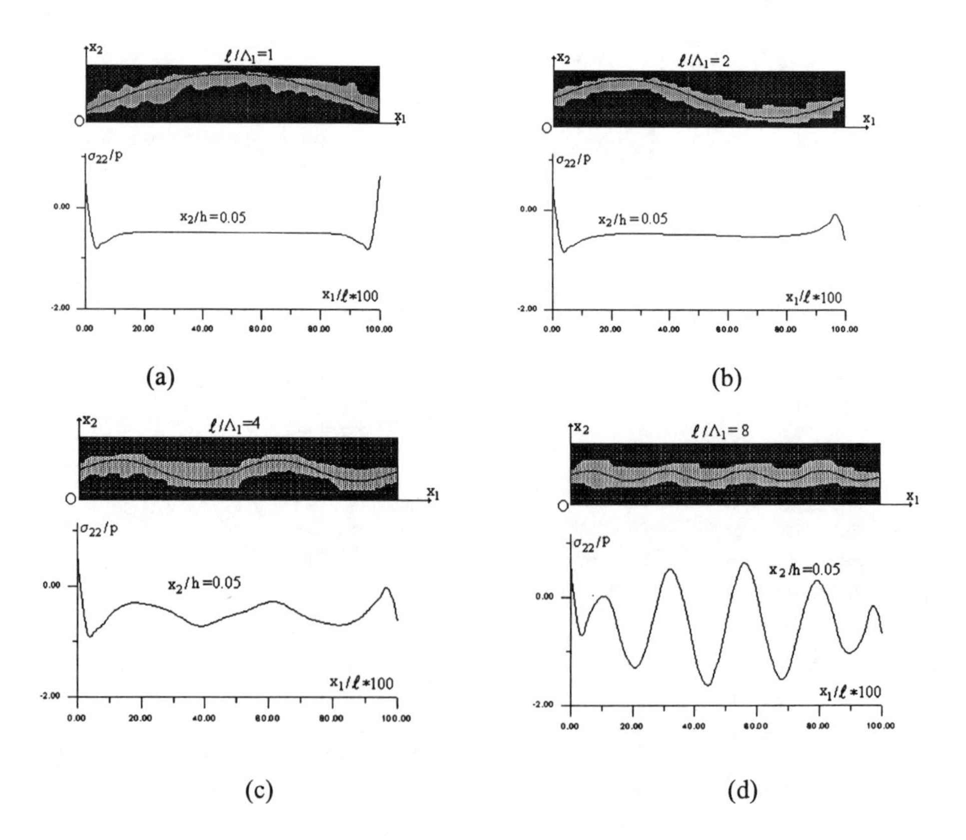

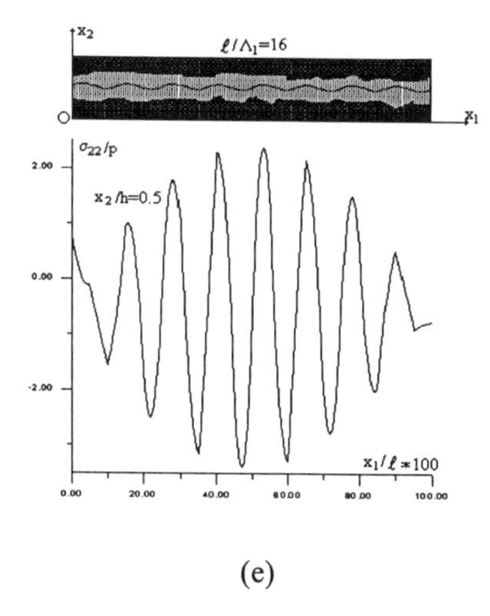

(e)

Fig.3.1.11. The influence of ℓ/Λ_1 to the distribution of σ_{22}/p in the case where $\delta=0$, h/ℓ=0.1, E_2/E_1=5; (a)-- ℓ/Λ_1=1, (b)-- ℓ/Λ_1=2, (c)-- ℓ/Λ_1=4, (d)-- ℓ/Λ_1=8, (e)-- ℓ/Λ_1=16.

Now we assume that $\delta=0$, $h/\ell=0.1$, $E_2/E_1=5$ and consider in more detail the influence of ℓ/Λ_1 on the distribution of σ_{22}/p with respect to x_1/ℓ. The graphs are given in Fig.3.1.11; ℓ/Λ_1 varies from 1 to 16. Note that for these problem parameters, the cases ℓ/Λ_1=1 or 2 can be taken as large-scale curving or as the case where the second inequality (2.10.2) is satisfied; if ℓ/Λ_1=4 or 8, the first relation of (2.10.2) is satisfied; ℓ/Λ_1=16 corresponds to small-scale curving. The results given in Fig.3.1.11 show that a significant perturbation in the distribution of σ_{22}/p arises when the curving is small-scale. Other numerical results, which are not given here, show that for $\ell/\Lambda_1 \geq 16$ the influence of increasing ℓ/Λ_1 on the stress distribution is insignificant. This situation is observed for all the problems that are considered in the present chapter; therefore, from now on, we take ℓ/Λ_1=16.

Numerical results obtained for Problem 2.

Problem 2 is governed by equations (3.1.1)-(3.1.4) and (3.1.7). First, we examine the

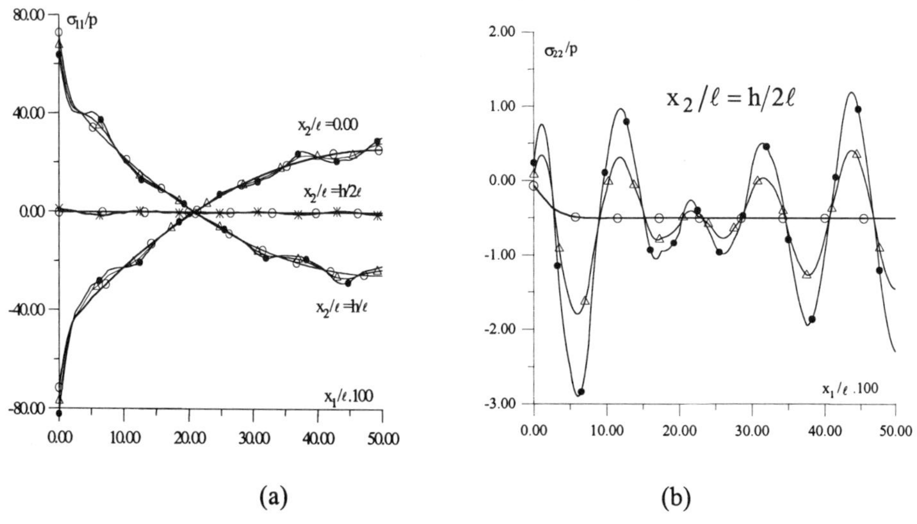

(a) (b)

Fig.3.1.12. The influence of the parameter ε on the distribution of stress σ_{11} (a) and σ_{22} (b) for $h/\ell=0.10$, $\ell/\Lambda_1=16$, $E_2/E_1=5$: ○ -- $\varepsilon=0.00$; △ -- $\varepsilon=0.05$; ● -- $\varepsilon=0.10$.

influence of ε on the distribution of the stresses σ_{11} and σ_{22} along the length of the

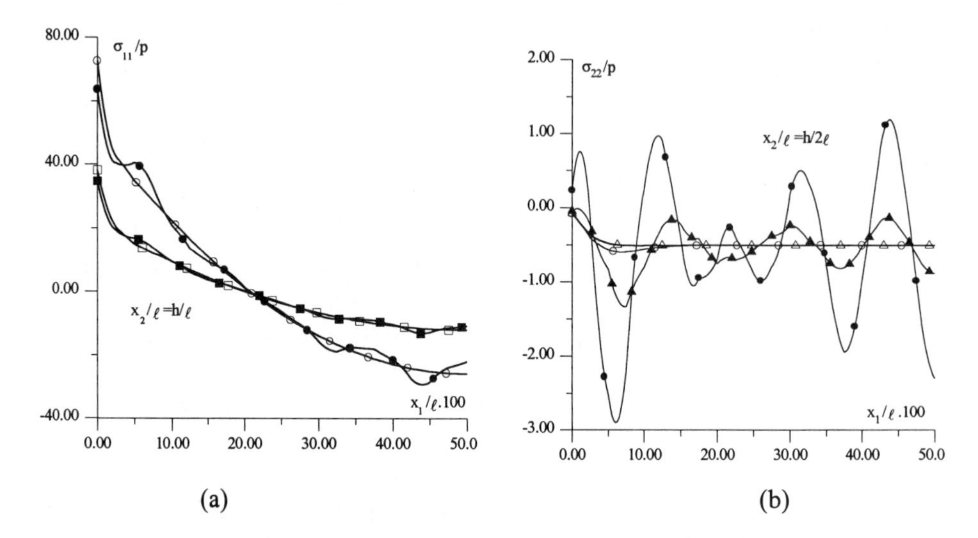

Fig.3.1.13. The influence of h/ℓ on the distribution of stress σ_{11} (a); σ_{22} (b) with respect to x_1 for $E_2/E_1 = 5$ $\ell/\Lambda_1 = 16$: ○ -- $h/\ell = 0.10$, $\varepsilon = 0.00$; ● -- $h/\ell = 0.10$, $\varepsilon = 0.05$;□ -- $h/\ell = 0.15$, $\varepsilon = 0.00$; ■ -- $h/\ell = 0.15$, $\varepsilon = 0.05$; △ -- $h/\ell = 0.20$, $\varepsilon = 0.00$;▲ -- $h/\ell = 0.20$, $\varepsilon = 0.05$.

strip. For this purpose we consider the dependence of σ_{11}/p on x_1/ℓ (Fig.3.1.12(a)), and of σ_{22}/p on x_1/ℓ (Fig.3.1.12(b)). The values of the other parameters are given in

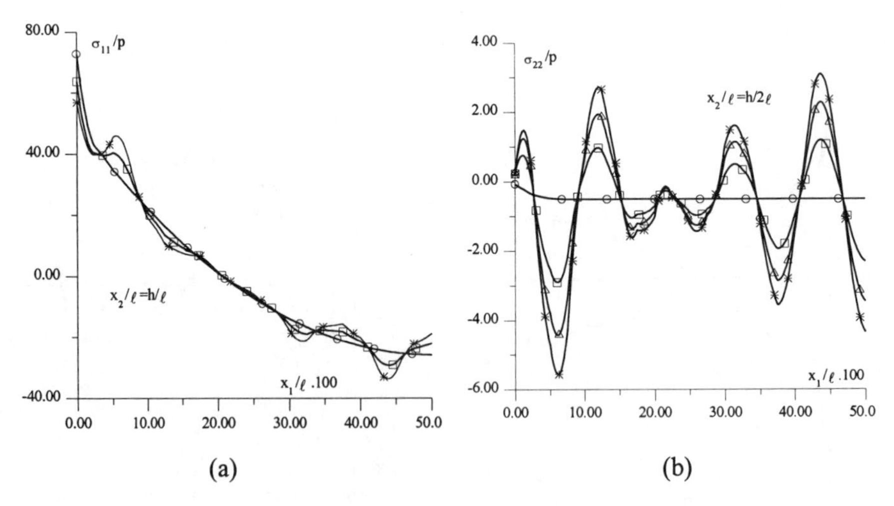

Fig.3.1.14. The influence of E_2 / E_1 on the distribution of stress σ_{11} (a); σ_{22} (b) with respect to x_1 . for $h/\ell = 0.10$, $\ell/\Lambda_1 = 16$: ○ -- $E_2/E_1 = 5$; $\varepsilon = 0.00$; while $\varepsilon = 0.05$ □ -- $E_2/E_1 = 5$;△ -- $E_2/E_1 = 10$ ✳ -- $E_2/E_1 = 20$.

the figure legends. These numerical results show that by increasing ε, the stress perturbations caused by the curving grow. A comparison of the results given in Fig.3.1.12 with those obtained for the Problem 1 and given in Fig.3.1.4 shows that as a result of the difference of the edge conditions, the influence of the curving on the stress distribution in the strip grows significantly near the edge of the strip. Consequently, the influence of the curving on the stress distributions depends significantly on the type of edge conditions. The influence of the strip thickness on the distribution of σ_{11} and σ_{22} is given in Fig.3.1.13. Fig. 3.1.14 shows the dependence of σ_{11}/p, σ_{22}/p on x_1/ℓ for various E_2/E_1 ratios.

These numerical results show that the perturbations of σ_{11} and σ_{22} decrease with increasing h/ℓ, and increase with increasing E_2/E_1.

The graphs of σ_{11}/p and σ_{22}/p for various ℓ/Λ_1 are given in Fig.3.1.15; they show that the dependence of σ_{11}, σ_{22} on ℓ/Λ_1 is non-monotonic. Other numerical results, which are not shown here, show that the characters of the variations of the stresses with respect to x_2/ℓ are the same as those for Problem1.

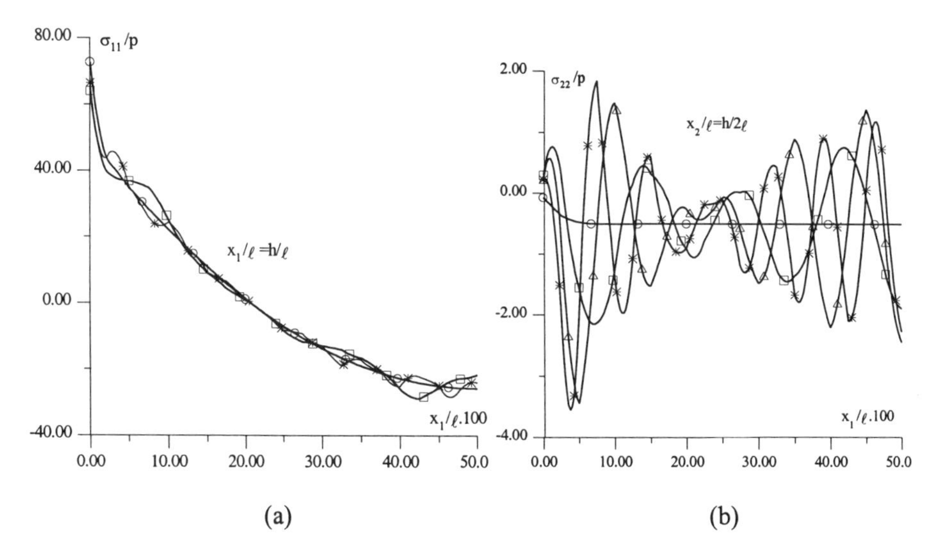

Fig.3.1.15. The influence of ℓ/Λ_1 on the distribution of stress σ_{11} (a); σ_{22} (b) with respect to x_1. for $h/\ell = 0.10$, $E_2/E_1 = 5$: ○ -- $\ell/\Lambda_1 = 00$; $\varepsilon = 0.00$; while $\varepsilon = 0.10$ □ -- $\ell/\Lambda_1 = 12$; △ -- $\ell/\Lambda_1 = 20$; ✳ -- $\ell/\Lambda_1 = 28$.

A comparison of the results obtained for Problems 1 and 2 shows that the character of the influence of the curving on the stress distribution is qualitatively the same for both problems. The numerical results agree with experimental results and coincide with the classical results when $\varepsilon = 0$. This confirms the reliability of the approach we have taken.

Numerical results obtained for Problem 3.

Now there is no symmetry; the region Ω occupied by the strip is divided into 80 finite elements (Fig.3.1.2) with 369 nodes and 720 NDOF.

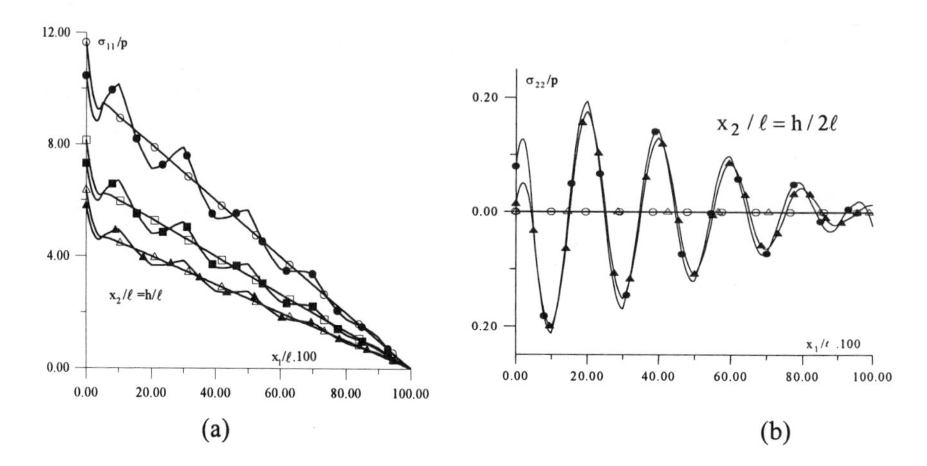

Fig.3.1.16. The influence of h/ℓ on the distribution of stress σ_{11} (a); σ_{22} (b) with respect to x_1 for $E_2/E_1 = 5$ $\ell/\Lambda_1 = 16$: ○ -- $h/\ell = 0.10$, $\varepsilon = 0.00$; ● -- $h/\ell = 0.10$, $\varepsilon = 0.05$ □ -- $h/\ell = 0.15$, $\varepsilon = 0.00$; ■ -- $h/\ell = 0.15$, $\varepsilon = 0.05$; △ -- $h/\ell = 0.20$, $\varepsilon = 0.00$; ▲ -- $h/\ell = 0.20$, $\varepsilon = 0.05$.

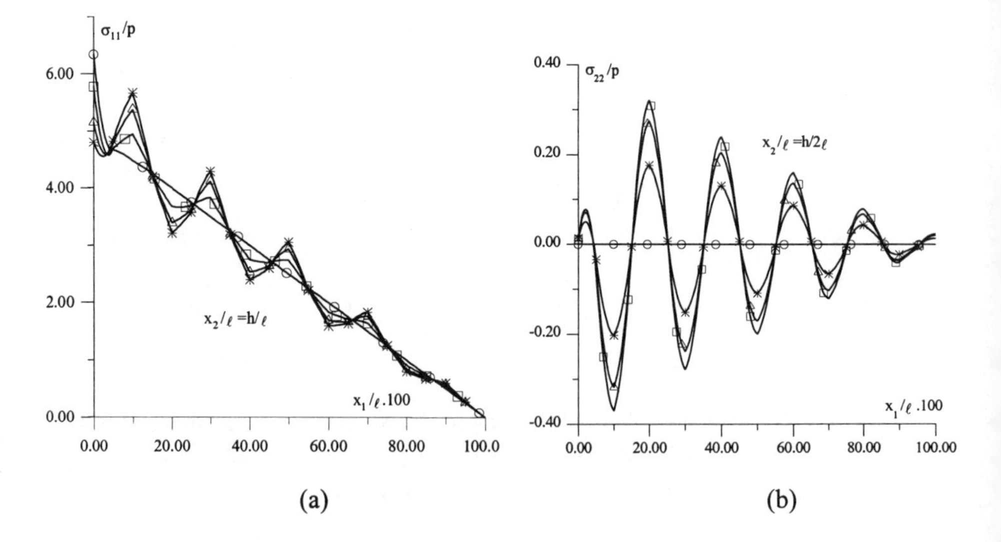

Fig.3.1.17. The influence of E_2/E_1 on the distribution of stress σ_{11} (a); σ_{22} (b) with respect to x_1 . for $h/\ell = 0.10$, $\ell/\Lambda_1 = 16$: ○ -- $E_2/E_1 = 5$; $\varepsilon = 0.00$; while $\varepsilon = 0.05$ □ -- $E_2/E_1 = 5$; △ -- $E_2/E_1 = 10$ ✳ -- $E_2/E_1 = 20$.

First, consider the influence of the problem parameters on the distribution of the stresses σ_{11} and σ_{22} in the cantilever strip. The distributions with respect to x_1/ℓ

are given in Fig.3.1.16 for various h/ℓ ; in Fig.3.1.17 for various E_2/E_1 and in Fig.3.1.18 for various ℓ/Λ_1 . These numerical results show that the influences of the curving on the stress distributions in the strip agree with those obtained for Problems 1 and 2.

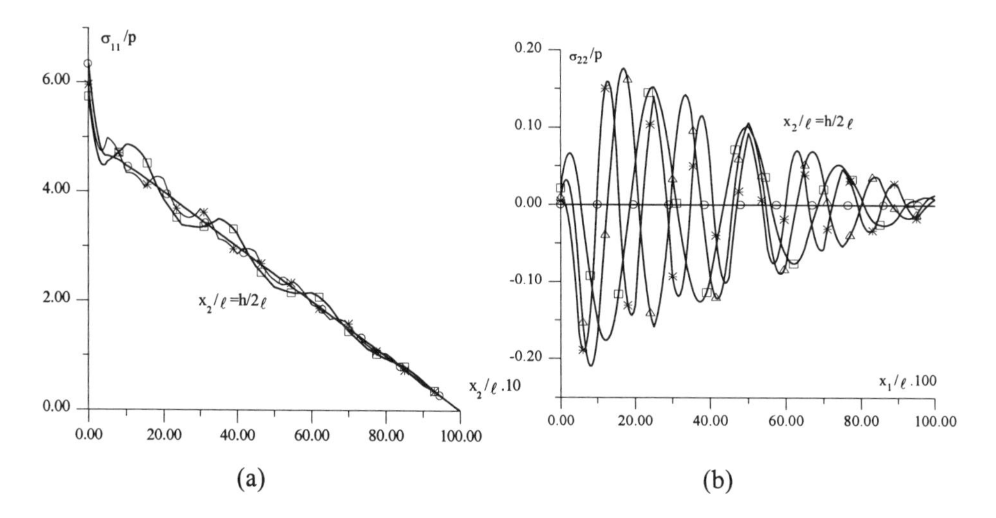

Fig.3.1.18. The influence of ℓ/Λ_1 on the distribution of stress σ_{11} (a); σ_{22} (b) with respect to x_1 . for $h/\ell = 0.10$, $E_2/E_1 = 5$: ○ -- $\ell/\Lambda_1 = 00$; $\varepsilon = 0.00$; while $\varepsilon = 0.10$ □ -- $\ell/\Lambda_1 = 12$; △ -- $\ell/\Lambda_1 = 20$; ✳ -- $\ell/\Lambda_1 = 28$.

3.1.4. ANALYSIS OF THE NUMERICAL RESULTS. LOCAL CURVING

So far we have supposed the curving to be periodic. However, as noted in the Introduction, the curving can be local; it can be regarded as *local damage*. Therefore the investigation of the influence of such a damage on the stress distributions is very significant.

First be note that observations of the cross sections of unidirectional composite materials indicate that the local curving form can be described by the following function with very high accuracy:

$$F(x_1) = \varepsilon f(x_1) = \begin{cases} \varepsilon\lambda^3\left(\dfrac{x_1}{\ell} - c\right)^2\left(\dfrac{x_1}{\ell} - d\right)^2 \exp\left(-\lambda^{2n}\left(\dfrac{x_1}{\ell} - \ell_0\right)^{2n}\right) \times \\ \cos\left(m\pi\lambda\left(\dfrac{x_1}{\ell} - \ell_0\right)\right), \quad \text{for} \quad \dfrac{x_1}{\ell} \in [c, d] \\ 0, \quad \text{for} \quad \dfrac{x_1}{\ell} \in [0, c) \cup (d, 1] \end{cases} \tag{3.1.32}$$

where $c = c_1/\ell$, $d = d_1/\ell$, $\Lambda_1 = d_1 - c_1$, $\lambda = 1/(d-c)$, $\varepsilon = H/\Lambda_1$, $\ell_0 = L/\ell$, H is the maximum height curving, L is the distance between the origin of the system of coordinates Ox_1x_2 and the point of maximum local curving. The parameters c and d characterize the size of the subregion in which the local curving exists; as (d-c) increases, the size of this subregion increases. Moreover the parameter m shows the "vibration" of the local curving form. Increasing m corresponds to increasing the number of vertices of the convexity. In some cases the graph of the function (3.1.32) is shown in Fig.3.1.19.

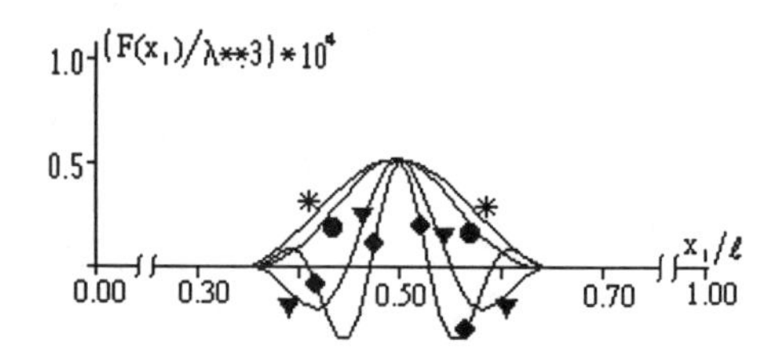

Fig.3.1.19. The form of the local curving for $\varepsilon = 0.1$, $c = 0.35$, $d = 0.65$, $\ell_0 = 0.5$, n=1 under various m: ✳ -- m=0, ● -- m=1, ▼ -- m=3, ◆ -- m=5.

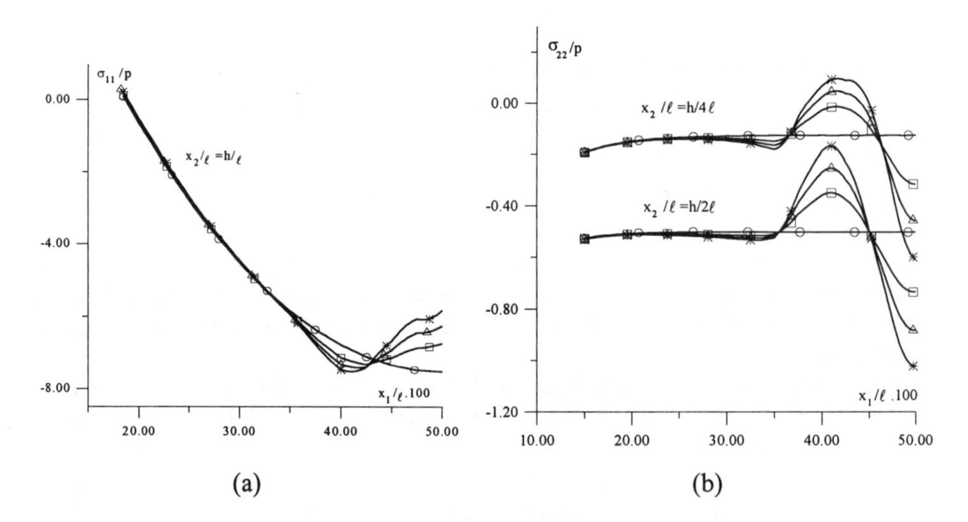

Fig.3.1.20. The influence of the parameter ε on the distribution of stress σ_{11} (a) and σ_{22} (b) for $h/\ell = 0.20$, $c = 0.35$, $m = 2$, $E_2/E_1 = 10$: ○ -- $\varepsilon = 0.00$; □ -- $\varepsilon = 0.3$; △ -- $\varepsilon = 0.5$ ✳ -- $\varepsilon = 0.7$.

Consider Problem 2 and suppose that c=1-d; this makes the problem symmetric. Saint-Venant's principle [126, 144] states that if $c > h/\ell$, the edge conditions do not affect the stress perturbations arising as a result of the local curving. Hence the numerical results which will be discussed can also be related to Problem 1.

Now we analyse the numerical results obtained for the distributions of the stresses σ_{11} and σ_{22} along Ox_1 axis. We fix the values of h/ℓ and E_2/E_1 (their values are taken as $h/\ell=0.2$ and $E_2/E_1=10$) because the influences of changes of these quantities on the stress distributions will be similar to those that obtained for

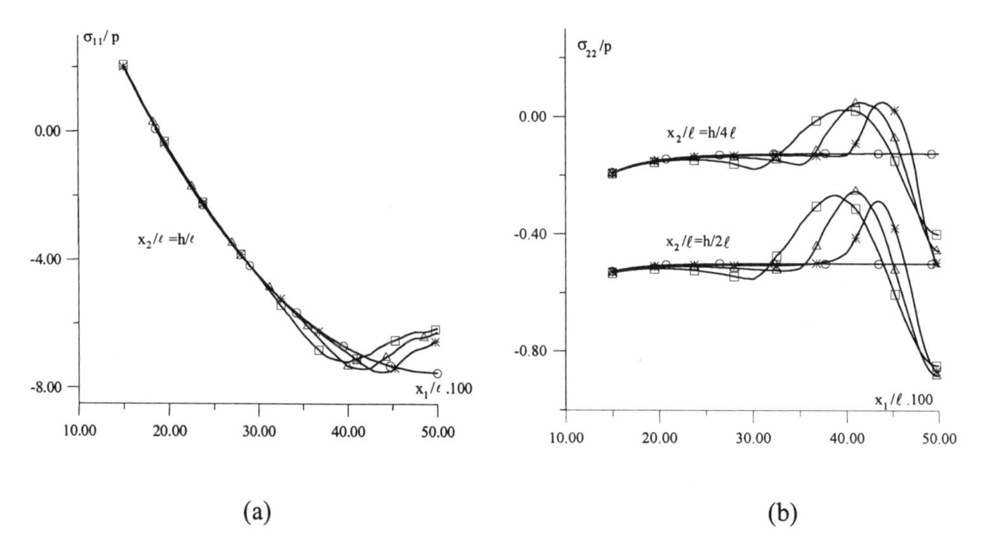

Fig.3.1.21. The influence of the parameter c to the distribution of stress σ_{11} (a) and σ_{22} (b) for $h/\ell=0.20$, $m=2$, $E_2/E_1=10$: ○ -- $\varepsilon=0.00$, $c=0$; while $\varepsilon=0.5$ □ -- $c=0.3$; △ -- $c=0.35$ ✳ -- $c=0.4$.

periodic curving. Moreover, we will take $n=1$, $\ell_0=1/2$ in equation (3.1.32).

Thus, consider the graphs given in Fig. 3.1.20 for $m=2$, $c=0.35$ (d=0.65); these show the dependence of σ_{11}/p, σ_{22}/p on x_1/ℓ for various ε. They show that the influence of the local curving on the stress distributions is also local. They also show that the influence of the local curving on the stress distributions increases monotonically with ε.

As the problem is symmetric with respect to $x_1=\ell/2$ we find that $\sigma_{12}=0$ at $x_1=\ell/2$. Moreover, as the local curving exists only near the section $x_1=\ell/2$, the influence of that on the distribution of σ_{12} can be neglected. Therefore we do not investigate the distribution of σ_{12}. Other numerical results, which are not given here, show that the effect of local curving on the character of the distributions of the stress σ_{11} with respect on x_2/ℓ is insignificant, and the distribution σ_{22} with respect to x_2/ℓ is the same as that was observed in Figs. 3.1.5, 3.1.7, 3.1.9 for the periodic curving; we do not consider the distributions of the stresses σ_{11} and σ_{22} with respect to x_2/ℓ.

The graphs given in Fig. 3.1.21 show the distributions of the stresses σ_{11}/p and σ_{22}/p with respect to x_1/ℓ, for various c. It is evident that by increasing c, the sub-region in which the influence of the local curving on the distributions of the stresses σ_{11} and σ_{22} is observed, must decrease in size also, as shown in Fig. 3.1.21.

Consider the influence of the change of the m on the stress distributions. Note that this influence is very considerable, as shown by the graphs given in Fig. 3.1.22 constructed for $\varepsilon = 0.5$ and c=0.3. These numerical results show that the perturbations of the stresses σ_{11} and σ_{22} increase monotonically with m.

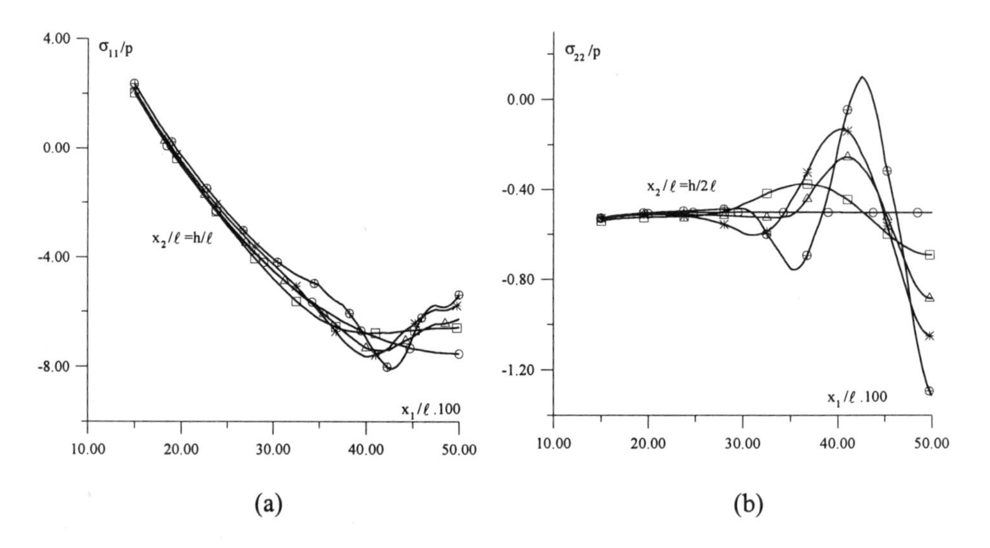

Fig. 3.1.22. The influence of the parameter m on the distribution of stress σ_{11} (a) and σ_{22} (b) in the case where $h/\ell = 0.20$, $c = 0.35$, $E_2/E_1 = 10$: ○ -- $\varepsilon = 0.00$, $m = 0$; while $\varepsilon = 0.5$ □ -- $m = 1$; △ -- $m = 2$; ✳ -- $m = 3$; ⊕ -- $m = 5$.

3.2. Bending of a Rectangular Plate

In this section we will consider some three-dimensional problems for the continuum theory presented in the previous chapter. We will study the stress distribution in the thick plate in the framework of the exact equations of the theory of elasticity; all numerical results will be obtained by employing semi-analytical FEM .

3.2.1. FORMULATION

Consider a plate occupying the region

$$V = \{0 \le x_1 \le \ell_1, 0 \le x_2 \le h, 0 \le x_3 \le \ell_3\} \tag{3.2.1}$$

with the geometrical parameters shown in Fig.3.2.1 and fabricated from composite material with curved structure. We assume that the reinforcing layers or fibers in this plate material are located in planes parallel to the plane Ox_1x_3 and the curving of these reinforcing elements exists only in the direction of the Ox_1 axis. In this case the parameters characterizing the plate material structure satisfy the limitations of the continuum theory proposed in the previous chapter. Suppose the plate is clamped at the edges $x_1 = 0; \ell_1$ and is simply supported at the edges $x_3 = 0; \ell_3$. Further, we assume that normal compressive forces continuously distributed with respect to x_1 and x_3 act on the upper plane of the plate. The formulation of the problem is as follows:

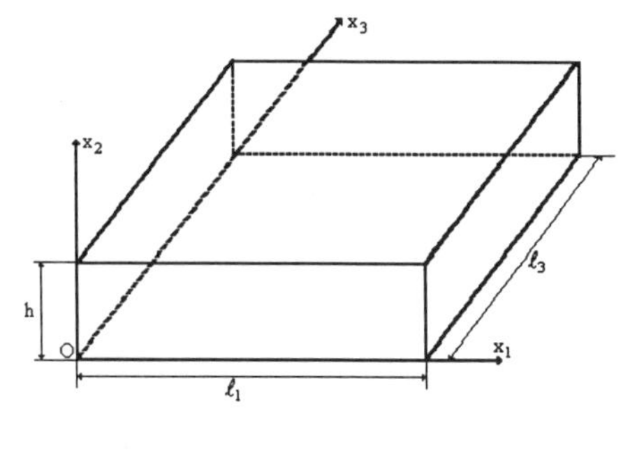

Fig.3.2.1. The geometry of the considered plate.

The equilibrium equations:

$$\frac{\partial \sigma_{ij}}{\partial x_j} = 0 \quad \text{for} \quad x_1, x_2, x_3 \in V, \quad i;j=1,2,3. \tag{3.2.2}$$

The stress-strain relations:

$$\sigma_{11} = A_{11}(x_1)\varepsilon_{11} + A_{12}(x_1)\varepsilon_{22} + A_{13}(x_1)\varepsilon_{33} + A_{16}(x_1)\varepsilon_{12},$$

$$\sigma_{22} = A_{12}(x_1)\varepsilon_{11} + A_{22}(x_1)\varepsilon_{22} + A_{23}(x_1)\varepsilon_{33} + A_{26}(x_1)\varepsilon_{12},$$

$$\sigma_{33} = A_{13}(x_1)\varepsilon_{11} + A_{23}(x_1)\varepsilon_{22} + A_{33}(x_1)\varepsilon_{33} + A_{36}(x_1)\varepsilon_{12},$$

$$\sigma_{23} = 2A_{44}(x_1)\varepsilon_{23} + 2A_{45}(x_1)\varepsilon_{13}, \quad \sigma_{13} = 2A_{45}(x_1)\varepsilon_{23} + 2A_{55}(x_1)\varepsilon_{13},$$

$$\sigma_{12} = A_{16}(x_1)\varepsilon_{11} + A_{26}(x_1)\varepsilon_{22} + A_{36}(x_1)\varepsilon_{33} + 2A_{66}(x_1)\varepsilon_{12}. \tag{3.2.3}$$

The geometrical relations:

$$\varepsilon_{\alpha\beta} = \frac{1}{2}\left(\frac{\partial u_\alpha}{\partial x_\beta} + \frac{\partial u_\beta}{\partial x_\alpha}\right), \qquad \alpha;\beta = 1,2,3. \tag{3.2.4}$$

The boundary conditions:

$$u_i\big|_{x_1=0;\ell_1} = 0, \quad i=1,2,3; \ \text{for} \ x_2 \in [0,h], \ x_3 \in [0,\ell_3], \tag{3.2.5}$$

$$u_2\big|_{x_3=0;\ell_3}=0,\quad \sigma_{33}\big|_{x_3=0;\ell_3}=0,\ \text{ for }\ x_2\in[0,h],\ \ x_1\in[0,\ell_1], \tag{3.2.6}$$

$$\sigma_{i2}\big|_{x_2=0}=0,\ i=1,2,3;\ \sigma_{22}\big|_{x_2=h}=-p(x_1,x_3),$$

$$\sigma_{12}\big|_{x_2=h}=\sigma_{32}\big|_{x_2=h}=0,\ \text{ for }\ x_1\in[0,\ell_1],\ \ x_3\in[0,\ell_3] \tag{3.2.7}$$

Note that the functions $A_{ij}(x_1)$ which enter the relations (3.2.3) are determined through the formulae (3.1.1) and (3.1.2). We will now consider the numerical solution procedure of particular problems in the framework of semi-analytical FEM.

3.2.2. SEMI-ANALYTICAL FEM MODELLING

There are many three-dimensional boundary-value problems whose solution can be expressed analytically with respect to one dimension; such problems can be reduced to two-dimensional problems which can be solved by employing FEM. According to [69, 156], in these and similar situations, the modelling is called the semi-analytical FEM modelling.

The problem formulated in the previous sub-section can be modelled by semi-analytical FEM. The function $p(x_1,x_3)$ in (3.2.7), may be expressed as follows:

$$p(x_1,x_3)=\sum_{n=1}^{N}p_n(x_1)\sin\left((2n-1)\pi\frac{x_3}{\ell_3}\right). \tag{3.2.8}$$

According to (3.2.8) the displacements are chosen as follows:

$$u_i=\sum_{n=1}^{N}v_{in}(x_1,x_3)\left(\delta_i^1\sin\left((2n-1)\pi\frac{x_3}{\ell_3}\right)+\delta_i^2\sin\left((2n-1)\pi\frac{x_3}{\ell_3}\right)+\right.$$

$$\left.\delta_i^3\cos\left((2n-1)\pi\frac{x_3}{\ell_3}\right)\right). \tag{3.2.9}$$

For the stresses and strains we obtain the following expressions from equations (3.2.9), (3.2.4) and (3.2.3):

$$\begin{pmatrix}\sigma_{ij}\\ \varepsilon_{ij}\end{pmatrix}=\sum_{n=1}^{N}\begin{pmatrix}S_{\underline{i}\underline{j}n}(x_1,x_2)\\ R_{\underline{i}\underline{j}n}(x_1,x_2)\end{pmatrix}\left(\delta_{\underline{i}}^{j}\sin\left((2n-1)\pi\frac{x_3}{\ell_3}\right)+\delta_{\underline{j}}^{2}\delta_{\underline{i}}^{1}\sin\left((2n-1)\pi\frac{x_3}{\ell_3}\right)+\right.$$

$$\delta_{\underline{i}}^{1}\delta_{\underline{j}}^{3}\cos\left((2n-1)\pi\frac{x_3}{\ell_3}\right)+\delta_{\underline{j}}^{3}\delta_{\underline{i}}^{2}\cos\left((2n-1)\pi\frac{x_3}{\ell_3}\right)\Bigg). \qquad (3.2.10)$$

In (3.2.9) and (3.2.10) δ_i^j are the Kronecker symbols, and $v_{in}(x_1,x_2)$ the unknown functions which must be determined. Note that the functions $S_{ijn}(x_1,x_2)$ and $R_{ijn}(x_1,x_2)$ are expressed through the functions $v_{in}(x_1,x_2)$ by the formulae which are obtained from (3.2.4) and (3.2.3). Substituting (3.2.8)-(3.2.10) into (3.2.2)-(3.2.7) we see that the equations and boundary conditions are satisfied automatically with respect to x_3. Consequently, the solution of the three-dimensional boundary-value problem (3.2.2)-(3.2.7) is reduced to the solution of N two-dimensional boundary-value problems for the functions $v_{in}(x_1,x_2)$. Note that the mathematical formulation of these two-dimensional problems can easily be obtained from the virtual work principle written separately for each approximation of (3.2.8)-(3.2.10); the total potential energy corresponding to the n-th approximation can be given as follows:

$$\Pi_n = \frac{1}{2}\iiint\limits_V \sigma_{\underline{ijn}}\varepsilon_{\underline{ijn}}dV - \iint\limits_{S_1} P_{\underline{in}}u_{\underline{in}}dS. \qquad (3.2.11)$$

According to the virtual work principle the first variation of the functional (3.2.11) must be zero, i.e.

$$\delta\Pi_n = \iiint\limits_V \sigma_{\underline{ijn}}\delta\varepsilon_{\underline{ijn}}dV - \iint\limits_{S_1} P_{\underline{in}}\delta u_{\underline{in}}dS = 0. \qquad (3.2.12)$$

After integrating (3.2.12) with respect to x_3 and carrying out some well-known transformations, we obtain the equations and boundary conditions for the unknown functions $v_{in}(x_1,x_2)$, $S_{ijn}(x_1,x_2)$. Also, taking into account (3.2.8)-(3.2.10) we obtain from (3.2.12) the semi-analytical FEM modelling for the n-th approximation. The region $\Omega=\{0\le x_1\le \ell_1, 0\le x_2\le h\}$ is divided into a finite number of elements Ω_k. We take the elements Ω_k to be nine noded rectangles (Fig.3.1.2), and use Lagrange second-order shape functions (3.1.13) and (3.1.14). Thus

$$\mathbf{v}_n^k \approx \mathbf{N}_{\underline{n}}^k \mathbf{a}_{\underline{n}}^k, \qquad (3.2.13)$$

where

$$\left(\mathbf{v}_n^k\right)^T = \left\{v_{1n}^k(x_1,x_2), v_{2n}^k(x_1,x_2), v_{3n}^k(x_1,x_2)\right\},$$

$$\left(\mathbf{a}_n^k\right)^T = \left\{v_{1n1}^k, v_{2n1}^k, v_{3n1}^k, \dots, v_{1n9}^k, v_{2n9}^k, v_{3n9}^k)\right\},$$

$$\mathbf{N}_n^k = \begin{pmatrix} N_{1n}^k & 0 & 0 & N_{2n}^k & 0 & 0 & \dots & N_{9n}^k & 0 & 0 \\ 0 & N_{1n}^k & 0 & 0 & N_{2n}^k & 0 & \dots & 0 & N_{9n}^k & 0 \\ 0 & 0 & N_{1n}^k & 0 & 0 & N_{2n}^k & .. & 0 & 0 & N_{9n}^k \end{pmatrix}. \quad (3.2.14)$$

Note that in (3.2.13) and (3.2.14) the upper index k relates to element Ω_k; the components of the vector $\mathbf{a}_n^k$ are the nodal values of the functions $v_{in}^k(x_1, x_2)$; the third subscript shows the number of the node (Fig.3.1.2).

We obtain the following equations for the nodal values of the functions $v_{in}(x_1, x_2)$ in the n-th approximation:

$$\mathbf{K}^{\underline{n}}\mathbf{a}_{\underline{n}} = \mathbf{r}_n, \quad (3.2.15)$$

where

$$(\mathbf{a}_n)^T = \left\{\mathbf{a}_n^1, \mathbf{a}_n^2, \dots, \mathbf{a}_n^M\right\}, \quad (3.2.16)$$

$\mathbf{r}_n$ is a vector whose components are determined through the given function $p_n(x_1)$ in (3.2.8), and $\mathbf{K}^n$ is the stiffness matrix:

$$\mathbf{K}^n = \sum_{k=1}^{M} \mathbf{K}^{kn}, \quad \mathbf{K}^{kn} = \begin{pmatrix} \mathbf{K}_{11}^{kn} & \mathbf{K}_{12}^{kn} & \dots & \mathbf{K}_{19}^{kn} \\ \mathbf{K}_{21}^{kn} & \mathbf{K}_{22}^{kn} & \dots & \mathbf{K}_{29}^{kn} \\ \dots & \dots & \dots & \dots \\ \mathbf{K}_{91}^{kn} & \mathbf{K}_{92}^{kn} & \dots & \mathbf{K}_{99}^{kn} \end{pmatrix}, \quad \mathbf{K}_{ij}^{kn} = \iint_{\Omega_k} \left(\mathbf{B}_{\underline{jn}}^{\underline{k}}\right)^T \mathbf{D}^{\underline{k}} \mathbf{B}_{\underline{in}}^{\underline{k}} \, d\Omega_{\underline{k}}, \quad (3.2.17)$$

where

$$\mathbf{D}^k = \left.\begin{pmatrix} A_{11}(x_1) & A_{12}(x_1) & A_{13}(x_1) & 2A_{16}(x_1) & 0 & 0 \\ A_{12}(x_1) & A_{22}(x_1) & A_{23}(x_1) & 2A_{26}(x_1) & 0 & 0 \\ A_{13}(x_1) & A_{23}(x_1) & A_{33}(x_1) & 2A_{36}(x_1) & 0 & 0 \\ A_{16}(x_1) & A_{26}(x_1) & A_{36}(x_1) & 2A_{66}(x_1) & 0 & 0 \\ 0 & 0 & 0 & 0 & 2A_{44}(x_1) & 2A_{45}(x_1) \\ 0 & 0 & 0 & 0 & 2A_{45}(x_1) & 2A_{55}(x_1) \end{pmatrix}\right|_{x_1 \in \Omega_k}$$

$$\mathbf{B}_{in}^{k} = \begin{pmatrix} \frac{\partial N_{in}^{k}}{\partial x_1} & 0 & 0 \\ 0 & \frac{\partial N_{in}^{k}}{\partial x_2} & 0 \\ 0 & 0 & -(2n-1)\pi\gamma N_{in}^{k} \\ \frac{\partial N_{in}^{k}}{\partial x_2} & \frac{\partial N_{in}^{k}}{\partial x_1} & 0 \\ (2n-1)\pi\gamma N_{in}^{k} & 0 & \frac{\partial N_{in}^{k}}{\partial x_1} \\ 0 & (2n-1)\pi\gamma N_{in}^{k} & \frac{\partial N_{in}^{k}}{\partial x_2} \end{pmatrix}, \quad \gamma = \frac{\ell_1}{\ell_3} \tag{3.2.18}$$

Thus the FEM modelling in the n-th approximation is similar to that for the two-dimensional problems considered in the previous section, except that there are three unknowns at each node instead of two in the two-dimensional problems. Since we use the displacement based FEM, the functions $S_{ijn}(x_1, x_2)$ calculated by the formulae

$$\mathbf{S}_n^d = \mathbf{D}\mathbf{B}_{\underline{n}}\mathbf{a}_{\underline{n}} \tag{3.2.19}$$

are discontinuous at the inter-element boundaries. Here $\mathbf{S}_n^d$ is a vector whose components are the functions $S_{ijn}(x_1, x_2)$, i.e.

$$\left(\mathbf{S}_n^d\right)^T = \{S_{11n}, S_{22n}, S_{33n}, S_{12n}, S_{13n}, S_{23n}\}(x_1, x_2), \tag{3.2.20}$$

and

$$(\mathbf{B}_n)^T = \left\{\mathbf{B}_n^1, \mathbf{B}_n^2, \ldots, \mathbf{B}_n^M\right\}; \left(\mathbf{B}_n^k\right)^T = \left\{\mathbf{B}_{1n}^k, \mathbf{B}_{2n}^k, \ldots, \mathbf{B}_{9n}^k\right\}; \quad 1 \le k \le M. \tag{3.2.21}$$

The superscript d in (3.2.19) and (3.2.20) means that the components of the vector $\mathbf{S}_n^d$ are discontinuous functions. For smoothing the functions $S_{ijn}(x_1, x_2)$ we use the approach

$$\mathbf{S}_n = \mathbf{N}_{\underline{n}}\overline{\mathbf{S}}_{\underline{n}}, \tag{3.2.22}$$

as in (3.1.23). In (3.2.22) the following notation is used:

$$\mathbf{N}_n = \left\{\mathbf{N}_n^1, \mathbf{N}_n^2, \ldots, \mathbf{N}_n^M\right\}, \quad \left(\overline{\mathbf{S}}_n^i\right)^T = \left\{\overline{\mathbf{S}}_{11n}^i, \overline{\mathbf{S}}_{22n}^i, \overline{\mathbf{S}}_{33n}^i, \overline{\mathbf{S}}_{12n}^i, \overline{\mathbf{S}}_{13n}^i, \overline{\mathbf{S}}_{23n}^i\right\}, \quad 1 \le i \le M;$$

$$
\mathbf{N}_n^i = \begin{pmatrix}
N_{1n}^i & 0 & 0 & 0 & 0 & 0 & \ldots & N_{9n}^i & 0 & 0 & 0 & 0 & 0 \\
0 & N_{1n}^i & 0 & 0 & 0 & 0 & \ldots & 0 & N_{9n}^i & 0 & 0 & 0 & 0 \\
0 & 0 & N_{1n}^i & 0 & 0 & 0 & \ldots & 0 & 0 & N_{9n}^i & 0 & 0 & 0 \\
0 & 0 & 0 & N_{1n}^i & 0 & 0 & \ldots & 0 & 0 & 0 & N_{9n}^i & 0 & 0 \\
0 & 0 & 0 & 0 & N_{1n}^i & 0 & \ldots & 0 & 0 & 0 & 0 & N_{9n}^i & 0 \\
0 & 0 & 0 & 0 & & N_{1n}^i & \ldots & 0 & 0 & 0 & 0 & 0 & N_{9n}^i
\end{pmatrix}
\tag{3.2.23}
$$

In (3.2.23) $\overline{\mathbf{S}}_{kjn}^i$ are the vector whose components are the values of the functions $S_{kjn}(x_1, x_2)$:

$$
\overline{\mathbf{S}}_{kjn}^i = \left\{ \overline{S}_{kjn1}^i, \ldots, \overline{S}_{kjn9}^i \right\}. \tag{3.2.24}
$$

The unknowns $\overline{S}_{kjn1}^i, \ldots, \overline{S}_{kjn9}^i$ are determined from the following equation

$$
\frac{\partial Q_n}{\partial \overline{S}_{kjnm}^i} = 0 \, , \; i=1,2,\ldots,M; \quad k,j=1,2,\ldots,6; \; m=1,2,\ldots,9, \tag{3.2.25}
$$

where

$$
Q_n = \iint_{\Omega} \left(\mathbf{S}_n - \mathbf{S}_n^d \right)^2 d\Omega . \tag{3.2.26}
$$

Thus, completes the semi-analytical FEM modelling of the three-dimensional problem. We now investigate some numerical examples.

3.2.3. NUMERICAL RESULTS

Consider the stress distributions in the plate shown in Fig.3.2.1, where function $p(x_1, x_3)$ in (3.2.7) is

$$
p(x_1, x_3) = p_1 \sin\left(\pi \frac{x_3}{\ell_3} \right) \tag{3.2.27}
$$

Now N=1, so that the semi-analytical FEM modelling procedure is made just once, and the shape functions entering (3.2.14) are as $N_{k1}^i = N_k$ (k=1,2,...,9; i=1,2,...,M) for every i, where the N_k are given in (3.1.14).

We accept the notation and limitations introduced in the sub-section 3.1.3, except that ℓ is replaced by ℓ_1. The form of the curving in the structure is given by (3.1.29). We write $\gamma = \ell_1/\ell_3$ and suppose that the problem is symmetrical with respect to $x_1 = \ell_1/2$. This means that we need model only the sub-domain $\{0 \le x_1/\ell_1 \le 1/2;\ \ 0 \le x_2/\ell_1 \le h/\ell_1\}$; we divide this into 80 rectangular Lagrange-family quadratic elements (Fig.3.1.2) with 369 nodes and 1107 NDOF. 4 rectangular elements are arranged in the direction of the Ox_2 space axis and 20 in the direction of the Ox_1 axis. The selection of the NDOF values follows from the requirements that the boundary conditions should be satisfied with very high accuracy, and the numerical results obtained for various NDOF should converge.

Numerical results related to the layered composite material.
As in sub-section 3.1.3, we assume that the composite material consists of alternating layers of two materials, and these layers are perpendicular to the Ox_2 axis in the absence of curving. This means that when $\varepsilon = 0$, the plate material can be taken as transversely isotropic about Ox_2; the normalized mechanical constants A_{ij}^{0}, in (3.1.1), are calculated through the formulae (1.9.1)-(1.9.3).

Periodic curving. The analyses of the numerical results obtained for various E_2/E_1, ε, h/ℓ_1, ℓ_1/Λ_1 show that the influence of these parameters on the stress distribution is

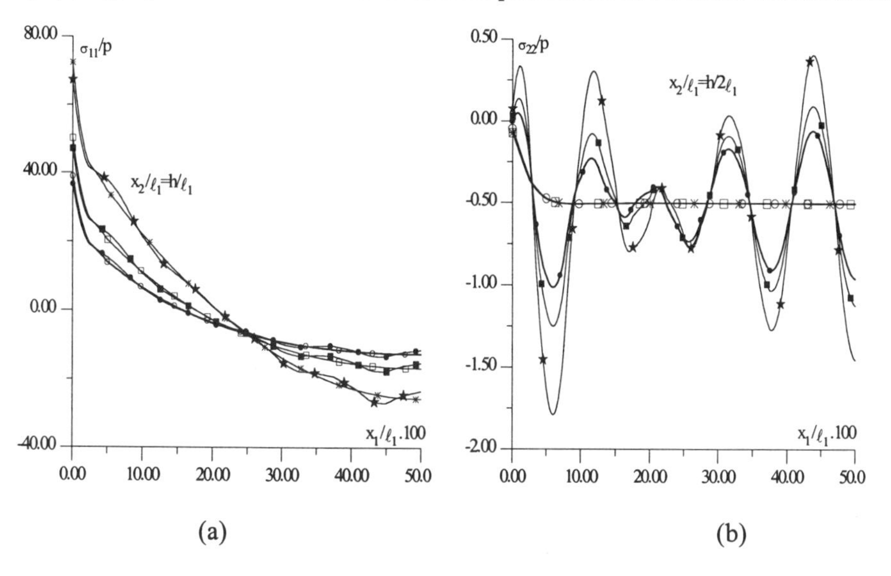

Fig.3.2.2. The influence of γ on the distribution of stress σ_{11} (a); σ_{22} (b) with respect to x_1 for $E_2/E_1 = 5$ $\ell/\Lambda_1 = 16$: ○ -- $\gamma = 1.3$, $\varepsilon = 0.00$; ● -- $\gamma = 1.3$, $\varepsilon = 0.05$ □-- $\gamma = 1.0$, $\varepsilon = 0.00$; ■ -- $\gamma = 1.0$, $\varepsilon = 0.05$; ✳-- $\gamma = 0.1$, $\varepsilon = 0.00$; ★ -- $\gamma = 0.1$, $\varepsilon = 0.05$; ★ -- Plane strain . Periodical curving.

qualitatively the same as that for the corresponding two-dimensional problems considered in section 3.1. Therefore, here we consider only the numerical results which relate to the influence of the parameter γ on the distribution of the stresses σ_{22} and σ_{11} with respect to x_1/ℓ_1 at the section $x_3/\ell_3 = 1/2$ for $E_2/E_1 = 5$, $\varepsilon = 0.05$, $h/\ell_1 = 0.1$ and $\ell_1/\Lambda_1 = 16$. The parameter γ characterizes the three-dimensionality of the problem; as $\gamma \to 0$, the results obtained must approach the corresponding results given in sub-section 3.1.3.

Consider the graphs given in Fig.3.2.2 which show the dependence of σ_{11}/p_1 on x_1/ℓ_1 (Fig.3.2.2 (a)), and σ_{22}/p_1 on x_1/ℓ_1 (Fig.3.2.2 (b)) for various γ. As γ decreases, stresses σ_{22} and σ_{11} approach the corresponding values obtained for the strip in the previous sub-section; see $\gamma = 0.1$. These stresses decrease monotonically with growing γ.

Local curving.

We analyse the results obtained for the stresses σ_{11} and σ_{22} for various γ. We assume that the form of the local curving is given by the formulae (3.1.32), where ℓ is replaced by ℓ_1. Again we consider the symmetric problem: $c = 1 - d$ and $\ell_0 = 1/2$; all

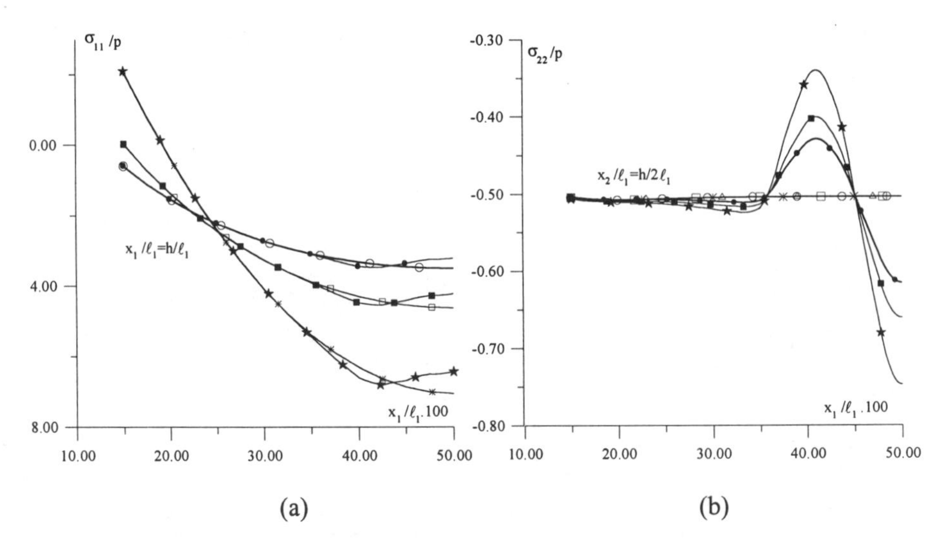

Fig.3.2.3. The influence of γ on the distribution of stress σ_{11} (a); σ_{22} (b) with respect to x_1 for $E_2/E_1 = 5$ $\ell/\Lambda_1 = 16$: ○ -- $\gamma = 1.3$, $\varepsilon = 0.00$; ● -- $\gamma = 1.3$, $\varepsilon = 0.05$; □ -- $\gamma = 1.0$, $\varepsilon = 0.00$; ■ -- $\gamma = 1.0$, $\varepsilon = 0.05$; ✳ -- $\gamma = 0.1$, $\varepsilon = 0.00$; ★ -- $\gamma = 0.1$, $\varepsilon = 0.05$; ★ -- Plane strain . Local curving.

numerical results are obtained for the region $\{0 \le x_1 \le \ell_1/2, 0 \le x_2 \le h\}$ with $x_3 = \ell_3/2$, and we take n=1 in (3.1.32).

The graphs in Fig.3.2.3 show the dependence of σ_{11}/p_1 (Fig.3.2.3 (a)), σ_{22}/p_1 (Fig.3.2.3 (b)) on x_1/ℓ_1 for various γ. Again, as γ increases, the perturbations of the stresses which arise from of the local curving decrease, and when γ is small, e.g. $\gamma = 0.1$, the stresses coincide with the plane strain values found in the previous section. Other numerous numerical results, not given here, show that the influences of the others problem parameters on the stress distributions are similar to those found in the plane strain case.

Fibrous-layered composite material.
So far all numerical results relate to a material consisting of alternating isotropic and homogeneous layers. Now we assume that these layers are composed of boron fibres in an epoxy matrix, with equal amounts of each. According to [74], the Young's modulus and the Poisson's coefficients for the boron and epoxy are $E_f = 400\text{GPa}$, $\nu_f = 0.21$ and $E_m = 4\text{GPa}$, $\nu_m = 0.4$, respectively. The orientations of these layers in the plate are assumed to be $\left(0^\circ/90^\circ\right)$, i.e., the fibres in the first (second) of these layers are directed along the $Ox_1(Ox_3)$ axis. In the continuum approach the material of each layer can be considered to be a homogeneous transversely-isotropic material, with the axis of isotropy lying along the unidirectional fibres. Consequently, in the absence of curving, the whole material can be considered to be orthotropic with principal axes $O(x_1, x_2, x_3)$ and with the normalized mechanical properties A_{ij}^0. The values of A_{ij}^0

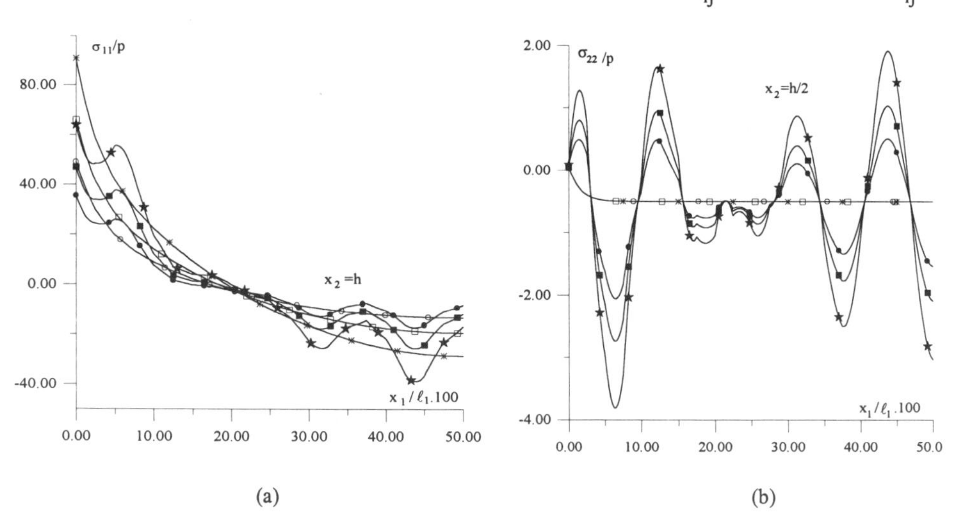

Fig.3.2.4. The influence of γ on the distribution of stress σ_{11} (a); σ_{22} (b) with respect to x_1 in the case where $\ell/\Lambda_1 = 16$: ○ -- $\gamma = 1.3$, $\varepsilon = 0.00$; ● -- $\gamma = 1.3$, $\varepsilon = 0.05$ □ -- $\gamma = 1.0$, $\varepsilon = 0.00$; ■ -- $\gamma = 1.0$, $\varepsilon = 0.05$; ✳-- $\gamma = 0.1$, $\varepsilon = 0.00$; ★ -- $\gamma = 0.1$, $\varepsilon = 0.05$; ★ -- Plane strain . Periodical curving. Layered-fibrous composite material

are calculated from the well-known formula given, for example, in [70]. Now we analyse the numerical results, and consider the influence of the problem parameters on σ_{11} and σ_{22} only; these are stresses most affected by the curving. Periodic and local curving will be considered separately.

Periodic curving. We investigate σ_{11} and σ_{22} as functions of x_1/ℓ_1 at the sections $\{x_3/\ell_3 = 1/2;\ x_2 = h\}$ and $\{x_3/\ell_3 = 1/2,\ x_2 = h/2\}$, respectively. The parameters δ, ℓ_1/Λ_1 which enter (3.1.29) and η_1, η_2 which are used for determination of the normalized mechanical properties A^0_{ij} are selected to be $\delta = \pi/2$, $\ell_1/\Lambda_1 = 16$, $\eta_1 = \eta_2 = 0.5$. Fig.3.2.4 a,b show the dependence of σ_{11}/p_1 and σ_{22}/p_1 on x_1/ℓ_1 for $\varepsilon = 0.05$, h/ℓ_1 =0.1, and various values of $\gamma = \ell_1/\ell_3$. The distributions of the stresses follows the form of the curving; as γ decreases, the amplitude of the dependence increases; the results for $\gamma = 0$ corresponds to those obtained for plane strain.

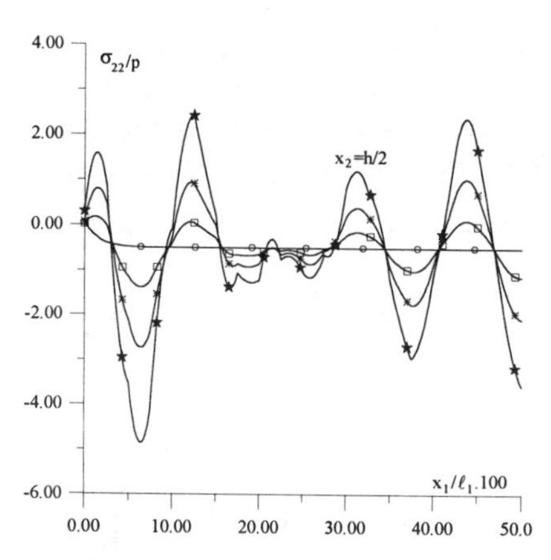

Fig.3.2.5. The influence of the ε on the distribution of the stress σ_{22} in the case where $h/\ell_1 = 0.1$; $\gamma = 1.0$; $\ell_1/\Lambda_1 = 16$: ○ -- $\varepsilon = 0.00$; □ -- $\varepsilon = 0.02$; ✳ -- $\varepsilon = 0.05$; ★ -- $\varepsilon = 0.10$

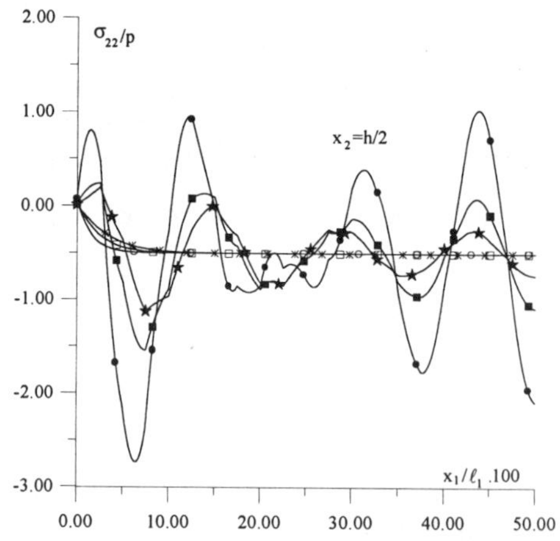

Fig.3.2.6. The influence of h/ℓ on the distribution of stress σ_{22} with respect to x_1 in the case where $\gamma = 1.0$; $\ell/\Lambda_1 = 16$: ○ -- $h/\ell = 0.10$, $\varepsilon = 0.00$; ● -- $h/\ell = 0.10$, $\varepsilon = 0.05$; □ -- $h/\ell = 0.15$, $\varepsilon = 0.00$; ■ -- $h/\ell = 0.15$, $\varepsilon = 0.05$; ✳ -- $h/\ell = 0.20$, $\varepsilon = 0.00$; ★ -- $h/\ell = 0.20$, $\varepsilon = 0.05$.

Fig.3.2.5 shows how ε affects the dependence of σ_{22}/p on x_1/ℓ_1. As ε increases the influence of the curving increases.

Fig. 3.2.6 shows the influence of the change of h/ℓ_1 on σ_{22}/p. When $\varepsilon = 0$, h/ℓ_1 has no effect. As h/ℓ_1 increases the "amplitude" of the dependence decreases.

We conclude that when the curving is periodic, a plate made from boron fibre-epoxy matrix composite behaves like one made from a composite with alternating isotropic homogeneous layers.

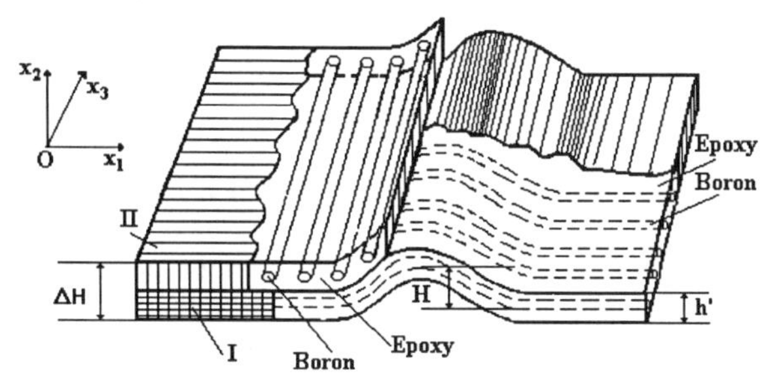

Fig.3.2.7. Schematic illustration of the representative packet-layer consisting of two mono-layer.

Local curving. We suppose the form of the local curving is given by the formulae (3.1.32), with ℓ replaced by ℓ_1, and again consider the symmetric problem with $c = 1 - d$ and $\ell_0 = 1/2$. Results are obtained for the region

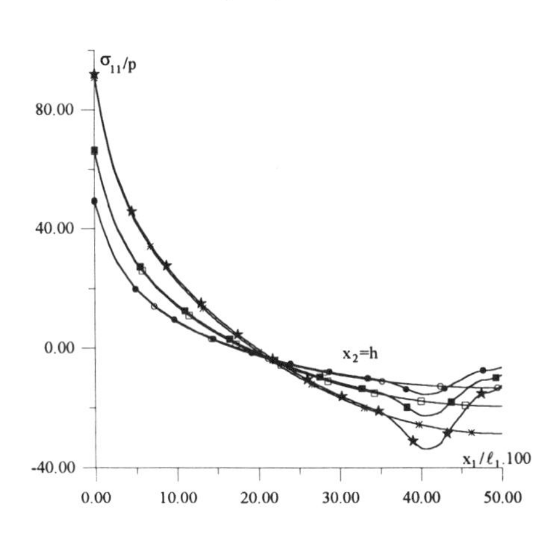

Fig.3.2.8. The influence of γ on the distribution of stress σ_{11} with respect to x_1 in the case where $h/\ell_1 = 0.1$, $\ell/\Lambda_1 = 16$, $c = 0.35$: ○ -- $\gamma = 1.3$, $\varepsilon = 0.00$; ● -- $\gamma = 1.3$, $\varepsilon = 0.5$; □ -- $\gamma = 1.0$, $\varepsilon = 0.00$; ■ -- $\gamma = 1.0$, $\varepsilon = 0.5$; ✳-- $\gamma = 0.1$, $\varepsilon = 0.00$; ★ -- $\gamma = 0.1$, $\varepsilon = 0.5$; ★ -- Plane strain. Local curving; layered-fibrous composite material.

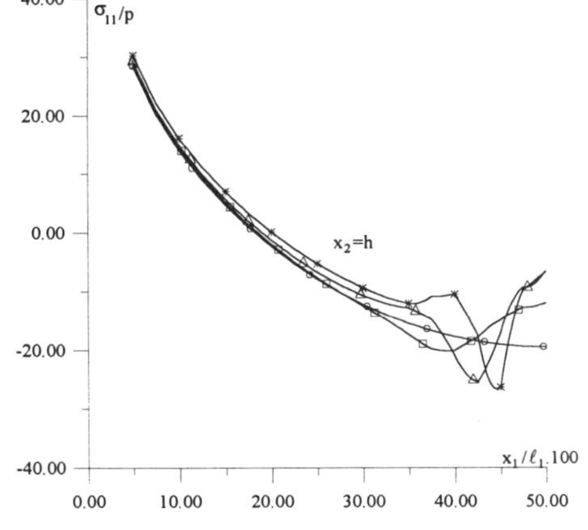

Fig.3.2.9. The influence of m on the distribution of stress σ_{11} with respect to x_1 in the case where $h/\ell_1 = 0.1$, $\ell/\Lambda_1 = 16$, $c = 0.35$, $\varepsilon = 0.5$, $\gamma = 1.0$: ○ -- ε=0; □ --m=1, △ -- $m = 3$; ✳ -- $m = 5$. Local curving; layered-fibrous composite material.

$\{0 \le x_1 \le \ell_1/2, 0 \le x_2 \le h\}$ with $x_3/\ell_3 = 1/2$. We assume that n=1 in (3.1.32) and explore the distributions of σ_{11} and σ_{22}.

Fig.3.2.7 shows the location of two single layers in a representative packet. We assume that the plate consists of many such packets. We apply a homogenisation procedure to each layer, i.e. each is modelled as a homogeneous transversely isotropic material with normalized mechanical characteristics. Further, we find the normalized mechanical properties of the representative packet-layer with thickness ΔH consisting of these two layers without curving. We suppose that the material of the representative packet layer is orthotropic, with mechanical constants A_{ij}^{0}, as in (3.1.1).

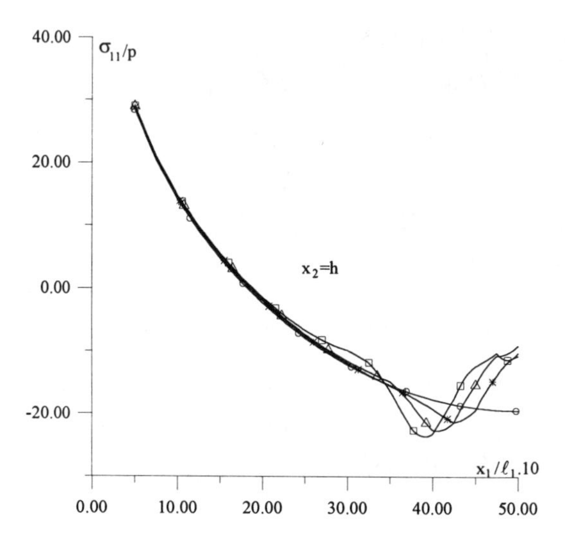

Fig.3.2.10. The influence of m on the distribution of stress σ_{11} with respect to x_1 in the case where $h/\ell_1 = 0.1$, $\ell/\Lambda_1 = 16$, m=2, $\varepsilon = 0.5$, $\gamma = 1.0$: ○ --c=0.5,; □ --c=0.30, △ --c=0.35; ✳.—c=0.40;. Local curving; layered-✳fibrous composite material.

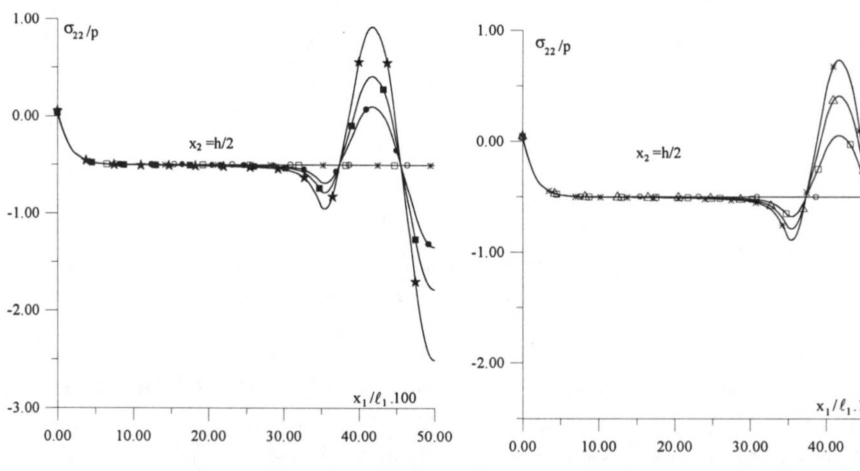

Fig.3.2.11. The influence of γ on the distribution of stress σ_{22} with respect to x_1 in the case where $h/\ell_1 = 0.1$, $c = 0.35$: ○ -- $\gamma = 1.3$, $\varepsilon = 0.00$; ● -- $\gamma = 1.3$, $\varepsilon = 0.5$; □ -- $\gamma = 1.0$, $\varepsilon = 0.00$; ■ -- $\gamma = 1.0$, $\varepsilon = 0.5$; ✳.-- $\gamma = 0.1$, $\varepsilon = 0.00$; ★ -- $\gamma = 0.1$, $\varepsilon = 0.5$; ★ -- Plane strain. Local curving; layered-fibrous composite material.

Fig.3.2.12. The influence of ε on the distribution of the stress σ_{22} in the case where $h/\ell_1 = 0.1$; $\gamma = 1.0$; c=0.35, m=2 : ○ -- $\varepsilon = 0.00$; □ -- $\varepsilon = 0.3$; △ -- $\varepsilon = 0.5$ ✳ -- $\varepsilon = 0.7$. Local curving; layered-fibrous composite material.

Fig.3.2.8 shows the dependence of σ_{11}/p_1 on x_1/ℓ_1 for various values of $\gamma = \ell_1/\ell_3$. The results show that, as γ decreases, the values of σ_{11} approach the plane strain values.

Fig. 3.2.9 shows how the parameter m in (3.1.32) affects the dependence of σ_{11}/p_1 on x_1/ℓ_1; here $\gamma = 1$. The numerical results show that, as m increases, the distortion in σ_{11} increases monotonically. Fig.3.2.10 shows the influence of the parameter c on the stress σ_{11}. The numerical results show that, as c decreases, the perturbations of the stress grow. Fig.3.2.11 shows the dependence σ_{22}/p_1 on x_1/ℓ_1 for various γ. These graphs show that, as $\gamma \to 0$, the values of σ_{22} at the local curving region significantly, increase and approach the corresponding plane strain values. The graphs given in Fig.3.2.12 show that as the local curving increases, i.e. ε increases, σ_{22} grows monotonically. In some parts of the local damage region the stress σ_{22} is positive; this can explain local delamination of the plate material.

3.3. Vibration Problems

In this section, we will investigate the natural and forced vibration of a strip and a rectangular plate fabricated from composite material with curved structure. We discuss the solution method, analyse the FEM results, and study the influence of the curving parameters on the natural frequencies and stress distributions in the structures.

3.3.1. FORMULATION AND SOLUTION

Consider a plate fabricated from composite material with curved structure. As in the previous section, we assume that the plate occupies the region (3.2.1), and the curving exists only in the direction of the Ox_1 axis. We assume that the relations (3.2.3) and (3.2.4) hold, and that instead of the equilibrium equation (3.2.2), the following equation of motion is satisfied:

$$\frac{\partial \sigma_{ij}}{\partial x_j} = \rho \frac{\partial^2 u_i}{\partial t^2}, \quad i;j=1,2,3. \tag{3.3.1}$$

Here t is the time and ρ is the volume density of the plate material. Suppose that on the upper plane of the above plate, i.e. at $x_2 = h$, normal forces act that are uniformly distributed with respect to x_1:

$$p(x_3,t) = p_1 \sin\left(\frac{\pi}{\ell_3} x_3\right) e^{i\omega t} \tag{3.3.2}$$

We will assume that the plate is simply supported along the edges $x_3 = 0$ and ℓ_3 (Fig.3.3.1), so that

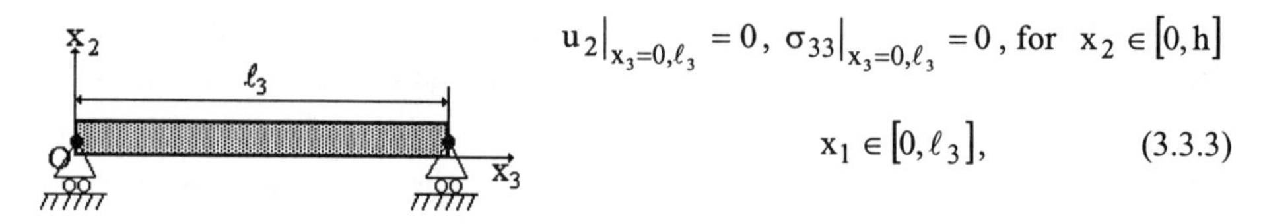

$$u_2\big|_{x_3=0,\ell_3} = 0,\ \sigma_{33}\big|_{x_3=0,\ell_3} = 0,\ \text{for}\ x_2 \in [0,h]$$

$$x_1 \in [0,\ell_3], \tag{3.3.3}$$

Fig.3.3.1. A diagram of the simply supported plate section at $x_1 = \text{const}$.

$$\sigma_{i2}\big|_{x_2=0} = 0;\ i=1,2,3,\quad \sigma_{22}\big|_{x_2=h} = -p_1 \sin\left(\frac{\pi}{\ell_3} x_3\right) e^{i\omega t},$$

$$\sigma_{12}\big|_{x_2=h} = \sigma_{32}\big|_{x_2=h} = 0,\ \text{for}\ x_1 \in [0,\ell_1],\ x_3 \in [0,\ell_3]. \tag{3.3.4}$$

Along the edges $x_1 = 0$, $x_1 = \ell_1$ we consider two possible end conditions:

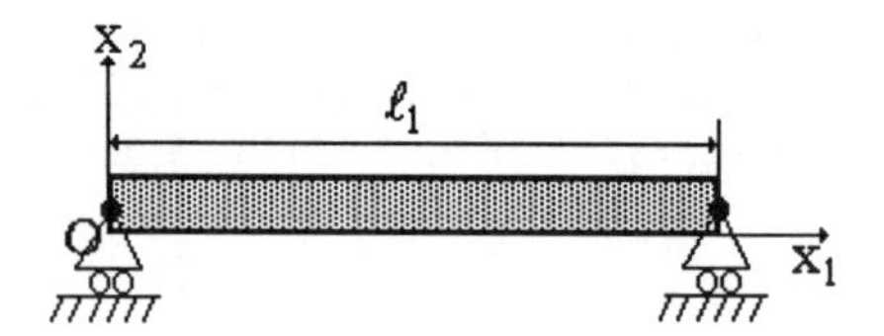

Fig.3.3.2. A diagram of the simply supported plate section at $x_1 = \text{const}$.

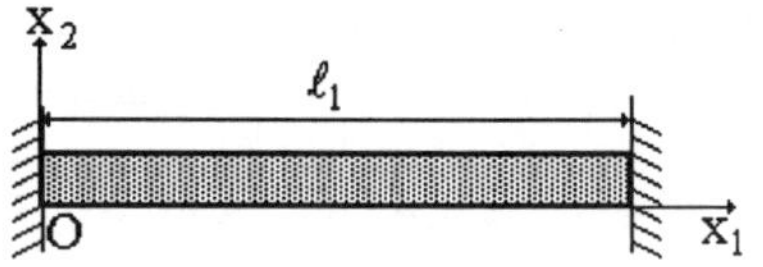

Fig.3.3.3. A diagram of the clamped plate section at $x_1 = \text{const}$.

1) simply supported (Fig.3.3.2)

$$u_2\big|_{x_1=0,\ell_1} = 0,\quad \sigma_{11}\big|_{x_1=0,\ell_1} = 0\ \ \text{for}\ x_2 \in [0,h],\ x_3 \in [0,\ell_3], \tag{3.3.5}$$

2) clamped (Fig.3.3.3)

$$u_i\big|_{x_1=0,\ell_1} = 0,\ i=1,2,3\ \ \text{for}\qquad x_2 \in [0,h],\ x_3 \in [0,\ell_3]. \tag{3.3.6}$$

This completes the formulation.

Now consider the solution of the vibration problems formulated in (3.3.1)-(3.3.6). Some vibration problems have been solved before; [45] develops the variational method presented in Section 1.6 for problems governed by (2.4.1)-(2.4.6), when conditions (2.3.1) are satisfied. The paper [123,124] use the refined plate theory developed in [120,121] and the variational method of [45] to investigate the natural frequencies of a plate fabricated from periodically curved composite. Here we will apply semi-analytical FEM to the equations (3.3.1)-(3.3.6).

The stresses are written

$$\left\{\sigma_{ij}, \varepsilon_{ij}, u_i\right\} = \left\{\bar{\sigma}_{ij}, \bar{\varepsilon}_{ij}, \bar{u}_i\right\} e^{i\omega t} \tag{3.3.7}$$

Adding to the functional (3.2.11) the term

$$-\omega^2 \rho \iiint_V \bar{u}_i dV \tag{3.3.8}$$

and carrying out the same operations performed in the previous section we obtain the following equation

$$\left(\mathbf{K} - \omega^2 \mathbf{M}\right) \bar{\mathbf{a}} = \bar{\mathbf{r}} \tag{3.3.9}$$

Here $\mathbf{K}$ is the stiffness matrix whose elements are determined through the relations (3.2.17)-(3.2.18) in the case n=1; $\bar{\mathbf{a}}$ is the vector whose components are the values of the $\bar{u}_i$ at the selected nodes; the components of the force vector $\bar{\mathbf{r}}$ are calculated like the force vector $\mathbf{r}_1$ which enters (3.2.15). In (3.3.9) matrix $\mathbf{M}$ is a mass matrix and

$$\mathbf{M} = \rho \iiint_V \mathbf{N}^T \mathbf{N} dV , \tag{3.3.10}$$

where $\mathbf{N}$ is determined by (3.2.14). After finding $\bar{\mathbf{a}}$ from the equation (3.3.9), we determine $\bar{\sigma}_{ij}$ by employing the procedure (3.2.22)-(3.2.26).

The natural frequencies ω_r are the values of ω for which the equation

$$\left(\mathbf{K} - \omega_r^2 \mathbf{M}\right) \bar{\mathbf{a}}_r = \mathbf{0} \tag{3.3.11}$$

has a non-trivial solution $\bar{\mathbf{a}}_r$. For investigating the forced vibrations of the plate we assume that the frequency ω is significantly less than the fundamental frequency ω_I.

3.3.2. FREE VIBRATION

First consider the free vibration of the simple supported strip shown in Fig.3.3.2. The end and boundary conditions are

$$u_2\big|_{x_1=0,\ell_1} = 0, \quad \sigma_{11}\big|_{x_1=0,\ell_1} = 0 \quad \text{for } x_2 \in [0,h], \tag{3.3.12}$$

$$\sigma_{i2}\big|_{x_2=0,h} = 0 \quad \text{for } x_1 \in (0, \ell_1) . \tag{3.3.13}$$

We introduce the dimensionless frequency $\overline{\omega}$ by

$$\overline{\omega}^2 = \frac{\omega^2 \rho \ell_1}{A_{22}^0} \tag{3.3.14}$$

We denote the values of $\overline{\omega}$ corresponding to ω_1, ω_2, ω_3 by ω_I, ω_{II}, ω_{III} respectively. We assume that the composite consists of alternating layers of two materials, and that these layers are perpendicular to the Ox_2 axis in the absence of curving. The normalized elastic constants A_{ij}^0 are determined from (1.9.1), (1.9.3) and the $A_{ij}(x_1)$ from (3.1.1), (3.1.2). We assume that the curving is periodic, and given by (3.1.29).

Now consider Tables 3.1-3.3 showing ω_r^2 (r = I, II, III) for various values of ε and $\alpha_1 = \ell_1/\Lambda$. These results are obtained for $E_2/E_1 = 50$, $\nu_1 = \nu_2 = 0.3$, $\eta^{(2)} = 0.5$, $h/\ell_1 = 0.1$, $\delta = 0$. The tables show that for fixed α_1, ω_r^2 decreases monotonically with increasing ε. However, for fixed ε, the influence of α_1 on ω_r^2 is complicated. The numerical results show that for $1 \le \alpha_1 \le r$, i.e. to the left of the broken vertical line, the dependence of ω_r^2 on α_1 is non-monotonic; for $\alpha_1 > r$, ω_r increases with α_1.

However these increases are insignificant for the cases when $\alpha_1 > r + 3$. Analogous results are obtained for the rigidly supported strip and for the plate with the boundary conditions described by (3.3.2)-(3.3.6). For this reason we will consider only $\alpha_1 >> r$; in fact we will take $\alpha_1 = 8$. Further numerical results show that the influence of the parameter δ on ω_r^2 is also negligible; we will take $\delta = \pi/2$.

TABLE 3.1. The values of ω_I^2 for various α_1 and ε ; ω_I^2 =0.62 for ε =0.00.

ε \ α_1	1	2	3	4	5	6	8	10	14	16
0.02	0.61	0.60	0.60	0.61	0.61	0.61	0.61	0.61	0.62	0.62
0.05	0.56	0.50	0.51	0.53	0.54	0.54	0.55	0.55	0.55	0.55
0.10	0.43	0.32	0.34	0.36	0.38	0.40	0.42	0.42	0.42	0.41

TABLE 3.2. The values of ω_{II}^2 for various α_1 and ε ; $\omega_{II}^2=5.62$ for $\varepsilon=0.00$

ε \ α_1	1	2	3	4	5	6	8	10	14	16
0.02	5.51	5.61	5.53	5.53	5.53	5.54	5.55	5.55	5.56	5.56
0.05	5.02	5.52	5.08	5.11	5.15	5.18	5.23	5.25	5.28	5.29
0.10	3.80	5.01	3.96	4.08	4.19	4.28	4.41	4.46	4.55	4.59

TABLE 3.3. The values of ω_{III}^2 for various α_1 and ε ; $\omega_{III}^2=16.57$ for $\varepsilon=0.00$

ε \ α_1	1	2	3	4	5	6	8	10	14	16
0.02	16.45	16.45	16.56	16.43	16.43	16.44	16.45	16.47	16.50	16.51
0.05	15.89	15.90	16.40	15.75	15.74	15.82	15.91	15.98	16.15	16.24
0.10	14.37	14.44	15.30	13.84	13.89	14.15	14.43	14.62	15.14	15.42

Now we analyse the numerical results for simply supported and clamped strips. These are given in Tables 3.4 – 3.6; the upper number refers to simply supported, the lower to clamped.

Table 3.4 shows the influence of $\varepsilon = H/\Lambda_1$; ε is the parameter which measures the height of the curving relation to its (half) wavelength. As ε increases, the material stiffness decreases and ω_r decreases significantly. On the other hand, Tables 3.5 and 3.6 show that as h/ℓ_1 or E_2/E_1 increase, the material stiffness increases, and ω_r increases.

TABLE 3.4. The influence of ε on values of ω_I^2, ω_{II}^2 and ω_{III}^2 under $h/\ell = 0.1$, $E_2/E_1 = 50$.

ε	ω^2		
	Mode I	Mode II	Mode III
0.00	0.62 / 1.58	5.62 / 6.98	16.57 / 17.98
1	2	3	4

Table 3.4 (Continuation)

1	2	3	4
0.05	0.55 / 1.50	5.22 / 6.76	15.89 / 17.76
0.07	0.50 / 1.43	4.92 / 6.59	15.34 / 17.53
0.10	0.41 / 1.31	4.40 / 6.28	14.38 / 17.08
0.15	0.30 / 1.10	3.55 / 5.73	12.66 / 16.24

TABLE 3.5. The influence of h/ℓ on values of ω_I^2, ω_{II}^2 and ω_{III}^2 for $E_2/E_1 = 50$.

h/ℓ \ ε	0.00			0.10		
	Mode I	Mode II	Mode III	Mode I	Mode II	Mode III
0.10	0.62 / 1.58	5.62 / 6.98	16.57 / 17.98	0.41 / 1.31	4.40 / 6.28	14.38 / 17.08
0.15	1.06 / 1.99	7.36 / 8.22	19.50 / 20.43	0.77 / 1.86	6.46 / 7.89	18.52 / 20.47
0.20	1.40 / 2.21	8.34 / 8.94	21.03 / 21.73	1.08 / 2.19	7.81 / 8.86	20.97 / 22.47

TABLE 3.6. The influence of E_2/E_1 on of ω_I^2, ω_{II}^2 and ω_{III}^2 for $h/\ell = 0.1$.

E_2/E_1 \ ε	0.00			0.10		
	Mode I	Mode II	Mode III	Mode I	Mode II	Mode III
1	0.06 / 0.30	0.93 / 1.96	4.12 / 6.38	0.06 / 0.30	0.93 / 1.96	4.12 / 6.38
20	0.31 / 1.05	3.58 / 5.32	12.33 / 14.51	0.24 / 0.91	2.97 / 4.89	10.97 / 13.94
50	0.62 / 1.58	5.62 / 6.98	16.57 / 17.98	0.41 / 1.31	4.40 / 6.28	14.38 / 17.08
100	0.98 / 1.93	7.10 / 8.04	19.10 / 20.09	0.59 / 1.61	5.47 / 7.15	16.47 / 19.00

Tables 3.7-3.9 relate to a rectangular plate, either simply supported (upper number) or clamped (lower number) along the Ox_1 axis. Table 3.7 shows that as $\gamma = \ell_1/\ell_3$ increases, ω_r increases monotonically, as one would expect; as $\gamma \to 0$ the value of ω_r approaches the value for the strip. Tables show that the influence of E_2/E_1 and ε on ω_r is similar to that for the strip (compare Table 3.9 and Table 3.4)

TABLE 3.7. The influence of $\gamma = \ell_1/\ell_3$ on values of ω_I^2, ω_{II}^2 and ω_{III}^2 for $h/\ell = 0.1$, $E_2/E_1 = 50$.

γ \ ε	0.00			0.10		
	Mode I	Mode II	Mode III	Mode I	Mode II	Mode III
1.5	3.95 / 4.60	9.98 / 11.27	21.49 / 22.98	3.72 / 4.49	8.99 / 11.03	19.91 / 23.09
1.3	2.94 / 3.65	8.80 / 10.14	20.22 / 21.71	2.67 / 3.47	7.70 / 9.73	18.42 / 21.46
1.0	1.82 / 2.82	7.42 / 8.79	18.69 / 20.16	1.54 / 2.38	6.23 / 8.23	16.67 / 19.54
0.5	0.85 / 1.77	6.04 / 7.41	17.09 / 18.52	0.62 / 1.50	4.81 / 6.74	14.92 / 17.61
0.3	0.70 / 1.64	5.76 / 7.13	16.75 / 18.18	0.48 / 1.38	4.54 / 6.45	14.57 / 17.25
0.1	0.63 / 1.58	5.63 / 6.99	16.59 / 18.00	0.42 / 1.32	4.42 / 6.30	14.40 / 17.10

TABLE 3.8. The influence of E_2/E_1 on ω_I^2, ω_{II}^2 and ω_{III}^2 for $\gamma = 0.1$, $h/\ell = 0.1$.

E_2/E_1 \ ε	0.00			0.10		
	Mode I	Mode II	Mode III	Mode I	Mode II	Mode III
1	0.23 / 0.49	1.38 / 2.43	4.91 / 7.16	0.23 / 0.49	1.38 / 2.43	4.91 / 7.16
20	1.02 / 1.71	4.92 / 6.61	14.11 / 16.28	0.90 / 1.56	4.28 / 6.22	12.80 / 15.78
50	1.82 / 2.62	7.42 / 8.79	18.69 / 20.16	1.54 / 2.38	6.23 / 8.23	16.67 / 19.54
100	2.58 / 3.32	9.17 / 10.20	21.41 / 22.48	0.59 / 1.61	5.47 / 7.15	16.47 / 19.00

TABLE 3.9. The influence of ε on values of ω_{I}^{2}, ω_{II}^{2} and ω_{III}^{2} for $\gamma = 1.0$, $h/\ell = 0.1$, $E_2/E_1 = 50$.

ε	ω^2		
	Mode I	Mode II	Mode III
0.00	1.82 / 2.62	7.42 / 8.79	18.69 / 20.16
0.05	1.72 / 2.55	7.03 / 8.62	18.06 / 20.01
0.07	1.65 / 2.49	6.73 / 8.48	17.55 / 19.85
0.10	1.54 / 2.38	6.23 / 8.23	16.67 / 19.54
0.15	1.37 / 2.18	5.38 / 7.75	15.10 / 18.97

3.3.3. FORCED VIBRATION

We consider a square plate ($\gamma = 1$) rigidly supported along the Ox_1 axis. We suppose the curving is periodic, as given by (3.1.29) with ℓ replaced by ℓ_1. We take $\ell_1/\Lambda = 16$, $\delta = \pi/2$, $\omega << \omega_I$ and consider $\overline{\sigma}_{22}/p_1$ along the Ox_1 axis at the mid section $x_3/\ell_3 = 0.5$, $x_2/\ell_1 = h/2\ell_1$. We choose to investigate σ_{22} because it has an important role in the adhesion strength of the plate material.

Fig.3.3.4 shows the dependence of $\overline{\sigma}_{22}/p_1$ on x_1/ℓ_1 for various ω^2. The figure shows that $\overline{\sigma}_{22}/p_1$ increases monotonically with ω^2; the influences of the other parameters are similar to those in the static problem, as shown by Figs. 3.3.5 and 3.3.6.

Analogous results are obtained for local curving given by (3.1.32) with ℓ replaced by ℓ_1. Fig.3.3.7 shows $\overline{\sigma}_{22}/p_1$ for $n = 1$, $d = 1 - c$ in (3.1.23), anf for various m and ω^2; in all cases $\overline{\sigma}_{22}/p_1$ increases monotonically with ω.

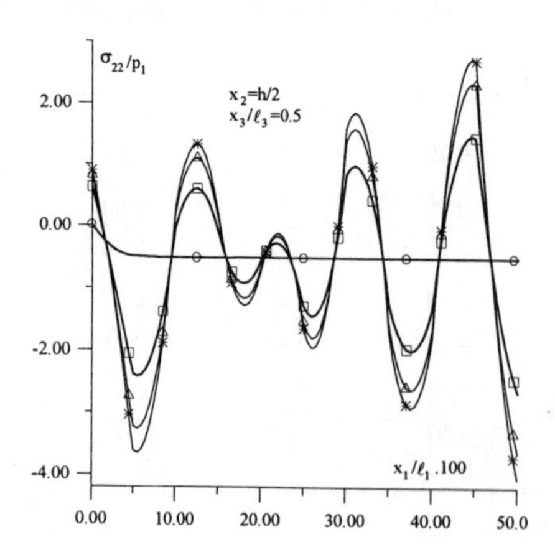

Fig.3.3.4. The relation between $\overline{\sigma}_{22}/p_1$ and x_1/ℓ_1 under $E_2/E_1 = 50$, $h/\ell_1 = 0.1$, $\varepsilon = 0.1$ with various ω^2: ○ -- $\varepsilon = 0.00$; □ -- $\omega^2 = 0.0$; △ -- $\omega^2 = 0.7$; ✳ -- $\omega^2 = 0.9$

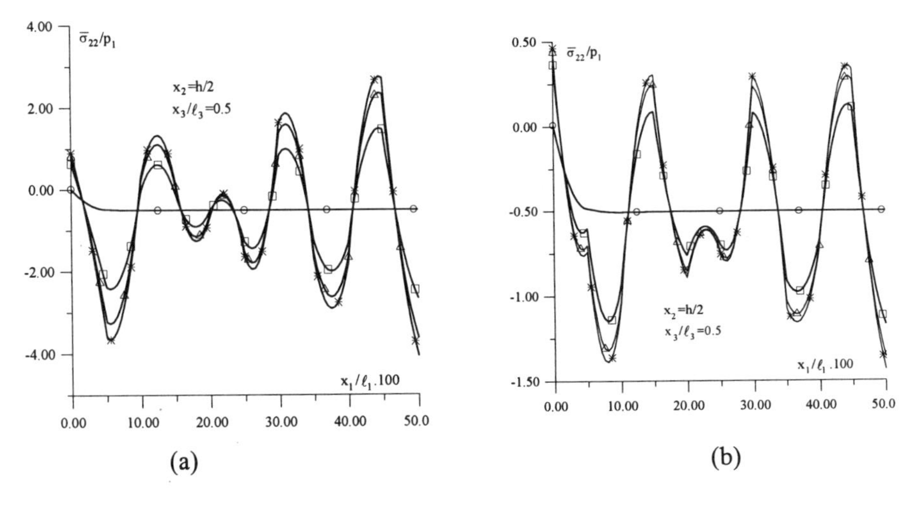

(a) (b)

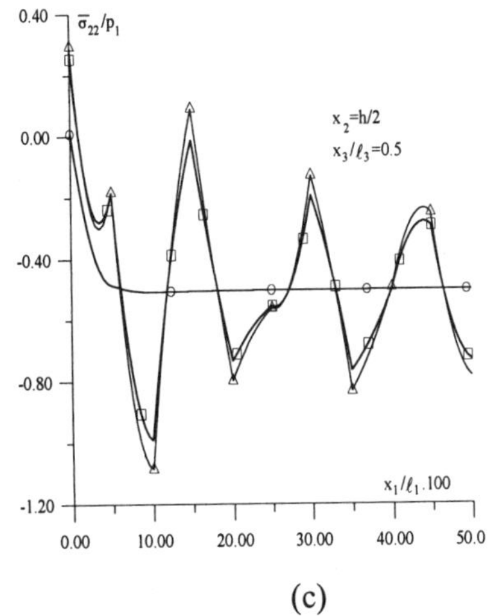

(c)

Fig.3.3.5. The influence of h/ℓ_1 ((a)-- h/ℓ_1 =0.1; (b)-- h/ℓ_1 =0.15; (c)-- h/ℓ_1 =0.2) on $\bar{\sigma}_{22}/p_1$ for $E_2/E_1 = 50$, $\varepsilon = 0.1$ and various ω^2 : ○ -- $\varepsilon = 0.00$; □ -- $\omega^2 = 0.0$; △ -- $\omega^2 = 0.7$; ✳ -- $\omega^2 = 0.9$.

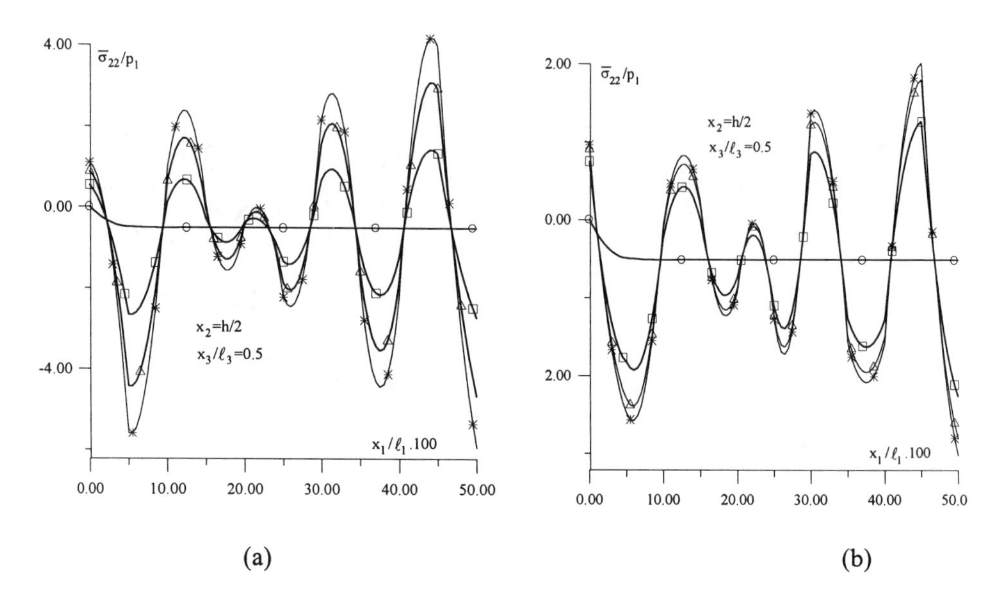

Fig.3.3.6. The influence of E_2/E_1 ((a)-- E_2/E_1 =20; (b)-- E_2/E_1 =100) on $\bar{\sigma}_{22}/p_1$ for $h/\ell_1 = 0.1$, $\varepsilon = 0.1$ and various ω^2 : ○ -- $\varepsilon = 0.00$; □ -- $\omega^2 = 0.0$; △ -- $\omega^2 = 0.7$; ✳ - $\omega^2 = 0.9$.

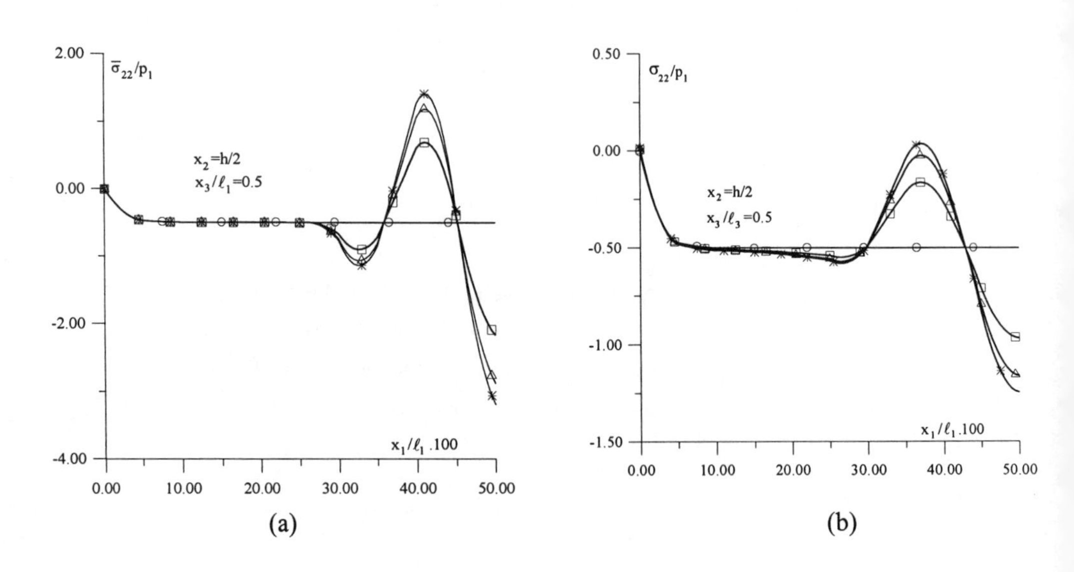

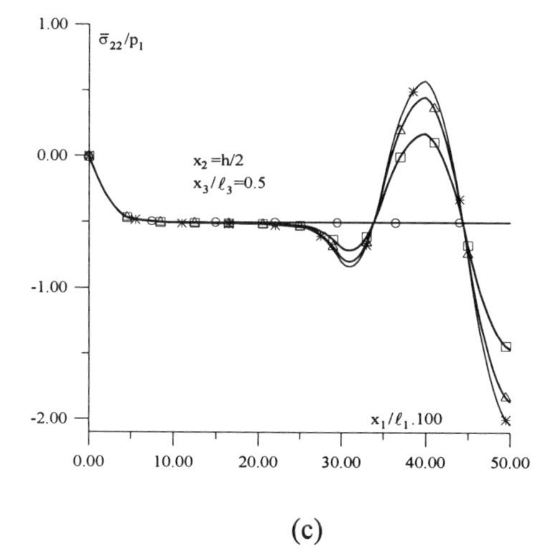

(c)

Fig.3.3.7. The influence of m ((a)-- m=1; (b)-- m=2; (c)-- m=3) on $\bar{\sigma}_{22}/p_1$ under $E_2/E_1 = 50$, c=0.3, $h/\ell_1 = 0.1$, $\varepsilon = 0.1$ with various ω^2 : ○ -- $\varepsilon = 0.00$; □ -- $\omega^2 = 0.0$; △ -- $\omega^2 = 0.7$; ✳ - $\omega^2 = 0.9$.

3.4. Bibliographical Notes

The results given in Section 3.1 and related to the stress distribution in the simply supported strip fabricated from periodic composite were obtained by S.D. Akbarov and N. Yahnioglu [50].

The numerical results related to the stress distribution in this strip under the boundary and edge conditions (3.1.7) and (3.1.8) were obtained by N.Yahnioglu [151]; the numerical investigations carried out for the stress distribution in the strip with the existence of a local curving in the structure, were also made by N. Yahnioglu [151].

The results given in Section 3.2 and related with the rectangular plate fabricated from periodically curved composite were obtained by N. Yahnioglu [152]; the investigations related to local curving, were made by S.D. Akbarov and N.Yahnioglu [51, 52].

The investigations presented in Section 3.3 were carried out by A.D. Zamanov [154].

CHAPTER 4

PLANE-STRAIN STATE IN PERIODICALLY CURVED COMPOSITES

It is evident that any continuum approach is an approximate one and the most accurate information on the stress distribution in the curved composites can be obtained only in the framework of the piecewise-homogeneous body model with the use of the exact equations of deformable solid body mechanics. Therefore, from now on, we begin study the problems of curved composites in the framework of the piecewise homogeneous body model. In this chapter we consider plane strain problems of periodically plane-curved composites. First, the method for investigation of the stress distribution in these composites is presented. Then this method is developed for cases in which the materials of the layers are: 1)viscoelastic; 2)rectilinearly anisotropic; 3) curvilinearly anisotropic. Many numerical results and their analyses are given for stress distributions in such composites.

4.1. Formulation

4.1.1. PLANE-STRAIN STATE

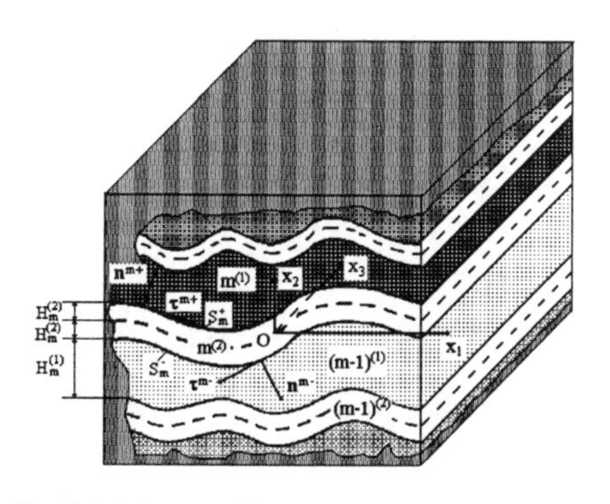

Fig. 4.1.1. A part of the composite material with plane curved structures at $x_1^{(k)}$ =const, $x_3^{(k)}$ =const.

We begin the general formulation of the problems on the plane-strain state in laminated composite materials with arbitrary plane-curved structures. This formulation will be made in the framework of the piecewise-homogeneous body model with the use of the exact equations of the theory of elasticity.

Thus, we consider an infinite elastic body, reinforced by an arbitrary number of non-intersecting curved filler layers. Values related to the matrix will be denoted by upper indices (1) ; those related to the filler by upper indices (2). The rectangular Cartesian system of coordinates $O_m^{(k)}x_{1m}^{(k)}x_{2m}^{(k)}x_{3m}^{(k)}$ which are obtained from the system of coordinates $Ox_1x_2x_3$ by parallel transfer along the Ox_2, will refer to each layer. We isolate from the composite the part at $x_{1m}^{(k)} = \text{const}$ and $x_{3m}^{(k)} = \text{const}$ shown in Fig.4.1.1. The reinforcing layers will be assumed to be located in planes $O_m^{(2)} x_{1m}^{(2)} x_{3m}^{(k)}$ and the thickness of each filler layer will be assumed constant. Furthermore, we assume that the reinforcing layers are curved only in the direction of the Ox_1 axis. It will also be assumed that the matrix and filler materials are homogeneous, isotropic and linearly elastic. We investigate the plain-strain state in this composite under the action at infinity of uniformly distributed forces, in the direction of the Ox_1 axis. We write the equilibrium equations, Hooke's law, and Cauchy relations within each layer

$$\frac{\partial \sigma_{ij}^{(k)\underline{m}}}{\partial x_{j\underline{m}}^{(\underline{k})}} = 0, \quad \sigma_{ij}^{(\underline{k})m} = \lambda^{(\underline{k})\underline{m}} \theta^{(\underline{k})\underline{m}} \delta_i^j + 2\mu^{(\underline{k})\underline{m}} \varepsilon_{ij}^{(\underline{k})\underline{m}}, \quad i, j, k=1,2;\ m=1,2,3,\ldots$$

$$\varepsilon_{ij}^{(\underline{k})\underline{m}} = \frac{1}{2}\left(\frac{\partial u_i^{(\underline{k})\underline{m}}}{\partial x_{j\underline{m}}^{(\underline{k})}} + \frac{\partial u_j^{(\underline{k})\underline{m}}}{\partial x_{i\underline{m}}^{(\underline{k})}} \right), \quad \theta^{(\underline{k})\underline{m}} = \frac{\partial u_i^{(\underline{k})\underline{m}}}{\partial x_{i\underline{m}}^{(\underline{k})}} . \tag{4.1.1}$$

In (4.1.1) the well-known notation is used.

We assume that the complete cohesion conditions are satisfied on the interface of the layers . Denoting the upper and lower surfaces of the $m^{(2)}$th filler layer by S_m^+ and S_m^-, respectively, and introducing the notation $m_1 = m-1$ we may write these conditions as follows:

$$\sigma_{ij}^{(1)\underline{m}}\Big|_{S_{\underline{m}}^+} n_j^{\underline{m},+} = \sigma_{ij}^{(2)\underline{m}}\Big|_{S_{\underline{m}}^+} n_j^{\underline{m},+}, \quad \sigma_{ij}^{(1)m_1}\Big|_{S_{\underline{m}}^-} n_j^{\underline{m},-} = \sigma_{ij}^{(2)\underline{m}}\Big|_{S_{\underline{m}}^-} n_j^{\underline{m},-},$$

$$u_i^{(1)\underline{m}}\Big|_{S_{\underline{m}}^+} = u_i^{(2)\underline{m}}\Big|_{S_{\underline{m}}^+}, \quad u_i^{(1)m_1}\Big|_{S_{\underline{m}}^-} = u_i^{(2)\underline{m}}\Big|_{S_{\underline{m}}^-}, \tag{4.1.2}$$

where $n_j^{m,\pm}$ are the components of the unit normal vector to the surface $S_m^{\pm}$.

We suppose that the equation of the middle surface of the $m^{(2)}$th filler layer is given in the form

$$x_{2\underline{m}}^{(2)} = F_{\underline{m}}(x_{1\underline{m}}^{(2)}) = \varepsilon f_{\underline{m}}(x_{1\underline{m}}^{2}) . \tag{4.1.3}$$

Here ε is a dimensionless small parameter and $0 \le \varepsilon < 1$. The geometrical meaning of this parameter will become apparent in a specific form of function (4.1.3).

This gives the general formulation of the problems. Now we consider some preparatory procedures for the exposition of the solution method.

4.1.2. THE PARAMETERISATION

First, using the condition of uniform filler layer thickness and the equations (4.1.3) we attempt to obtain the equation for the interface surfaces $S_m^{\pm}$. For this purpose we introduce the parameter t_{1m}, where $t_{1m} \in (-\infty, +\infty)$ and rewrite the equation (4.1.3) in the following form

$$x_{1m}^{(2)} = t_{1m}, \quad x_{2\underline{m}}^{(2)} = F_{\underline{m}}(x_{1\underline{m}}^{(2)}) = \varepsilon f_{\underline{m}}(x_{1\underline{m}}^{(2)}). \tag{4.1.4}$$

We assume that the functions $F_{\underline{m}}(t_{1\underline{m}})$ and their first order derivatives are continuous and from (4.1.4) we obtain the expression

$$\frac{dx_{2\underline{m}}^{(2)}}{dx_{1\underline{m}}^{(2)}} = \frac{dF_{\underline{m}}(t_{1\underline{m}})}{dt_{1\underline{m}}} \tag{4.1.5}$$

for the slope of the tangent line to the surface (4.1.4) at $x_{3m}^{(2)} = \text{const}$. Taking (4.1.5) into account, we can write the equation of the line passing through the point $(t_{1\underline{m}}, F_{\underline{m}}(t_{1\underline{m}}))$ and perpendicular to this tangent line as follows:

$$x_{1\underline{m}}^{(2)} - t_{1\underline{m}} = -\frac{dF_{\underline{m}}(t_{1\underline{m}})}{dt_{1\underline{m}}}(x_{2\underline{m}}^{(2)} - F_{\underline{m}}(t_{1\underline{m}})). \tag{4.1.6}$$

As the surfaces $S_m^{\pm}$ and the line (4.1.6) have common points, the coordinates the intersections with these surfaces must satisfy equation (4.1.6). Moreover, taking into account the uniformity of the filler layer thickness we obtain the following system of equation for the coordinates of the points of the surfaces $S_m^{\pm}$.

$$x_{1\underline{m}}^{(2)\pm} - t_{1\underline{m}} = -\frac{dF_{\underline{m}}(t_{1\underline{m}})}{dt_{1\underline{m}}}(x_{2\underline{m}}^{(2)\pm} - F_{\underline{m}}(t_{1\underline{m}})),$$

$$\left(x_{1\underline{m}}^{(2)\pm} - t_{1\underline{m}}\right)^2 + \left(x_{2\underline{m}}^{(2)\pm} - F_{\underline{m}}(t_{1\underline{m}})\right)^2 = \left(H_{\underline{m}}^{(2)}\right)^2. \tag{4.1.7}$$

Here $x_{1m}^{(2)\pm}, x_{2m}^{(2)\pm}$ are the coordinates of the points of the surfaces $S_m^{\pm}$; $H_m^{(2)}$ is the half-thickness of the $m^{(2)}$ th filler layer.

Thus, from (4.1.7) we have the following equation for the surfaces $S_m^{\pm}$.

$$x_{1\underline{m}}^{(2)\pm} = t_{1\underline{m}} \mp H_{\underline{m}}^{(2)} \frac{dF_{\underline{m}}(t_{1\underline{m}})}{dt_{1\underline{m}}} T(t_{1\underline{m}}), \quad x_{2\underline{m}}^{(2)\pm} = F_{\underline{m}}(t_{1\underline{m}}) \pm H_{\underline{m}}^{(2)} T(t_{1\underline{m}}), \qquad (4.1.8)$$

where

$$T(t_{1\underline{m}}) = \left[1 + \left(\frac{dF_{\underline{m}}(t_{1\underline{m}})}{dt_{1\underline{m}}}\right)^2\right]^{-\frac{1}{2}}. \qquad (4.1.9)$$

Using the equations (4.1.8) and (4.1.9) and doing some calculations we obtain the following expressions for the components of the unit normal vector of the surfaces $S_m^{\pm}$.

$$n_1^{\underline{m},\pm} = -\frac{dx_{2\underline{m}}^{(2)\pm}}{dt_{1\underline{m}}} V^{\pm}(t_{1\underline{m}}), \quad n_2^{\underline{m},\pm} = \frac{dx_{1\underline{m}}^{(2)\pm}}{dt_{1\underline{m}}} V^{\pm}(t_{1\underline{m}}), \qquad (4.1.10)$$

where

$$V^{\pm}(t_{1\underline{m}}) = \left[\left(\frac{dx_{1\underline{m}}^{(2)\pm}}{dt_{1\underline{m}}}\right)^2 + \left(\frac{dx_{2\underline{m}}^{(2)\pm}}{dt_{1\underline{m}}}\right)^2\right]^{-\frac{1}{2}}. \qquad (4.1.11)$$

In all further investigations we will assume that the condition

$$\left|\varepsilon \frac{df(t_{1\underline{m}})}{dt_{1\underline{m}}}\right| < 1 \qquad (4.1.12)$$

is satisfied. Taking (4.1.12), (4.1.9) and (4.1.11) into account we represent the expressions (4.1.8), (4.1.10) as power series form in the small parameter ε as follows:

$$x_{1\underline{m}}^{(2)\pm} = t_{1\underline{m}} \mp \varepsilon f_{\underline{m}}'(t_{1\underline{m}}) H_{\underline{m}}^{(2)} \pm \varepsilon^3 \frac{1}{2}(f_{\underline{m}}'(t_{1\underline{m}}))^3 \mp \varepsilon^5 \frac{3}{8}\left(f_{\underline{m}}'(t_{1\underline{m}})\right)^5 H_{\underline{m}}^{(2)} + \ldots =$$

$$t_{1\underline{m}} + \sum_{k=1}^{\infty} (-1)^k \varepsilon^{2k-1} a_{1k}\left(f'_{\underline{m}}(t_{1\underline{m}}), \pm H_{\underline{m}}^{(2)}\right),$$

$$x_{2\underline{m}}^{(2)\pm} = \pm H_{\underline{m}}^{(2)} + \varepsilon f_{\underline{m}}(t_{1\underline{m}}) \mp \varepsilon^2 \frac{1}{2} H_{\underline{m}}^{(2)} \left(f'_{\underline{m}}(t_{1\underline{m}})\right)^2 \pm \varepsilon^4 \frac{3}{8} H_{\underline{m}}^{(2)} \left(f'_{\underline{m}}(t_{1\underline{m}})\right)^4 + \ldots =$$

$$\pm H_{\underline{m}}^{(2)} + \varepsilon f_{\underline{m}}(t_{1\underline{m}}) + \sum_{k=1}^{\infty} (-1)^k \varepsilon^{2k} a_{2k}\left(f'_{\underline{m}}(t_{1\underline{m}}), \pm H_{\underline{m}}^{(2)}\right),$$

$$n_1^{\underline{m},+} = n_1^{\underline{m},-} = -\varepsilon f'_{\underline{m}}(t_{1\underline{m}}) + \varepsilon^3 \frac{1}{2}\left(f'_{\underline{m}}(t_{1\underline{m}})\right)^3 - \varepsilon^5 \frac{3}{8}\left(f'_{\underline{m}}(t_{1\underline{m}})\right)^5 + \ldots =$$

$$\sum_{k=1}^{\infty} \varepsilon^{2k-1} b_{1k}\left(f'_{\underline{m}}(t_{1\underline{m}})\right),$$

$$n_2^{\underline{m},+} = n_2^{\underline{m},-} = 1 - \varepsilon^2 \frac{1}{2}\left(f'_{\underline{m}}(t_{1\underline{m}})\right)^2 + \varepsilon^4 \frac{3}{8}\left(f'_{\underline{m}}(t_{1\underline{m}})\right)^4 + \ldots = 1 + \sum_{k=1}^{\infty} \varepsilon^{2k} b_{2k}\left(f'_{\underline{m}}(t_{1\underline{m}})\right), \tag{4.1.13}$$

where

$$f'_{\underline{m}}(t_{1\underline{m}}) = \frac{df_{\underline{m}}(t_{1\underline{m}})}{dt_{1\underline{m}}}. \tag{4.1.14}$$

4.1.3. THE PRESENTATION OF THE STRESS-STRAIN VALUES IN SERIES FORM

It is difficult or impossible to obtain the exact solutions of the problems formulated in subsection 4.1.1 for composite materials with curved structures, because the contact conditions are given on the curved surfaces which do not coincide with the coordinate surfaces. To solve these problems we must use series expansions in the small parameter ε; the condition (4.1.12) limits the degree of the curving of the reinforcing layers in the structure of the composite materials.

We present the quantities characterizing the stress-strain state of the $m^{(k)}$th layer in the form of series in the parameter ε as follows:

$$\left\{\sigma_{ij}^{(k)\underline{m}}; \varepsilon_{ij}^{(k)\underline{m}}; u_i^{(k)\underline{m}}\right\}\left(x_{1\underline{m}}^{(k)}, x_{2\underline{m}}^{(k)}\right) = \sum_{q=0}^{\infty} \varepsilon^q \left\{\sigma_{ij}^{(k)\underline{m},q}; \varepsilon_{ij}^{(k)\underline{m},q}; u_i^{(k)\underline{m},q}\right\}\left(x_{1\underline{m}}^{(k)}, x_{2\underline{m}}^{(k)}\right). \tag{4.1.15}$$

Because of linearity, the equations (4.1.1) will be satisfied for each approximation separately. In other words, substituting (4.1.15) in (4.1.1) and grouping by equal powers of ε we obtain

$$\frac{\partial \sigma_{ij}^{(\underline{k})\underline{m},q}}{\partial x_{j\underline{m}}^{(\underline{k})}} = 0, \quad \sigma_{ij}^{(\underline{k})m,q} = \lambda^{(\underline{k})\underline{m}} \theta^{(\underline{k})\underline{m},q} \delta_i^j + 2\mu^{(\underline{k})\underline{m}} \varepsilon_{ij}^{(\underline{k})\underline{m},q}, \quad i, j, k=1,2; m=1,2,3,\ldots$$

$$\varepsilon_{ij}^{(\underline{k})\underline{m},q} = \frac{1}{2}\left(\frac{\partial u_i^{(\underline{k})\underline{m},q}}{\partial x_{j\underline{m}}^{(\underline{k})}} + \frac{\partial u_j^{(\underline{k})\underline{m},q}}{\partial x_{i\underline{m}}^{(\underline{k})}} \right), \quad \theta^{(\underline{k})\underline{m},q} = \frac{\partial u_i^{(\underline{k})\underline{m},q}}{\partial x_{i\underline{m}}^{(\underline{k})}}. \tag{4.1.16}$$

Now we consider obtaining the contact conditions for the q-th approximation.

4.1.4. CONTACT CONDITIONS FOR EACH APPROXIMATION (4.1.15)

We rewrite the equation of the surfaces S_m^+ and S_m^- in the system of coordinates $O_m^{(1)} x_{1m}^{(1)} x_{2m}^{(1)} x_{3m}^{(1)}$ and $O_{m_1}^{(1)} x_{1m_1}^{(1)} x_{2m_1}^{(1)} x_{3m_1}^{(1)}$, respectively. For this purpose we substitute the relations $x_{1m}^{(1)} = x_{1m_1}^{(1)} = x_{1m}^{(2)}$, $x_{2m}^{(2)} = x_{2m}^{(1)} - H_m^{(1)} - H_m^{(2)}$, $x_{2m}^{(2)} = x_{2m_1}^{(1)} + H_{m_1}^{(1)} + H_m^{(2)}$ in equations (4.1.13) and obtain:

the equation of the surface S_m^+ in the system of coordinates $O_m^{(1)} x_{1m}^{(1)} x_{2m}^{(1)} x_{3m}^{(1)}$:

$$x_{1\underline{m}}^{(1)+} = t_{1\underline{m}} + \sum_{k=1}^{\infty} (-1)^k \varepsilon^{2k-1} a_{1k}\left(f'_{\underline{m}}(t_{1\underline{m}}), H_{\underline{m}}^{(2)}\right),$$

$$x_{2\underline{m}}^{(1)+} = -H_{\underline{m}}^{(1)} + \varepsilon f_{\underline{m}}(t_{1\underline{m}}) + \sum_{k=1}^{\infty} (-1)^k \varepsilon^{2k} a_{2k}\left(f'_{\underline{m}}(t_{1\underline{m}}), H_{\underline{m}}^{(2)}\right); \tag{4.1.17}$$

the equation of the surface S_m^- in the system of coordinates $O_{m_1}^{(1)} x_{1m_1}^{(1)} x_{2m_1}^{(1)} x_{3m_1}^{(1)}$:

$$x_{1m_1}^{(1)-} = t_{1\underline{m}} + \sum_{k=1}^{\infty} (-1)^k \varepsilon^{2k-1} a_{1k}\left(f'_{\underline{m}}(t_{1\underline{m}}), -H_{\underline{m}}^{(2)}\right),$$

$$x_{2m_1}^{(1)-} = H_{m_1}^{(1)} + \varepsilon f_m(t_{1\underline{m}}) + \sum_{k=1}^{\infty}(-1)^k \varepsilon^{2k} a_{2k}\left(f_{\underline{m}}^{'}(t_{1\underline{m}}), -H_{\underline{m}}^{(2)}\right). \tag{4.1.18}$$

The expressions for the components $n_i^{m,+}$ and $n_i^{m,-}$ in the system of coordinates $O_m^{(1)}x_{1m}^{(1)}x_{2m}^{(1)}x_{3m}^{(1)}$ and $O_{m_1}^{(1)}x_{1m_1}^{(1)}x_{2m_1}^{(1)}x_{3m_1}^{(1)}$, respectively, remain as in (4.1.13). Thus, taking into account the equations (4.1.13), (4.1.17) and (4.1.18) we rewrite the contact conditions (4.1.2) in the following form.

$$\sigma_{ij}^{(1)\underline{m}}\left(x_{1\underline{m}}^{(1)+}, x_{2\underline{m}}^{(1)+}\right) n_j^{\underline{m},+} = \sigma_{ij}^{(2)\underline{m}}\left(x_{1\underline{m}}^{(2)+}, x_{2\underline{m}}^{(2)+}\right) n_j^{\underline{m},+},$$

$$\sigma_{ij}^{(1)\underline{m}_1}\left(x_{1\underline{m}_1}^{(1)-}, x_{2\underline{m}_1}^{(1)-}\right) n_j^{m,+} = \sigma_{ij}^{(2)\underline{m}}\left(x_{1\underline{m}}^{(2)-}, x_{2\underline{m}}^{(2)-}\right) n_j^{\underline{m},-},$$

$$u_i^{(1)\underline{m}}\left(x_{1\underline{m}}^{(1)+}, x_{2\underline{m}}^{(1)+}\right) = u_i^{(2)\underline{m}}\left(x_{1\underline{m}}^{(2)+}, x_{2\underline{m}}^{(2)+}\right),$$

$$u_i^{(1)\underline{m}_1}\left(x_{1\underline{m}_1}^{(1)-}, x_{2\underline{m}_1}^{(1)-}\right) = u_i^{(2)\underline{m}}\left(x_{1\underline{m}}^{(2)-}, x_{2\underline{m}}^{(2)-}\right). \tag{4.1.19}$$

Substituting (4.1.15) in (4.1.19) we obtain

$$\sum_{q=1}^{\infty}\varepsilon^q \sigma_{ij}^{(1)\underline{m},q}\left(x_{1\underline{m}}^{(1)+}, x_{2\underline{m}}^{(1)+}\right) n_j^{\underline{m},+} = \sum_{q=1}^{\infty}\varepsilon^q \sigma_{ij}^{(2)\underline{m},q}\left(x_{1\underline{m}}^{(2)+}, x_{2\underline{m}}^{(2)+}\right) n_j^{\underline{m},+},$$

$$\sum_{q=1}^{\infty}\varepsilon^q \sigma_{ij}^{(1)\underline{m}_1,q}\left(x_{1\underline{m}_1}^{(1)-}, x_{2\underline{m}_1}^{(1)-}\right) n_j^{m,+} = \sum_{q=1}^{\infty}\varepsilon^q \sigma_{ij}^{(2)\underline{m},q}\left(x_{1\underline{m}}^{(2)-}, x_{2\underline{m}}^{(2)-}\right) n_j^{\underline{m},-},$$

$$\sum_{q=1}^{\infty}\varepsilon^q u_i^{(1)\underline{m},q}\left(x_{1\underline{m}}^{(1)+}, x_{2\underline{m}}^{(1)+}\right) = \sum_{q=1}^{\infty}\varepsilon^q u_i^{(2)\underline{m},q}\left(x_{1\underline{m}}^{(2)+}, x_{2\underline{m}}^{(2)+}\right),$$

$$\sum_{q=1}^{\infty}\varepsilon^q u_i^{(1)\underline{m}_1,q}\left(x_{1\underline{m}_1}^{(1)-}, x_{2\underline{m}_1}^{(1)-}\right) = \sum_{q=1}^{\infty}\varepsilon^q u_i^{(2)\underline{m},q}\left(x_{1\underline{m}}^{(2)-}, x_{2\underline{m}}^{(2)-}\right). \tag{4.1.20}$$

Expanding the values $\sigma_{ij}^{(1)\underline{m},q}\left(x_{1\underline{m}}^{(1)+}, x_{2\underline{m}}^{(1)+}\right)$, $u_i^{(1)\underline{m},q}\left(x_{1\underline{m}}^{(1)+}, x_{2\underline{m}}^{(1)+}\right)$ in a Taylor series in the vicinity of $\left(t_{1\underline{m}}, -H_{\underline{m}}^{(1)}\right)$, the values $\sigma_{ij}^{(2)\underline{m},q}\left(x_{1\underline{m}}^{(2)\pm}, x_{2\underline{m}}^{(2)\pm}\right)$, $u_i^{(2)\underline{m},q}\left(x_{1\underline{m}}^{(2)\pm}, x_{2\underline{m}}^{(2)\pm}\right)$ in the vicinity $\left(t_{1\underline{m}}, \pm H_{\underline{m}}^{(2)}\right)$, and the values $\sigma_{ij}^{(1)\underline{m}_1,q}\left(x_{1\underline{m}_1}^{(1)-}, x_{2\underline{m}_1}^{(1)-}\right)$, $u_i^{(1)\underline{m}_1,q}\left(x_{1\underline{m}_1}^{(1)-}, x_{2\underline{m}_1}^{(1)-}\right)$

in the vicinity $\left(t_{1\underline{m}},+H^{(1)}_{\underline{m}_1}\right)$, taking into account the expressions of $n_j^{m,\pm}$ given in (4.1.13) and grouping in (4.1.20) by equal power of ε we obtain the following representations.

$$\left\{\sigma_{ij}^{(2)\underline{m}};u_i^{(2)\underline{m}}\right\}\left(x_{1\underline{m}}^{(2)\pm},x_{2\underline{m}}^{(2)\pm}\right)=\sum_{n=0}^{\infty}\varepsilon^n\left\{P_{ij}^{(2)\underline{m},n};U_i^{(2)\underline{m},n}\right\}\left(t_{1\underline{m}},\pm H_{\underline{m}}^{(2)}\right),$$

$$\left\{\sigma_{ij}^{(1)\underline{m}};u_i^{(1)\underline{m}}\right\}\left(x_{1\underline{m}}^{(1)+},x_{2\underline{m}}^{(1)+}\right)=\sum_{n=0}^{\infty}\varepsilon^n\left\{P_{ij}^{(1)\underline{m},n};U_i^{(1)\underline{m},n}\right\}\left(t_{1\underline{m}},-H_{\underline{m}}^{(1)}\right),$$

$$\left\{\sigma_{ij}^{(1)\underline{m}_1};u_i^{(1)\underline{m}_1}\right\}\left(x_{1\underline{m}_1}^{(1)-},x_{2\underline{m}_1}^{(1)-}\right)=\sum_{n=0}^{\infty}\varepsilon^n\left\{P_{ij}^{(1)\underline{m}_1,n};U_i^{(1)\underline{m}_1,n}\right\}\left(t_{1\underline{m}_1},+H_{\underline{m}_1}^{(1)}\right). \quad (4.1.21)$$

We write the formulae for the expressions of the $P_{ij}^{(k)m,q}$ and $U_i^{(k)m,q}$ at $\left(t_{1m},\pm H_m^{(k)}\right)$ in the case where the zeroth approximation of (4.1.15) corresponds to the homogeneous stress state:

the formulae for $P_{ij}^{(k)m,q}$:

$$P_{ij}^{(k)m,0}=\sigma_{ij}^{(k)m,0},\quad P_{ij}^{(k)m,1}=\sigma_{ij}^{(k)m,1},$$

$$P_{ij}^{(k)m,q}=\sigma_{ij}^{(k)m,q}+\sum_{s=1}^{q-1}L_s\sigma_{ij}^{(k)m,q-s}\quad \text{for } q\geq 2; \qquad (4.1.22)$$

the formulae for $U_i^{(k)m,q}$:

$$U_i^{(k)m,0}=u_i^{(k)m,0},\quad U_i^{(k)m,1}=u_i^{(k)m,1}+L_1u_i^{(k)m,0},$$

$$U_i^{(\underline{k})\underline{m},2}=u_i^{(\underline{k})\underline{m},2}+L_1u_i^{(\underline{k})\underline{m},1}\mp\frac{H_{\underline{m}}^{(2)}}{2}\left(\frac{df_{\underline{m}}}{dx_{1\underline{m}}^{(\underline{k})}}\right)^2\frac{\partial u_i^{(\underline{k})\underline{m},0}}{\partial x_{1\underline{m}}^{(\underline{k})}},$$

$$U_i^{(k)m,q}=u_i^{(k)m,q}+\sum_{s=1}^{q-1}L_su_i^{(k)m,q-s}\quad \text{for } q\geq 3. \qquad (4.1.23)$$

In (4.1.22) and (4.1.23) L_s are linear differential operators. Here we present the expressions of these operators for the first five approximations. Note that $H_m^{(2)}$, $x_{1m}^{(k)}$,

$x_{2m}^{(k)}$, f_m are denoted as H, x_1, x_2, f, respectively, in these expressions.

$$L_1 = \pm Hf'\frac{\partial}{\partial x_1} + f\frac{\partial}{\partial x_2};$$

$$L_2 = \mp\frac{H}{2}(f')^2\frac{\partial}{\partial x_2} + \frac{H^2}{2}(f')^2\frac{\partial^2}{\partial x_1^2} \mp Hff'\frac{\partial^2}{\partial x_1 \partial x_2} + \frac{1}{2}(f)^2\frac{\partial^2}{\partial x_2^2};$$

$$L_3 = \frac{1}{2}H^2(f')^3\frac{\partial^2}{\partial x_1\partial x_2} + \frac{1}{2}H^2 f(f')^2\frac{\partial^3}{\partial x_1^2\partial x_2} \mp \frac{1}{2}Hf^2 f'\frac{\partial^3}{\partial x_1\partial x_2^2} + \frac{1}{6}f^3\frac{\partial^3}{\partial x_2^3} \pm$$

$$\frac{1}{2}H(f')^3\frac{\partial}{\partial x_1} \mp \frac{1}{2}Hf(f')^2\frac{\partial^2}{\partial x_2^2} \mp \frac{1}{6}H^3(f')^3\frac{\partial^3}{\partial x_1^3};$$

$$L_4 = L_{41} + L_{42} + L_{43} + L_{44}, \tag{4.1.24}$$

where

$$L_{41} = \pm\frac{3}{8}H(f')^4\frac{\partial}{\partial x_2}; \quad L_{42} = -\frac{H^2}{2}(f')^4\frac{\partial^2}{\partial x_1^2} \pm \frac{H}{2}f(f')^3\frac{\partial^2}{\partial x_1\partial x_2} + \frac{H^2}{8}(f')^4\frac{\partial^2}{\partial x_2^2};$$

$$L_{43} = \mp\frac{H^3}{4}(f')^4\frac{\partial^3}{\partial x_1^2\partial x_2} + \frac{H^2}{2}f(f')^3\frac{\partial^3}{\partial x_1\partial x_2^2} \mp \frac{H}{4}f^2(f')^2\frac{\partial^3}{\partial x_2^3};$$

$$L_{44} = \mp\frac{H}{6}f^3 f'\frac{\partial^4}{\partial x_1\partial x_2^3} + \frac{f^4}{24}\frac{\partial^4}{\partial x_2^4} + \frac{H^4}{24}(f')^4\frac{\partial^4}{\partial x_1^4} + \frac{H^2}{4}(ff')^2\frac{\partial^4}{\partial x_1^2\partial x_2^2} \mp$$

$$\frac{H^3}{6}f(f')^3\frac{\partial^4}{\partial x_1^3\partial x_2} \mp \frac{H}{6}f^3 f'\frac{\partial^4}{\partial x_1\partial x_2^3}; \quad f' = \frac{df}{dx_1}. \tag{4.1.25}$$

It should be noted that the formulae for calculating $P_{ij}^{(1)m,q}$, $U_i^{(1)m,q}$ at $\left(t_{1m}, -H_m^{(1)}\right)$ and $P_{ij}^{(1)m_1,q}$, $U_i^{(1)m_1,q}$ at $\left(t_{1m}, H_{m_1}^{(1)}\right)$ are obtained from the formulae (4.1.22)-(4.1.25) by replacing $\pm H_m^{(2)}$, $\mp H_m^{(2)}$, $\pm H_m^{(k)}$ with $-H_m^{(2)}$, $H_m^{(2)}$, $-H_m^{(1)}$ and with $H_m^{(2)}$, $-H_m^{(2)}$, $H_{m_1}^{(1)}$, respectively.

Thus, taking (4.1.13), (4.1.21)-(4.1.23) into account from the (4.1.20) we obtain the following contact conditions for the q-th approximation of the series presentation (4.1.15).

$$\left.\left(-\frac{df_{\underline{m}}}{dx^{(2)}_{1\underline{m}}}P^{(1)\underline{m},q-1}_{i1}+P^{(1)\underline{m},q}_{i2}\right)\right|_{\left(t_{1\underline{m}},-H^{(1)}_{\underline{m}}\right)}=\left.\left(-\frac{df_{\underline{m}}}{dx^{(2)}_{1\underline{m}}}P^{(2)\underline{m},q-1}_{i1}+P^{(2)\underline{m},q}_{i2}\right)\right|_{\left(t_{1\underline{m}},H^{(2)}_{\underline{m}}\right)},$$

$$\left.\left(-\frac{df_{\underline{m}}}{dx^{(2)}_{1\underline{m}}}P^{(2)\underline{m},q-1}_{i1}+P^{(2)\underline{m},q}_{i2}\right)\right|_{\left(t_{1\underline{m}},-H^{(2)}_{\underline{m}}\right)}=\left.\left(-\frac{df_{\underline{m}}}{dx^{(2)}_{1\underline{m}}}P^{(1)\underline{m}_1,q-1}_{i1}+P^{(2)\underline{m}_1,q}_{i2}\right)\right|_{\left(t_{1\underline{m}},H^{(1)}_{\underline{m}_1}\right)},$$

$$U^{(1)\underline{m},q}_{i}\Big|_{\left(t_{1\underline{m}},-H^{(1)}_{\underline{m}}\right)}=U^{(2)\underline{m},q}_{i}\Big|_{\left(t_{1\underline{m}},H^{(2)}_{\underline{m}}\right)},\quad U^{(2)\underline{m},q}_{i}\Big|_{\left(t_{1\underline{m}},-H^{(2)}_{\underline{m}}\right)}=U^{(1)\underline{m}_1,q}_{i}\Big|_{\left(t_{1\underline{m}},H^{(1)}_{\underline{m}_1}\right)}. \tag{4.1.26}$$

It follows from (4.1.26) that the contact conditions related to the q-th approximation involve all previous approximations. Moreover, it follows from the (4.1.26) that these contact conditions are independent of the character of the mechanical relations of the components of the composite materials.

Hence we have a closed system of equations and corresponding contact conditions for each approximation for plane-strain in the composite material with arbitrary curving of the reinforcing layers of the material. In the next sections we will consider the solution of these contact problems for particular combinations of periodical curving layers in the structure of the composite material; we will then investigate the stress distribution in the material.

4.2. Method of Solution

In this section, with concrete problems as examples, we consider the determination of the values of each approximation (4.1.15). As concrete problems we select plane-strain problems in the composite material with 1) *co-phase* ; 2) *anti-phase* periodically curved layers. We investigate these problems separately.

4.2.1. CO-PHASE PERIODICALLY CURVED COMPOSITE

We investigate the plane-strain state in a laminated composite material with an infinite number of co-phase plane-curved layers alternating in the direction of the Ox_2 axis. We suppose that these layers are periodically curved in the direction of the Ox_1 axis, and the plane-strain takes place in the Ox_1x_2 plane under the action, at infinity, of uniformly distributed normal forces of intensity p in the direction of the Ox_1 axis.

The structure of this material is shown schematically in Fig.4.2.1. The periodicity of the composite structure in the direction of the Ox_2 axis has period $2\left(H^{(2)}+H^{(1)}\right)$, where $2H^{(1)}$ is a thickness of the matrix layer and $2H^{(2)}$ is a thickness of the filler layer. Among the layers we single out two of them, i.e. $1^{(1)}, 1^{(2)}$ (Fig.4.2.1), and discuss them below. The equation (4.1.3) for the middle surface of the $1^{(2)}$-th layer we take as follows:

$$x_2^{(2)} = F\left(x_1^{(2)}\right) = \varepsilon f\left(x_1^{(2)}\right) = L\sin\left(\frac{2\pi x_1^{(2)}}{\ell}\right) = \varepsilon\ell\sin\left(\frac{2\pi x_1^{(2)}}{\ell}\right), \tag{4.2.1}$$

where L is the curving rise, ℓ is the wavelength of the form of the middle surface curving. With the assumption $L < \ell$, the value of the small parameter is selected as

$$\varepsilon = \frac{L}{\ell}. \tag{4.2.2}$$

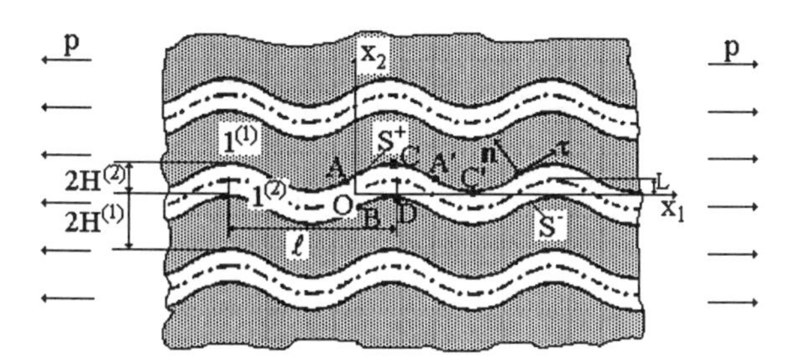

Fig.4.2.1. The structure of the composite material with cophasically periodically plane-curved layers

We will omit the index m which enter the expression given in the previous section. Taking into account the periodicity of the material structure in the direction of the Ox_2 axis we can rewrite the contact relations (4.1.26) for this problem as follows:

$$\left.\left(-\frac{df}{dx_1^{(2)}}P_{i1}^{(1),q-1} + P_{i2}^{(1),q}\right)\right|_{\left(t_1,\pm H^{(1)}\right)} = \left.\left(-\frac{df}{dx_1^{(2)}}P_{i1}^{(2),q-1} + P_{i2}^{(2),q}\right)\right|_{\left(t_1,\mp H^{(2)}\right)}$$

$$\left.U_i^{(1),q}\right|_{\left(t_1,\pm H^{(1)}\right)} = \left.U_i^{(2),q}\right|_{\left(t_1,\mp H^{(2)}\right)}. \tag{4.2.3}$$

We consider the determination of each approximation separately.

The zeroth approximation. The values of this approximation correspond to the stress-strain state in the composite material with ideal (uncurved) layout of the layers under

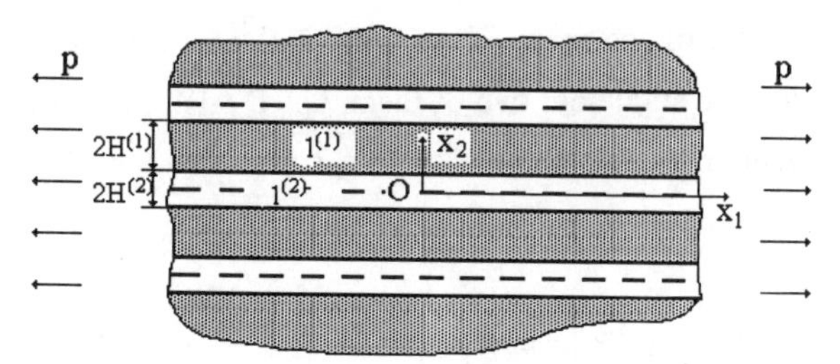

Fig.4.2.2. An ideal layout of the layers in the structure of the composite material

the action of the same external forces (Fig.4.2.2). As we assume that the body is infinite in the direction Ox_1 axis, we do not consider the effect of the edges on which the external forces act on the stress distribution.

Thus for this approximation we obtain the following contact conditions from (4.1.22), (4.1.23), (4.2.3):

$$\sigma_{i2}^{(1),0}\Big|_{\left(t_1,\mp H^{(1)}\right)}=\sigma_{i2}^{(2),0}\Big|_{\left(t_1,\pm H^{(2)}\right)}\,,\quad u_i^{(1),0}\Big|_{\left(t_1,\mp H^{(1)}\right)}=u_{i2}^{(2),0}\Big|_{\left(t_1,\pm H^{(2)}\right)} \tag{4.2.4}$$

According to the statement of the problem in the zeroth approximation (Fig.4.2.2) we can write

$$\sigma_{12}^{(k),0}=\sigma_{22}^{(k),0}=0\,,\;\; k=1,2. \tag{4.2.5}$$

Further, remembering (4.2.5) we obtain

$$\sigma_{11}^{(1),0}\eta^{(1)}+\sigma_{11}^{(2),0}\eta^{(2)}=p\,, \tag{4.2.6}$$

and using Hooke's law and the relations (4.2.4)-(4.2.6), we find the values of the zeroth approximation as follows:

$$\sigma_{11}^{(1),0}=p\left(\eta^{(1)}+\eta^{(2)}\frac{E^{(2)}}{E^{(1)}}\frac{1-\left(\nu^{(1)}\right)^2}{1-\left(\nu^{(2)}\right)^2}\right)^{-1},\quad \eta^{(k)}=\frac{H^{(k)}}{H^{(1)}+H^{(2)}}\,,$$

$$\sigma_{11}^{(2),0}=\frac{E^{(2)}}{E^{(1)}}\frac{1-(\nu^{(1)})^2}{1-(\nu^{(2)})^2}\sigma_{11}^{(1),0}\,,\quad u_1^{(k),0}=\frac{1-(\nu^{(k)})^2}{E^{(k)}}\sigma_{11}^{(k),0}x_1^{(k)}\,,$$

$$u_2^{(k),0} = -\frac{\nu^{(k)}\left(1+\nu^{(k)}\right)}{E^{(k)}}\sigma_{11}^{(k),0} x_2^{(k)} + C^{(k)}, \quad k=1,2, \quad C^{(k)} = \text{const}, \quad (4.2.7)$$

where $E^{(k)}, \nu^{(k)}$ are Young's moduli and Poisson's ratio, respectively.

(4.2.7) gives the zeroth approximation. If follows from the expressions (4.2.7) that the stress state corresponding to the zeroth approximation balances the external forces p. Therefore the values of the first and subsequent approximations will correspond to self-balanced stresses arising as a result of the curving of the reinforcing layers.

The first approximation. For this approximation we obtain the following contact conditions from the expressions (4.1.22)-(4.1.25), (4.2.1), (4.2.3), (4.2.5) and (4.2.7):

$$\sigma_{i2}^{(1),1}\Big|_{\left(t_1,\mp H^{(1)}\right)} - \sigma_{i2}^{(2),1}\Big|_{\left(t_1,\pm H^{(2)}\right)} = 2\pi\cos(\alpha t_1)\left(\sigma_{11}^{(1),0} - \sigma_{11}^{(2),0}\right)\delta_i^1,$$

$$u_i^{(1),1}\Big|_{\left(t_1,\mp H^{(1)}\right)} - u_i^{(2),1}\Big|_{\left(t_1,\pm H^{(2)}\right)} = \ell\sin(\alpha t_1)\left(\frac{\partial u_2^{(2),0}}{\partial x_2^{(2)}} - \frac{\partial u_2^{(1),0}}{\partial x_2^{(1)}}\right)\delta_i^2, \quad (4.2.8)$$

where δ_i^1, δ_i^2 are Kronecker symbols, and $\alpha = 2\pi/\ell$.

Since the equations (4.1.16) are satisfied for each approximation separately, therefore the values of the first and subsequent approximations can be sought with the use of *Papkovich-Neuber* representations. These representations for plain-strain displacements can be written as follows:

$$2G^{(\underline{k})}u_1^{(\underline{k}),q} = -\frac{\partial\Phi_0^{(\underline{k}),q}}{\partial x_1^{(\underline{k})}} - x_2^{(\underline{k})}\frac{\partial\Phi_2^{(\underline{k}),q}}{\partial x_1^{(\underline{k})}},$$

$$2G^{(\underline{k})}u_2^{(\underline{k}),q} = \left(3-4\nu^{(\underline{k})}\right)\Phi_2^{(\underline{k}),q} - \frac{\partial\Phi_0^{(\underline{k}),q}}{\partial x_2^{(\underline{k})}} - x_2^{(\underline{k})}\frac{\partial\Phi_2^{(\underline{k}),q}}{\partial x_2^{(\underline{k})}}, \quad (4.2.9)$$

where $G^{(k)}$ is a shear modulus and the functions $\Phi_0^{(\underline{k}),q}(x_1^{(\underline{k})}, x_2^{(\underline{k})})$, $\Phi_2^{(\underline{k}),q}(x_1^{(\underline{k})}, x_2^{(\underline{k})})$ are harmonic, i.e., these functions satisfy the following equation

$$\left(\frac{\partial^2}{\partial\left(x_1^{(k)}\right)^2}+\frac{\partial^2}{\partial\left(x_2^{(k)}\right)^2}\right)\Phi_0^{(k),q}=0, \quad \left(\frac{\partial^2}{\partial\left(x_1^{(k)}\right)^2}+\frac{\partial^2}{\partial\left(x_2^{(k)}\right)^2}\right)\Phi_2^{(k),q}=0. \quad (4.2.10)$$

Taking into account the contact conditions (4.2.8), we choose the harmonic functions $\Phi_0^{(k),1}$, $\Phi_2^{(k),1}$ in the form

$$\Phi_0^{(k),1}=\left[C_{01}^{(k),1}\cosh\left(\alpha x_2^{(k)}\right)+C_{02}^{(k),1}\sinh\left(\alpha x_2^{(k)}\right)\right]\sin\left(\alpha x_1^{(k)}\right),$$

$$\Phi_2^{(k),1}=\left[C_{21}^{(k),1}\cosh\left(\alpha x_2^{(k)}\right)+C_{22}^{(k),1}\sinh\left(\alpha x_2^{(k)}\right)\right]\sin\left(\alpha x_1^{(k)}\right). \quad (4.2.11)$$

The relations (4.2.9), (4.1.16), (4.2.11) and the contact conditions (4.2.8) give a system of non-homogeneous algebraic equations for the unknown constants $C_{n1}^{(k),1}, C_{n2}^{(k),1}$ (n=0,2) which enter (4.2.11); these constants give the first approximation. In this case the stresses and displacements can be presented as follows:

$$\sigma_{ii}^{(k),1}=\psi_{ii}^{(k),1}\left(x_2^{(k)}\right)\sin\left(\alpha x_1^{(k)}\right), \quad \sigma_{12}^{(k),1}=\psi_{12}^{(k),1}\left(x_2^{(k)}\right)\cos\left(\alpha x_1^{(k)}\right)$$

$$u_i^{(k),1}=\varphi_i^{(k),1}\left(x_2^{(k)}\right)\left(\delta_i^1\cos\left(\alpha x_1^{(k)}\right)+\delta_i^2\sin\left(\alpha x_1^{(k)}\right)\right), \quad i=1,2 \quad (4.2.12)$$

where the expressions of the functions $\psi_{ij}^{(k),1}, \varphi_i^{(k),1}$, $(k,i,j=1,2)$ can be easily determined from the relations (4.2.11), (4.2.9) and (4.1.16).

The second approximation. From the (4.2.3), (4.1.12) we obtain the following contact conditions for this approximation:

$$P_{i2}^{(1),2}\Big|_{\left(t_1,\mp H^{(1)}\right)}-P_{i2}^{(2),2}\Big|_{\left(t_1,\pm H^{(2)}\right)}=2\pi\cos\left(\alpha t_1\right)\left(\sigma_{i1}^{(1),1}\Big|_{\left(t_1,\mp H^{(1)}\right)}-\sigma_{i1}^{(2),1}\Big|_{\left(t_1,\pm H^{(2)}\right)}\right),$$

$$U_i^{(1),2}\Big|_{\left(t_1,\mp H^{(1)}\right)}-U_i^{(2),2}\Big|_{\left(t_1,\pm H^{(2)}\right)}=0. \quad (4.2.13)$$

Taking into account the expressions (4.2.12), (4.1.22)-(4.1.25) and carrying out some transformations we reduce the contact conditions (4.2.13) to the following:

$$\sigma_{12}^{(1),2}\Big|_{\left(t_1,\mp H^{(1)}\right)}-\sigma_{12}^{(2),2}\Big|_{\left(t_1,\pm H^{(2)}\right)}=2\pi\sin\left(2\alpha t_1\right)\left[\psi_{11}^{(1),1}\Big|_{\left(\mp H^{(1)}\right)}-\psi_{11}^{(2),1}\Big|_{\left(\pm H^{(2)}\right)}\pm\frac{1}{2}\alpha H^{(2)}\times\right.$$

$$\left(\psi_{12}^{(2),1}\Big|_{(\pm H^{(2)})}-\psi_{12}^{(1),1}\Big|_{(\mp H^{(1)})}\right)\Bigg],$$

$$\sigma_{22}^{(1),2}\Big|_{(t_1,\mp H^{(1)})}-\sigma_{22}^{(2),2}\Big|_{(t_1,\pm H^{(2)})}=2\pi\cos(2\alpha t_1)\left(\psi_{12}^{(1),1}\Big|_{(\mp H^{(1)})}-\psi_{12}^{(2),1}\Big|_{(\pm H^{(2)})}\right),$$

$$u_1^{(1),2}\Big|_{(t_1,\mp H^{(1)})}-u_1^{(2),2}\Big|_{(t_1,\pm H^{(2)})}=\frac{\ell}{2}\sin(2\alpha t_1)\left(\frac{d\varphi_1^{(2),1}}{dx_2^{(2)}}\Bigg|_{(\pm H^{(2)})}-\frac{d\varphi_1^{(1),1}}{dx_2^{(1)}}\Bigg|_{(\mp H^{(1)})}\right),$$

$$u_2^{(1),2}\Big|_{(t_1,\mp H^{(1)})}-u_2^{(2),2}\Big|_{(t_1,\pm H^{(2)})}=\pi\cos(2\alpha t_1)\Bigg(\mp\alpha H^{(2)}\left(\varphi_2^{(2),1}\Big|_{(\pm H^{(2)})}+\varphi_2^{(1),1}\Big|_{(\mp H^{(1)})}\right)-$$

$$\frac{\ell}{2}\left(\frac{d\varphi_1^{(2),1}}{dx_2^{(2)}}\Bigg|_{(\pm H^{(2)})}-\frac{d\varphi_1^{(1),1}}{dx_2^{(1)}}\Bigg|_{(\mp H^{(1)})}\right)\Bigg)\mp\alpha H^{(2)}\left(\varphi_2^{(2),1}\Big|_{(\pm H^{(2)})}-\varphi_2^{(1),1}\Big|_{(\mp H^{(1)})}\right)+$$

$$\frac{\ell}{2}\left(\frac{d\varphi_1^{(2),1}}{dx_2^{(2)}}\Bigg|_{(\pm H^{(2)})}-\frac{d\varphi_1^{(1),1}}{dx_2^{(1)}}\Bigg|_{(\mp H^{(1)})}\right). \qquad (4.2.14)$$

Considering the form the contact conditions (4.2.14), we select the unknown harmonic functions $\Phi_0^{(k),2},\Phi_2^{(k),2}$ as follows:

$$\Phi_0^{(k),2}=\left[C_{01}^{(k),2}\cosh\left(2\alpha x_2^{(k)}\right)+C_{02}^{(k)}\sinh\left(2\alpha x_2^{(k)}\right)\right]\cos\left(2\alpha x_1^{(k)}\right),$$

$$\Phi_2^{(k),2}=\left[C_{21}^{(k),2}\cosh\left(2\alpha x_2^{(k)}\right)+C_{22}^{(k)}\sinh\left(2\alpha x_2^{(k)}\right)\right]\cos\left(2\alpha x_1^{(k)}\right)+B^{(k)},\ B^{(k)}=\text{const} \qquad (4.2.15)$$

The relations (4.2.9), (4.1.16) and the contact conditions (4.2.14) give the equations for the unknown constants $C_{ni}^{(k),2},B^{(k)}$ (n=0,2; i,k=1,2); following the solution to these equations we determine the values of the second approximation in the form:

$$\sigma_{ii}^{(k),2}=\psi_{ii}^{(k),2}\left(x_2^{(k)}\right)\cos\left(2\alpha x_1^{(k)}\right),\quad \sigma_{12}^{(k),2}=\psi_{12}^{(k),2}\left(x_2^{(k)}\right)\sin\left(2\alpha x_1^{(k)}\right),$$

$$u_i^{(k),2}=\varphi_i^{(k),2}\left(x_2^{(k)}\right)\left(\delta_i^1\sin\left(2\alpha x_1^{(k)}\right)+\delta_i^2\cos\left(2\alpha x_1^{(k)}\right)\right)+\widetilde{B}^{(k)}\delta_i^2,\quad i=1,2, \qquad (4.2.16)$$

where the expressions for the functions $\psi_{ij}^{(k),2}, \varphi_i^{(k),2}$, $(k,i,j=1,2)$ can easily be determined from the relations (4.2.15), (4.2.9) and (4.1.16).

We can determine subsequent approximations in a similar way.

4.2.2. ANTI-PHASE PERIODICALLY CURVED COMPOSITE

We now investigate plane-strain in a composite material with alternating anti-phase periodically plane-curved layers, as shown in Fig.4.2.3.

The periodicity of the composite structure in the direction of the Ox_2 axis has period $4\left(H^{(2)}+H^{(1)}\right)$ among the layers we single out four of them, i.e. $1^{(1)}, 1^{(2)}, 2^{(1)}, 2^{(2)}$ (Fig.4.2.3), and discuss them below. The equation (4.1.3) for the middle surface of the $1^{(2)}$-th layer we take as

$$x_{21}^{(2)} = F_1\left(x_{11}^{(2)}\right) = \varepsilon f_1\left(x_{11}^{(2)}\right) = L\sin\left(\frac{2\pi}{\ell}x_{11}^{(2)}\right) = \varepsilon\ell\sin\left(\frac{2\pi}{\ell}x_1\right), \quad (4.2.17)$$

but the equation for the middle surface of the $2^{(2)}$-th layer we take as

$$x_{22}^{(2)} = F_2\left(x_{12}^{(2)}\right) = \varepsilon f_2\left(x_{12}^{(2)}\right) = -L\sin\left(\frac{2\pi}{\ell}x_{12}^{(2)}\right) = -\varepsilon\ell\sin\left(\frac{2\pi}{\ell}x_1\right), \quad (4.2.18)$$

where $x_{11}^{(2)} = x_{12}^{(2)} = x_{11}^{(1)} = x_{12}^{(1)} = x_1$ and $L<\ell$; the small parameter ε is determined by the relation (4.2.2).

Using the periodicity of the material structure with period $4\left(H^{(2)}+H^{(1)}\right)$ and taking $t_{11}=t_{12}=t_1$, we can write the explicit form of the contact conditions (4.1.26) as follows:

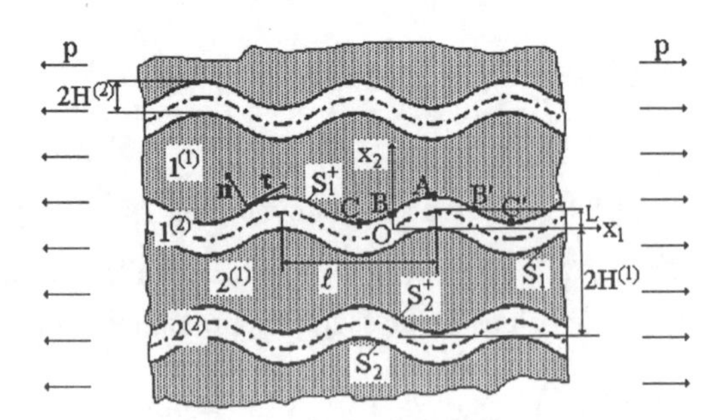

Fig.4.2.3. The structure of the composite material with anti-phase periodically plane-curved layers.

$$\left(-\frac{df_1}{dx_1}P_{i1}^{(1)1,q-1}+P_{i2}^{(1)1,q}\right)\Bigg|_{(t_1,-H^{(1)})}=\left(-\frac{df_1}{dx_1}P_{i1}^{(2)1,q-1}+P_{i2}^{(2)1,q}\right)\Bigg|_{(t_1,H^{(2)})},$$

$$\left(-\frac{df_1}{dx_1}P_{i1}^{(2)1,q-1}+P_{i2}^{(2)1,q}\right)\Bigg|_{(t_1,-H^{(2)})}=\left(-\frac{df_1}{dx_1}P_{i1}^{(1)2,q-1}+P_{i2}^{(1)2,q}\right)\Bigg|_{(t_1,H^{(1)})},$$

$$U_i^{(1)1,q}\Big|_{(t_1,-H^{(1)})}=U_i^{(2)1,q}\Big|_{(t_1,H^{(2)})},\quad U_i^{(2)1,q}\Big|_{(t_1,-H^{(2)})}=U_i^{(1)2,q}\Big|_{(t_1,H^{(1)})},$$

$$\left(-\frac{df_2}{dx_1}P_{i1}^{(1)2,q-1}+P_{i2}^{(1)2,q}\right)\Bigg|_{(t_1,-H^{(1)})}=\left(-\frac{df_2}{dx_1}P_{i1}^{(2)2,q-1}+P_{i2}^{(2)2,q}\right)\Bigg|_{(t_1,H^{(2)})},$$

$$\left(-\frac{df_2}{dx_1}P_{i1}^{(2)2,q-1}+P_{i2}^{(2)2,q}\right)\Bigg|_{(t_1,-H^{(2)})}=\left(-\frac{df_2}{dx_1}P_{i1}^{(1)1,q-1}+P_{i2}^{(1)1,q}\right)\Bigg|_{(t_1,H^{(1)})},$$

$$U_i^{(1)2,q}\Big|_{(t_1,-H^{(1)})}=U_i^{(2)2,q}\Big|_{(t_1,H^{(2)})},\quad U_i^{(2)2,q}\Big|_{(t_1,-H^{(2)})}=U_i^{(1)1,q}\Big|_{(t_1,H^{(1)})} \tag{4.2.19}$$

The values of the zeroth approximation, except $u_2^{(k)m,0}$, are determined as in (4.2.5), (4.2.7) and

$$\sigma_{ij}^{(k)m,0}=\sigma_{ij}^{(k),0},\quad u_1^{(k)m,0}=u_1^{(k),0}; \tag{4.2.20}$$

however for the $u_2^{(k)m,0}$ we obtain the following expression

$$u_2^{(k)m,0}=-\frac{\nu^{(k)}\left(1+\nu^{(k)}\right)}{E^{(k)}}\sigma_{11}^{(k),0}x_{2m}^{(k)}+C_m^{(k)},\quad C_m^{(k)}=\text{const}. \tag{4.2.21}$$

Now we consider first and second approximations.

The first approximation. From the (4.2.5), (4.2.7), (4.2.17)-(4.2.21) and (4.1.22)-(4.1.25) we obtain the following contact conditions:

$$\sigma_{i2}^{(1)1,1}\Big|_{(t_1,-H^{(1)})}-\sigma_{i2}^{(2)1,1}\Big|_{(t_1,H^{(2)})}=2\pi\cos(\alpha t_1)\left(\sigma_{i1}^{(1),0}-\sigma_{i1}^{(2),0}\right),$$

$$\sigma_{i2}^{(2)1,1}\Big|_{(t_1,-H^{(2)})} - \sigma_{i2}^{(1)2,1}\Big|_{(t_1,H^{(1)})} = 2\pi\cos(\alpha t_1)\left(\sigma_{i1}^{(2),0} - \sigma_{i1}^{(1),0}\right),$$

$$\sigma_{i2}^{(1)2,1}\Big|_{(t_1,-H^{(1)})} - \sigma_{i2}^{(2)2,1}\Big|_{(t_1,H^{(2)})} = 2\pi\cos(\alpha t_1)\left(\sigma_{i1}^{(1),0} - \sigma_{i1}^{(2),0}\right),$$

$$\sigma_{i2}^{(2)2,1}\Big|_{(t_1,-H^{(2)})} - \sigma_{i2}^{(1)1,1}\Big|_{(t_1,H^{(1)})} = 2\pi\cos(\alpha t_1)\left(\sigma_{i1}^{(1),0} - \sigma_{i1}^{(2),0}\right),$$

$$u_i^{(1)1,1}\Big|_{(t_1,-H^{(1)})} - u_i^{(2)1,1}\Big|_{(t_1,H^{(2)})} = \ell\sin(\alpha t_1)\left(\frac{\partial u_2^{(2)1,0}}{\partial x_{21}^{(2)}} - \frac{\partial u_2^{(1)1,0}}{\partial x_{21}^{(1)}}\right)\delta_i^2,$$

$$u_i^{(2)1,1}\Big|_{(t_1,-H^{(2)})} - u_i^{(1)2,1}\Big|_{(t_1,H^{(1)})} = \ell\sin(\alpha t_1)\left(\frac{\partial u_2^{(2)2,0}}{\partial x_{21}^{(2)}} - \frac{\partial u_2^{(1)2,0}}{\partial x_{22}^{(1)}}\right)\delta_i^2,$$

$$u_i^{(1)2,1}\Big|_{(t_1,-H^{(1)})} - u_i^{(2)2,1}\Big|_{(t_1,H^{(2)})} = \ell\sin(\alpha t_1)\left(\frac{\partial u_2^{(1)2,0}}{\partial x_{22}^{(2)}} - \frac{\partial u_2^{(2)2,0}}{\partial x_{22}^{(1)}}\right)\delta_i^2,$$

$$u_i^{(2)2,1}\Big|_{(t_1,-H^{(2)})} - u_i^{(1)1,1}\Big|_{(t_1,H^{(1)})} = \ell\sin(\alpha t_1)\left(\frac{\partial u_2^{(2)2,0}}{\partial x_{22}^{(2)}} - \frac{\partial u_2^{(1)1,0}}{\partial x_{21}^{(1)}}\right)\delta_i^2. \quad (4.2.22).$$

As in the previous subsection, using Papkovich-Neuber representations (4.2.9) and taking into account the form of the expressions (4.2.22) we take the unknown harmonic functions (4.2.10) for the selected four layers (Fig.4.2.3) as follows:

$$\Phi_0^{(\underline{k})\underline{m},1} = \left[C_{01}^{(\underline{k})\underline{m},1}\cosh\left(\alpha x_{2\underline{m}}^{(\underline{k})}\right) + C_{02}^{(\underline{k})\underline{m},1}\sinh\left(\alpha x_{2\underline{m}}^{(\underline{k})}\right)\right]\sin(\alpha x_1),$$

$$\Phi_2^{(\underline{k})\underline{m},1} = \left[C_{21}^{(\underline{k})\underline{m},1}\cosh\left(\alpha x_{2\underline{m}}^{(\underline{k})}\right) + C_{22}^{(\underline{k})\underline{m},1}\sinh\left(\alpha x_{2\underline{m}}^{(\underline{k})}\right)\right]\sin(\alpha x_1), \qquad (4.2.23)$$

where $C_{n2}^{(k)m,1}$ (n=0,1; k,m=1,2) are the unknown constants which are determined from the contact conditions (4.2.22) by employing the procedure described in the previous subsection. Further, the approximation is taken in the following form:

$$\sigma_{ii}^{(\underline{k})\underline{m},1} = \psi_{ii}^{(\underline{k})\underline{m},1}\left(x_{2\underline{m}}^{(\underline{k})}\right)\sin(\alpha x_1), \quad \sigma_{12}^{(\underline{k})\underline{m},1} = \psi_{12}^{(\underline{k})\underline{m},1}\left(x_{2\underline{m}}^{(\underline{k})}\right)\cos(\alpha x_1),$$

$$u_i^{(\underline{k})\underline{m},1} = \varphi_i^{(\underline{k})\underline{m},1}\left(x_{2\underline{m}}^{(\underline{k})}\right)\left(\delta_i^1 \cos(\alpha x_1) + \delta_i^2 \sin(\alpha x_1)\right), \tag{4.2.24}$$

where the expressions of the functions $\psi_{ij}^{(k)m,1}, \varphi_i^{(k)m,1}$ can easily be determined from the (4.2.9), (4.1.16). This gives the first approximation.

The second approximation. Using the expressions (4.2.24), (4.1.22)-(4.1.25), (4.2.5), (4.2.7), (4.2.20), (4.2.21) we obtain the following contact conditions for the second approximation from (4.2.19).

$$\sigma_{12}^{(2)1,2}\Big|_{(t_1,H^{(2)})} - \sigma_{12}^{(1)1,2}\Big|_{(t_1,-H^{(1)})} = \Psi\left(\psi_{12}^{(1)1,1}, \psi_{12}^{(2)1,1}, \psi_{11}^{(1)1,1}, \psi_{11}^{(2)1,1}, -H^{(1)}, H^{(2)}\right)\sin(2\alpha t_1)$$

$$\sigma_{22}^{(2)1,2}\Big|_{(t_1,H^{(2)})} - \sigma_{22}^{(1)1,2}\Big|_{(t_1,-H^{(1)})} = -2\pi\left(\psi_{12}^{(1)1,1}\Big|_{(-H^{(1)})} - \psi_{12}^{(2)1,1}\Big|_{(H^{(2)})}\right)\cos(2\alpha t_1),$$

$$u_1^{(2)1,2}\Big|_{(t_1,H^{(2)})} - u_1^{(1)1,2}\Big|_{(t_1,-H^{(1)})} = \sin(2\alpha t_1)D_{\varphi_1}\left(\varphi_1^{(1)1,1}, \varphi_1^{(2)1,1}, -H^{(1)}, H^{(2)}\right),$$

$$u_2^{(2)1,2}\Big|_{(t_1,H^{(2)})} - u_2^{(1)1,2}\Big|_{(t_1,-H^{(1)})} = \cos(2\alpha t_1)D_{\varphi_2}\left(\varphi_2^{(1)1,1}, \varphi_2^{(2)1,1}, -H^{(1)}, H^{(2)}\right) +$$

$$d_{\varphi_2}\left(\varphi_2^{(1)1,1}, \varphi_2^{(2)1,1}, -H^{(1)}, H^{(2)}\right), \quad \sigma_{12}^{(2)1,2}\Big|_{(t_1,-H^{(2)})} - \sigma_{12}^{(1)2,2}\Big|_{(t_1,H^{12)})} =$$

$$\Psi\left(\psi_{12}^{(1)2,1}, \psi_{12}^{(2)1,1}, \psi_{11}^{(1)2,1}, \psi_{11}^{(2)1,1}, H^{(1)}, -H^{(2)}\right)\sin(2\alpha t_1),$$

$$\sigma_{22}^{(2)1,2}\Big|_{(t_1,-H^{(2)})} - \sigma_{22}^{(1)2,2}\Big|_{(t_1,H^{(1)})} = -2\pi\left(\psi_{12}^{(1)2,1}\Big|_{(H^{(1)})} - \psi_{12}^{(2)1,1}\Big|_{(-H^{(2)})}\right)\cos(2\alpha t_1),$$

$$u_1^{(2)1,2}\Big|_{(t_1,-H^{(2)})} - u_1^{(1)2,2}\Big|_{(t_1,H^{(1)})} = \sin(2\alpha t_1)D_{\varphi_1}\left(\varphi_1^{(1)2,1}, \varphi_1^{(2)1,1}, H^{(1)}, -H^{(2)}\right),$$

$$u_2^{(2)1,2}\Big|_{(t_1,-H^{(2)})} - u_2^{(1)2,2}\Big|_{(t_1,H^{(1)})} = \cos(2\alpha t_1)D_{\varphi_2}\left(\varphi_2^{(1)2,1}, \varphi_2^{(2)1,1}, H^{(1)}, -H^{(2)}\right) +$$

$$d_{\varphi_2}\left(\varphi_2^{(1)2,1}, \varphi_2^{(2)1,1}, H^{(1)}, -H^{(2)}\right), \quad \sigma_{12}^{(2)2,2}\Big|_{(t_1,H^{(2)})} - \sigma_{12}^{(1)2,2}\Big|_{(t_1,-H^{12)})} =$$

$$\Psi\left(\psi_{12}^{(1)2,1}, \psi_{12}^{(2)2,1}, \psi_{11}^{(1)2,1}, \psi_{11}^{(2)2,1}, -H^{(1)}, H^{(2)}\right)\sin(2\alpha t_1)$$

$$\sigma_{22}^{(2)2,2}\Big|_{(t_1,H^{(2)})}-\sigma_{22}^{(1)2,2}\Big|_{(t_1,-H^{(1)})}=-2\pi\left(\psi_{12}^{(1)2,1}\Big|_{(-H^{(1)})}-\psi_{12}^{(2)2,1}\Big|_{(H^{(2)})}\right)\cos(2\alpha t_1),$$

$$u_1^{(2)2,2}\Big|_{(t_1,H^{(2)})}-u_1^{(1)2,2}\Big|_{(t_1,-H^{(1)})}=\sin(2\alpha t_1)D_{\varphi_1}\left(\varphi_1^{(1)2,1},\varphi_1^{(2)2,1},-H^{(1)},H^{(2)}\right),$$

$$u_2^{(2)2,2}\Big|_{(t_1,H^{(2)})}-u_2^{(1)2,2}\Big|_{(t_1,-H^{(1)})}=\cos(2\alpha t_1)D_{\varphi_2}\left(\varphi_2^{(1)2,1},\varphi_2^{(2)2,1},-H^{(1)},H^{(2)}\right)+$$

$$d_{\varphi_2}\left(\varphi_2^{(1)2,1},\varphi_2^{(2)2,1},-H^{(1)},H^{(2)}\right),\quad \sigma_{12}^{(2)2,2}\Big|_{(t_1,-H^{(2)})}-\sigma_{12}^{(1)1,2}\Big|_{(t_1,H^{12)})}=$$

$$\Psi\left(\psi_{12}^{(1)1,1},\psi_{12}^{(2)2,1},\psi_{11}^{(1)1,1},\psi_{11}^{(2)2,1},H^{(1)},-H^{(2)}\right)\sin(2\alpha t_1)$$

$$\sigma_{22}^{(2)2,2}\Big|_{(t_1,-H^{(2)})}-\sigma_{22}^{(1)1,2}\Big|_{(t_1,H^{(1)})}=-2\pi\left(\psi_{12}^{(1)1,1}\Big|_{(H^{(1)})}-\psi_{12}^{(2)2,1}\Big|_{(-H^{(2)})}\right)\cos(2\alpha t_1),$$

$$u_1^{(2)2,2}\Big|_{(t_1,-H^{(2)})}-u_1^{(1)1,2}\Big|_{(t_1,H^{(1)})}=\sin(2\alpha t_1)D_{\varphi_1}\left(\varphi_1^{(1)1,1},\varphi_1^{(2)2,1},H^{(1)},-H^{(2)}\right),$$

$$u_2^{(2)2,2}\Big|_{(t_1,-H^{(2)})}-u_2^{(1)1,2}\Big|_{(t_1,H^{(1)})}=\cos(2\alpha t_1)D_{\varphi_2}\left(\varphi_2^{(1)1,1},\varphi_2^{(2)2,1},H^{(1)},-H^{(2)}\right)+$$

$$d_{\varphi_2}\left(\varphi_2^{(1)1,1},\varphi_2^{(2)2,1},H^{(1)},-H^{(2)}\right), \tag{4.2.25}$$

In (4.2.25) the following notation is used:

$$\Psi\left(\psi_{12}^{(1)n,1},\psi_{12}^{(2)m,1},\psi_{11}^{(1)n,1},\psi_{11}^{(2)m,1},H^{(1)},H^{(2)}\right)=\pi\left[\alpha H^{(2)}\left(\psi_{12}^{(1)n,1}\Big|_{(H^{(1)})}-\right.\right.$$

$$\left.\left.\psi_{12}^{(1)m,1}\Big|_{(H^{(2)})}\right)-2\left(\psi_{11}^{(1)n,1}\Big|_{(H^{(1)})}-\psi_{11}^{(2)m,1}\Big|_{(H^{(2)})}\right)\right],$$

$$D_{\varphi_1}\left(\varphi_1^{(1)n,1},\varphi_1^{(2)m,1},H^{(1)},H^{(2)}\right)=\frac{\ell}{2}\left(\frac{d\varphi_1^{(1)n,1}}{dx_{2n}^{(1)}}\Bigg|_{(H^{(1)})}-\frac{d\varphi_1^{(2)m,1}}{dx_{2m}^{(2)}}\Bigg|_{(H^{(2)})}\right),$$

$$D_{\varphi_2}\left(\varphi_2^{(1)n,1},\varphi_2^{(2)m,1},H^{(1)},H^{(2)}\right)=-\pi\left(-\alpha H^{(2)}\left(\varphi_2^{(2)m,1}\Big|_{\left(H^{(2)}\right)}+\varphi_2^{(1)n,1}\Big|_{\left(H^{(1)}\right)}\right)-\right.$$

$$\left.\frac{\ell}{2}\left(\frac{d\varphi_2^{(2)m,1}}{dx_{2m}^{(2)}}\Bigg|_{\left(H^{(2)}\right)}-\frac{d\varphi_2^{(1)n,1}}{dx_{2n}^{(1)}}\Bigg|_{\left(H^{(1)}\right)}\right)\right),$$

$$d_{\varphi_2}\left(\varphi_2^{(1)n,1},\varphi_2^{(2)m,1},H^{(1)},H^{(2)}\right)=\alpha H^{(2)}\left(\varphi_2^{(2)m,1}\Big|_{\left(H^{(2)}\right)}-\varphi_2^{(1)n,1}\Big|_{\left(H^{(1)}\right)}\right)-$$

$$\left.\frac{\ell}{2}\left(\frac{d\varphi_2^{(2)m,1}}{dx_{2m}^{(2)}}\Bigg|_{\left(H^{(2)}\right)}-\frac{d\varphi_2^{(1)n,1}}{dx_{2n}^{(1)}}\Bigg|_{\left(H^{(1)}\right)}\right)\right). \quad (4.2.26)$$

With the form of the contact conditions we take the solutions to the harmonic equations (4.2.10) for the approximation as follows:

$$\Phi_0^{(k)m,2}=\left[C_{01}^{(k)m,2}\cosh\left(2\alpha x_{2m}^{(k)}\right)+C_{02}^{(k)m,2}\sinh\left(2\alpha x_{2m}^{(k)}\right)\right]\cos\left(2\alpha x_1\right),$$

$$\Phi_2^{(k)m,2}=\left[C_{21}^{(k)m,2}\cosh\left(2\alpha x_{2m}^{(k)}\right)+C_{22}^{(k)m,2}\sinh\left(2\alpha x_{2m}^{(k)}\right)\right]\cos\left(2\alpha x_1\right)+B^{(k)m}. \quad (4.2.27)$$

As before, we determine the unknown constants entering (4.2.27) from the contact conditions (4.2.25). We can use this procedure for all subsequent approximations.

The solution method presented in the last two sections can be called *the boundary form disturbance method* based on the piecewise-homogeneous body model, with the application to composite materials with curved structures. Note that in the following sections this method will be developed for composite materials with more complex structures.

4.3. Stress Distribution in Composites with Alternating Layers

In this section we will investigate the distribution of the stresses which arise as a result of the curving of the reinforcing layers shown in Figs. 4.2.1 and 4.2.3. We will consider the distribution of the stresses at the interface between the filler and matrix layers under the action, at infinity, of uniformly distributed normal forces of intensity p in Ox_1 direction. The materials of the matrix and filler layers will be taken to be homogeneous and isotropic.

We introduce the parameter $\chi=2\pi H^{(2)}/\ell$. Here $2H^{(2)}$ is the thickness of the filler layer, ℓ is the wavelength of forms of curvature of the middle surface of the filler layer. We call χ the parameter of the wave generation of the curving. We assume the Poisson's ratios $\nu^{(1)} = \nu^{(2)} = 0.3$. The numerical results which will be analysed in the present section have been obtained from the zeroth and the first five approximations of (4.1.15).

4.3.1. LOWER FILLER CONCENTRATION

First consider the case in which the concentration of the reinforcing layers in is very small, and the interaction between them can be neglected. In other words, we assume that $H^{(1)} >> H^{(2)}$ and neglect the interaction between the curved reinforcing layers. In this case the problems shown in Figs. 4.2.1 and 4.2.3 are reduced to problems for the infinite body with a single periodic curved reinforcing layer. The values of the zeroth approximation are determined by the expressions (4.2.5), (4.2.7) with $\eta^{(2)} = 0$. The values of the first and subsequent approximations are determined with the use of the appropriate decay conditions at $|x_2| \to +\infty$; the functions $\cosh(n\alpha x_2)$, $\sinh(n\alpha x_2)$, which enter the expressions of the harmonic functions $\Phi_0^{(1),n}$ and $\Phi_2^{(1),n}$, are replaced by $\exp(-n\alpha x_2)$ (for the upper half-plane) or by $\exp(n\alpha x_2)$ (for the lower half-plane).

We analyse the numerical results related to the stress distribution at the interface between the filler and matrix material. Taking into account the periodicity of the reinforcing layer curving, we consider only the part AC on surface S^+ and the part BD on S^- (Fig.4.2.1). The values related to the surfaces S^+ and S^- will be denoted by the upper symbols + and - , respectively. Note that the parts AC and BD correspond to the variation of the parameter αt_1 in the interval $[0, \pi/2]$. In other words, accounting (4.2.1) and taking $\alpha t_1 = 0$ $(\alpha t_1 = \pi/2)$ in (4.1.13) we obtain the coordinates of the point A (C) on the surface S^+ and of the point B (D) on the surface S^- (Fig.4.2) from (4.1.8). Analogously, changing the parameter αt_1 from 0 up to $\pi/2$ we obtain from (4.1.8) the coordinates of all points of the parts AC and BD.

Thus we consider the graphs given in Fig. 4.3.1 which show the distributions of the $\overset{+}{\sigma}_{nn}/\sigma_{11}^{(1),0}$ $\left(\overset{-}{\sigma}_{nn}/\sigma_{11}^{(1),0}\right)$ and $\overset{+}{\sigma}_{n\tau}/\sigma_{11}^{(1),0}$ $\left(\overset{-}{\sigma}_{n\tau}/\sigma_{11}^{(1),0}\right)$ on the part AC (BD) for $\varepsilon = 0.015$, $\chi = 0.3$ and various values of the ratio $E^{(2)}/E^{(1)}$. Here $\overset{+}{\sigma}_{nn}$ $\left(\overset{-}{\sigma}_{nn}\right)$ is the normal stress on the surface S^+ $\left(S^-\right)$ in the direction of the normal vector $\mathbf{n}$ to this surface; $\overset{+}{\sigma}_{n\tau}\left(\overset{-}{\sigma}_{n\tau}\right)$ is the tangential stress on the surface S^+ $\left(S^-\right)$ in the direction of the tangential vector $\boldsymbol{\tau}$ to this surface; $\sigma_{11}^{(1),0}$ is the normal stress in the matrix in the direction of the Ox_1 axis in the zeroth approximation, i.e. $\sigma_{11}^{(1),0}$ is the normal stress in the matrix in the direction of the Ox_1 axis in the composite material with the ideal (uncurved) structure shown in Fig.4.2.2.

These graphs show that the normal stress $\sigma_{nn}^{+}\left(\sigma_{nn}^{-}\right)$ which arises as a result of the curving of the reinforcing layers has its minimal value at $\alpha t_1 = 0$, i.e. at point A

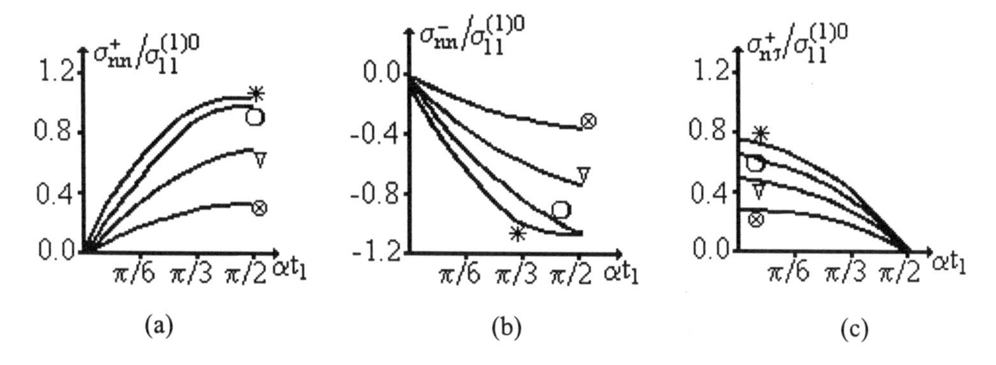

Fig.4.3.1. The distribution of the stresses σ_{nn}^{+} (a), σ_{nn}^{-} (b) and $\sigma_{n\tau}^{+}$ (c)with respect to αt_1 for $\varepsilon = 0.015$, $\chi = 0.3$, $E^{(2)}/E^{(1)} = 20$ (⊗); 50 (∇); 100 (O); 150 (✱).

(B) (Fig.4.2.1). With increasing αt_1 from 0 up to $\pi/2$, i.e. by approaching the point C (D) from point A (B) on the surface $S^{+}(S^{-})$, the values of $\sigma_{nn}^{+}\left(\sigma_{nn}^{-}\right)$ increase monotonically and this stress has its absolute maximal value at point C (D); $\sigma_{nn}^{+}\left(\sigma_{nn}^{-}\right)$ is the tension (compression) stress on the AC (BD) surface under $p > 0$.

Consider the distribution of the tangential stress $\sigma_{n\tau}^{+}$ on the surface S^{+}. Note that the distribution of the stress $\sigma_{n\tau}^{-}$ on the surface S^{-} almost coincide with that obtained on the surface S^{+}; we therefore consider only $\sigma_{n\tau}^{+}$. Fig. 4.3.1. shows that the stress $\sigma_{n\tau}^{+}$ has its maximal value at the point A , i.e. at $\alpha t_1 = 0$, and decreases monotonically as it approaches the point C; $\sigma_{n\tau}^{+}=0$ at C.

Fig.4.3.1 also shows that the absolute values of the stresses σ_{nn}^{+}, σ_{nn}^{-}, $\sigma_{n\tau}^{+}$ grow monotonically with $E^{(2)}/E^{(1)}$. These stresses arise only as a result of the curving of the reinforcing layer; if there is no curving, they are equal to zero and therefore are self-balanced.

We now consider the distribution of $\sigma_{\tau\tau}^{(1)\pm}/\sigma_{11}^{(1),0}$ and $\sigma_{\tau\tau}^{(2)\pm}/\sigma_{11}^{(2),0}$ with respect to αt_1 in the part AC (BD) on the surface $S^{+}(S^{-})$, where $\sigma_{\tau\tau}^{(1)\pm}\left(\sigma_{\tau\tau}^{(2)\pm}\right)$ is the normal stress in the direction of the tangential vector $\boldsymbol{\tau}$ (Fig.4.2.1) in the matrix (filler) layer at points of $S^{\pm}$; $\sigma_{11}^{(2),0}$ is the normal stress in the filler layer in the direction of the Ox_1 axis in the zeroth approximation, i.e. the normal stress in the filler layer in the composite material with the ideal (uncurved) structure (Fig.4.2.2). Note that the graphs of these distributions for fixed ε, χ and with various $E^{(2)}/E^{(1)}$ are given in Fig. 4.3.2.

For $\varepsilon = 0$ the stresses $\sigma_{\tau\tau}^{(1)\pm}\left(\sigma_{\tau\tau}^{(2)\pm}\right)$ coincide with the stresses $\sigma_{11}^{(1),0}\left(\sigma_{11}^{(2),0}\right)$. The stresses $\sigma_{\tau\tau}^{(1)\pm}\left(\sigma_{\tau\tau}^{(2)\pm}\right)$ consist of two parts: the first balances the external forces p, and the second part is self-balanced. Fig.4.3.2 shows that this latter part of $\sigma_{\tau\tau}^{(1)+}$, $\sigma_{\tau\tau}^{(2)+}$ is compressive. However, the self-balanced part of the $\sigma_{\tau\tau}^{(1)-}$, $\sigma_{\tau\tau}^{(2)-}$ has a complex character: it is compressive for $t_1 < t_1'$, tensile for $t_1 > t'_1$. At $t_1 = t_1'$ we obtain $\sigma_{\tau\tau}^{(1)-} = \sigma_{11}^{(1),0}$, $\sigma_{\tau\tau}^{(2)-} = \sigma_{11}^{(2),0}$; the value of t_1' depends on the ratio $E^{(2)}/E^{(1)}$. Fig. 4.3.2. shows that the absolute values of the self-balanced parts of the stresses $\sigma_{\tau\tau}^{(1)\pm}, \sigma_{\tau\tau}^{(2)\pm}$ grow monotonically with $E^{(2)}/E^{(1)}$.

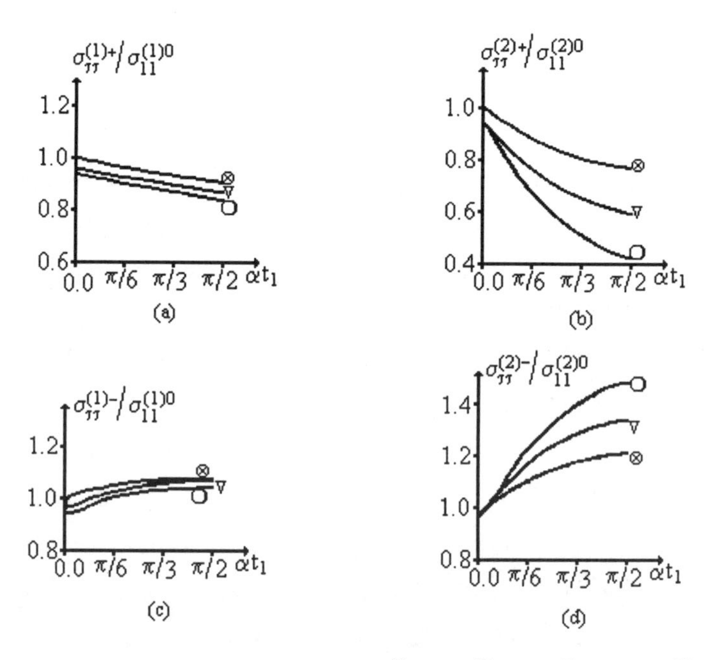

Fig.4.3.2. The distribution of the stresses $\sigma_{\tau\tau}^{(1)+}$ (a), $\sigma_{\tau\tau}^{(2)+}$ (b), $\sigma_{\tau\tau}^{(1)-}$ (c) and $\sigma_{\tau\tau}^{(2)-}$ (d) with respect to αt_1 in the case where $\varepsilon = 0.015$, $\chi = 0.3$, $E^{(2)}/E^{(1)} = 20$ (⊗); 50 (∇); 100 (○).

Consider the relation between the parameter χ and these self-balanced stresses. We assume that $\chi \leq 1.5$ in order to avoid the contact between different points of the contact surfaces $S^{\pm}$. Fig.4.3.3 shows the relations between $\sigma_{nn}^{\pm}/\sigma_{11}^{(1),0}$, $\sigma_{n\tau}^{+}/\sigma_{11}^{(1),0}$ and the parameter χ. In these graphs the values of the normal stresses $\sigma_{nn}^{+}\left(\sigma_{nn}^{-}\right)$ are calculated at the point C(D) (Fig.4.2.1), while the values of the tangential stress $\sigma_{n\tau}^{+}$ are calculated at point A (Fig.4.2.1). Fig.4.3.3 shows the (non monotonic) dependence of $\sigma_{nn}^{+}/\sigma_{11}^{(1),0}$ $\left(\sigma_{nn}^{-}/\sigma_{11}^{(1),0}\right)$ at the point C(D) on the parameter χ, and the (non monotonic) dependence of $\sigma_{n\tau}^{+}/\sigma_{11}^{(1),0}$ at point A on the parameter χ. A similar dependence was observed at all points of the contact surfaces $S^{\pm}$. Consequently, it can be concluded that *for each fixed* $E^{(2)}/E^{(1)}$ *there is such a value of the parameter* χ *(say* χ'*) for which the self-balanced stresses* σ_{nn}^{+}, σ_{nn}^{-}, $\sigma_{n\tau}^{+}$ *have their maxima; as* $E^{(2)}/E^{(1)}$ *increases,* χ' *decreases*. Similar conclusions can be drawn for the dependence of $\sigma_{\tau\tau}^{(1)\pm}/\sigma_{11}^{(1),0}$, $\sigma_{\tau\tau}^{(2)\pm}/\sigma_{11}^{(2),0}$ on χ. The graphics are shown in Fig.4.3.4; the values of $\sigma_{\tau\tau}^{(1)+}$ are calculated at C, while $\sigma_{\tau\tau}^{(1)-}$, $\sigma_{\tau\tau}^{(2)+}$, $\sigma_{\tau\tau}^{(2)-}$ are calculated at D (Fig.4.2.1).

Table 4.3.1 shows $\sigma_{nn}^{\pm}/\sigma_{11}^{(1),0}$, $\sigma_{n\tau}^{+}/\sigma_{11}^{(1),0}$, $\sigma_{\tau\tau}^{(k)\pm}/\sigma_{11}^{(k),0}$ (k=1,2) for various ε and $E^{(2)}/E^{(1)}$. Note that the stresses σ_{nn}^{+}, $\sigma_{\tau\tau}^{(1)+}$, $\sigma_{\tau\tau}^{(2)+}$ are calculated at point C, σ_{nn}^{-}, $\sigma_{\tau\tau}^{(1)-}$, $\sigma_{\tau\tau}^{(2)-}$ at D, and $\sigma_{n\tau}^{+}$ at A (Fig.4.2.1). Moreover note that for

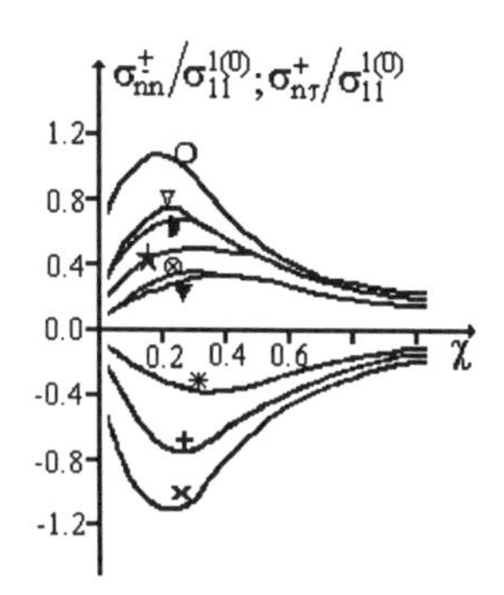

Fig.4.3.3. The graphs of the relations between $\sigma_{nn}^{\pm}$, $\sigma_{n\tau}^{+}$ and χ for $\varepsilon = 0.015$, $E^{(2)}/E^{(1)} = 20, 50, 100$: ⊗, ∇, ○ -- $\sigma_{nn}^{+}/\sigma_{11}^{(1),0}$;✳, ✚, ✖ -- $\sigma_{nn}^{-}/\sigma_{11}^{(1),0}$; ▼, ★, ◗-- $\sigma_{n\tau}^{+}/\sigma_{11}^{(1),0}$

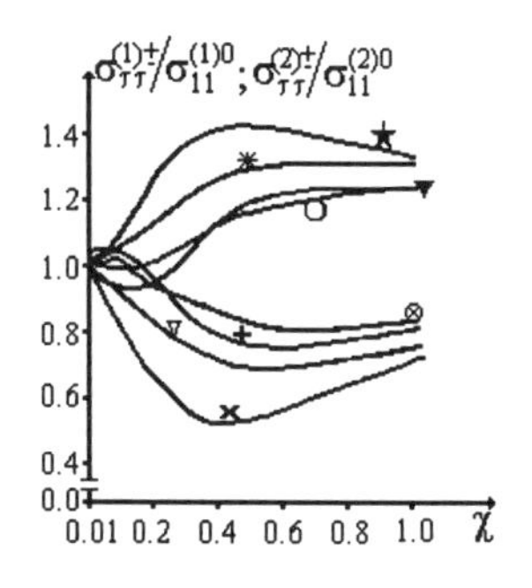

Fig.4.3.4. The graphs of the relations between the stresses $\sigma_{\tau\tau}^{(1)\pm}$, $\sigma_{\tau\tau}^{(2)\pm}$ and χ for $\varepsilon = 0.015$, $E^{(2)}/E^{(1)} = 20, 50$: ⊗, ✚ -- $\sigma_{\tau\tau}^{(1)+}/\sigma_{11}^{(1),0}$;∇,✖ -- $\sigma_{\tau\tau}^{(1)-}/\sigma_{11}^{(1),0}$; ○,▼ -- $\sigma_{\tau\tau}^{(2)+}/\sigma_{11}^{(2),0}$;✳, ★ -- $\sigma_{\tau\tau}^{(2)-}/\sigma_{11}^{(2),0}$.

$E^{(2)}/E^{(1)}$ =20, 50 the stresses $\sigma_{nn}^{\pm}$, $\sigma_{n\tau}^{+}$ are calculated at $\chi = 0.3 \approx \chi'$, for $E^{(2)}/E^{(1)}$ =100 at $\chi = 0.2 \approx \chi'$. However for all the selected $E^{(2)}/E^{(1)}$ the values of the stresses $\sigma_{\tau\tau}^{(k)\pm}$ are calculated at $\chi = 0.5$. Table 4.3.1 shows that among the considered values of ε, χ, $E^{(2)}/E^{(1)}$ there are some for which the self-balanced stress

σ_{nn}^{+} is greater than the stress $\sigma_{11}^{(1),0}$ which acts in the matrix and balances the external forces p (Fig.4.2.1).

Table 4.1. The values of the relations $\sigma_{nn}^{\pm}/\sigma_{11}^{(1),0}$, $\sigma_{n\tau}^{+}/\sigma_{11}^{(1),0}$, $\sigma_{\tau\tau}^{(k),\pm}/\sigma_{11}^{(k),0}$ (k=1,2) with various $E^{(2)}/E^{(1)}$, ε.

$\frac{E^{(2)}}{E^{(1)}}$	ε	$\frac{\sigma_{nn}^{+}}{\sigma_{11}^{(1),0}}$	$\frac{\sigma_{nn}^{-}}{\sigma_{11}^{(1),0}}$	$\frac{\sigma_{n\tau}^{+}}{\sigma_{11}^{(1),0}}$	$\frac{\sigma_{\tau\tau}^{(1)+}}{\sigma_{11}^{(1),0}}$	$\frac{\sigma_{\tau\tau}^{(2)+}}{\sigma_{11}^{(1),0}}$	$\frac{\sigma_{\tau\tau}^{(1)-}}{\sigma_{11}^{(1),0}}$	$\frac{\sigma_{\tau\tau}^{(2)-}}{\sigma_{11}^{(1),0}}$
20	0.010	0.228	-0.224	-0.200	0.885	0.802	1.102	1.184
	0.015	0.337	-0.327	-0.294	0.824	0.702	1.147	1.268
	0.020	0.437	-0.418	-0.383	0.761	0.602	1.189	1.347
	0.025	0.523	-0.490	-0.465	0.695	0.501	1.230	1.421
50	0.010	0.453	-0.445	-0.318	0.849	0.719	1.130	1.260
	0.015	0.662	-0.638	-0.465	0.770	0.578	1.186	1.377
	0.020	0.853	-0.789	-0.600	0.689	0.439	1.237	1.485
	0.025	0.913	-0.876	-0.714	0.603	0.299	1.286	1.586
100	0.010	0.750	-0.738	-0.416	0.829	0.672	1.147	1.304
	0.015	1.045	-1.021	-0.595	0.740	0.508	1.208	1.439
	0.020	1.212	-1.179	-0.754	0.649	0.348	1.263	1.562
	0.025	1.260	-1.226	-0.907	0.553	0.186	1.317	1.679

4.3.2. COMPOSITE MATERIAL WITH CO-PHASE CURVED LAYERS

Consider composite material that alternates in the direction of the Ox_2 axis

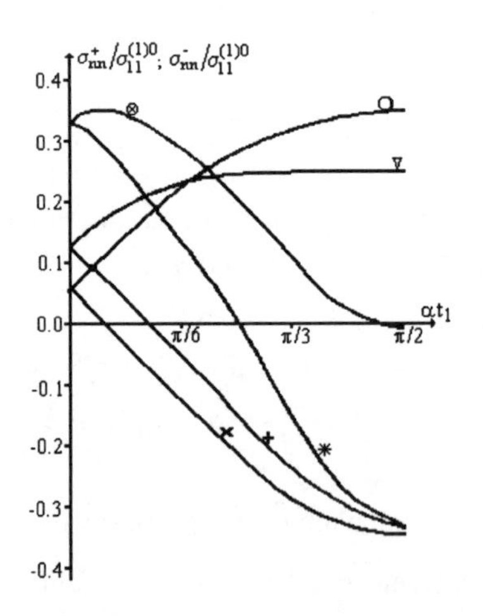

Fig. 4.3.5. The relations between the $\sigma_{nn}^{\pm}/\sigma_{11}^{(1),0}$ and αt_1 for $\varepsilon = 0.015$, $\chi = 0.1$, $E^{(2)}/E^{(1)}$ =50 with $\eta^{(2)}$=0.5, 0.2, 0.1: ⊗, ∇, O -- $\sigma_{nn}^{+}/\sigma_{11}^{(1),0}$; ✱, ✚, ✖ -- $\sigma_{nn}^{-}/\sigma_{11}^{(1),0}$

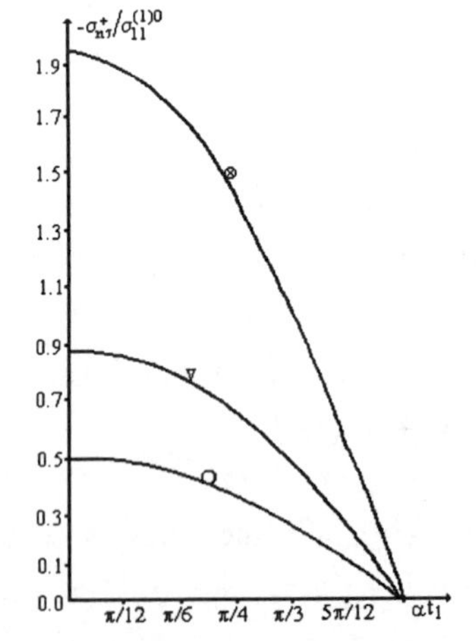

Fig. 4.3.6. The relations between the $\sigma_{n\tau}^{+}/\sigma_{11}^{(1),0}$ and αt_1 for $\varepsilon = 0.015$, $\chi = 0.1$, $E^{(2)}/E^{(1)}$ =50 with $\eta^{(2)}$=0.5, 0.2, 0.1: ⊗, ∇, O -- $\sigma_{n\tau}^{+}/\sigma_{11}^{(1),0}$

with periodic co-phase curving in the direction of the Ox_1 axis (Fig.4.2.1). We investigate the distribution of stresses $\sigma_{nn}^{\pm}$, $\sigma_{n\tau}^{+}$, $\sigma_{\tau\tau}^{(k)\pm}$ (k=1,2) on the surfaces $S^{\pm}$ under the action at infinity of normal uniformly distributed forces with intensity p in the direction of the Ox_1 axis. We use the notation and restrictions described in the previous subsection. The graphs given in Figs. 4.3.5 and 4.3.6 show the dependence of $\sigma_{nn}^{\pm}/\sigma_{11}^{(1),0}$ (Fig.4.3.5), $\sigma_{n\tau}^{+}/\sigma_{11}^{(1),0}$ (Fig.4.3.6) on αt_1 under fixed ε, χ, $E^{(2)}/E^{(1)}$ for various $\eta^{(2)}$; $\eta^{(2)}$ is the concentration of the filler layers in the composite.

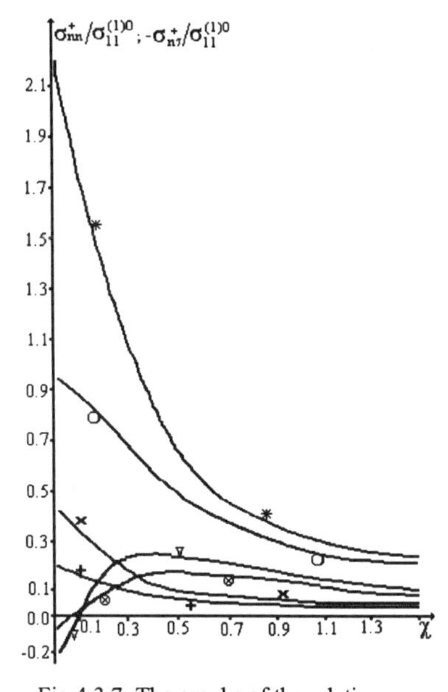

Fig.4.3.7. The graphs of the relation between σ_{nn}^{+}, $\sigma_{n\tau}^{+}$ and χ for $\eta^{(2)}=0.5$, $E^{(2)}/E^{(1)}=20{,}50$:

⊗, ∇ -- $\sigma_{nn}^{+}/\sigma_{11}^{(1),0}$; at point C (Fig.4.2.1)

❍, ✳ -- $\sigma_{n\tau}^{+}/\sigma_{11}^{(1),0}$; at point A (Fig.4.2.1)

✚, ✖ -- $\sigma_{n\tau}^{+}/\sigma_{11}^{(1),0}$; at point A (Fig.4.2.1)

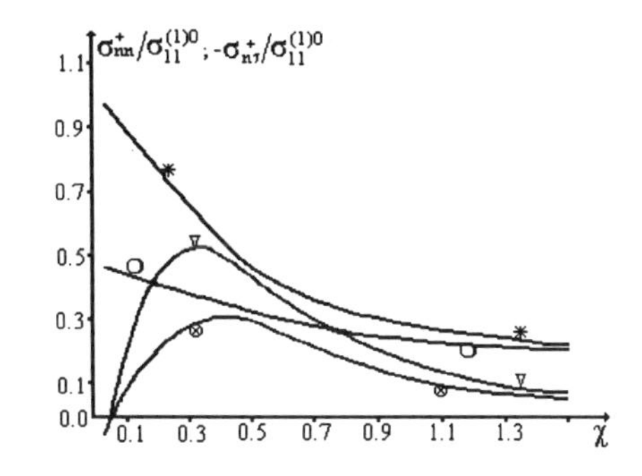

Fig.4.3.8. The graphs of the relation between σ_{nn}^{+}, $\sigma_{n\tau}^{+}$ and χ for $\eta^{(2)}=0.2$, $E^{(2)}/E^{(1)}=20{,}50$:

⊗, ∇ -- $\sigma_{nn}^{+}/\sigma_{11}^{(1),0}$; at point C (Fig.4.2.1)

❍, ✳ -- $\sigma_{n\tau}^{+}/\sigma_{11}^{(1),0}$; at point A (Fig.4.2.1)

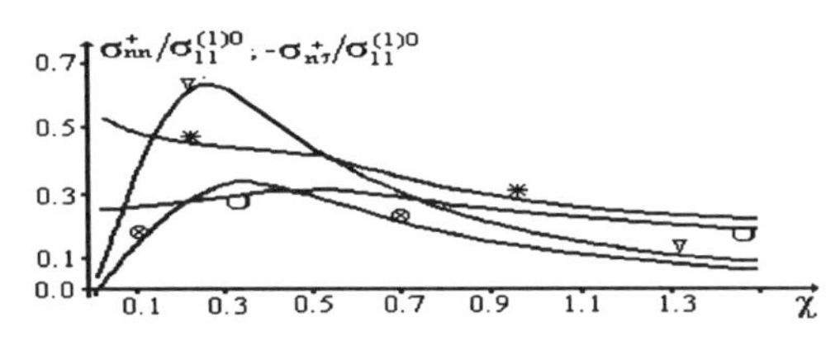

Fig.4.3.9. The graphs of the relation between σ_{nn}^{+}, $\sigma_{n\tau}^{+}$ and χ for $\eta^{(2)}=0.1$, $E^{(2)}/E^{(1)}=20{,}50$:

⊗, ∇ -- $\sigma_{nn}^{+}/\sigma_{11}^{(1),0}$; at point C Fig.4.2.1)

❍, ✳ -- $\sigma_{n\tau}^{+}/\sigma_{11}^{(1),0}$; at point A Fig.4.2.1)

These graphs show that the filler concentration $\eta^{(2)}$ significantly influences the distributions of the stresses $\sigma_{nn}^{\pm}/\sigma_{11}^{(1),0}$, $\sigma_{n\tau}^{+}/\sigma_{11}^{(1),0}$ on the surfaces $S^{\pm}$. As a result of the mutual interaction between curved filler layers the point at which the stress σ_{nn}^{+} has its maximal value moves from C to A with increasing $\eta^{(2)}$. This influence of $\eta^{(2)}$ on the distribution of σ_{nn}^{+} on the inter-medium surface $S^{\pm}$ is observed more clearly when $\chi \leq 0.3$. Note that the character of the dependence of σ_{nn}^{-}, $\sigma_{n\tau}^{+}$ on αt_1 does not change with increasing $\eta^{(2)}$. However, the values of the $\sigma_{nn}^{-}/\sigma_{11}^{(1),0}$, $\sigma_{n\tau}^{+}/\sigma_{11}^{(1),0}$ (specially the values of $\sigma_{n\tau}^{+}/\sigma_{11}^{(1),0}$) grow significantly with increasing $\eta^{(2)}$. Note that as the distribution of the stress $\sigma_{n\tau}^{-}$ on the surface S^{-} is virtually the same as that for $\sigma_{n\tau}^{+}$ on the surface S^{+}, we do not show the results related to $\sigma_{n\tau}^{-}$ here.

We now investigate the influence of $\eta^{(2)}$ on the character of the relation between $\sigma_{nn}^{+}/\sigma_{11}^{(1),0}$, $\sigma_{n\tau}^{+}/\sigma_{11}^{(1),0}$ and the parameter χ. Consider the graphs given in Figs. 4.3.7-4.3.10 which show these dependences with various $\eta^{(2)}$. These graphs show that the non-monotonic character of the relation between $\sigma_{nn}^{+}(C)/\sigma_{11}^{(1),0}$ (*the notation* $\sigma_{ij}^{\pm}(\circ)$ *means the value of* $\sigma_{ij}^{\pm}$ *at point* $(\circ)$) and χ remains for all values of $\eta^{(2)}$. This situation does not hold for the relation between $\sigma_{nn}^{+}(A)/\sigma_{11}^{(1),0}$ and χ, and between $\sigma_{n\tau}^{+}(A)/\sigma_{11}^{(1),0}$ and χ. The numerical results given in Figs.4.3.7-4.3.10 reveal that with increasing $\eta^{(2)}$ and decreasing χ the values of $\sigma_{nn}^{+}(A)/\sigma_{11}^{(1),0}$ and $\sigma_{n\tau}^{+}(A)/\sigma_{11}^{(1),0}$ grow monotonically. However, for smaller $\eta^{(2)}$ the non-monotonic character of these dependences reappears.

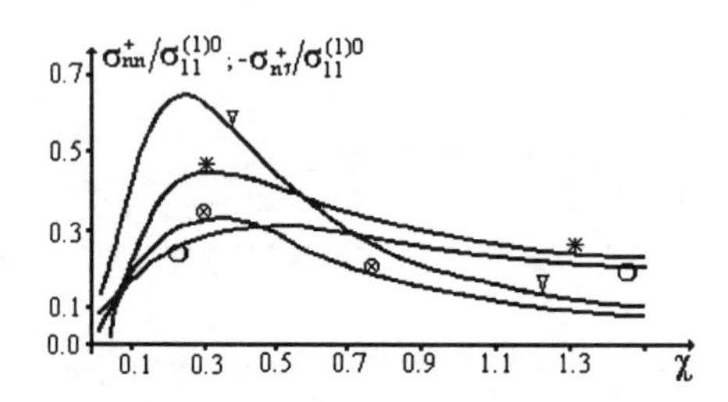

Fig.4.3.10. The graphs of the relation between σ_{nn}^{+}, $\sigma_{n\tau}^{+}$ and χ for $\eta^{(2)}=0.02$, $E^{(2)}/E^{(1)} = 20{,}50$:
$\otimes$, ∇ -- $\sigma_{nn}^{+}/\sigma_{11}^{(1),0}$; at point C Fig.4.2.1)
O, $*$ -- $\sigma_{n\tau}^{+}/\sigma_{11}^{(1),0}$; at point A Fig.4.2.1)

Note that with decreasing $\eta^{(2)}$ the values of $\sigma_{nn}^{+}(A)/\sigma_{11}^{(1),0}$ become insignificant. Moreover note that the relation between $-\sigma_{nn}^{-}(B)/\sigma_{11}^{(1),0}$, $-\sigma_{nn}^{-}(D)/\sigma_{11}^{(1),0}$ and χ coincide with those between $\sigma_{nn}^{+}(A)/\sigma_{11}^{(1),0}$, $\sigma_{nn}^{+}(C)/\sigma_{11}^{(1),0}$ and χ, respectively. Therefore here we do not show the former.

Table 4.2. The values of $\sigma_{nn}^{+}(C)/\sigma_{11}^{(1),0}$ for various approximations and $\eta^{(2)}$ for $\chi = 0.3$, $\varepsilon = 0.02$, $E^{(2)}/E^{(1)} = 50$.

$\eta^{(2)}$	Number of approximations (4.1.15)				
	1	2	3	4	5
0.50	0.437	0.285	0.221	0.250	0.257
0.20	0.782	0.772	0.675	0.680	0.695
0.10	0.911	0.922	0.821	0.821	0.839
0.02	0.924	0.937	0.836	0.835	0.853
0.00	0.924	0.937	0.836	0.835	0.853

Table 4.3. The values of $\sigma_{nn}^{-}(D)/\sigma_{11}^{(1),0}$ for various approximations and $\eta^{(2)}$ for $\chi = 0.3$, $\varepsilon = 0.02$, $E^{(2)}/E^{(1)} = 50$.

$\eta^{(2)}$	Number of approximations (4.1.15)				
	1	2	3	4	5
0.50	-0.437	-0.588	-0.524	-0.495	-0.485
0.20	-0.782	-0.793	-0.697	-0.692	-0.672
0.10	-0.911	-0.899	-0.798	-0.798	-0.778
0.02	-0.924	-0.910	-0.809	-0.810	-0.789
0.00	-0.924	-0.910	-0.809	-0.810	-0.789

Table 4.4. The values of $\sigma_{n\tau}^{+}(A)/\sigma_{11}^{(1),0}$ for various approximations and $\eta^{(2)}$ for $\chi = 0.3$, $\varepsilon = 0.02$, $E^{(2)}/E^{(1)} = 50$.

$\eta^{(2)}$	Number of approximations (4.1.15)				
	1	2	3	4	5
0.50	-1.629	-1.629	-1.394	-1.394	-1.450
0.20	-0.880	-0.880	-0.768	-0.768	-0.796
0.10	-0.673	-0.673	-0.598	-0.598	-0.619
0.02	-0.651	-0.651	-0.579	-0.579	-0.600
0.00	-0.651	-0.651	-0.579	-0.579	-0.600

In Tables 4.2-4.4 give the values of $\sigma_{nn}^{+}(C)/\sigma_{11}^{(1),0}$, $\sigma_{nn}^{-}(D)/\sigma_{11}^{(1),0}$, $\sigma_{n\tau}^{+}(A)/\sigma_{11}^{(1),0}$ for fixed χ, ε and $E^{(2)}/E^{(1)}$, for various approximations (4.1.15) and $\eta^{(2)}$. These Tables show that for $\eta^{(2)} \geq 0.2$ the self-balanced tangential stress $\sigma_{n\tau}^{+}(A)$ is more significant than the self-balanced normal stresses $\sigma_{nn}^{+}(C)$, $\sigma_{nn}^{-}(D)$. But, when $\eta^{(2)} \leq 0.1$ the normal stresses dominate the tangential ones. This may be explained by the increase of the reinforcement of the filler layers in the direction of the normal vector **n** (Fig.4.2.1) as $\eta^{(2)}$ increases. When $\eta^{(2)}$ is large the shear deformation dominates, and the self-balanced tangential stress is greater than the self-balanced normal stress.

Table 4.5. The values of $\sigma^{+}_{n\tau}(A)/\sigma^{(1),0}_{11}$ for $\chi=0.3$, $\eta^{(2)}=0.5$ for various $E^{(2)}/E^{(1)}$, ε and the approximation (4.1.15).

ε	Num. of appr. (4.1.15)	$E^{(2)}/E^{(1)}$		
		20	50	100
	1	-0.483	-0.814	-1.066
0.010	2	-0.483	-0.814	-1.066
	3	-0.472	-0.785	-1.010
	4	-0.472	-0.785	-1.010
	5	-0.473	-0.787	-1.020
	1	-0.967	-1.625	-2.133
0.020	2	-0.967	-1.625	-2.133
	3	-0.880	-1.391	-1.747
	4	-0.880	-1.391	-1.747
	5	-0.886	-1.450	-1.866
	1	-1.209	-2.031	-2.665
0.025	2	-1.209	-2.031	-2.665
	3	-1.039	-1.570	-1.898
	4	-1.039	-1.570	-1.898
	5	-1.089	-1.761	-2.277

Table 4.6. The values of $\sigma^{+}_{nn}(C)/\sigma^{(1),0}_{11}$ for $\chi=0.3$, with various $E^{(2)}/E^{(1)}$, ε and $\eta^{(2)}$.

$\eta^{(2)}$	ε	$E^{(2)}/E^{(1)}$			
		20	50	100	150
	0.010	0.226	0.446	0.641	0.750
0.10	0.015	0.333	0.651	0.907	1.047
	0.020	0.430	0.839	1.086	1.220
	0.025	0.513	0.897	1.121	1.323
	0.010	0.228	0.453	0.653	0.970
0.02	0.015	0.337	0.662	0.924	1.311
	0.020	0.437	0.853	1.107	1.417
	0.025	0.523	0.913	1.242	1.498

We now consider the influence of $E^{(2)}/E^{(1)}$ and ε on the values of $\sigma^{+}_{n\tau}(A)/\sigma^{(1),0}_{11}$ and $\sigma^{+}_{nn}(C)/\sigma^{(1),0}_{11}$. Tables 4.5 and 4.6 show that the values of $\sigma^{+}_{nn}(C)$ and $\sigma^{+}_{n\tau}(A)$ increase monotonically with growing $E^{(2)}/E^{(1)}$ and $\eta^{(2)}$. For some values of the parameters these stresses are significantly greater than the stress $\sigma^{(1),0}_{11}$ which balances the external forces p (Fig.4.2.1).

The comparison of the numerical results obtained for various approximations and given in Tables 4.2-4.5 show that the solution method is highly effective. The numerical results have been obtained for the case in which the Poisson's ratios of the matrix and filler materials are equal, i.e. $\nu^{(1)}=\nu^{(2)}=0.3$. Other numerical results show that the influence of the difference between $\nu^{(1)}$ and $\nu^{(2)}$ on the distribution of the self-balanced stresses is insignificant. For instance, if $\eta^{(2)}=0.5$, $E^{(2)}/E^{(1)}=50$, $\chi=0.3$, $\varepsilon=0.015$, $\nu^{(2)}=0.3$ we obtain $\sigma_{n\tau}^{+}(A)/\sigma_{11}^{(1),0}=1.239$, 1.200, 1.137, 1.049 for $\nu^{(1)}=0.1$, 0.2, 0.3, 0.4, respectively.

We now consider the numerical results related to the normal stresses $\sigma_{\tau\tau}^{(k)\pm}$ that act in the direction of the tangential vector $\boldsymbol{\tau}$ (Fig.4.2.1) at the points of the surfaces $S^{\pm}$. Consider the graphs given in Fig.4.3.11 which show the dependence of $\sigma_{\tau\tau}^{(k)\pm}/\sigma_{11}^{(k),0}$ on αt_1 for various $\eta^{(2)}$ and with fixed ε, χ and $E^{(2)}/E^{(1)}$. These graphs show that the mutual interaction between the curved reinforcing layers does not change the character of these dependences. In this case the self-balanced part of the stresses $\sigma_{\tau\tau}^{(k)\pm}$ increases with growing $\eta^{(2)}$. This conclusion corroborated by Table

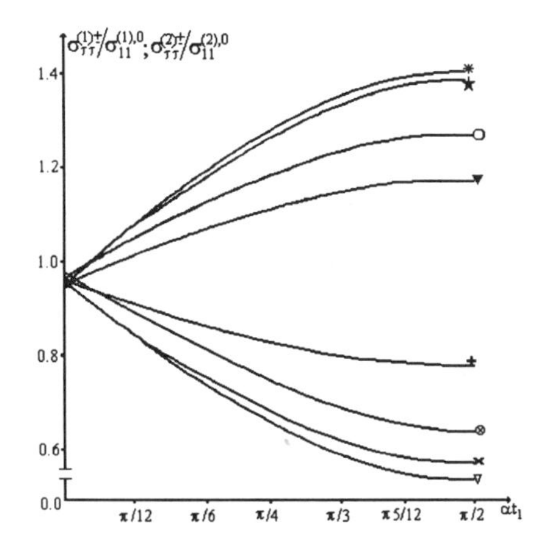

Fig.4.3.11. The distribution of the stresses $\sigma_{\tau\tau}^{(1)\pm},\sigma_{\tau\tau}^{(2)\pm}$ with respect to αt_1 for $\varepsilon=0.015$, $\chi=0.4$, $E^{(2)}/E^{(1)}=50$, $\eta^{(2)}=0.5, 0.2$: ⊗, ✚ -- $\sigma_{\tau\tau}^{(1)+}/\sigma_{11}^{(1),0}$; ∇, ✖ -- $\sigma_{\tau\tau}^{(1)-}/\sigma_{11}^{(1),0}$; ○, ▼ -- $\sigma_{\tau\tau}^{(2)+}/\sigma_{11}^{(2),0}$; ✳, ★ -- $\sigma_{\tau\tau}^{(2)-}/\sigma_{11}^{(2),0}$.

4.7. Table 4.7 as shows that as $\eta^{(2)}\to 0$ the values of $\sigma_{\tau\tau}^{(k)\pm}/\sigma_{11}^{(k),0}$ approach the values obtained in the previous subsection.

Investigate the influence of $\eta^{(2)}$ on the character of the relation between $\sigma_{\tau\tau}^{(k)\pm}/\sigma_{11}^{(k),0}$ and χ. For this purpose consider the graphs indicated in Fig.4.3.12 which are constructed with various $\eta^{(2)}$. These results show that the mutual interaction between the curved reinforcing layers does not change the non-monotonic character of these dependences. However with increasing $\eta^{(2)}$ this character of the considered dependence is observed more clearly.

Table 4.7. The values of $\sigma_{\tau\tau}^{(k)\pm}/\sigma_{11}^{(k),0}$ (k=1,2) for $\chi=0.3$, $\varepsilon=0.015$ for various $E^{(2)}/E^{(1)}$, and $\eta^{(2)}$.

$E^{(2)}/E^{(1)}$	$\eta^{(2)}$	$\sigma_{\tau\tau}^{(1)+}(C)/\sigma_{11}^{(1),0}$	$\sigma_{\tau\tau}^{(1)-}(D)/\sigma_{11}^{(1),0}$	$\sigma_{\tau\tau}^{(2)+}(C)/\sigma_{11}^{(121,0}$	$\sigma_{\tau\tau}^{(2)-}(D)/\sigma_{11}^{(2),0}$
50	0.50	0.624	1.263	0.521	1.405
	0.20	0.763	1.165	0.549	1.380
	0.10	0.797	1.137	0.560	1.341
	0.02	0.798	1.136	0.560	1.370
	0.00	0.798	1.136	0.560	1.370
100	0.50	0.546	1.312	0.420	1.488
	0.20	0.716	1.193	0.443	1.468
	0.10	0.760	1.115	0.452	1.460
	0.02	0.761	1.115	0.452	1.460
	0.00	0.761	1.115	0.452	1.460

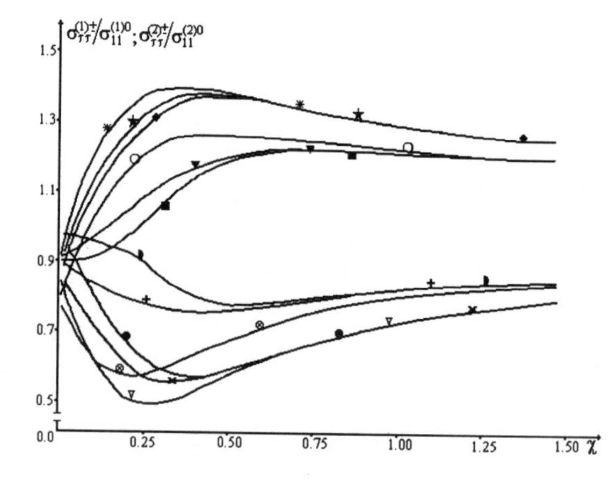

Fig.4.3.12. The graphs of the dependence of $\sigma_{\tau\tau}^{(k)\pm}$ on χ for $\varepsilon=0.015$, $E^{(2)}/E^{(1)}=50$, $\eta^{(2)}$=0.5, 0.2, 0.1: ⊗, ✚, ◗ -- $\sigma_{\tau\tau}^{(1)+}/\sigma_{11}^{(1),0}$; at point C (Fig.4.2.1), ∇, ✖, ● -- $\sigma_{\tau\tau}^{(2)+}/\sigma_{11}^{(2),0}$; at point C (Fig.4.2.1), ❍, ▼, ■ -- $\sigma_{\tau\tau}^{(1)-}/\sigma_{11}^{(1),0}$; at point D (Fig.4.2.1) ✳, ★, ◆ -- $\sigma_{\tau\tau}^{(2)-}/\sigma_{11}^{(2),0}$; at point D (Fig.4.2.1)

Multi-layered composites. We compare the values of σ_{nn}^{+} obtained in the framework of the piecewise-homogeneous material with those obtained from the continuum approaches presented in Chapters 1 and 2. For this purpose we consider a multi-layered composite material with

$$\ell > L, \quad \ell >> H^{(2)}, \quad \ell >> H^{(1)}, \quad L >> H^{(2)}, \quad L >> H^{(1)}, \quad H^{(2)} \approx H^{(1)}. \tag{4.3.1}$$

The numerical investigation carried out in the framework of the piecewise homogeneous body model shows that in the case (4.3.1) for fixed L and ℓ (i.e. for fixed ε) and with $\chi \le 0.05$; $0.2 \le \eta^{(2)} \le 0.5$ the maximal values of σ_{nn}^{+} are obtained at points which are determined by the conditions $\alpha t_1 = n\pi$ and $\alpha t_1 = \pi/2 + 2n\pi$, n=0, 1, 2, ... In this case this stress is tensile at $\alpha t_1 = n\pi$, and compressive at $\alpha t_1 = \pi/2 + 2n\pi$. The character of the stress σ_{nn}^{+} agrees with that obtained from the continuum approach presented in Chapter 1 and written by expression (1.9.8).

At points $\alpha t_1 = n\pi$ the stress σ_{nn}^{+} in piecewise-homogeneous body model can be presented up to ε^2 as

$$\sigma_{nn}^{+} = Kp\varepsilon^2, \qquad K = K\left(\eta^{(2)}, \chi, E^{(2)}/E^{(1)}\right). \tag{4.3.2}$$

Table 4.8 gives the values of K for $E^{(2)}/E^{(1)}$ =50, $\eta^{(2)}$=0.5 and various χ. Table 4.9 gives the values of K for various $E^{(2)}/E^{(1)}$ and χ=0.01, $\eta^{(2)}$=0.5. These tables show that with decreasing χ and increasing $E^{(2)}/E^{(1)}$ the influence of χ and $E^{(2)}/E^{(1)}$ on the values of K decreases. For the continuum approaches the following

Table 4.8. The values of K with various χ in the case where $E^{(2)}/E^{(1)} = 50$

χ	0.05	0.04	0.03	0.02	0.01
K	82.4	83.6	84.4	84.4	85.1

Table 4.9. The values of K with various $E^{(2)}/E^{(1)}$ in the case where χ=0.01

$E^{(2)}/E^{(1)}$	20	50	100
K	81.1	85.1	86.0

values are obtained for K: K $\approx$ 45.5 in the continuum approach presented in Chapter 1.; K $\approx$ 68.95 in that presented in Chapter 2. The continuum approach of Chapter 2 is more accurate than that of Chapter 1, although both agree qualitatively with the piecewise-homogeneous body model.

4.3.3. COMPOSITE MATERIAL WITH ANTI-PHASE CURVED LAYERS

Consider a layered composite material with an infinity of alternating layers in the direction of the Ox_2 axis, periodically curving in anti-phase in the direction Ox_1, and assume that this composite is loaded at infinity by the normal forces with intensity p in the direction of the Ox_1 axis (Fig. 4.2.3). We retain the notation and restrictions given in the previous subsection and suppose $L < H^{(1)}$, i.e. neighbouring reinforcing layers do not contact each other. We investigate the plain-strain state in this composite material, and taking into account the periodicity of the curving of the reinforcing layers, we consider only the stress distribution in the part CBA on the surface S^{+}; this corresponds to αt_1 in $[-\pi/2, \pi/2]$; the values $\alpha t_1 = -\pi/2, 0, \pi/2$ correspond to C, B, A, respectively. We take $\nu^{(1)} = \nu^{(2)} = 0.3$.

First we investigate the distributions of the self-balanced stresses σ_{nn} and $\sigma_{n\tau}$. Fig.4.3.13 gives the relation between $\sigma_{nn}/\sigma_{11}^{(1),0}$, $\sigma_{n\tau}/\sigma_{11}^{(1),0}$ and αt_1 for

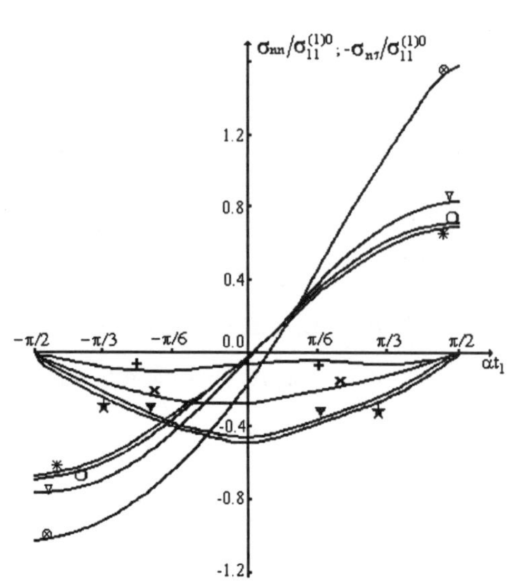

Fig.4.3.13. The graphs of the dependence of $\sigma_{nn}/\sigma_{11}^{(1),0}$, $\sigma_{n\tau}/\sigma_{11}^{(1),0}$ on αt_1 for ε=0.015, $E^{(2)}/E^{(1)}$=50, χ=0.3 with $\eta^{(2)}$=0.5, 0.2, 0.1, 0.02 : ⊗, ∇, O, ✳ -- $\sigma_{nn}/\sigma_{11}^{(1),0}$, ✚, ✖, ▼, ★ -- $\sigma_{n\tau}/\sigma_{11}^{(1),0}$.

various $\eta^{(2)}$.

This figure shows that as the concentration of the filler layers, i.e. $\eta^{(2)}$, increases, the self-balanced normal stress σ_{nn} grows monotonically, and the self-balanced tangential stress $\sigma_{n\tau}$ decreases. In this case σ_{nn} has its absolute maxima at C and A (Fig.4.2.3); at A σ_{nn} is tensile, at C it is compressive. The sizes of the parts of CBA on which σ_{nn} is tensile or compressive depends on $\eta^{(2)}$. The numerical results show that for comparatively small values of $\eta^{(2)}$, i.e. $\eta^{(2)} \leq 0.2$, the tangential self-balanced stress $\sigma_{n\tau}$ has its maximum at B; while $\eta^{(2)}$=0.5 the maximum occurs near the middle point of CB (Fig.4.2.3).

The comparison of numerical results for σ_{nn} and $\sigma_{n\tau}$ when $\eta^{(2)}$=0.02 and those for $\eta^{(2)}$=0 given in sub-section 4.3.1 shows that these results differently by terms of order $10^{-3}-10^{-5}$. Consequently, $\eta^{(2)} \leq 0.02$ we can neglect the mutual interaction between the reinforcing layers.

Now we consider the relations between σ_{nn}, $\sigma_{n\tau}$ and parameter χ, as shown in Fig. 4.3.14 for various $\eta^{(2)}$. Again the results for $\eta^{(2)}$=0.02 and $\eta^{(2)}$=0.00 coincide.

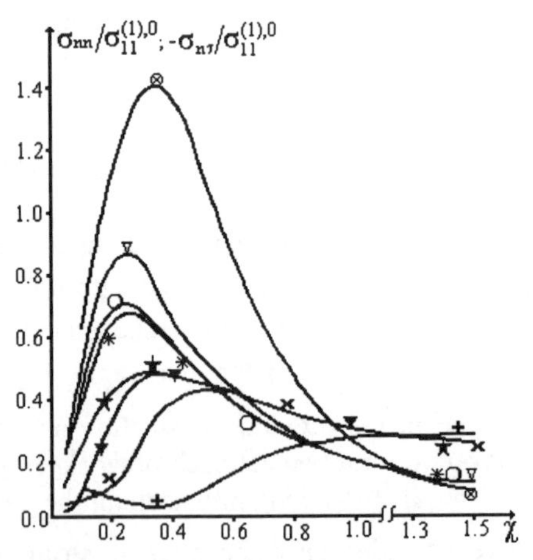

Fig.4.3.14. The graphs of the dependence of $\sigma_{nn}/\sigma_{11}^{(1),0}$, $\sigma_{n\tau}/\sigma_{11}^{(1),0}$ on χ for ε=0.015, $E^{(2)}/E^{(1)}$=50 with $\eta^{(2)}$=0.5, 0.2, 0.1, 0.02 : ⊗, ∇, O, ✳ -- $\sigma_{nn}/\sigma_{11}^{(1),0}$ at point A (Fig.4.2.3), ✚, ✖, ▼, ★ -- $\sigma_{n\tau}/\sigma_{11}^{(1),0}$ at point B (Fig.4.2.3).

These graphs show that the mutual interaction between the reinforcing layers does not change the character of the dependence of σ_{nn} on χ; the dependence is always non-monotonic; σ_{nn} always increases with $\eta^{(2)}$.

However, this mutual interaction does significantly change the relations between $\sigma_{n\tau}$ and χ. The numerical results show that if $\eta^{(2)} \geq 0.1$ $\sigma_{n\tau}$ is significantly less than σ_{nn} and thus has no practical interest.

Table 4.9 gives the values of $\sigma_{nn}(A)$ for various $\eta^{(2)}$, ε and $E^{(2)}/E^{(1)}$ (Fig.4.2.3.). The numerical results for $E^{(2)}/E^{(1)}$ =20 are calculated with χ=0.4; $E^{(2)}/E^{(1)}$ =50,100,150 with χ=0.3. Table 4.9 shows that the values of the self-balanced normal stress σ_{nn} grow monotonically with ε and $E^{(2)}/E^{(1)}$. Also, when $\varepsilon \geq 0.01$ the values of σ_{nn} can be significantly greater than the values $\sigma_{11}^{(1),0}$ which balance the external forces p.

Table 4.10 illustrates the convergence of the various approximations (4.1.15) for $\sigma_{nn}(A)/\sigma_{11}^{(1),0}$; the convergence is acceptable.

We now consider the distribution of the normal stresses $\sigma_{\tau\tau}^{(k)}$ (k=1,2) which act in the direction of the vector $\boldsymbol{\tau}$. Taking the periodicity of the curving into account,

Table 4.9. The values of $\sigma_{nn}(A)/\sigma_{11}^{(1),0}$ for various $E^{(2)}/E^{(1)}$, ε and $\eta^{(2)}$.

$\eta^{(2)}$	ε	$E^{(2)}/E^{(1)}$			
		20	50	100	150
	0.010	0.399	0.874	1.602	2.204
0.5	0.015	0.621	1.382	2.526	3.465
	0.020	0.859	1.946	3.548	4.853
	0.025	1.116	2.578	4.695	6.413
	0.030	1.395	3.296	6.011	8.222
	0.010	0.247	0.554	0.845	1.020
0.2	0.015	0.368	0.820	1.239	1.488
	0.020	0.484	1.074	1.546	1.927
	0.025	0.595	1.323	1.978	2.377
	0.030	0.704	1.583	2.396	2.926

we consider only the part CBA on the surface S_1^+. Fig. 4.3.15 shows the relation between $\sigma_{\tau\tau}^{(k)}/\sigma_{11}^{(k),0}$ (k=1,2) and αt_1 for various $\eta^{(2)}$; the graphs for $\eta^{(2)}$=0.02 and $\eta^{(2)}$=0.00 coincide. Fig.4.3.15 shows that when $\eta^{(2)} \geq 0.2$ the stress $\sigma_{\tau\tau}^{(1)}$ is compressive on BA and tensile on CB (Fig.4.2.3). For all $\eta^{(2)}$, $\sigma_{\tau\tau}^{(2)}$ is a tensile on CB

Table 4.10. The values of $\sigma_{nn}(A)/\sigma_{11}^{(1),0}$ in the first five approximations (4.1.15) for various $E^{(2)}/E^{(1)}$ for $\eta^{(2)}=0.5$, $\chi=0.3$.

ε	Numb. of appr.	$E^{(2)}/E^{(1)}$		
		50	100	150
0.020	1	1.573	2.892	3.988
	2	1.573	3.523	4.859
	3	1.916	3.497	4.773
	4	1.923	3.515	4.788
	5	1.946	3.548	4.853
0.030	1	2.360	4.338	5.982
	2	3.132	5.759	7.942
	3	3.154	5.669	7.652
	4	3.224	5.769	7.726
	5	3.296	6.011	8.222

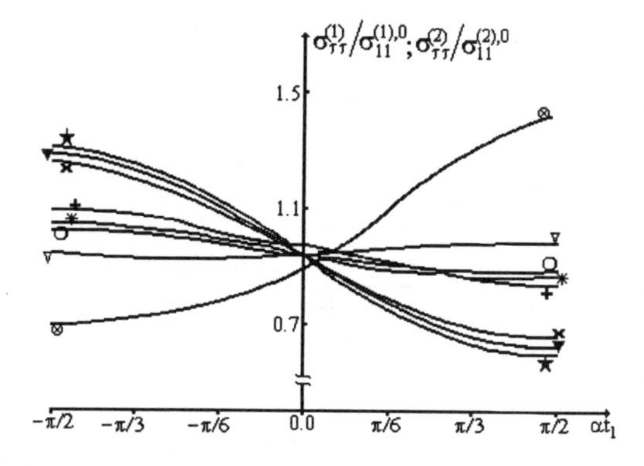

Fig.4.3.15. The dependence of $\sigma_{\tau\tau}^{(1)}/\sigma_{11}^{(1),0}$, $\sigma_{\tau\tau}^{(2)}/\sigma_{11}^{(1),0}$ on αt_1 for $\varepsilon=0.015$, $E^{(2)}/E^{(1)}=50$, $\chi=0.3$ with $\eta^{(2)}=0.5, 0.2, 0.1, 0.02$:
⊗, ∇, O, ✳ -- $\sigma_{\tau\tau}^{(1)}/\sigma_{11}^{(1),0}$,
✚, ✖, ▼, ★ -- $\sigma_{\tau\tau}^{(2)}/\sigma_{11}^{(1),0}$

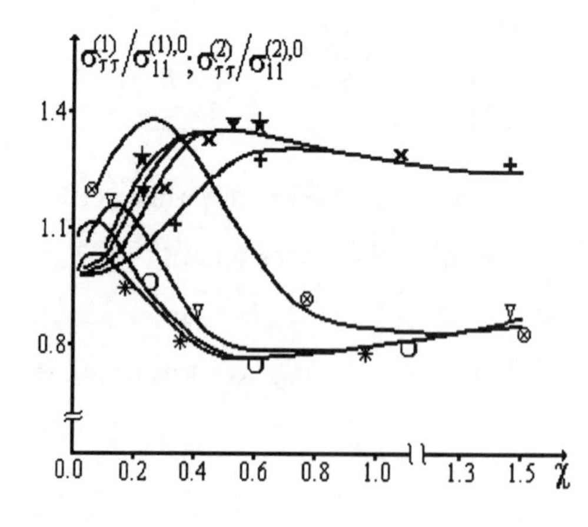

Fig.4.3.16. The dependence of $\sigma_{\tau\tau}^{(1)}/\sigma_{11}^{(1),0}$, $\sigma_{\tau\tau}^{(2)}/\sigma_{11}^{(1),0}$ on χ for $\varepsilon=0.015$, $E^{(2)}/E^{(1)}=50$ with $\eta^{(2)}=0.5, 0.2, 0.1, 0.02$:
⊗, ∇, O, ✳ -- $\sigma_{\tau\tau}^{(1)}/\sigma_{11}^{(1),0}$ at point A (Fig.4.2.3),
✚, ✖, ▼, ★ -- $\sigma_{\tau\tau}^{(2)}/\sigma_{11}^{(1),0}$ at point C Fig.4.2.3).

and compressive stress on BA. As $\eta^{(2)}$ increases the absolute value of the self-balanced part of the stress $\sigma_{\tau\tau}^{(1)}$ increase; but that of $\sigma_{\tau\tau}^{(2)}$ decreases.

Fig.4.3.16 shows the dependence of $\sigma_{\tau\tau}^{(k)}/\sigma_{11}^{(k),0}$ (k=1,2) on χ for various $\eta^{(2)}$; $\sigma_{\tau\tau}^{(1)}$ is calculated at A, and $\sigma_{\tau\tau}^{(2)}$ at C. These graphs show that the interaction between reinforcing layers qualitatively changes the character of the dependence of $\sigma_{\tau\tau}^{(1)}$ on χ, but does not change the character of the dependence of $\sigma_{\tau\tau}^{(2)}$ on χ.

Table 4.11 gives the values of $\sigma_{\tau\tau}^{(1)}(A)/\sigma_{11}^{(1),0}$ and $\sigma_{\tau\tau}^{(2)}(C)/\sigma_{11}^{(2),0}$ calculated for various ε and $E^{(2)}/E^{(1)}$; that the absolute values of self-balanced part of the stresses $\sigma_{\tau\tau}^{(k)}$ (k=1,2) grow monotonically with increasing ε and $E^{(2)}/E^{(1)}$.

Table 4.11. The values of $\sigma_{\tau\tau}^{(k)}(A)/\sigma_{11}^{(k),0}(k=1,2)$ with various $E^{(2)}/E^{(1)}$, ε for $\eta^{(2)}$=0.5, χ=0.3.

	$E^{(2)}/E^{(1)}$	ε			
		0.015	0.020	0.025	0.030
$\frac{\sigma_{\tau\tau}^{(1)}(A)}{\sigma_{11}^{(1),0}}$	20	1.081	1.126	1.175	1.237
	50	1.413	1.596	1.811	2.070
	100	1.827	2.181	2.597	3.095
	150	2.166	2.659	3.236	3.930
$\frac{\sigma_{\tau\tau}^{(2)}(C)}{\sigma_{11}^{(2),0}}$	20	1.142	1.181	1.216	1.246
	50	1.221	1.258	1.280	1.370
	100	1.305	1.384	1.455	1.522
	150	1.360	1.452	1.534	1.616

4.4. Stress Distribution in Composites with Partially Curved Layers

The analysis of sections of various composite materials shows that some cases occur when some of the reinforcing elements, relative to others, with sufficient accuracy, may be taken as ideal (non-curved). For these reasons, in the present section the stress-strain state is investigated in laminated composites with various types of the partial curving in the structure.

4.4.1. ALTERNATING STRAIGHT AND CO-PHASE CURVED LAYERS

Fig.4.4.1 shows the structure. We assume that there is an infinity of layers, the curving is in the direction of the Ox_1 axis, and is periodic. Using the analysis in the previous sections we investigate the plane-strain state for $\sigma_{11} = p$ at infinity.

Solution method.

Taking into account the period $4\left(H^{(1)}+H^{(2)}\right)$ in the direction of the Ox_2 we single out four layers, i.e. $1^{(1)},1^{(2)},2^{(1)},2^{(2)}$ (Fig.4.4.1). The equation of the middle surface of layer $2^{(2)}$ we take as $x_{22}^{(2)}=L\sin(2\pi x_1/\ell)$; we assume that $L<\ell$ and introduce the small parameter $\varepsilon=L/\ell$.

We consider approximation (4.1.15). For the zeroth approximation we obtain the following expressions.

$$\sigma_{ij}^{(k)m,0}=\sigma_{ij}^{(k),0},\quad u_1^{(k)m,0}=u_1^{(k),0},$$

$$u_2^{(k)m,0}=-\frac{\nu^{(k)}\left(1+\nu^{(k)}\right)}{E^{(k)}}\sigma_{11}^{(k),0}x_{2m}^{(k)}+C^{(k)m},\quad C^{(k)m}=\text{const}, \tag{4.4.1}$$

where $\sigma_{ij}^{(k),0}$ and $u_1^{(k),0}$ are determined by the formulae (4.2.5), (4.2.7).

We now consider the determination of the values of the first approximation. In since $f_m\equiv 0$ for the straight layers we obtain the following contact condition for the first approximation from (4.1.22)-(4.1.26).

$$\sigma_{i2}^{(2)1,1}\Big|_{\left(t_1,-H^{(2)}\right)}-\sigma_{i2}^{(1)1,1}\Big|_{\left(t_1,H^{(1)}\right)}=0,\quad u_i^{(2)1,1}\Big|_{\left(t_1,-H^{(2)}\right)}-u_i^{(1)1,1}\Big|_{\left(t_1,H^{(1)}\right)}=0,$$

$$\sigma_{i2}^{(2)2,1}\Big|_{\left(t_1,H^{(2)}\right)}-\sigma_{i2}^{(1)1,1}\Big|_{\left(t_1,-H^{(1)}\right)}=2\pi\delta_i^1\left(\sigma_{11}^{(2),0}-\sigma_{11}^{(1),0}\right)\cos(\alpha t_1),$$

$$u_i^{(2)2,1}\Big|_{\left(t_1,H^{(2)}\right)}-u_i^{(1)1,1}\Big|_{\left(t_1,-H^{(1)}\right)}=\ell\delta_i^2\left(\frac{\nu^{(2)}(1+\nu^{(2)})}{E^{(2)}}\sigma_{11}^{(2),0}-\frac{\nu^{(1)}(1+\nu^{(1)})}{E^{(1)}}\sigma_{11}^{(1),0}\right)\times$$

$$\sin(\alpha t_1),\quad \sigma_{i2}^{(2)2,1}\Big|_{\left(t_1,-H^{(2)}\right)}-\sigma_{i2}^{(1)2,1}\Big|_{\left(t_1,H^{(1)}\right)}=2\pi\delta_i^1\left(\sigma_{11}^{(2),0}-\sigma_{11}^{(1),0}\right)\cos(\alpha t_1),$$

$$u_i^{(2)2,1}\Big|_{\left(t_1,-H^{(2)}\right)}-u_i^{(1)2,1}\Big|_{\left(t_1,H^{(1)}\right)}=\ell\delta_i^2\left(\frac{\nu^{(2)}(1+\nu^{(2)})}{E^{(2)}}\sigma_{11}^{(2),0}-\frac{\nu^{(1)}(1+\nu^{(1)})}{E^{(1)}}\sigma_{11}^{(1),0}\right)\times$$

$$\sin(\alpha t_1),\ \sigma_{i2}^{(2)1,1}\Big|_{\left(t_1,H^{(2)}\right)}-\sigma_{i2}^{(1)2,1}\Big|_{\left(t_1,-H^{(1)}\right)}=0,\quad u_i^{(2)1,1}\Big|_{\left(t_1,H^{(2)}\right)}-u_i^{(1)2,1}\Big|_{\left(t_1,-H^{(1)}\right)}=0,$$

$$i=1,2,\quad \delta_1^1=\delta_2^2=1,\quad \delta_2^1=\delta_1^2=0. \tag{4.4.2}$$

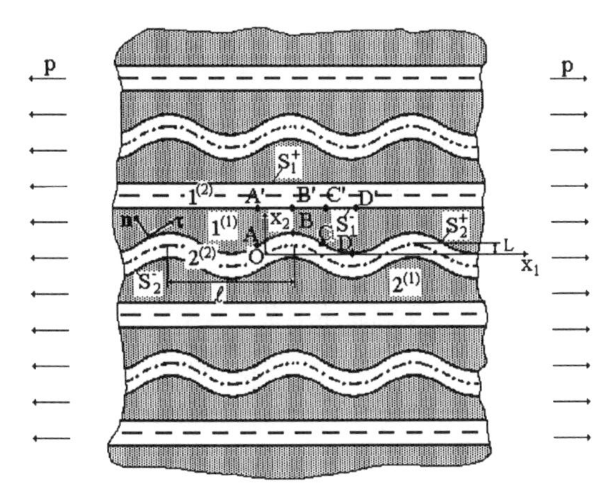

Fig.4.4.1. Structure of the composite material with alternating every second noncurved layer and co-phase periodically curved layers.

Taking the form of the expressions entering the right side of the conditions (4.4.2) into account, we take the harmonic functions (4.2.10) as in (4.2.23). Using the relations (4.2.9), (4.1.16) we obtain from (4.4.2) the system of equations for unknown constants entering in the expressions (4.2.23). After this procedure we determine the values of the first approximation and present they as (4.2.24).

Consider the second approximation. In this case we obtain the following contact conditions from the relations (4.4.1), (4.2.24), (4.1.22)-(4.1.26).

$$\sigma_{i2}^{(2)1,2}\Big|_{(t_1,-H^{(2)})} - \sigma_{i2}^{(1)1,2}\Big|_{(t_1,H^{(1)})} = 0, \quad u_i^{(2)1,2}\Big|_{(t_1,-H^{(2)})} - u_i^{(1)1,2}\Big|_{(t_1,H^{(1)})} = 0$$

$$\sigma_{12}^{(2)2,2}\Big|_{(t_1,H^{(2)})} - \sigma_{12}^{(1)1,2}\Big|_{(t_1,-H^{(1)})} = \Bigg[\pi H^{(2)}\left(\psi_{12}^{(1)1,1}\Big|_{(-H^{(1)})} - \psi_{12}^{(2)2,1}\Big|_{(H^{(2)})}\right) +$$

$$\pi\left(\frac{d\psi_{12}^{(1)1,1}}{dx_{21}^{(1)}}\Bigg|_{(-H^{(1)})} - \frac{d\psi_{12}^{(2)2,1}}{dx_{22}^{(2)}}\Bigg|_{(H^{(2)})}\right) - \pi\left(\psi_{11}^{(2)2,1}\Big|_{(H^{(2)})} - \psi_{11}^{(1)1,1}\Big|_{(-H^{(1)})}\right)\Bigg]\sin(2\alpha t_1),$$

$$\sigma_{22}^{(2)2,2}\Big|_{(t_1,H^{(2)})} - \sigma_{22}^{(1)1,2}\Big|_{(t_1,-H^{(1)})} = 2\pi\left(\psi_{12}^{(2)2,1}\Big|_{(H^{(2)})} - \psi_{12}^{(1)1,1}\Big|_{(-H^{(1)})}\right)\cos(2\alpha t_1),$$

$$u_1^{(2)2,2}\Big|_{\left(t_1,H^{(2)}\right)} - u_1^{(1)1,2}\Big|_{\left(t_1,-H^{(1)}\right)} = \frac{\ell}{2}\left(\frac{d\varphi_1^{(1)1,1}}{dx_{21}^{(1)}}\Bigg|_{\left(-H^{(1)}\right)} - \frac{d\varphi_1^{(2)2,1}}{dx_{22}^{(2)}}\Bigg|_{\left(H^{(2)}\right)}\right)\sin(2\alpha t_1) +$$

$$\pi H^{(2)}\left(\varphi_1^{(1)1,1}\Big|_{\left(-H^{(1)}\right)} - \varphi_1^{(2)2,1}\Big|_{\left(H^{(2)}\right)}\right)\sin(2\alpha t_1),$$

$$u_2^{(2)2,2}\Big|_{\left(t_1,H^{(2)}\right)} - u_2^{(1)1,2}\Big|_{\left(t_1,-H^{(1)}\right)} = \ell\sin^2(\alpha t_1)\left(\frac{d\varphi_2^{(1)1,1}}{dx_{21}^{(1)}}\Bigg|_{\left(-H^{(1)}\right)} - \frac{d\varphi_2^{(2)2,1}}{dx_{22}^{(2)}}\Bigg|_{\left(H^{(2)}\right)}\right) +$$

$$2\pi H^{(2)}\cos^2(\alpha t_1)\left(\varphi_2^{(1)1,1}\Big|_{\left(-H^{(1)}\right)} - \varphi_2^{(2)2,1}\Big|_{\left(H^{(2)}\right)}\right)$$

$$\sigma_{12}^{(2)2,2}\Big|_{\left(t_1,-H^{(2)}\right)} - \sigma_{12}^{(1)2,2}\Big|_{\left(t_1,H^{(1)}\right)} = \left[-\pi H^{(2)}\left(\psi_{12}^{(1)2,1}\Big|_{\left(H^{(1)}\right)} - \psi_{12}^{(2)2,1}\Big|_{\left(-H^{(2)}\right)}\right) +\right.$$

$$\left.\pi\left(\frac{d\psi_{12}^{(1)2,1}}{dx_{22}^{(1)}}\Bigg|_{\left(H^{(1)}\right)} - \frac{d\psi_{12}^{(2)2,1}}{dx_{22}^{(2)}}\Bigg|_{\left(-H^{(2)}\right)}\right) - \pi\left(\psi_{11}^{(2)2,1}\Big|_{\left(-H^{(2)}\right)} - \psi_{11}^{(1)2,1}\Big|_{\left(H^{(1)}\right)}\right)\right]\sin(2\alpha t_1),$$

$$\sigma_{22}^{(2)2,2}\Big|_{\left(t_1,-H^{(2)}\right)} - \sigma_{22}^{(1)2,2}\Big|_{\left(t_1,H^{(1)}\right)} = 2\pi\left(\psi_{12}^{(2)2,1}\Big|_{\left(-H^{(2)}\right)} - \psi_{12}^{(1)2,1}\Big|_{\left(H^{(1)}\right)}\right)\cos(2\alpha t_1),$$

$$u_1^{(2)2,2}\Big|_{\left(t_1,-H^{(2)}\right)} - u_1^{(1)2,2}\Big|_{\left(t_1,H^{(1)}\right)} = \frac{\ell}{2}\left(\frac{d\varphi_1^{(1)2,1}}{dx_{22}^{(1)}}\Bigg|_{\left(H^{(1)}\right)} - \frac{d\varphi_1^{(2)2,1}}{dx_{22}^{(2)}}\Bigg|_{\left(-H^{(2)}\right)}\right)\sin(2\alpha t_1) -$$

$$\pi H^{(2)}\left(\varphi_1^{(1)2,1}\Big|_{\left(H^{(1)}\right)} - \varphi_1^{(2)2,1}\Big|_{\left(-H^{(2)}\right)}\right)\sin(2\alpha t_1),$$

$$u_2^{(2)2,2}\Big|_{\left(t_1,-H^{(2)}\right)} - u_2^{(1)2,2}\Big|_{\left(t_1,H^{(1)}\right)} = \ell\sin^2(\alpha t_1)\left(\frac{d\varphi_2^{(1)2,1}}{dx_{22}^{(1)}}\Bigg|_{\left(H^{(1)}\right)} - \frac{d\varphi_2^{(2)2,1}}{dx_{22}^{(2)}}\Bigg|_{\left(-H^{(2)}\right)}\right) -$$

$$2\pi H^{(2)} \cos^2(\alpha t_1)\left(\varphi_2^{(1)2,1}\Big|_{\left(H^{(1)}\right)} - \varphi_2^{(2)2,1}\Big|_{\left(-H^{(2)}\right)}\right),$$

$$\sigma_{i2}^{(2)1,2}\Big|_{\left(t_1,H^{(2)}\right)} - \sigma_{i2}^{(1)2,2}\Big|_{\left(t_1,-H^{(1)}\right)} = 0\,, \qquad u_i^{(2)1,2}\Big|_{\left(t_1,H^{(2)}\right)} - u_i^{(1)2,2}\Big|_{\left(t_1,-H^{(1)}\right)} = 0\,. \quad (4.4.3)$$

As in the first approximation, taking into account the form of the expressions entering the right side of (4.4.3) we select the unknown harmonic functions (4.2.10) in the form (4.2.27). Employing the well-known procedure we obtain corresponding algebraic equations for the constants which enter the expressions (4.2.27). So, continuing we can determine all subsequent approximations (4.1.15).

Numerical results and its analysis.

Consider some numerical results related to the stress distribution in the composite material with structure shown in Fig.4.4.1. We use the notation and restriction given in

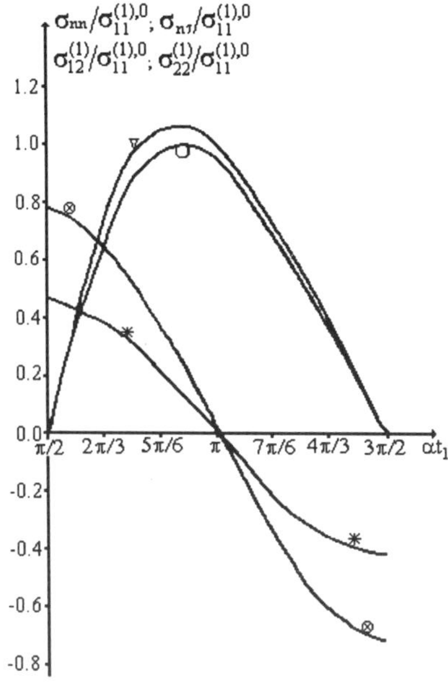

Fig.4.4.2. The distributions of self-balanced normal and tangential stresses on the inter-medium surface for $\varepsilon = 0.015$, $E^{(2)}/E^{(1)} = 50$, $\eta^{(2)} = 0.5$; ⊗ -- $\sigma_{nn}/\sigma_{11}^{(1),0}$ at $\chi = 0.4$, ∇ -- $\sigma_{n\tau}/\sigma_{11}^{(1),0}$ at $\chi = 0.2$, O -- $\sigma_{12}^{(1)}/\sigma_{11}^{(1),0}$ at $\chi = 0.1$, ∗ -- $\sigma_{22}^{(1)}/\sigma_{11}^{(1),0}$ at $\chi = 0.4$.

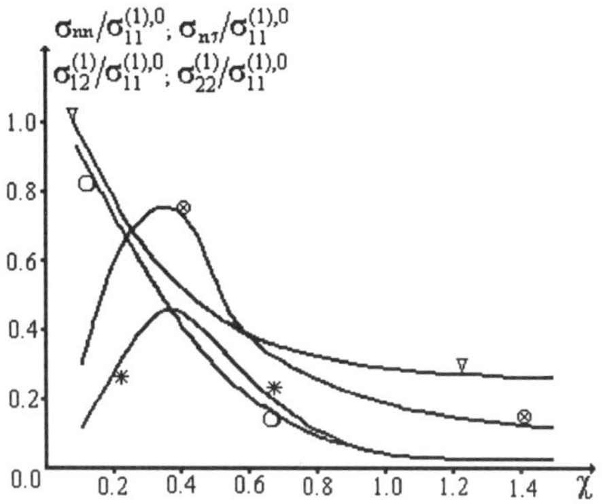

Fig.4.4.3. The distributions of self-balanced normal and tangential stresses with respect to the parameter χ for $\varepsilon = 0.015$, $E^{(2)}/E^{(1)} = 50$, $\eta^{(2)} = 0.5$; ⊗ -- $\sigma_{nn}/\sigma_{11}^{(1),0}$ at point B (Fig.4.4.1), ∇ -- $\sigma_{n\tau}/\sigma_{11}^{(1),0}$ at point C (Fig.4.4.1), O -- $\sigma_{12}^{(1)}/\sigma_{11}^{(1),0}$ at point C' (Fig.4.4.1), ∗ -- $\sigma_{22}^{(1)}/\sigma_{11}^{(1),0}$ at point B' (Fig.4.4.1).

the previous sections. Moreover, we introduce the following notation: $\sigma_{22}^{(1)}$, $\sigma_{12}^{(1)}$ are the normal and tangential stress, respectively, which act on the surface S_1^-.

Since the curving is periodic, we consider only the stress distribution on BCD on the surface S_2^+ and on B'C'D' on the surface S_1^- (Fig.4.4.1). Further we assume that the neighbouring filler layers do not contact each other, i.e., $L < 2H^{(1)}$.

First we analyse the distribution of the stresses on the inter-medium surface, as shown in Fig.4.4.2. We compare these results with the corresponding ones obtained in subsection 4.3.2 for composite material with alternating co-phase periodically curved layers; they are very similar. Fig.4.4.2 shows that $|\sigma_{nn}| \geq |\sigma_{22}^{(1)}|$ and $|\sigma_{n\tau}| \geq |\sigma_{12}^{(1)}|$. Similar results hold for the distributions of these stresses with respect to parameter χ, as shown in Fig.4.4.3.

Table 4.12. The values of self-balanced stresses with various $\eta^{(2)}$, ε for $E^{(2)}/E^{(1)}=50$.

$\eta^{(2)}$	ε	$\dfrac{\sigma_{nn}(B)}{\sigma_{11}^{(1),0}}$	$\dfrac{\sigma_{n\tau}(C)}{\sigma_{11}^{(1),0}}$	$\dfrac{\sigma_{12}^{(1)}(C')}{\sigma_{11}^{(1),0}}$	$\dfrac{\sigma_{22}^{(1)}(B')}{\sigma_{11}^{(1),0}}$
	0.015	0.662	0.464	0.214	0.029
0.1	0.020	0.855	0.599	0.277	0.039
	0.025	1.035	0.724	0.336	0.048
	0.030	1.217	0.845	0.393	0.057
	0.015	0.679	0.473	0.430	0.131
0.2	0.020	0.878	0.606	0.553	0.171
	0.025	1.064	0.724	0.668	0.208
	0.030	1.251	0.834	0.781	0.244
	0.015	0.761	0.798	0.766	0.463
0.5	0.020	1.001	1.020	0.984	0.614
	0.025	1.236	1.217	1.185	0.763
	0.030	1.472	1.402	1.384	0.918

Table 4.13. The values of $\sigma_{nn}(B)/\sigma_{11}^{(1),0}$ with various approximation (4.1.15) and ε for $E^{(2)}/E^{(1)}=50$, $\eta^{(2)}=0.5$

Numb. of	ε			
Appr.	0.015	0.020	0.025	0.030
1	0.764	1.019	1.274	1.529
2	0.789	1.064	1.344	1.629
3	0.758	0.989	1.198	1.378
4	0.757	0.988	1.194	1.369
5	0.761	1.001	1.236	1.472

Table 4.12 shows the influence of the parameters ε and $\eta^{(2)}$ on the values of these stresses; σ_{nn}, $\sigma_{12}^{(1)}$ are obtained for χ=0.3 and 0.1, respectively. However, the data corresponding to the case $\eta^{(2)}$=0.1 and related to the stresses $\sigma_{n\tau}$, $\sigma_{12}^{(1)}$ are obtained for χ=0.3, 0.1, respectively, but those corresponding to the cases $\eta^{(2)} \geq 0.2$ are obtained for χ=0.1. These numerical results show that the stresses increase monotonically as ε and $\eta^{(2)}$ increase. The comparison of these results with those obtained in section 4.3 shows that $\tilde{\sigma}_{nn} < \sigma_{nn} < \breve{\sigma}_{nn}$, $\tilde{\sigma}_{n\tau} > \sigma_{n\tau} > \breve{\sigma}_{nn}$; here $\tilde{\sigma}_{nn}$, $\tilde{\sigma}_{n\tau}$ are the self-balanced normal and tangential stresses obtained for the composite depicted in Fig.4.2.1, and $\breve{\sigma}_{nn}$, $\breve{\sigma}_{n\tau}$ are those obtained for the composite shown in Fig.4.2.3; these inequalities are to be expected from mechanical considerations.

All numerical results presented in this subsection have been obtained by using the first five approximations. Table 4.13 shows the numerical convergence of $\sigma_{nn}/\sigma_{11}^{(1),0}$ for various values of ε; convergence is acceptable.

4.4.2. ALTERNATING STRAIGHT AND ANTI-PHASE CURVED LAYERS

The structure is shown in Fig.4.4.4. We investigate the plane strain state for $\sigma_{11}(\infty) = p$.

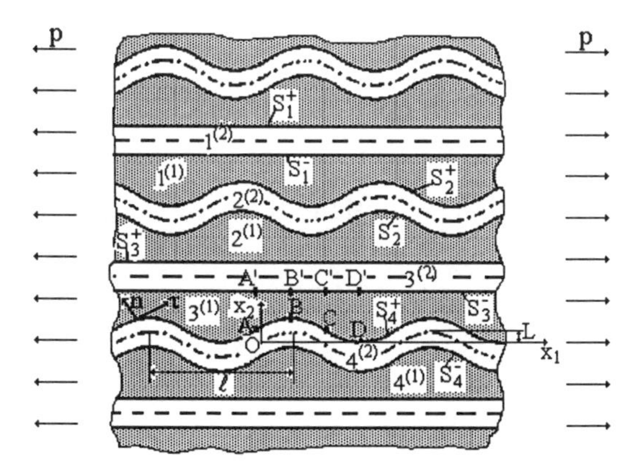

Fig.4.4.4. Structure of the composite material with alternating straight and anti-phase periodically curved layers

Solution method.

The solution procedure is the same as that described in the previous subsection. Now the period in the Ox_2 direction is $8\left(H^{(2)} + H^{(1)}\right)$; we single out eight layers, i.e. $1^{(1)}$, $1^{(2)}$, $2^{(1)}$, $2^{(2)}$, $3^{(1)}$, $3^{(2)}$, $4^{(1)}$, $4^{(2)}$ (Fig.4.4.4), and discuss them below.

The equation of the middle surface of the curved layers $2^{(2)}$ and $4^{(2)}$ are $x_{22}^{(2)} = -L\sin(2\pi x_1/\ell)$ and $x_{24}^{(2)} = L\sin(2\pi x_1/\ell)$, respectively. Assume that $L < \ell$ and introduce a small parameter $\varepsilon = L/\ell$.

The analysis proceeds as before, with slight changes. We obtain the contact conditions for first approximation from (4.4.1), (4.1.22)-(4.1.26) in the following form:

$$\sigma_{i2}^{(2)1,1}\Big|_{(t_1,-H^{(2)})} - \sigma_{i2}^{(1)1,1}\Big|_{(t_1,H^{(1)})} = 0, \quad u_i^{(2)1,1}\Big|_{(t_1,-H^{(2)})} - u_i^{(1)1,1}\Big|_{(t_1,H^{(1)})} = 0,$$

$$\sigma_{i2}^{(2)2,1}\Big|_{(t_1,H^{(2)})} - \sigma_{i2}^{(1)1,1}\Big|_{(t_1,-H^{(1)})} = 2\pi\delta_i^1\left(\sigma_{11}^{(2),0} - \sigma_{11}^{(1),0}\right)\cos(\alpha t_1),$$

$$u_i^{(2)2,1}\Big|_{(t_1,H^{(2)})} - u_i^{(1)1,1}\Big|_{(t_1,-H^{(1)})} = -\ell\delta_i^2\sin(\alpha t_1)\left(\frac{\nu^{(2)}\left(1+\nu^{(2)}\right)}{E^{(2)}}\sigma_{11}^{(2),0} - \frac{\nu^{(1)}\left(1+\nu^{(1)}\right)}{E^{(1)}}\sigma_{11}^{(1),0}\right),$$

$$\sigma_{i2}^{(2)2,1}\Big|_{(t_1,-H^{(2)})} - \sigma_{i2}^{(1)2,1}\Big|_{(t_1,H^{(1)})} = 2\pi\delta_i^1\left(\sigma_{11}^{(2),0} - \sigma_{11}^{(1),0}\right)\cos(\alpha t_1),$$

$$u_i^{(2)2,1}\Big|_{(t_1,-H^{(2)})} - u_i^{(1)2,1}\Big|_{(t_1,H^{(1)})} = -\ell\delta_i^2\sin(\alpha t_1)\left(\frac{\nu^{(2)}\left(1+\nu^{(2)}\right)}{E^{(2)}}\sigma_{11}^{(2),0} - \frac{\nu^{(1)}\left(1+\nu^{(1)}\right)}{E^{(1)}}\sigma_{11}^{(1),0}\right),$$

$$\sigma_{i2}^{(2)3,1}\Big|_{(t_1,H^{(2)})} - \sigma_{i2}^{(1)2,1}\Big|_{(t_1,-H^{(1)})} = 0, \quad u_i^{(2)3,1}\Big|_{(t_1,H^{(2)})} - u_i^{(1)2,1}\Big|_{(t_1,-H^{(1)})} = 0,$$

$$\sigma_{i2}^{(2)3,1}\Big|_{(t_1,-H^{(2)})} - \sigma_{i2}^{(1)3,1}\Big|_{(t_1,H^{(1)})} = 0, \quad u_i^{(2)3,1}\Big|_{(t_1,-H^{(2)})} - u_i^{(1)3,1}\Big|_{(t_1,H^{(1)})} = 0,$$

$$\sigma_{i2}^{(2)4,1}\Big|_{(t_1,H^{(2)})} - \sigma_{i2}^{(1)3,1}\Big|_{(t_1,-H^{(1)})} = -2\pi\delta_i^1\left(\sigma_{11}^{(2),0} - \sigma_{11}^{(1),0}\right)\cos(\alpha t_1),$$

$$u_i^{(2)4,1}\Big|_{(t_1,H^{(2)})} - u_i^{(1)3,1}\Big|_{(t_1,-H^{(1)})} = \ell\delta_i^2\sin(\alpha t_1)\left(\frac{\nu^{(2)}\left(1+\nu^{(2)}\right)}{E^{(2)}}\sigma_{11}^{(2),0} - \right.$$

$$\left.\frac{\nu^{(1)}\left(1+\nu^{(1)}\right)}{E^{(1)}}\sigma_{11}^{(1),0}\right),$$

$$\sigma_{i2}^{(2)4,1}\Big|_{\left(t_1,-H^{(2)}\right)} - \sigma_{i2}^{(1)3,1}\Big|_{\left(t_1,H^{(1)}\right)} = -2\pi\delta_i^1\left(\sigma_{11}^{(2),0} - \sigma_{11}^{(1),0}\right)\cos(\alpha t_1),$$

$$u_i^{(2)4,1}\Big|_{\left(t_1,-H^{(2)}\right)} - u_i^{(1)3,1}\Big|_{\left(t_1,H^{(1)}\right)} = -\ell\delta_i^2\sin(\alpha t_1)\left(\frac{\nu^{(2)}\left(1+\nu^{(2)}\right)}{E^{(2)}}\sigma_{11}^{(2),0} - \right.$$

$$\left.\frac{\nu^{(1)}\left(1+\nu^{(1)}\right)}{E^{(1)}}\sigma_{11}^{(1),0}\right),$$

$$\sigma_{i2}^{(2)1,1}\Big|_{\left(t_1,H^{(2)}\right)} - \sigma_{i2}^{(1)4,1}\Big|_{\left(t_1,-H^{(1)}\right)} = 0, \quad u_i^{(2)1,1}\Big|_{\left(t_1,H^{(2)}\right)} - u_i^{(1)4,1}\Big|_{\left(t_1,-H^{(1)}\right)} = 0. \quad (4.4.4)$$

Thus, the values of the first approximation are determined as before and proceeding this procedure we find the values of every subsequent approximation.

Numerical results and its analysis.
Taking into account the periodicity of the curving we consider only BCD and B'C'D' on the surfaces S_4^+ and S_3^- (Fig.4.4.4), respectively. Fig.4.4.5 shows the dependence of σ_{nn}, $\sigma_{n\tau}$, $\sigma_{22}^{(1)}$, $\sigma_{12}^{(1)}$ on αt_1; stress distributions on the inter-medium surfaces are very similar to those in the previous subsection.

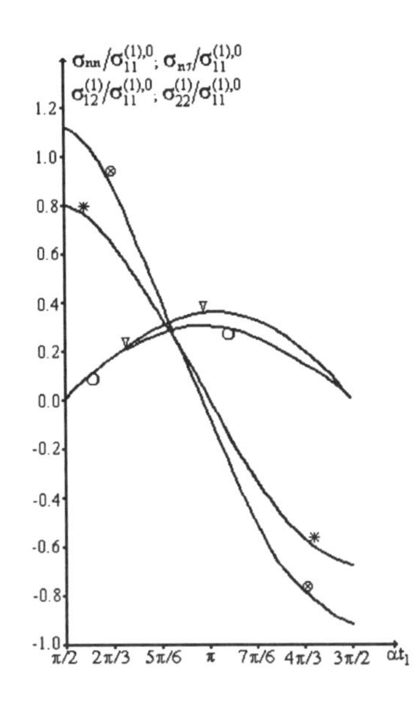

Fig.4.4.5. The distributions of the self-balanced normal and tangential stresses on the inter-medium surface for $\varepsilon = 0.015$, $E^{(2)}/E^{(1)} = 50$, $\eta^{(2)} = 0.5$; ⊗-- $\sigma_{nn}/\sigma_{11}^{(1),0}$ at $\chi = 0.3$, ∇ -- $\sigma_{n\tau}/\sigma_{11}^{(1),0}$ at $\chi = 0.6$, ❍ -- $\sigma_{12}^{(1)}/\sigma_{11}^{(1),0}$ at $\chi = 0.3$, ✱ -- $\sigma_{22}^{(1)}/\sigma_{11}^{(1),0}$ at $\chi = 0.4$.

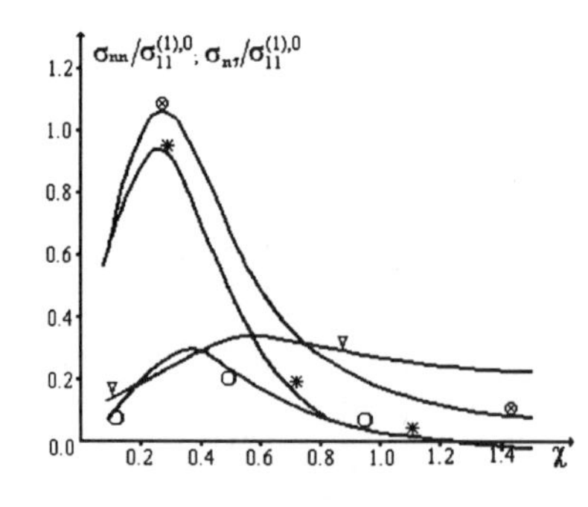

Fig.4.4.6. The distributions of the self-balanced normal and tangential stresses with respect to the parameter χ for $\varepsilon=0.015$, $E^{(2)}/E^{(1)}=50$, $\eta^{(2)}=0.5$; ⊗-- $\sigma_{nn}/\sigma_{11}^{(1),0}$ at point B (Fig.4.4.4), ∇ -- $\sigma_{n\tau}/\sigma_{11}^{(1),0}$ at point C (Fig.4.4.4), O -- $\sigma_{12}^{(1)}/\sigma_{11}^{(1),0}$ at point C' (Fig.4.4.4), ✳ -- $\sigma_{22}^{(1)}/\sigma_{11}^{(1),0}$ at point B' (Fig.4.4.4).

Fig.4.4.6 shows the influence of χ; the dependence of $\sigma_{nn}, \sigma_{n\tau}, \sigma_{12}^{(1)}, \sigma_{22}^{(1)}$ on χ is non-monotonic; the self-balanced stresses arising on the curved surface are significantly greater than those obtained on the ideal (straight) contact plane S_3^-.

Table 4.14 shows the influence of ε and $\eta^{(2)}$ on the stresses; it follows from the data shown in this table that the self-balanced stresses increase monotonically with ε. As $\eta^{(2)}$ increases, the stresses σ_{nn}, $\sigma_{12}^{(1)}$ and $\sigma_{22}^{(1)}$ increase, but $\sigma_{n\tau}$ decreases. A comparison with the data given in the previous subsection shows that $\sigma_{nn} > \hat{\sigma}_{nn}$, $\sigma_{22}^{(1)} > \hat{\sigma}_{22}^{(1)}$, $\sigma_{n\tau} < \hat{\sigma}_{n\tau}$ and $\sigma_{12}^{(1)} < \hat{\sigma}_{12}^{(1)}$; here σ and $\hat{\sigma}$ are the stresses given in Tables 4.14, 4.12, respectively. The inequalities $\tilde{\sigma}_{nn} < \sigma_{nn} < \breve{\sigma}_{nn}$, $\tilde{\sigma}_{n\tau} > \sigma_{n\tau} > \breve{\sigma}_{n\tau}$ also hold, here $\tilde{\sigma}_{nn}$, $\breve{\sigma}_{n\tau}$ are the self-balanced normal and tangential stresses obtained for the composite shown in Fig.4.2.1, and $\breve{\sigma}_{nn}$, $\breve{\sigma}_{n\tau}$ are those obtained for the composite shown in Fig.4.2.3.

Table 4.14. The values of the self-balanced stresses with various $\eta^{(2)}$, ε for $E^{(2)}/E^{(1)}=50$.

$\eta^{(2)}$	ε	$\frac{\sigma_{nn}(B)}{\sigma_{11}^{(1),0}}$	$\frac{\sigma_{n\tau}(C)}{\sigma_{11}^{(1),0}}$	$\frac{\sigma_{12}^{(1)}(C')}{\sigma_{11}^{(1),0}}$	$\frac{\sigma_{22}^{(1)}(B')}{\sigma_{11}^{(1),0}}$
	0.015	0.662	0.464	0.107	0.238
0.1	0.020	0.855	0.599	0.140	0.302
	0.025	1.035	0.724	0.170	0.381
	0.030	1.217	0.845	0.199	0.448
	0.015	0.722	0.448	0.198	0.172
0.2	0.020	0.941	0.583	0.255	0.542
	0.025	1.147	0.710	0.309	0.659
	0.030	1.348	0.798	0.360	0.773
	0.015	1.078	0.329	0.284	0.774
0.5	0.020	1.446	0.435	0.376	1.030
	0.025	1.817	0.540	0.463	1.284
	0.030	2.193	0.647	0.524	1.543

Table 4.15 gives the first five approximations for $\sigma_{nn}(B)/\sigma_{11}^{(1),0}$, for various ε; the convergence is acceptable.

Table 4.15. The values of $\sigma_{nn}(B)/\sigma_{11}^{(1),0}$ with various approximation (4.1.15) and ε for $E^{(2)}/E^{(1)}=50$, $\eta^{(2)}=0.5$

Numb. of approx	ε			
	0.015	0.020	0.025	0.030
1	1.029	1.373	1.716	2.059
2	1.105	1.507	1.926	2.362
3	1.077	1.440	1.795	2.136
4	1.075	1.435	1.784	2.112
5	1.078	1.446	1.816	2.193

4.4.3. COMPOSITE MATERIAL WITH A SINGLE PERIODICALLY CURVED LAYER

Fig.4.4.7 shows the structure; there is just one curved layer; all the others are straight. We assume that the curving is periodic in the direction of the Ox_1 axis and does not depend on x_3.

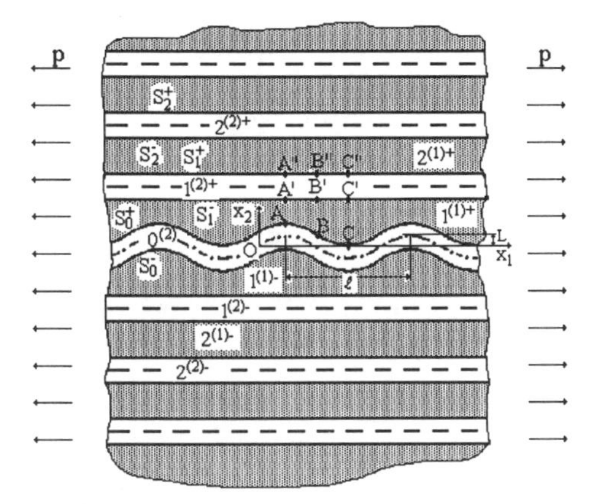

Fig.4.4.7. The structure of the composite material with a single periodically curved layer.

We investigate the plane strain stress distribution for $\sigma_{11}(\infty)=p$.

Solution method.

In previous problems the structure in the Ox_2 direction has been periodic; this one is not periodic, and this complicates the analysis. In the periodic any problems investigated so far we obtained a closed system of equations for the unknown constants entering the harmonic functions (4.2.10); now we obtain an infinite system of equations. We propose an approximate solution method.

The effect of the single curved layer on the stresses on far layers will be the same as that of a straight replacement layer. But when the curved layer is replaced by a straight one, the structure is that considered in the zeroth approximation.This means that

$$\overline{\sigma}_{ij}^{(k)m\pm} = \sigma_{ij}^{(k)m\pm} - \sigma_{ij}^{(k)m\pm,0} \to 0 \quad \text{as} \quad m^{(k)\pm} \to \infty, \tag{4.4.5}$$

where $m^{(k)\pm}$ is the number of the layer; $\sigma_{ij}^{(k)m\pm,0}$ are the components of the stress tensor in the zeroth approximation.

Taking (4.4.5) into account, we assume that, after some specific layer with number $m_*^{(k)\pm}$ all components of $\overline{\sigma}_{ij}^{(k)m\pm}$ are zero, i.e.

$$\overline{\sigma}_{i2}^{(k)m_*\pm}\Big|_{S^{\pm}_{m_*^{(k)\pm}}} = 0. \tag{4.4.6}$$

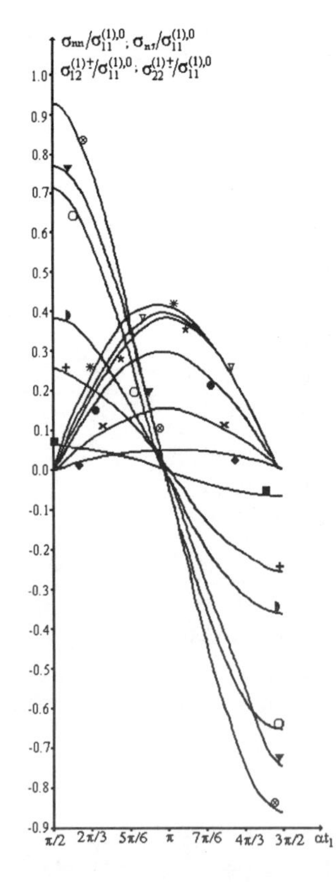

Now there is only a finite number of layers. Since the problem is linear, equation (4.4.6) will be satisfied for in each approximation separately. Thus, adding the condition (4.4.6) to the corresponding contact relations we obtain a finite system of linear inhomogeneous algebraic equations; the appropriate value of $m_*^{(k)\pm}$ for a required accuracy is obtained by comparing numerical results for various $m_*^{(k)\pm}$.

Fig.4.4.8. Graphs illustrating the changes of self-equilibrating stresses in their dependence on αt_1 in the composite material with a single curved layer for $\varepsilon=0.015$, $\chi=0.3$, $E^{(2)}/E^{(1)}=50$ and $\eta^{(2)}=0.5;0.3$:
⊗, ▼ -- $\sigma_{nn}/\sigma_{11}^{(1),0}$; ∇, ★ -- $\sigma_{n\tau}/\sigma_{11}^{(1),0}$;
○, ◗ -- $\sigma_{22}^{(1)-}/\sigma_{11}^{(1),0}$; ✳, ● -- $\sigma_{12}^{(1)-}/\sigma_{11}^{(1),0}$;
✚, ■ -- $\sigma_{22}^{(1)+}/\sigma_{11}^{(1),0}$; ✖, ◆ -- $\sigma_{12}^{(1)+}/\sigma_{11}^{(1),0}$.

Some numerical results and its analyses.
In the numerical results we consider only the distribution of the self-balanced normal and tangential stresses on the surfaces S_0^+, S_1^-, S_1^+ shown in Fig.4.4.7. The equation of the middle surface of the $0^{(2)}$th layer is taken as $x_{20}^{(2)} = L\sin(2\pi x_1/\ell)$; we consider only the stresses on ABC, A'B'C' and A''B''C'' on the surfaces S_0^+, S_1^- and S_1^+, respectively; these correspond to αt_1 in $[\pi/2, 3\pi/2]$. Again we assume that $L < \ell$, and take the small parameter $\varepsilon = L/\ell$. The normal (tangential) stresses acting on the surfaces S_0^+, S_1^- and S_1^+ we denote by $\sigma_{nn}(\sigma_{n\tau})$, $\sigma_{22}^{(1)-}$ $\left(\sigma_{12}^{(1)-}\right)$ and $\sigma_{22}^{(1)+}$ $\left(\sigma_{12}^{(1)+}\right)$, respectively. As before we assume that the Poisson ratio of the matrix and filler are equal, i.e., $\nu^{(1)} = \nu^{(2)} = 0.3$.

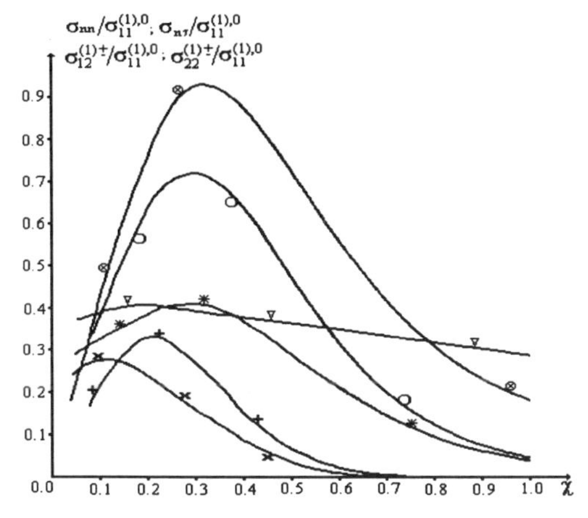

Fig.4.4.9. Graphs illustrating the changes of self-equilibrating stresses in their dependence on χ in the composite material with a single curved layer for ε =0.015, χ =0.3, $E^{(2)}/E^{(1)}$ =50 and $\eta^{(2)}$ =0.5

⊗ -- $\sigma_{nn}/\sigma_{11}^{(1),0}$ at point A (Fig.4.4.7);

∇-- $\sigma_{n\tau}/\sigma_{11}^{(1),0}$ at point B (Fig.4.4.7);

O -- $\sigma_{22}^{(1)-}/\sigma_{11}^{(1),0}$ at point A' (Fig.4.4.7);

✳ -- $\sigma_{12}^{(1)-}/\sigma_{11}^{(1),0}$ at point B' (Fig.4.4.7);

✚ -- $\sigma_{22}^{(1)+}/\sigma_{11}^{(1),0}$ at point A'' (Fig.4.4.7);

✖ -- $\sigma_{12}^{(1)+}/\sigma_{11}^{(1),0}$ at point B'' (Fig.4.4.7)

Fig.4.4.8 shows the distribution of the stresses on the surfaces S_0^+, S_1^- and S_1^+; the self-balanced normal stresses have absolute maximal values for $\alpha t_1 = \pi/2; 3\pi/2$, yet, the tangential self-balanced stresses have near $\alpha t_1 = \pi$; the stresses $\sigma_{22}^{(1)+}$, $\sigma_{12}^{(1)+}$ become insignificant for small $\eta^{(2)}$.

Fig.4.4.9 shows dependence of the self-balanced stresses on χ; the dependence is similar to that obtained for the previous problems.

Table 4.16 shows the influence of $\eta^{(2)}$, $E^{(2)}/E^{(1)}$ and ε on the stresses; all the stress components increase monotonically with ε, $E^{(2)}/E^{(1)}$. With an increase in $\eta^{(2)}$ these stresses (except $\sigma_{n\tau}$) increase monotonically. However, the relation between $\sigma_{n\tau}/\sigma_{11}^{(1),0}$ and $\eta^{(2)}$ is non-monotonic. Note that the influence of $\eta^{(2)}$ on the stresses acting on the surfaces S_1^-, S_1^+ is more significant than its influence on the stress acting at the surface S_0^+.

Table 4.16. The self-balanced stresses for various $\eta^{(2)}$, ε, $E^{(2)}/E^{(1)}$: $E^{(2)}/E^{(1)}=50$, and $\chi=0.3$; $E^{(2)}/E^{(1)}=100$ and $\chi=0.2$.

$E^{(2)}/E^{(1)}$	$\eta^{(2)}$	ε	$\frac{\sigma_{nn}(A)}{\sigma_{11}^{(1),0}}$	$\frac{\sigma_{n\tau}(B)}{\sigma_{11}^{(1),0}}$	$\frac{\sigma_{12}^{(1)-}(B')}{\sigma_{11}^{(1),0}}$	$\frac{\sigma_{22}^{(1)-}(A')}{\sigma_{11}^{(1),0}}$	$\frac{\sigma_{12}^{(1)+}(B'')}{\sigma_{11}^{(1),0}}$	$\frac{\sigma_{22}^{(1)+}(A'')}{\sigma_{11}^{(1),0}}$
50	0.1	0.020	0.828	0.592	0.017	0.016	0.000	0.000
		0.025	0.993	0.706	0.021	0.019	0.000	0.000
		0.030	1.153	0.806	0.024	0.022	0.000	0.000
	0.3	0.020	0.971	0.493	0.380	0.484	0.061	0.084
		0.025	1.176	0.593	0.457	0.585	0.080	0.101
		0.030	1.374	0.644	0.529	0.679	0.092	0.117
	0.5	0.020	1.203	0.501	0.529	0.938	0.182	0.326
		0.025	1.487	0.611	0.641	1.163	0.224	0.396
		0.030	1.770	0.715	0.746	1.387	0.261	0.463
100	0.1	0.020	1.347	0.725	0.142	0.126	0.005	0.004
		0.025	1.595	0.833	0.164	0.146	0.006	0.005
		0.030	1.860	0.915	0.185	0.164	0.007	0.006
	0.3	0.020	1.646	0.588	0.731	1.090	0.255	0.365
		0.025	1.981	0.695	0.860	1.307	0.301	0.432
		0.030	2.315	0.784	0.973	1.514	0.341	0.494
	0.5	0.020	1.954	0.711	0.805	1.678	0.429	0.806
		0.025	2.426	0.856	0.956	2.092	0.513	0.979
		0.030	2.911	0.987	1.083	2.518	0.588	1.146

Table 4.17. The self-balanced normal stresses in the layered composite shown in Fig.4.4.7 with various approximations and $m_*^{(k)+}$ for $\eta^{(2)}=0.5$, $\varepsilon=0.020$, $\chi=0.2$, $E^{(2)}/E^{(1)}=100$.

Numb. of layers (equations)	Numb. of approxm. (4.1.15)	$\frac{\sigma_{nn}(A)}{\sigma_{11}^{(1),0}}$	$\frac{\sigma_{22}^{(1)-}(A')}{\sigma_{11}^{(1),0}}$	$\frac{\sigma_{22}^{(1)+}(A'')}{\sigma_{11}^{(1),0}}$
6 (26)	1	1.919	1.606	0.774
	2	2.119	1.801	0.802
	3	1.944	1.665	0.732
	4	1.930	1.651	0.729
	5	1.962	1.677	0.741
7 (30)	1	1.921	1.609	0.781
	2	2.121	1.804	0.809
	3	1.946	1.668	0.739
	4	1.933	1.654	0.736
	5	1.964	1.680	0.748
8 (34)	1	1.935	1.633	0.852
	2	2.096	1.794	0.874
	3	1.920	1.653	0.795
	4	1.927	1.656	0.795
	5	1.954	1.678	0.806

Table 4.16 also shows that, with identical values of the problem parameters, it is always the case that $\sigma_{12}^{(1)-} > \sigma_{12}^{(1)+}$, $\sigma_{nn} > \sigma_{22}^{(1)-} > \sigma_{22}^{(1)+}$; on moving away from the distorted layer in the direction of the axis Ox_2 (Fig.4.4.7), the influence of the curved layer $0^{(2)}$ on the distribution of the given stress components declines.

Table 4.17 shows normal self-equilibrium stress for various approximations and various $m_*^{(k)+}$; the method is effective.

4.5. Viscoelastic Composites

So far we have assumed that all the materials in the composite are elastic. However, under prolonged loading the mechanical behaviour of a composite material with one or several polymer components varies substantially on time; the viscosity of the matrix and filler material must be taken into account in studying the influence of curvature on the stress-strain state. We present a solution method for the composite material with viscoelastic curved layers.

For simplicity we investigate a quasi-static stress-strain problem in a composite material with an infinity of co-phase periodically plane-curved viscoelastic layers alternating in the direction of the Ox_2 axis (Fig.4.2.1), under the loading $\sigma_{11}(\infty) = p(t)$. We use assumptions and notation, developed in subsection 4.2.1. We need to replace the elastic stress-strain equations by viscoelastic ones:

$$\sigma_{ij}^{(k)} = \lambda^{(k)*}\theta^{(k)}\delta_i^j + \mu^{(k)*}\varepsilon_{ij}^{(k)*}, \tag{4.5.1}$$

where $\lambda^{(k)*}$ and $\mu^{(k)*}$ are the following operators

$$\lambda^{(k)*}g(t) = \lambda_0^{(k)}g(t) + \int_0^t \lambda_1^{(k)}(t-\tau)g(\tau)d\tau,$$

$$\mu^{(k)*}g(t) = \mu_0^{(k)}g(t) + \int_0^t \mu_1^{(k)}(t-\tau)g(\tau)d\tau. \tag{4.5.2}$$

Here $\lambda_0^{(k)}$ and $\mu_0^{(k)}$ are time independent Lamé's constants; $\lambda_1^{(k)}(t)$ and $\mu_1^{(k)}(t)$ are the kernels of the bounded operators $\lambda^{(k)*}$, $\mu^{(k)*}$. As with the pure elastic problem, the zeroth approximation corresponds to the stress state in the composite material with ideal (straight) structure shown in Fig.4.2.2, under the loading $\sigma_{11}(\infty) = p(t)$. The stresses in the zeroth approximation balance the external forces p(t); the first and subsequent approximations of (4.1.15) correspond to the self-balanced stress state caused by the curving of the reinforcing layers. Each approximation is reduced to the solution of the corresponding quasistatic problem of the theory of linear viscoelasticity; we will use Schapery's methods [136,137] according to which we apply the Laplace transform with real parameter s>0, i.e.

$$\overline{V}(s)=\int_0^{\infty}V(t)e^{-st}dt \tag{4.5.3}$$

to all the relationships involved in the approximation. After applying the Laplace transform (4.5.3) to the mechanical relations (4.5.1), (4.5.2) and using the convolution theorem we obtain

$$\overline{\sigma}_{ij}^{(k)}(s)=\overline{\lambda}^{(k)}(s)\overline{\theta}^{(k)}(s)+\overline{\mu}^{(k)}(s)\overline{\varepsilon}_{ij}^{(k)}(s),$$

$$\overline{\lambda}^{(k)*}(s)=\lambda_0^{(k)}+\overline{\lambda}_1^{(k)}(s);\quad \overline{\mu}^{(k)*}(s)=\mu_0^{(k)}+\overline{\mu}_1^{(k)}(s). \tag{4.5.4}$$

Under Laplace transform, all others relationships retain their form; $\sigma_{ij}^{(k),q}$, $\varepsilon_{ij}^{(k),q}$, $u_i^{(k),q}$ are replaced by Laplace transforms $\overline{\sigma}_{ij}^{(k),q}$, $\overline{\varepsilon}_{ij}^{(k),q}$, $\overline{u}_i^{(k),q}$, respectively; latter values are determined by the method presented in section 4.1. Thus for the zeroth approximation we obtain

$$\overline{\sigma}_{11}^{(1),0}(s)=\overline{p}(s)\left(\eta^{(1)}+\frac{\overline{E}^{(2)}(s)}{\overline{E}^{(1)}(s)}\frac{1-\left(\nu^{(1)}(s)\right)^2}{1-\left(\nu^{(2)}(s)\right)^2}\eta^{(2)}\right)^{-1},$$

$$\overline{\sigma}_{11}^{(2),0}(s)=\frac{\overline{E}^{(2)}(s)}{\overline{E}^{(1)}(s)}\frac{1-\left(\overline{\nu}^{(1)}(s)\right)^2}{1-\left(\overline{\nu}^{(2)}(s)\right)^2}\overline{\sigma}_{11}^{(1),0}(s),\quad \overline{\sigma}_{12}^{(k),0}=\overline{\sigma}_{22}^{(k),0}=0;$$

$$k=1,2,\quad \overline{u}_1^{(k),0}=\frac{1-\left(\overline{\nu}^{(k)}\right)^2}{\overline{E}^{(k)}(s)}\overline{\sigma}_{11}^{(k),0}(s)x_1,$$

$$\overline{u}_2^{(k),0}=-\frac{\overline{\nu}^{(k)}(s)\left(1+\overline{\nu}^{(k)}(s)\right)}{\overline{E}^{(k)}(s)}\overline{\sigma}_{11}^{(k),0}x_2^{(k)}+\frac{C^{(k)}}{s}, \tag{4.5.5}$$

where

$$\overline{\nu}^{(k)}(s)=\frac{\overline{\lambda}^{(k)}(s)}{2\left(\overline{\lambda}^{(k)}(s)+\overline{\mu}^{(k)}(s)\right)},\quad \overline{E}^{(k)}(s)=\frac{\overline{\mu}^{(k)}(s)\left(3\overline{\lambda}^{(k)}(s)+2\overline{\mu}^{(k)}(s)\right)}{\overline{\lambda}^{(k)}(s)+\overline{\mu}^{(k)}(s)}. \tag{4.5.6}$$

According to the correspondence principle the Laplace transforms of the first and subsequent approximations, are determined as in the pure elastic case by changing $C_{0i}^{(k),q}$, $C_{2i}^{(k),q}$ (i=1,2; k=1,2; q=1,2,...) to $C_{0i}^{(k),q}(s)$, $C_{2i}^{(k),q}(s)$, respectively; the unknown constants entering the expressions of the harmonic functions (4.2.10) will now depend on the transform parameter s. After determination $\bar{\sigma}_{ij}^{(k),q}$, $\bar{\varepsilon}_{ij}^{(k),q}$, $\bar{u}_{i}^{(k),q}$ we must find their inverse Laplace transforms; we use of Schapery's method [136,137]. This method is based on the approximate inversion of the integral equation (4.5.3) for V(t). First, we calculate of $C_{0i}^{(k),q}(s)$, $C_{2i}^{(k),q}(s)$ at $s=s_n$ (n=1,2,..., N) from the corresponding contact conditions; from these we determine the values of $\bar{\sigma}_{ij}^{(k),q}$, $\bar{\varepsilon}_{ij}^{(k),q}$, $\bar{u}_{i}^{(k),q}$ at $s=s_n$. After these calculations we construct the graphs of the relations between $s\bar{\sigma}_{ij}^{(k),q}(s)$, $s\bar{u}_{i}^{(k),q}(s)$ and $\log s$. If the curvatures of these graphs are sufficiently small, in the sense described in [136,137], then we find the inverse transforms by using the approximate relationships

$$\sigma_{ij}^{(k),q}(x_1,x_2,x_3,t) \approx s\bar{\sigma}_{ij}^{(k),q}(x_1,x_2,x_3,s)\big|_{s=1/(2t)};$$

$$u_{i}^{(k),q}(x_1,x_2,x_3,t) \approx s\bar{u}_{i}^{(k),q}(x_1,x_2,x_3,s)\big|_{s=1/(2t)}. \tag{4.5.7}$$

If the smallness condition is not satisfied then the relations (4.5.7) are not applicable; we proceed as follows. First by using the limiting $\lambda^{(k)*}$, $\mu^{(k)*}$ as $t\to\infty$, instead of the operators, we determine the protracted (long-time) stresses (we denote them by $\sigma_{ijpr.}^{(k),q}$, $u_{ipr.}^{(k),q}$). The stress tensor and displacement vector components are represented as

$$\sigma_{ij}^{(k),q} = \delta\sigma_{ij}^{(k),q} + \sigma_{ijpr.}^{(k),q}, \quad u_{i}^{(k),q} = \delta u_{i}^{(k),q} + u_{ipr.}^{(k),q}. \tag{4.5.8}$$

To determine $\delta\sigma_{ij}^{(k),q}$, $\delta u_{i}^{(k),q}$ we apply the collocation method described in [136,137], based on the values of $\bar{\sigma}_{ij}^{(k),q}$, $\bar{u}_{i}^{(k),q}$ at $s=s_q$. As an example, we consider the determination of the values $\delta\sigma_{ij}^{(k),q}$; we assume that it can be approximated by the finite Dirichlet series

$$\delta\sigma_{ij}^{(k),q} \cong \sum_{i=1}^{N} S_i e^{-\lambda_i t}, \tag{4.5.9}$$

λ_i are chosen real constants, and S_i are real unknowns. For the determination of these unknowns we obtain the system of equations

$$\sum_{j=1}^{N}\left(1+\frac{\lambda_j}{\lambda_i}\right)^{-1} S_j = s\delta\bar{\sigma}_{nm}^{(k),q}(s)\Big|_{s=\lambda_i}, \tag{4.5.10}$$

which are obtained from the minimisation of the square deviation

$$E=\int_0^{\infty}\left(\delta\sigma_{nm}^{(k),q}-\sum_{i=1}^{N}\left(1+\frac{\lambda_j}{\lambda_i}\right)^{-1} S_i\right)^2 dt \tag{4.5.11}$$

with respect to each S_i. In (4.5.10), the notation is

$$\delta\bar{\sigma}_{nm}^{(k),q}(s)=\bar{\sigma}_{nm}^{(k),q}(s)-\sigma_{nmpr.}^{(k),q} \tag{4.5.12}$$

In [136,137] Schapery shows how the values for the parameter λ_i can be selected from the graphs of $s\bar{\sigma}_{nm}^{(k),q}(s)$ against $\log s$. We can often restrict ourselves to equally spaced λ_i, we can improve accuracy by increasing N. The accuracy of the numerical results depends on the character of the function p(t) describing the external forces with respect to time t. If p(t) is a slowly changing continuous function, Schapery's method gives results with very high accuracy.

4.6. Stress Distribution in Composites with Viscoelastic Layers

Using the approach presented in the previous section we investigate the stress distribution in the composite materials with co-phase (Fig.4.2.1) and anti-phase (Fig.4.2.3) periodically curved layers. We take $\sigma_{11}(\infty)=p(t)=p=\text{constant}$. We suppose the layers of filler are purely elastic with Young's modulus $E^{(2)}$, and Poisson's ratio $\nu^{(2)}$. The material of the layers of the matrix is assumed to be linearly viscoelastic with operators

$$E^{(1)*}=E_0^{(1)}\left[1-\omega_0\Pi_{\alpha'}^{*}(-\omega_0-\omega_\infty\right),$$

$$\nu^{(1)*}=\nu_0^{(1)}\left[1+\frac{1-2\nu_0^{(1)}}{2\nu_0^{(1)}}\omega_0\Pi_{\alpha'}^{*}(-\omega_0-\omega_\infty)\right], \tag{4.6.1}$$

where $E_0^{(1)}$ and $\nu_0^{(1)}$ are time independent values of the Young's modulus and Poisson's ratio, respectively; ω_0, ω_∞, α' are rheological parameters of the matrix material; $\Pi^*_{\alpha'}$ is Rabotnov's fractional-exponential operator [131]:

$$\Pi^*_{\alpha'}(\beta) = \int_0^\infty \Pi_{\alpha'}(\beta, t-\tau)d\tau , \tag{4.6.2}$$

where

$$\Pi_{\alpha'}(\beta, t) = t^{\alpha'} \sum_{n=0}^{\infty} \frac{\beta^n t^{n(1+\alpha')}}{\Gamma[(n+1)(1+\alpha')]}, \quad -1 < \alpha' \le 0 . \tag{4.6.3}$$

In equation (4.6.3) $\Gamma(x)$ is the gamma function.

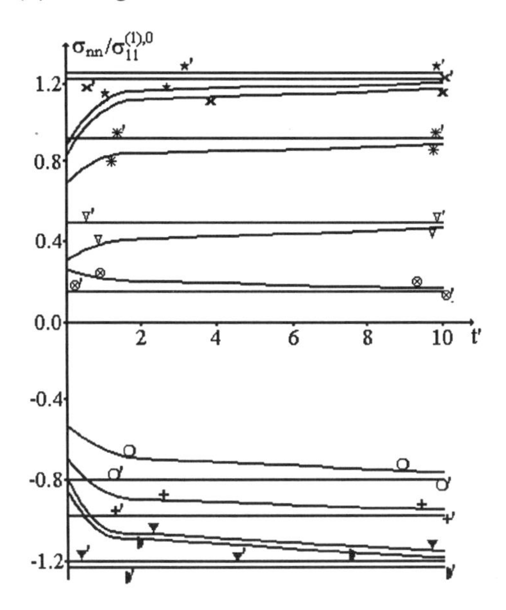

Fig.4.6.1. The graphs of the relations between $\sigma_{nn}/\sigma_{11}^{(1),0}$ and t' for ω =0.5, α' = - 0.3 with $\eta^{(2)}$ =0.5; 0.2; 0.1; 0.02; ⊗, ✳, ✖, ★ -- $\sigma_{nn}/\sigma_{11}^{(1),0}$ at point C (Fig.4.2.1), ❍, ✚, ▼, ◗ -- $\sigma_{nn}/\sigma_{11}^{(1),0}$ at point D (Fig.4.2.1), ∇-- $\sigma_{nn}/\sigma_{11}^{(1),0}$ at point A (Fig.4.2.1) under $\eta^{(2)}$ =0.5, ⊗', ✳', ✖', ★', ❍', ✚', ▼', ◗', ∇' – long – time values of $\sigma_{nn}/\sigma_{11}^{(1),0}$ corresponding to the cases shown by the symbols ⊗, ✳, ✖, ★, ❍, ✚, ▼, ◗, ∇, respectively.

Taking into account the numerical results obtained in section 4.3 we investigate the influence of the rheological parameters of the matrix material on the stress distributions in the composite materials with periodically plane curved alternating layers (Fig.4.2.1 and 4.2.3). For this purpose we introduce the dimensionless rheological parameter $\omega = \omega_\infty / \omega_0$ and dimensionless time $t' = \omega_0^{\frac{1}{1+\alpha'}} t$, and study the stress distribution with respect to these parameters, i.e., with the parameters α', ω, t'. We will consider only the values of the normal σ_{nn} and of the tangential $\sigma_{n\tau}$ self-balanced stresses at points of the inter-layer surfaces for which these stresses obtain their absolute

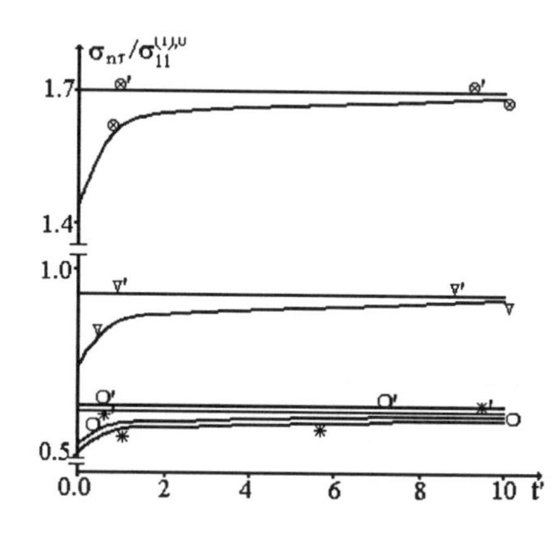

Fig.4.6.2. The graphs of the relations between $\sigma_{n\tau}/\sigma_{11}^{(1),0}$ and t' for ω =0.5, α'= - 0.3 with $\eta^{(2)}$ =0.5; 0.2; 0.1; 0.02; ⊗, ∇, O, ✱ -- $\sigma_{n\tau}/\sigma_{11}^{(1),0}$ at point A (Fig.4.2.1), ⊗', ∇', O', ✱' – long – time values of $\sigma_{n\tau}/\sigma_{11}^{(1),0}$ corresponding to the cases shown by the symbols ⊗, ∇, O, ✱, respectively.

maxima. We will assume that $E^{(2)}/E_0^{(1)}$ =50, χ =0.3, $\nu^{(2)} = \nu_0^{(1)}$ =0.3, ε =0.02.

Fig.4.6.1 shows the relation between $\sigma_{nn}/\sigma_{11}^{(1),0}$ and t' for fixed values of the rheological parameters α' and ω for the composite with alternating co-phase periodically curved layers. Fig.4.6.2 shows similar results for $\sigma_{n\tau}/\sigma_{11}^{(1),0}$; the values of $\sigma_{n\tau}/\sigma_{11}^{(1),0}$ are calculated at point A (Fig.4.2.1). These graphs show in the co-phase curving case with $\eta^{(2)} = 0.5$ the absolute values of $\sigma_{nn}/\sigma_{11}^{(1),0}$ at point C (Fig.4.2.1) decrease with time t', while those $\sigma_{nn}/\sigma_{11}^{(1),0}$ at points A and D (Fig.4.2.1)

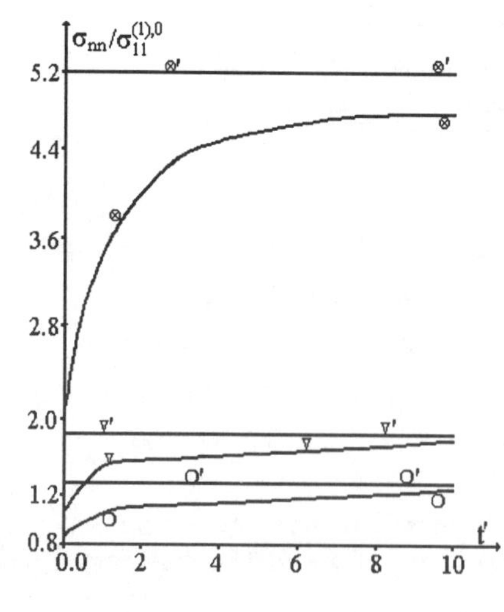

Fig.4.6.3. The graphs of the relations between $\sigma_{nn}/\sigma_{11}^{(1),0}$ and t' for ω =0.5, α'= - 0.3 with $\eta^{(2)}$ =0.5; 0.2; 0.1; ⊗, ∇, O -- $\sigma_{nn}/\sigma_{11}^{(1),0}$ at point A (Fig.4.2.3), ⊗', ∇', O' – long – time values of $\sigma_{nn}/\sigma_{11}^{(1),0}$ corresponding to the cases shown by the symbols ⊗, ∇, O, respectively.

increase. However, for comparatively small $\eta^{(2)}$, e.g. $\eta^{(2)}$=0.2; 0.1; 0.02, the absolute values of $\sigma_{nn}/\sigma_{11}^{(1),0}$ increase with time t'. Fig.4.6.2 shows also that for all selected values of $\eta^{(2)}$ the values of $\sigma_{n\tau}/\sigma_{11}^{(1),0}$ increase with t'. In all cases the significant change in $\sigma_{nn}/\sigma_{11}^{(1),0}$, $\sigma_{n\tau}/\sigma_{11}^{(1),0}$ takes place for $0 < t' \leq 3$. For t'>3 the values of $\sigma_{nn}/\sigma_{11}^{(1),0}$, $\sigma_{n\tau}/\sigma_{11}^{(1),0}$ change slowly with t'. From the comparison of $\sigma_{nn}/\sigma_{11}^{(1),0}$ with $\sigma_{n\tau}/\sigma_{11}^{(1),0}$ for various $\eta^{(2)}$ it follows that for co-phase curving (Fig.4.2.1) the effect of the viscosity of the matrix material on $\sigma_{n\tau}/\sigma_{11}^{(1),0}$ ($\sigma_{nn}/\sigma_{11}^{(1),0}$) increases (decreases) with growing $\eta^{(2)}$. Note that numerical results obtained for $\eta^{(2)}$=0.02 coincide with those obtained for $\eta^{(2)}$=0, with an accuracy of the order of $10^{-3}-10^{-4}$.

We now consider the numerical results obtained for the composite material with anti-phase periodically curved layers (Fig.4.2.3). Fig.4.6.3 shows the graphs of the relation between $\sigma_{nn}/\sigma_{11}^{(1),0}$ and t' for fixed rheological parameters α', ω and various $\eta^{(2)}$; with growing $\eta^{(2)}$ and t' the values of $\sigma_{nn}/\sigma_{11}^{(1),0}$ increase. For $\eta^{(2)}$=0.2, 0.1 the main rise in the values of $\sigma_{nn}/\sigma_{11}^{(1),0}$ takes place in the time interval $0 < t' \leq 3$; however for $\eta^{(2)}$=0.5 this rise appears in the time period $0 < t' \leq 6$; with time, the values of $\sigma_{nn}/\sigma_{11}^{(1),0}$ can become significantly greater than their corresponding values at t'=0. The values of $\sigma_{n\tau}/\sigma_{11}^{(1),0}$ are insignificant and neglect them.

Table 4.18. The values of self-balanced normal and tangential stresses in layered viscoelastic composite shown in Fig.4.2.1 with various $\eta^{(2)}$ and α' for ω=0.5, t'=10.

$\eta^{(2)}$	α'	$\sigma_{nn}(C)/\sigma_{11}^{(1),0}$	$\sigma_{n\tau}(A)/\sigma_{11}^{(1),0}$
	-0.3	0.178	-1.674
0.5	-0.5	0.190	-1.660
	-0.7	0.285	-1.638
	-0.3	0.905	-0.932
0.2	-0.5	0.888	-0.923
	-0.7	0.864	-0.908
	-0.3	1.183	-0.674
0.1	-0.5	1.153	-0.671
	-0.7	1.111	-0.667
0.02	-0.3	1.216	-0.641
	-0.5	1.184	-0.639
	-0.7	1.139	-0.639

Consider the influence of the parameters α' and ω on the values of the stresses under co-phase curving (Fig.4.2.1). Tables 4.18 and 4.19 give $\sigma_{nn}/\sigma_{11}^{(1),0}$ and $\sigma_{n\tau}/\sigma_{11}^{(1),0}$ for fixed t' and various $\eta^{(2)}$, α' and ω.

Table 4.19. The values of self-balanced normal and tangential stresses in the layered viscoelastic composite shown in Fig.4.2.1 with various $\eta^{(2)}$ and ω, for α'=-0.3, t'=10.

$\eta^{(2)}$	ω	$\frac{\sigma_{nn}(C)}{\sigma_{11}^{(1),0}}$	$\frac{\sigma_{n\tau}(A)}{\sigma_{11}^{(1),0}}$
	0.5	0.178	-1.674
0.5	1.0	0.217	-1.617
	2.0	0.241	-1.562
	0.5	0.905	-0.932
0.2	1.0	0.842	-0.895
	2.0	0.789	-0.861
	0.5	1.183	-0.674
0.1	1.0	1.073	-0.662
	2.0	0.984	-0.649
0.02	0.5	1.216	-0.641
	1.0	1.099	-0.633
	2.0	1.005	-0.624

For each $\eta^{(2)}$, $\sigma_{n\tau}/\sigma_{11}^{(1),0}$ decreases with growing α' and ω. When $\eta^{(2)}$=0.5, the absolute values of $\sigma_{nn}/\sigma_{11}^{(1),0}$ grow with α' and ω, but for $\eta^{(2)}$=0.2, 0.1, 0.01, they decrease. Tables 4.20 and 4.21 show the corresponding results for anti-phase curving. For $\eta^{(2)}$=0.5, $\sigma_{nn}/\sigma_{11}^{(1),0}$ decreases as α' and ω increase; the influence of ω is more significant than that of α'. Similar results occur for the viscoelastic composite with co-phase curved layers (Fig.4.2.1). Moreover, the comparison of Tables 4.20 and 4.21 show that for given $\eta^{(2)}$, the stresses $\sigma_{nn}(A)/\sigma_{11}^{(1),0}$, $\sigma_{n\tau}(C)/\sigma_{11}^{(1),0}$ (Fig.4.2.3) decrease as ω increase. This conclusion is explained by the growth of the long-time modulus of elasticity of the matrix material with ω. All numerical results have been obtained in the framework of the first five approximations of the presentation (4.1.15). To illustrate of the convergence of these numerical results in Table 4.22 shows the convergence of $\sigma_{n\tau}(A)/\sigma_{11}^{(1),0}$ for the co-phase curving (Fig.4.2.1) and $\sigma_{nn}(A)/\sigma_{11}^{(1),0}$ for the anti-phase curving (Fig.4.2.3) for fixed problem parameters; the convergence is acceptaible.

Table 4.20. The values of self-balanced normal tresses in the layered viscoelastic composite shown in Fig.4.2.3 with various α' and ω; $\eta^{(2)}=0.5$, $t'=10$.

ω	α'	$\sigma_{nn}(A)/\sigma_{11}^{(1),0}$
	-0.3	4.770
0.5	-0.5	4.425
	-0.7	3.918
	-0.3	3.413
1.0	-0.5	3.296
	-0.7	3.122
	-0.3	2.742
2.0	-0.5	2.712
	-0.7	2.657

Table 4.21. The values of self-balanced normal and tangential stresses in layered viscoelastic composite shown in Fig.4.2.3 with various $\eta^{(2)}$ and ω for $\alpha'=-0.3$, $t'=10$.

$\eta^{(2)}$	ω	$\sigma_{nn}(A)/\sigma_{11}^{(1),0}$	$\sigma_{n\tau}(B)/\sigma_{11}^{(1),0}$
	0.5	1.768	-0.179
0.2	1.0	1.517	-0.253
	2.0	1.288	-0.314
	0.5	1.251	-0.605
0.1	1.0	1.127	-0.603
	2.0	1.028	-0.598
0.02	0.5	1.216	-0.641
	1.0	1.099	-0.633
	2.0	1.005	-0.624

Table 4.22. The values of $\sigma_{n\tau}(A)/\sigma_{11}^{(1),0}$ (Fig.4.2.1) and $\sigma_{nn}(A)/\sigma_{11}^{(1),0}$ (Fig.4.2.3) for various approximations (4.1.15) for $\eta^{(2)}=0.5$, $\alpha'=-0.3$, $\omega=0.5$, $t'=10$,

Numb.of approx.	$\sigma_{n\tau}(A)/\sigma_{11}^{(1),0}$	$\sigma_{nn}(A)/\sigma_{11}^{(1),0}$
1	-1.953	3.775
2	-1.953	4.691
3	-1.553	4.711
4	-1.553	4.737
5	-1.674	4.770

Table 4.23 gives the values of $\sigma_{nn}(A)/\sigma_{11}^{(1),0}$ in the anti-phase curving case (Fig.4.2.3) obtained for various N in (4.5.9) and various t', and the corresponding values obtained for (4.5.7). In the collocation method (4.5.9), the values of λ_i are

selected as $\lambda_1=0.02$, $\lambda_i=\lambda_{i-1}+i\times 0.098$, ($i\geq 2$). The comparison of these data reveals that the difference between the results obtained by the collocation method and by the formula (4.5.7) is not greater than 2-6 %; Consequently, with respect to the problems considered we can conclude that in the framework of this accuracy the studied quasistatic problems may also be investigated by the use of the inverse transform formula (4.5.7).

Table 4.23. The values $\sigma_{nn}(A)/\sigma_{11}^{(1),0}$ (Fig.4.2.3) with various N (4.5.9) and t' for $\eta^{(2)}=0.5$, $\alpha'=-0.3$, $\omega=0.5$.

N (4.5.9)	t'				
	1	11	21	31	51
10	3.262	4.686	4.865	4.974	4.994
15	3.257	4.667	4.863	4.960	5.046
the results obtained by (4.5.7)	3.412	4.581	4.770	4.857	4.944

4.7. Composite Materials with Anisotropic Layers

In this section we assume that the material of the layers from which composite materials are constructed, is anisotropic (orthotropic). The following two cases will be investigated: 1) the materials of the layers are rectilinearly anisotropic; 2) the materials of the layers are curvilinearly anisotropic, in the latter we will suppose that the curvilinear anisotropy is caused by curving of these layers. The solution method presented in section 4.2 will be developed for these cases.

4.7.1. RECTILINEAR ANISOTROPY

For simplicity we consider stress-strain distributions in a composite with alternating co-phase periodically plane-curved layers (Fig.4.2.1) under loading $\sigma_{11}(\infty)=p$. We assume that the layers are orthotropic with principal symmetry axes Ox_1, Ox_2 and Ox_3. We will also assume that all suppositions and restrictions accepted in section 4.2 hold. Thus, instead of the mechanical relations given in (4.1.1) we take the following ones for the plane-strain state.

$$\varepsilon_{11}^{(k)}=\frac{1}{E_1^{(k)}}\left[\left(1-\left(\nu_{13}^{(k)}\right)^2\frac{E_3^{(k)}}{E_1^{(k)}}\right)\sigma_{11}^{(k)}-\left(\nu_{12}^{(k)}+\nu_{13}^{(k)}\nu_{32}^{(k)}\right)\sigma_{22}^{(k)}\right],$$

$$\varepsilon_{22}^{(k)}=\frac{1}{E_2^{(k)}}\left[\left(-\nu_{12}^{(k)}-\nu_{13}^{(k)}\nu_{32}^{(k)}\right)\sigma_{11}^{(k)}+\frac{E_1^{(k)}}{E_2^{(k)}}\left(1-\frac{E_2^{(k)}}{E_3^{(k)}}\left(\nu_{32}^{(k)}\right)^2\right)\sigma_{22}^{(k)}\right],$$

$$\varepsilon_{12}^{(k)} = \frac{1}{G_{12}^{(k)}} \sigma_{12}^{(k)}, \qquad k=1,2\,. \tag{4.7.1}$$

In (4.7.1) $E_1^{(k)}$, $E_2^{(k)}$, $E_3^{(k)}$ are the moduli of elasticity in the directions of the Ox_1, Ox_2, Ox_3 axes, $G_{12}^{(k)}$ is the shear modulus in the Ox_1x_2 plane, $\nu_{13}^{(k)}, \nu_{12}^{(k)}, \nu_{32}^{(k)}$ are the Poisson's ratios. The relations (4.7.1) will be satisfied by each approximation (4.1.15) separately.

Thus using the solution procedure described in subsection 4.2.1 we obtain the following zeroth approximation:

$$\sigma_{11}^{(1),0} = p\left[\eta^{(1)} + \eta^{(2)}\kappa\right]^{-1}, \qquad \sigma_{11}^{(2),0} = \kappa\sigma_{11}^{(1),0},$$

$$u_1^{(k),0} = \frac{1}{E_1^{(k)}}\left(1 - \left(\nu_{13}^{(k)}\right)^2 \frac{E_3^{(k)}}{E_1^{(k)}}\right)\sigma_{11}^{(k),0} x_1,$$

$$u_2^{(k),0} = -\frac{1}{E_1^{(k)}}\left(\nu_{13}^{(k)} + \nu_{13}^{(k)}\nu_{32}^{(k)}\right)\sigma_{11}^{(k),0} x_2^{(k)} + C^{(k)},$$

$$\kappa = E_1^{(2)}\left(1 - \left(\nu_{13}^{(1)}\right)^2 \frac{E_3^{(1)}}{E_1^{(1)}}\right)\left(E_1^{(1)}\left(1 - \left(\nu_{13}^{(2)}\right)^2 \frac{E_3^{(2)}}{E_1^{(2)}}\right)\right)^{-1}, C^{(k)} = \text{const}\,, \sigma_{12}^{(k)} = \sigma_{22}^{(k)} = 0\,. \tag{4.7.2}$$

Now consider the first approximation. In this case the contact conditions (4.2.8) hold. Introduce the stress potential $\Phi^{(k),1}$ through which the stresses are expressed as

$$\sigma_{11}^{(k),1} = \frac{\partial^2 \Phi^{(k),1}}{\partial (x_2^{(k)})^2}, \quad \sigma_{22}^{(k),1} = \frac{\partial^2 \Phi^{(k),1}}{\partial (x_1^{(k)})^2}, \quad \sigma_{12}^{(k),1} = -\frac{\partial^2 \Phi^{(k),1}}{\partial x_1^{(k)} \partial x_2^{(k)}}. \tag{4.7.3}$$

Using (4.7.1) and (4.7.3) and the deformation compatibility conditions we obtain the following equation for the function $\Phi^{(k),1}$.

$$\rho_1^{(k)} \frac{\partial^4 \Phi^{(k),1}}{\partial (x_2^{(k)})^4} + \left[\frac{E_1^{(k)}}{G_{12}^{(k)}} - 2\left(\nu_{12}^{(k)} + \nu_{13}^{(k)}\nu_{32}^{(k)}\right)\right] \frac{\partial^4 \Phi^{(k),1}}{\partial^2 (x_1^{(k)})^2 \partial^2 (x_2^{(k)})^2} +$$

$$\rho_2^{(k)} \frac{\partial^4 \Phi^{(k),1}}{\partial (x_1^{(k)})^4} = 0 .$$

$$\rho_1^{(k)} = \left[1 - \left(\nu_{13}^{(k)}\right)^2 \frac{E_3^{(k)}}{E_1^{(k)}} \right], \rho_2^{(k)} = \frac{E_1^{(k)}}{E_2^{(k)}} \left[1 - \frac{E_2^{(k)}}{E_3^{(k)}} \left(\nu_{32}^{(k)}\right)^2 \right] \tag{4.7.4}$$

Taking the form of the expressions in the right hand side of the contact conditions (4.2.8) into account, we select the particular solution to the equation (4.7.4) as

$$\Phi^{(k),1} = C_1^{(k)} \exp\left(\kappa^{(k)} \alpha x_2^{(k)}\right) \sin(\alpha x_1) , \ \alpha = \frac{2\pi}{\ell} . \tag{4.7.5}$$

Substituting (4.7.5) in (4.7.4) we obtain the following characteristic equation for $\kappa^{(k)}$.

$$\left(\kappa^{(k)}\right)^4 - Q^{(k)} \left(\kappa^{(k)}\right)^2 + R^{(k)} = 0 , \tag{4.7.6}$$

where

$$Q^{(k)} = \left(\frac{E_1^{(k)}}{G_{12}^{(k)}} - 2\left(\nu_{12}^{(k)} + \nu_{13}^{(k)} \nu_{32}^{(k)}\right) \right) \left(1 - \left(\nu_{13}^{(k)}\right)^2 \right)^{-1} , \ R^{(k)} = \rho_2^{(k)} \left(\rho_1^{(k)}\right)^{-1} . \tag{4.7.7}$$

In [125] it had been proven that the equation

$$\left(\kappa^{(k)}\right)^4 + Q^{(k)} \left(\kappa^{(k)}\right)^2 + R^{(k)} = 0 , \tag{4.7.8}$$

that is obtained from (4.7.6) by the replacement $Q^{(k)}$ with $-Q^{(k)}$, cannot have a real root. This conclusion corresponds to the following three cases for the coefficients of the equation (4.7.8):

$$\text{I}: \ R^{(k)} > 0 , \ -2\sqrt{R^{(k)}} < Q^{(k)} < 2\sqrt{R^{(k)}} ;$$

$$\text{II}: \ R^{(k)} > 0 , \ Q^{(k)} > 2\sqrt{R^{(k)}} ;$$

$$\text{III}: \ R^{(k)} > 0 , \ Q^{(k)} = 2\sqrt{R^{(k)}} . \tag{4.7.9}$$

The case where $Q^{(k)} \le -2\sqrt{R^{(k)}}$ cannot occur. According to (4.7.9), the roots of equation (4.7.6) and the solution of equation (4.7.4) are determined in the following form.

Case I : the roots are complex

$$\kappa_1^{(k)} = \alpha^{(k)} + i\beta^{(k)},\ \kappa_2^{(k)} = \alpha^{(k)} - i\beta^{(k)},\ k_3^{(k)} = -k_1^{(k)},\ \kappa_4^{(k)} = -\kappa_2^{(k)},$$

$$\alpha^{(k)} = \left[\frac{R^{(k)}}{2}\left(1 + \frac{Q^{(k)}}{2\sqrt{R^{(k)}}}\right)\right]^{\frac{1}{2}}, \qquad \beta^{(k)} = \left[\frac{R^{(k)}}{2}\left(1 - \frac{Q^{(k)}}{2\sqrt{R^{(k)}}}\right)\right]^{\frac{1}{2}}; \tag{4.7.10}$$

$$\Phi^{(k),1} = \left[C_1^{(k),1}\cosh\left(\alpha^{(k)}\alpha x_2^{(k)}\right)\cos\left(\beta^{(k)}\alpha x_2^{(k)}\right) + C_2^{(k),1}\sinh\left(\alpha^{(k)}\alpha x_2^{(k)}\right)\cos\left(\beta^{(k)}\alpha x_2^{(k)}\right) + C_3^{(k),1}\cosh\left(\alpha^{(k)}\alpha x_2^{(k)}\right)\sin\left(\beta^{(k)}\alpha x_2^{(k)}\right) + C_4^{(k),1}\sinh\left(\alpha^{(k)}\alpha x_2^{(k)}\right)\sin\left(\beta^{(k)}\alpha x_2^{(k)}\right)\right]\sin(\alpha x_1) \tag{4.7.11}$$

Case II : the roots are real

$$\kappa_{1,2}^{(k)} = \left[\frac{Q^{(k)}}{2} \pm \left(\frac{\left(Q^{(k)}\right)^2}{4} - R^{(k)}\right)^{\frac{1}{2}}\right]^{\frac{1}{2}},\ \kappa_3^{(k)} = -\kappa_1^{(k)},\ \kappa_4^{(k)} = -\kappa_2^{(k)}. \tag{4.7.12}$$

$$\Phi^{(k),1} = \left[C_1^{(k),1}\sinh\left(\kappa_1^{(k)}\alpha x_2^{(k)}\right) + C_2^{(k),1}\cosh\left(\kappa_1^{(k)}\alpha x_2^{(k)}\right) + C_3^{(k),1}\sinh\left(\kappa_2^{(k)}\alpha x_2^{(k)}\right) + C_4^{(k),1}\cosh\left(\kappa_2^{(k)}\alpha x_2^{(k)}\right)\right]\sin(\alpha x_1). \tag{4.7.13}$$

We omit case III which corresponds to the case in which there is isotropy in the Ox_1x_2 plane. Using the relations (4.7.3), (4.7.11), (4.7.13) and (4.7.1) we obtain the closed system equation from the contact conditions (4.2.8) to determine the unknown constants. So, we find the values of the first approximations. The second and subsequent approximations are considered similarly; each has the characteristic equation (4.7.6).

4.7.2. CURVILINEAR ANISOTROPY

The elasticity relations for every approximation

We begin with the construction of the relations between $\sigma_{ij}^{(k)m,q}$ and $\varepsilon_{ij}^{(k)m,q}$ in the case where the material of the curving reinforcing layers are curvilinear anisotropic. In this case we will assume that the curvilinearity of the material anisotropy is caused by curving of the reinforcing layer.

For simplicity, we choose an individual curved layer of filler having a constant thickness over its entire length (Fig.4.7.1).

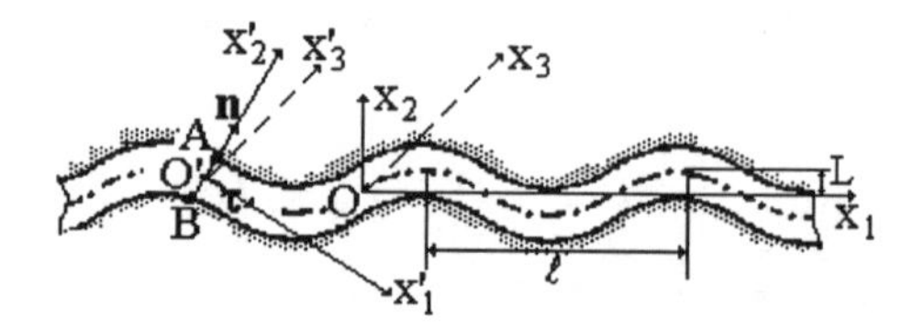

Fig. 4.7.1. The selected curvilinear anisotropic reinforcing layer.

We use the rectangular coordinate system $Ox_1x_2x_3$ and assume that the middle surface of this layer is described by the equation

$$x_2 = \varepsilon f(x_1), \tag{4.7.14}$$

where $\varepsilon \in [0,1)$ is a dimensionless small parameter.

We take each band AB (Fig.4.7.1), located perpendicular to the tangent vector $\boldsymbol{\tau}$ of the middle surface, and associate it with the local coordinate system $O'x'_1\,x'_2\,x'_3$. Assume that the $O'x'_1$ and $O'x'_2$ axes of this system coincide with tangential $\boldsymbol{\tau}$ and normal $\mathbf{n}$ vectors to the middle surface (4.7.14), and the $O'x'_3$ axis coincides with Ox_3 axis. We suppose that within each band AB the material of the chosen layer is homogeneous and orthotropic with principal elastic symmetry axes $O'x'_1$, $O'x'_2$ and $O'x'_3$. Hence in the global coordinate system $Ox_1x_2x_3$ the material of the chosen layer can be considered as an anisotropic material with continuously changing elastic symmetry axes $O'x'_1$ and $O'x'_2$. In other words, if the material of the chosen filler layer in the uncurved case is homogeneous and rectilinearly orthotropic with principal elastic symmetry axes Ox_1, Ox_2 and Ox_3, then in the curved case the material of this layer will be curvilinearly anisotropic in the coordinate system $Ox_1x_2x_3$.

Suppose the material of the chosen layer is transversely isotropic at each point of the band AB, with the axis of elastic symmetry coinciding with the axis $O'x'_1$. We write the expressions for the compliance coefficient in terms of engineering constants in the $O'x'_1\,x'_2\,x'_3$ coordinate system here retaining the notation used in the previous subsection:

$$a'_{14} = a'_{24} = a'_{34} = a'_{46} = a'_{15} = a'_{25} = a'_{35} = a'_{16} = a'_{26} = a'_{36} = a'_{45} = 0,$$

$$a'_{12} = a'_{13} = -\frac{\nu_1}{E_1}, \quad a'_{11} = \frac{1}{E}, \quad a'_{22} = \frac{1}{E}, \quad a'_{23} = -\frac{\nu}{E},$$

$$a'_{33} = a'_{22}, \quad a'_{55} = a'_{66} = \frac{1}{G_{12}}, \quad a'_{44} = \frac{2(1+\nu)}{E}. \tag{4.7.15}$$

Denoting the angle between the tangent vector $\boldsymbol{\tau}$ and the Ox_1 (Fig.4.7.1) axis by φ and using the transformation (1.3.20) of compliance constants under the rotation of the axes and taking (4.7.15) into account we obtain the following expressions for the compliance constants in the global coordinate system $Ox_1x_2x_3$:

$$a_{11}=\frac{1}{E_1}\cos^4\varphi+\left(-\frac{2\nu_1}{E_1}+\frac{1}{G_{12}}\right)\sin^2\varphi\cos^2\varphi+\frac{1}{E}\sin^4\varphi,$$

$$a_{22}=\frac{1}{E_1}\sin^4\varphi+\left(-\frac{2\nu_1}{E_1}+\frac{1}{G_{12}}\right)\sin^2\varphi\cos^2\varphi+\frac{1}{E}\cos^4\varphi,$$

$$a_{12}=-\frac{\nu_1}{E_1}+\left(\frac{1}{E_1}+\frac{1}{E}+\frac{2\nu}{E_1}-\frac{1}{G_{12}}\right)\sin^2\varphi\cos^2\varphi,$$

$$a_{66}=\frac{1}{G_{12}}+4\left(\frac{1}{E_1}+\frac{1}{E}+\frac{2\nu_1}{E_1}-\frac{1}{G_{12}}\right)\sin^2\varphi\cos^2\varphi, \tag{4.1.16}$$

$$a_{16}=\left[\frac{2}{E}\sin^2\varphi-\frac{2}{E_1}\cos^2\varphi+\left(-\frac{2\nu_1}{E_1}+\frac{1}{G_{12}}\right)\left(\cos^2\varphi-\sin^2\varphi\right)\right]\sin\varphi\cos\varphi,$$

$$a_{26}=\left[\frac{2}{E}\cos^2\varphi-\frac{2}{E_1}\sin^2\varphi-\left(-\frac{2\nu_1}{E_1}+\frac{1}{G_{12}}\right)\left(\cos^2\varphi-\sin^2\varphi\right)\right]\sin\varphi\cos\varphi,$$

$$a_{44}=\frac{2(1+\nu)}{E}\cos^2\varphi+\frac{1}{G_{12}}\sin^2\varphi,\quad a_{45}=\left(\frac{2(1+\nu)}{E}-\frac{1}{G_{12}}\right)\sin\varphi\cos\varphi,$$

$$a_{55}=\frac{2(1+\nu)}{E}\sin^2\varphi+\frac{1}{G_{12}}\cos^2\varphi,\quad a_{13}=-\frac{\nu_1}{E_1}\cos^2\varphi-\frac{\nu}{E}\sin^2\varphi,$$

$$a_{36}=2\left(\frac{\nu_1}{E_1}-\frac{\nu}{E}\right)\sin\varphi\cos\varphi,\quad a_{23}=-\frac{\nu_1}{E_1}\sin^2\varphi-\frac{\nu}{E}\cos^2\varphi,\quad a_{33}=\frac{1}{E}$$

It follows from equation (4.7.14) that

$$\tan\varphi=\varepsilon f'(x_1). \tag{4.7.17}$$

Consider the case where the curving of the reinforcing layer is periodic and the function (4.7.14) is given in the form (4.2.1); in this case

$$\tan\varphi = \varepsilon 2\pi\cos(\alpha x_1),\ \alpha = \frac{2\pi}{\ell},\quad \varepsilon = \frac{L}{\ell} \tag{4.7.18}$$

Thus, assuming that $(\varepsilon 2\pi) < 1$, from (4.7.16) and (4.7.18) and doing some transformations we derive the following expressions for the compliance constants a_{ij}.

$$a_{ij} = \begin{cases} a'_{ij} + \sum_{n=1}^{\infty} \varepsilon^{2n} a_{ijn} \cos^{2n}(\alpha x_1) & \text{for combinations } ij = 11; 22; 12; 66; 44; 55; 13; 23 \\ \sum_{n=1}^{\infty} \varepsilon^{2n-1} a_{ijn} \cos^{2n-1}(\alpha x_1) & \text{for combinations } ij = 16; 26; 45; 36 \\ a'_{ij} & \text{for combination } ij = 33 \end{cases}$$

(4.7.19)

Note that these relations are the inverse ones of the relations given by (2.3.2) in the case where the curving is described by the function (4.2.1). Now we write the expressions for the a_{ijn} for $n = 1$ and 2.

$$a_{111} = 4\pi^2\left(\frac{1}{G_{12}} - \frac{2}{E_1} - \frac{2\nu_1}{E_1}\right),\quad a_{112} = 16\pi^4\left(\frac{3}{E_1} + \frac{4\nu_1}{E_1} - \frac{2}{G_{12}} + \frac{1}{E}\right),$$

$$a_{221} = 4\pi^2\left(\frac{1}{G_{12}} - \frac{2\nu_1}{E_1} - \frac{2}{E}\right),\quad a_{222} = 8\pi^4\left(\frac{6}{E} + \frac{8\nu_1}{E_1} - \frac{4}{G_{12}} + \frac{2}{E_1}\right),$$

$$a_{121} = 4\pi^2\left(\frac{1}{E_1} + \frac{1}{E} + \frac{2\nu_1}{E_1} - \frac{1}{G_{12}}\right);\quad a_{121} = 32\pi^4\left(\frac{1}{E_1} + \frac{1}{E} + \frac{2\nu_1}{E_1} - \frac{1}{G_{12}}\right);$$

$$a_{661} = 16\pi^2\left(\frac{1}{E_1} + \frac{1}{E} + \frac{2\nu_1}{E_1} - \frac{1}{G_{12}}\right);\quad a_{662} = -128\pi^4\left(\frac{1}{E_1} + \frac{2}{E} + \frac{2\nu_1}{E_1} - \frac{1}{G_{12}}\right);$$

$$a_{441} = 4\pi^2\left(\frac{1}{G_{12}} - \frac{2(1+\nu)}{E}\right);\quad a_{442} = -16\pi^4\left(\frac{1}{G_{12}} - \frac{2(1+\nu)}{E}\right);$$

$$a_{551} = -4\pi^2\left(\frac{1}{G_{12}} - \frac{2(1+\nu)}{E}\right);\quad a_{552} = 16\pi^4\left(\frac{1}{G_{12}} - \frac{2(1+\nu)}{E}\right);$$

$$a_{131} = 4\pi^2\left(\frac{\nu_1}{E_1} - \frac{\nu}{E}\right); \qquad a_{132} = -16\pi^4\left(\frac{\nu_1}{E_1} - \frac{\nu}{E}\right);$$

$$a_{231} = -4\pi^2\left(\frac{\nu_1}{E_1} - \frac{\nu}{E}\right); \qquad a_{232} = 16\pi^4\left(\frac{\nu_1}{E_1} - \frac{\nu}{E}\right);$$

$$a_{161} = 2\pi\left(\frac{1}{G_{12}} - \frac{2(1+\nu_1)}{E_1}\right); \quad a_{162} = 8\pi^3\left(-\frac{3}{G_{12}} + \frac{2(2+3\nu_1)}{E_1} + \frac{2}{E}\right);$$

$$a_{261} = 2\pi\left(\frac{2}{E} + \frac{2\nu_1}{E_1} - \frac{1}{G_{12}}\right); \quad a_{262} = 8\pi^3\left(-\frac{4}{E} - \frac{2+6\nu_1}{E_1} + \frac{3}{G_{12}}\right);$$

$$a_{451} = 2\pi\left(\frac{2(1+\nu)}{E} - \frac{1}{G_{12}}\right); \quad a_{452} = -8\pi^3\left(\frac{2(1+\nu)}{E} - \frac{1}{G_{12}}\right);$$

$$a_{361} = 4\pi\left(-\frac{\nu}{E} + \frac{\nu_1}{E_1}\right); \quad a_{362} = -16\pi^3\left(-\frac{\nu}{E} + \frac{\nu_1}{E_1}\right). \tag{4.7.20}$$

Note that the relations (4.7.20) are the inverse ones of the relations (2.3.3) in the case (4.7.18).

Now we write the elasticity relations for the plane strain state for the anisotropic body having one elastic symmetry plane which is perpendicular to the Ox_3 axis.

$$\varepsilon_{11} = \left(a_{11} - \frac{a_{13}^2}{a_{33}}\right)\sigma_{11} + \left(a_{12} - \frac{a_{12}a_{23}}{a_{33}}\right)\sigma_{22} + \left(a_{16} - \frac{a_{13}a_{36}}{a_{33}}\right)\sigma_{12},$$

$$\varepsilon_{22} = \left(a_{22} - \frac{a_{22}a_{13}}{a_{33}}\right)\sigma_{11} + \left(a_{22} - \frac{a_{23}a_{23}}{a_{33}}\right)\sigma_{22} + \left(a_{26} - \frac{a_{23}a_{36}}{a_{33}}\right)\sigma_{12},$$

$$2\varepsilon_{12} = \left(a_{16} - \frac{a_{36}a_{13}}{a_{33}}\right)\sigma_{11} + \left(a_{26} - \frac{a_{36}a_{23}}{a_{33}}\right)\sigma_{22} + \left(a_{66} - \frac{a_{36}a_{36}}{a_{33}}\right)\sigma_{12}. \tag{4.7.21}$$

Using (4.7.19) we derive the following expressions for the coefficients in (4.7.21):

$$a_{11} - \frac{a_{13}^2}{a_{33}} = \frac{1}{E_1}\left(\alpha_{110} + \sum_{n=1}^{\infty}\varepsilon^{2n}\alpha_{11n}\cos^{2n}(\alpha x_1)\right);$$

$$a_{12} - \frac{a_{13}a_{23}}{a_{33}} = \frac{1}{E_1}\left(\alpha_{120} + \sum_{n=1}^{\infty}\varepsilon^{2n}\alpha_{12n}\cos^{2n}(\alpha x_1)\right);$$

$$a_{22} - \frac{a_{23}a_{23}}{a_{33}} = \frac{1}{E_1}\left(\alpha_{220} + \sum_{n=1}^{\infty}\varepsilon^{2n}\alpha_{22n}\cos^{2n}(\alpha x_1)\right);$$

$$a_{66} - \frac{a_{36}a_{36}}{a_{33}} = \frac{1}{E_1}\left(\alpha_{660} + \sum_{n=1}^{\infty}\varepsilon^{2n}\alpha_{66n}\cos^{2n}(\alpha x_1)\right);$$

$$a_{16} - \frac{a_{13}a_{36}}{a_{33}} = \frac{1}{E_1}\sum_{n=1}^{\infty}\varepsilon^{2n-1}\alpha_{16n}\cos^{2n-1}(\alpha x_1);$$

$$a_{26} - \frac{a_{23}a_{36}}{a_{33}} = \frac{1}{E_1}\sum_{n=1}^{\infty}\varepsilon^{2n-1}\alpha_{26n}\cos^{2n-1}(\alpha x_1). \tag{4.7.22}$$

Introducing the notation

$$e = \frac{E_1}{E}, \quad g = \frac{E_1}{G_{12}} \qquad e_1 = \nu_1 - \frac{\nu}{e} \tag{4.7.23}$$

we write the expressions for some α_{ijn} which will be used for further investigations.

$$\alpha_{110} = 1 - e\nu_1^2, \quad \alpha_{111} = 4\pi^2[g - 2(1+\nu_1) + 2\nu_1 e e_1],$$

$$\alpha_{112} = 16\pi^4\left[3 + 4\nu_1 + \frac{1}{e} - 2g - 2\nu_1 e e_1\right], \quad \alpha_{120} = -\nu_1(1+\nu),$$

$$\alpha_{121} = 4\pi^2\left[1 + 2\nu_1 + \frac{1}{e} - g - e e_1^2\right],$$

$$\alpha_{122} = 16\pi^4\left[2g - \frac{2}{e} - 2(1+2\nu_1) + 2e e_1^2\right],$$

$$\alpha_{161} = 2\pi[g - 2(1+\nu_1) + 2\nu_1 e e_1],$$

$$\alpha_{162} = 8\pi^3\left[2(2+3\nu_1) + \frac{2}{e} - 3g - 2\nu_1 e e_1 - 2e e_1^2\right],$$

$$\alpha_{220} = \left(1 - \nu^2\right)e, \quad \alpha_{221} = 4\pi^2\left[g - \frac{2}{e} - 2\nu_1 - 2\nu e_1^2\right],$$

$$\alpha_{222} = 16\pi^4\left[\frac{3}{e} + 4\nu_1 + 1 - 2g + 2\nu e_1 - e_1^2 e\right],$$

$$\alpha_{261} = 2\pi\left[\frac{2}{e} + 2\nu_1 - g + 2\nu e_1\right],$$

$$\alpha_{262} = 8\pi^3\left[\frac{3}{g} - \frac{4}{e} - (2 + 6\nu_1) - 2\nu e_1 + 2ee_1^2\right],$$

$$\alpha_{660} = g, \quad \alpha_{661} = 16\pi^2\left[1 + 2\nu_1 + \frac{1}{e} - g - ee_1^2\right],$$

$$\alpha_{662} = 128\pi^4\left[g - (1 + 2\nu_1) - \frac{1}{e} + ee_1^2\right]. \tag{4.7.24}$$

Note that for $\varepsilon = 0$ the relations (4.7.21) are the relations for a transversally-isotropic body with symmetry axis Ox_1.

Thus, substituting the presentations (4.1.15) and (4.7.22) in (4.7.21) we obtain the following mechanical relations for each approximation.

$$E_1\varepsilon_{11}^k = \alpha_{110}\,\sigma_{11}^k + \alpha_{120}\sigma_{12}^k + \sum_{n=1}^{[k/2]}\left(\alpha_{11n}\sigma_{11}^{k-2n} + \alpha_{12n}\sigma_{22}^{k-2n}\right)\cos^{2n}(\alpha x_1) +$$

$$\sum_{n=1}^{[(k+1)/2]}\alpha_{16n}\sigma_{12}^{k+1-2n}\cos^{2n-1}(\alpha x_1),$$

$$E_1\varepsilon_{22}^k = \alpha_{120}\,\sigma_{11}^k + \alpha_{220}\sigma_{12}^k + \sum_{n=1}^{[k/2]}\left(\alpha_{26n}\sigma_{11}^{k-2n} + \alpha_{22n}\sigma_{22}^{k-2n}\right)\cos^{2n}(\alpha x_1) +$$

$$\sum_{n=1}^{[(k+1)/2]}\alpha_{26n}\sigma_{12}^{k+1-2n}\cos^{2n-1}(\alpha x_1),$$

$$2E_1\varepsilon_{12}^k = \alpha_{660}\,\sigma_{12}^k + \sum_{n=1}^{[(k+1)/2]}\left(\alpha_{16n}\sigma_{11}^{k+1-2n} + \alpha_{26n}\sigma_{22}^{k+1-2n}\right)\cos^{2n-1}(\alpha x_1) +$$

$$\sum_{n=1}^{[k/2]} \alpha_{66n} \sigma_{12}^{k+1-2n} \cos^{2n}(\alpha x_1). \tag{4.7.25}$$

Note that in (4.7.25) the upper indices on the stress and strain components show the order of the approximation (4.1.15) and the $[k/2]$ gives the integer part of $k/2$. It follows from (4.7.25) that for curvilinear anisotropy the mechanical relations written for the k-th approximation contain the values of all previous approximations. Note that by replacing $2\pi\cos(\alpha x_1)$ with $f'(x_1)$ in (4.7.25) we obtain the mechanical relations for each approximation under local curving of the anisotropic filler layer.

The determination of the values of the first and second approximations.
We assume that the material of the matrix layers are isotropic. Note that the values of the zeroth approximation is determined as (4.7.2) in the case where,

$$E_2^{(2)} = E_3^{(2)} = E^{(2)}, \quad \nu_{13}^{(2)} = \nu_1^{(2)}, \quad \nu_{32}^{(2)} = \nu^{(2)}, \tag{4.7.26}$$

$$E_2^{(1)} = E_3^{(1)} = E_3^{(2)} = E^{(1)}, \quad \nu_{13}^{(1)} = \nu_{32}^{(1)} = \nu^{(1)}. \tag{4.7.27}$$

Further we will consider the determination of the values related to the filler layers because the values related to the matrix layers are determined as in section 4.2.

The first approximation. For this approximation we have the equilibrium equation, geometrical relations given in (4.1.16), contact condition (4.2.8) and the compatibility condition for the components of the deformations, i.e.,

$$\frac{\partial^2 \varepsilon_{11}^{(k),1}}{\partial (x_2^{(k)})^2} + \frac{\partial^2 \varepsilon_{22}^{(k),1}}{\partial (x_1^{(k)})^2} = 2\frac{\partial^2 \varepsilon_{12}^{(k),1}}{\partial x_1^{(k)} \partial x_2^{(k)}}. \tag{4.7.28}$$

To these equations we must add the elasticity relations which are obtained from (4.7.25) as

$$E_1^{(2)} \varepsilon_{11}^{(2),1} = \alpha_{110}^{(2)} \sigma_{11}^{(2),1} + \alpha_{120}^{(2)} \sigma_{22}^{(2),1}, \quad E_1^{(2)} \varepsilon_{22}^{(2),1} = \alpha_{120}^{(2)} \sigma_{11}^{(2),1} + \alpha_{220}^{(2)} \sigma_{22}^{(2),1},$$

$$2E_1^{(2)} \varepsilon_{12}^{(2),1} = \alpha_{660}^{(2)} \sigma_{12}^{(2),1} + \alpha_{161}^{(2)} \sigma_{11}^{(2),0} \cos(\alpha x_1). \tag{4.7.29}$$

As before we introduce the function $\Phi^{(2),1}$ through which the stresses are determined by (4.7.3). Thus using (4.7.3) and (4.7.29) we obtain from (4.7.28) the equation (4.7.4) for the function $\Phi^{(2),1}$. Therefore, in the first approximation the function $\Phi^{(2),1}$ is determined as (4.7.11), (4.7.13); using the corresponding relations we obtain the following expressions for the stresses and displacements:

$$\sigma_{11}^{(2),1} = s_{11}(x_2^{(2)})\sin(\alpha x_1),\ \sigma_{22}^{(2),1} = s_{22}(x_2^{(2)})\sin(\alpha x_1),\ \sigma_{12}^{(2),1} = s_{12}(x_2^{(2)})\cos(\alpha x_1),$$

$$u_1^{(2),1} = u_{11}\left(x_2^{(2)}\right)\cos(\alpha x_1),\ u_2^{(2),1} = u_{21}(x_2^{(2)})\sin(\alpha x_1) + \frac{1}{E_1^{(2)}}\alpha_{161}\sigma_{11}^{(2),0}\sin(\alpha x_1), \tag{4.7.30}$$

where

$$s_{11}\left(x_2^{(2)}\right) = \frac{d^2 B\left(x_2^{(k)}\right)}{d(x_2^{(2)})^2},\quad s_{12}\left(x_2^{(2)}\right) = -\frac{dB\left(x_2^{(k)}\right)}{dx_2^{(2)}},\quad s_{22}\left(x_2^{(2)}\right) = -B\left(x_2^{(2)}\right),$$

$$B\left(x_2^{(2)}\right) = C_1^{(2),1} b_{11}\left(x_2^{(2)}\right) + C_2^{(2),1} b_{12}\left(x_2^{(2)}\right) + C_3^{(2),1} b_{21}\left(x_2^{(2)}\right) + C_4^{(2),1} b_{22}\left(x_2^{(2)}\right),$$

$$u_{11}\left(x_2^{(2)}\right) = \frac{1}{E_1^{(2)}}\left(\alpha_{110}\frac{d^2}{d(x_2^{(2)})^2} - \alpha_{120}\right)B\left(x_2^{(2)}\right)$$

$$u_{21}\left(x_2^{(2)}\right) = \frac{1}{E_1^{(2)}}\left(\alpha_{110}\frac{d^2}{d(x_2^{(2)})^2} - \alpha_{120}\frac{d}{dx_2^{(2)}} - \alpha_{660}\right)B\left(x_2^{(2)}\right). \tag{4.7.31}$$

In (4.7.31) the following notation has been used.

For case I (4.7.10):

$$b_{11}\left(x_2^{(2)}\right) = \cosh\left(\alpha^{(2)}\alpha x_2^{(2)}\right)\cos\left(\beta^{(2)}\alpha x_2^{(2)}\right),$$

$$b_{12}\left(x_2^{(2)}\right) = \sinh\left(\alpha^{(2)}\alpha x_2^{(2)}\right)\cos\left(\beta^{(2)}\alpha x_2^{(2)}\right)$$

$$b_{21}\left(x_2^{(2)}\right) = \cosh\left(\alpha^{(2)}\alpha x_2^{(2)}\right)\sin\left(\beta^{(2)}\alpha x_2^{(2)}\right),$$

$$b_{22}\left(x_2^{(2)}\right) = \sinh\left(\alpha^{(2)}\alpha x_2^{(2)}\right)\sin\left(\beta^{(2)}\alpha x_2^{(2)}\right). \tag{4.7.32}$$

For case II (4.7.12):

$$b_{11}\left(x_2^{(2)}\right) = \sinh\left(\kappa_1^{(2)}\alpha x_2^{(2)}\right),\ b_{12}\left(x_2^{(2)}\right) = \cosh\left(\kappa_1^{(2)}\alpha x_2^{(2)}\right)$$

$$b_{21}\left(x_2^{(2)}\right) = \sinh\left(\kappa_2^{(2)}\alpha x_2^{(2)}\right),\ b_{22}\left(x_2^{(2)}\right) = \cosh\left(\kappa_2^{(2)}\alpha x_2^{(2)}\right). \tag{4.7.33}$$

Thus considering the equations (4.1.16), (4.2.9), (4.2.11) and (4.7.30)-(4.7.33) we derive the corresponding equation from the contact condition (4.2.8) for the unknown constants $C_i^{(2),1}(i=1,2,3,4)$ and $C_{n1}^{(1),1}, C_{n2}^{(1),1}(n=0,2)$ which are in the expressions of the values of the first approximation. It follows from the expressions and equations for the first approximation, that the curvilinearity of the filler layer changes the stress and displacement distribution in the framework of this approximation.

The second approximation. For this approximation we obtain the following elasticity relations for the filler layer material from (4.7.25).

$$E_1^{(2)}\varepsilon_{11}^{(2),2} = \alpha_{110}\,\sigma_{11}^{(2),2} + \alpha_{120}\,\sigma_{22}^{(2),2} + \frac{1}{2}\alpha_{112}\,\sigma_{11}^{(2),0} + \frac{1}{2}\alpha_{112}\,\sigma_{11}^{(2),0}\cos(2\alpha x_1) +$$

$$\alpha_{161}\sigma_{12}^{(2),1}\cos(\alpha x_1),$$

$$E_1^{(2)}\varepsilon_{22}^{(2),2} = \alpha_{120}\,\sigma_{11}^{(2),2} + \alpha_{220}\,\sigma_{22}^{(2),2} + \frac{1}{2}\alpha_{122}\,\sigma_{11}^{(2),0} + \frac{1}{2}\alpha_{122}\,\sigma_{11}^{(2),0}\cos(2\alpha x_1) +$$

$$\alpha_{261}\sigma_{12}^{(2),1}\cos(\alpha x_1),$$

$$E_1^{(2)}\varepsilon_{11}^{(2),2} = \alpha_{660}\sigma_{12}^{(2),2} + \left(\alpha_{161}\sigma_{11}^{(2),1} + \alpha_{261}\sigma_{22}^{(2),1}\right)\cos(\alpha x_1). \qquad (4.7.34)$$

The contact conditions are derived from (4.2.13) as

$$\sigma_{12}^{(2),2}\Big|_{(t_1,\pm H^{(2)})} - \sigma_{12}^{(1),2}\Big|_{(t_1,\mp H^{(1)})} = \pi\Bigg[\alpha H^{(2)}\left(\psi_{12}^{(1),1}\Big|_{(\mp H^{(1)})} - s_{12}^{(2),1}\Big|_{(\pm H^{(2)})}\right) +$$

$$\left.\frac{d\psi_{12}^{(1),1}}{dx_2^{(1)}}\right|_{(\mp H^{(1)})} - \left.\frac{ds_{12}^{(2),1}}{dx_2^{(2)}}\right|_{(\pm H^{(2)})} + s_{11}^{(2),1}\Big|_{(\pm H^{(2)})} - \psi_{11}^{(1),1}\Big|_{(\mp H^{(1)})}\Bigg]\sin(2\alpha x_1),$$

$$\sigma_{22}^{(2),2}\Big|_{(t_1,\pm H^{(2)})} - \sigma_{22}^{(1),2}\Big|_{(t_1,\mp H^{(1)})} = 2\pi\left(s_{12}^{(2),1}\Big|_{(\pm H^{(2)})} - \psi_{12}^{(1),1}\Big|_{(\mp H^{(1)})}\right)\cos(2\alpha x_1),$$

$$u_1^{(2),2}\Big|_{(t_1,\pm H^{(2)})} - u_1^{(1),2}\Big|_{(t_1,\mp H^{(1)})} = 2\pi\ell\left(\frac{C^{(1),1}}{E^{(1)}} - \frac{C^{(2),1}}{E_1^{(2)}}\right)\sin(2\alpha x_1) +$$

$$\pi\ell\left(\left.\frac{d\varphi_1^{(1),1}}{E^{(1)}dx_2^{(1)}}\right|_{(\mp H^{(1)})}-\left.\frac{du_{11}^{(2),1}}{E_1^{(2)}dx_2^{(2)}}\right|_{(\pm H^{(2)})}\right)\sin(2\alpha x_1),$$

$$\left.u_2^{(2),2}\right|_{(t_1,\pm H^{(2)})}-\left.u_2^{(1),2}\right|_{(t_1,\mp H^{(1)})}=\pi\ell\left[\alpha H^{(2)}\left.\frac{u_{21}^{(2),1}}{E_1^{(2)}}\right|_{(\pm H^{(2)})}-\left.\frac{\varphi_2^{(1),1}}{E^{(1)}}\right|_{(\mp H^{(1)})}+\right.$$

$$\left.\ell\left(\left.\frac{du_{21}^{(2),1}}{E_1^{(2)}dx_2^{(2)}}\right|_{(\pm H^{(2)})}-\left.\frac{d\varphi_2^{(1),1}}{E^{(1)}dx_2^{(1)}}\right|_{(\mp H^{(1)})}\right)\cos(2\alpha x_1)\right]. \tag{4.7.35}$$

Rewriting the representation (4.7.3) and the equation (4.7.28) for the second approximation we derive the following equation for $\Phi^{(2),2}$ from (4.7.34), (4.7.28).

$$\alpha_{110}\frac{\partial^4\Phi^{(2),2}}{\partial(x_2^{(2)})^4}+(2\alpha_{120}+\alpha_{660})\frac{\partial^4\Phi^{(2),2}}{\partial(x_1)^2\partial(x_2^{(2)})^2}+\alpha_{220}\frac{\partial^4\Phi^{(2),2}}{\partial(x_1)^4}=$$

$$-\frac{1}{2}\alpha_{161}\frac{d^2s_{12}^{(2),1}}{d\left(x_2^{(2)}\right)^2}+\cos(2\alpha x_1)\left(2\alpha_{122}\sigma_{11}^{(2),0}-\frac{1}{2}\alpha_{161}\frac{d^2s_{12}^{(2),1}}{d\left(x_2^{(2)}\right)^2}+\right.$$

$$\left.2\alpha_{261}s_{12}^{(2),1}+\alpha_{161}\frac{ds_{11}^{(2),1}}{dx_2^{(2)}}+\alpha_{261}\frac{ds_{22}^{(2),1}}{dx_2^{(2)}}\right). \tag{4.7.36}$$

The equation (4.7.36) is inhomogeneous; taking into account the form of the expressions on the right hand side of the contact conditions (4.7.35), we seek the general solution to this equation as

$$\Phi_g^{(2),2}=\phi_g\left(x_2^{(2)}\right)\cos(2\alpha x_1). \tag{4.7.37}$$

By employing the solution procedure described in subsection 4.6.1 we obtain the following expression for the function ϕ_g.

For case I (4.7.10):

$$\phi_g=C_{12}^{(2),2}\cosh\left(2\alpha^{(2)}\alpha x_2^{(2)}\right)\cos\left(2\beta^{(2)}\alpha x_2^{(2)}\right)+C_{22}^{(2),2}\sinh\left(2\alpha^{(2)}\alpha x_2^{(2)}\right)\cos\left(2\beta^{(2)}\alpha x_2^{(2)}\right)$$

$$+C_{32}^{(2),2}\cosh\left(2\alpha^{(2)}\alpha x_2^{(2)}\right)\sin\left(2\beta^{(2)}\alpha x_2^{(2)}\right)+C_{42}^{(2),2}\sinh\left(2\alpha^{(2)}\alpha x_2^{(2)}\right)\sin\left(2\beta^{(2)}\alpha x_2^{(2)}\right), \tag{4.7.38}$$

For case II (4.7.12):

$$\phi_g = C_{12}^{(2),2}\cosh\left(2\kappa_1^{(2)}\alpha x_2^{(2)}\right)+C_{22}^{(2),2}\sinh\left(2\kappa_1^{(2)}\alpha x_2^{(2)}\right)+$$

$$C_{32}^{(2),2}\cosh\left(2\kappa_2^{(2)}\alpha x_2^{(2)}\right)+C_{42}^{(2),2}\sinh\left(2\kappa_2^{(2)}\alpha x_2^{(2)}\right). \tag{4.7.39}$$

We now consider the determination of the partial solutions to the equation (4.7.36). To the simplify the right hand side of this equation we use the following relations

$$\frac{d^2 s_{12}^{(2),1}}{d\left(x_2^{(2)}\right)^2}=-\frac{ds_{11}^{(2),1}}{dx_2^{(2)}}, \qquad s_{12}^{(2),1}=\frac{ds_{22}^{(2),1}}{dx_2^{(2)}}, \tag{4.7.40}$$

which may be verified directly. Taking these relations into account we can write the right hand side of the equation (4.7.40) as follows.

$$-\frac{1}{2}\alpha_{161}\frac{d^2 s_{12}^{(2),1}}{d(x_2^{(2)})^2}+\cos(2\alpha x_1)\left(2\alpha_{122}\sigma_{11}^{(2),0}+\frac{3}{2}\alpha_{161}\frac{ds_{11}^{(2),1}}{dx_2^{(2)}}+3\alpha_{261}s_{12}^{(2),1}\right). \tag{4.7.41}$$

The partial solution corresponding to the first term of (4.7.41) we denote by $\Phi_{pr.1}^{(2),2}\left(x_2^{(2)}\right)$; for this the following relation is obtained from (4.7.36):

$$\frac{d^2\Phi_{pr.1}^{(2)}\left(x_2^{(2)}\right)}{d(x_2^{(2)})^2}=-\frac{1}{2}\frac{\alpha_{161}}{\alpha_{110}}s_{12}^{(2),1}\left(x_2^{(2)}\right). \tag{4.7.42}$$

It follows from (4.7.42), (4.7.3) that the partial solution $\Phi_{pr.1}^{(2),2}\left(x_2^{(2)}\right)$ gives only one additional term to the stress $\sigma_{11}^{(2),2}$ and this term is equal to the right hand side of the relation (4.7.42). Thereby it is not necessary to obtain the explicit expression of this partial solution. Thus consider the determination of the partial solution (let us denote it by $\Phi_{pr.2}^{(2),2}(x_1)$) of the equation (4.7.36) corresponding to the term $\left(2\alpha_{122}\sigma_{11}^{(2),0}\right)$ $\times\cos(2\alpha x_1)$. After some operations we obtain:

$$\Phi_{pr.2}^{(2),2}(x_1)=F_0^{(2)}\cos(2\alpha x_1), \tag{4.7.43}$$

where

$$F_0^{(2)} = 2\alpha_{122}\sigma_{11}^{(2),0}\left(16\alpha_{220}\right)^{-1}. \tag{4.7.44}$$

Consider the determination of the partial solution of the equation (4.7.36) corresponding to the remaining part of the expression (4.7.41), i.e., to

$$\cos(2\alpha x_1)\left(\frac{3}{2}\alpha_{161}\frac{ds_{11}^{(2),1}}{dx_2^{(2)}} + 3\alpha_{361}s_{12}^{(2),1}\right). \tag{4.7.45}$$

This has the following form

$$\Phi_{pr.3}^{(2),2} = \left(F_1^{(2)}b_{11}^{(2)}(x_2^{(2)}) + F_2^{(2)}b_{12}^{(2)}(x_2^{(2)}) + F_3^{(2)}b_{21}^{(2)}(x_2^{(2)}) + F_4^{(2)}b_{22}^{(2)}(x_2^{(2)})\right)\cos(2\alpha x_1). \tag{4.7.46}$$

Note that the expressions of the functions $b_{ij}^{(2)}(x_2^{(2)})$ are given by (4.7.32), (4.7.33).

Let us consider the determination of the constants $F_i^{(2)}(i = 1,2,3,4)$ in the cases I (4.7.10) and II (4.7.12).

Case I (4.7.10). In this case by direct verification we prove that the following relations hold.

$$\frac{db_{11}^{(2)}}{dx_2^{(2)}} = -\beta^{(2)}b_{21}^{(2)} + \alpha^{(2)}b_{12}^{(2)}, \quad \frac{db_{12}^{(2)}}{dx_2^{(2)}} = -\beta^{(2)}b_{22}^{(2)} + \alpha^{(2)}b_{11}^{(2)},$$

$$\frac{db_{21}^{(2)}}{dx_2^{(2)}} = \beta^{(2)}b_{11}^{(2)} + \alpha^{(2)}b_{22}^{(2)}, \quad \frac{db_{22}^{(2)}}{dx_2^{(2)}} = \beta^{(2)}b_{12}^{(2)} + \alpha^{(2)}b_{21}^{(2)},$$

$$\frac{d^2b_{11}^{(2)}}{d\left(x_2^{(2)}\right)^2} = \gamma_1^{(2)}b_{11}^{(2)} - \gamma_2^{(2)}b_{22}^{(2)}, \quad \frac{d^2b_{12}^{(2)}}{d\left(x_2^{(2)}\right)^2} = \gamma_1^{(2)}b_{12}^{(2)} - \gamma_2^{(2)}b_{21}^{(2)},$$

$$\frac{d^2b_{21}^{(2)}}{d\left(x_2^{(2)}\right)^2} = \gamma_1^{(2)}b_{21}^{(2)} + \gamma_2^{(2)}b_{12}^{(2)}, \quad \frac{d^2b_{22}^{(2)}}{d\left(x_2^{(2)}\right)^2} = \gamma_1^{(2)}b_{22}^{(2)} + \gamma_2^{(2)}b_{11}^{(2)},$$

$$\frac{d^3b_{11}^{(2)}}{d\left(x_2^{(2)}\right)^3} = -\gamma_3^{(2)}b_{21}^{(2)} + \gamma_4^{(2)}b_{12}^{(2)}, \quad \frac{d^3b_{12}^{(2)}}{d\left(x_2^{(2)}\right)^3} = -\gamma_3^{(2)}b_{22}^{(2)} + \gamma_4^{(2)}b_{21}^{(2)},$$

$$\frac{d^3b_{21}^{(2)}}{d\left(x_2^{(2)}\right)^3}=-\gamma_3^{(2)}b_{11}^{(2)}+\gamma_4^{(2)}b_{22}^{(2)},\quad \frac{d^3b_{22}^{(2)}}{d\left(x_2^{(2)}\right)^3}=\gamma_3^{(2)}b_{12}^{(2)}+\gamma_4^{(2)}b_{21}^{(2)},$$

$$\frac{d^4b_{11}^{(2)}}{d\left(x_2^{(2)}\right)^4}=\gamma_5^{(2)}b_{11}^{(2)}+\gamma_6^{(2)}b_{22}^{(2)},\quad \frac{d^4b_{12}^{(2)}}{d\left(x_2^{(2)}\right)^4}=\gamma_5^{(2)}b_{12}^{(2)}+\gamma_6^{(2)}b_{21}^{(2)},$$

$$\frac{d^4b_{21}^{(2)}}{d\left(x_2^{(2)}\right)^4}=\gamma_5^{(2)}b_{21}^{(2)}-\gamma_6^{(2)}b_{12}^{(2)},\quad \frac{d^4b_{22}^{(2)}}{d\left(x_2^{(2)}\right)^4}=\gamma_5^{(2)}b_{22}^{(2)}-\gamma_6^{(2)}b_{11}^{(2)},\quad (4.7.47)$$

where

$$\gamma_1^{(2)}=\left(\alpha^{(2)}\right)^2-\left(\beta^{(2)}\right)^2,\quad \gamma_2^{(2)}=2\alpha^{(2)}\beta^{(2)},\quad \gamma_3^{(2)}=\gamma_1^{(2)}\beta^{(2)}+\gamma_2^{(2)}\alpha^{(2)},$$

$$\gamma_4^{(2)}=\gamma_1^{(2)}\alpha^{(2)}-\gamma_2^{(2)}\beta^{(2)},\ \gamma_5^{(2)}=-\gamma_3^{(2)}\beta^{(2)}+\gamma_4^{(2)}\alpha^{(2)},\ \gamma_6^{(2)}=-\gamma_3^{(2)}\alpha^{(2)}-\gamma_4^{(2)}\beta^{(2)}.\quad (4.7.48)$$

Taking account of the relations (4.7.47) and (4.7.48) we represent the expression (4.7.45) in the following form.

$$\cos(2\alpha x_1)\left(\frac{3}{2}\alpha_{161}\frac{ds_{11}^{(2),1}}{dx_2^{(2)}}+3\alpha_{361}s_{12}^{(2),1}\right)=$$

$$\left[\overline{C}_1^{(2),1}b_{11}^{(2)}+\overline{C}_2^{(2),1}b_{12}^{(2)}+\overline{C}_3^{(2),1}b_{21}^{(2)}+\overline{C}_4^{(2),1}b_{22}^{(2)}\right]\quad (4.7.49)$$

where

$$\overline{C}_1^{(2),1}=\frac{3}{2}\alpha_{161}\left(C_2^{(2),1}\gamma_4^{(2)}+C_3^{(2),1}\gamma_3^{(2)}\right)+3\alpha_{261}\left(-C_2^{(2),1}\alpha^{(2)}-C_3^{(2),1}\beta^{(2)}\right),$$

$$\overline{C}_2^{(2),1}=\frac{3}{2}\alpha_{161}\left(C_1^{(2),1}\gamma_4^{(2)}+C_4^{(2),1}\gamma_3^{(2)}\right)+3\alpha_{261}\left(-C_1^{(2),1}\alpha^{(2)}-C_4^{(2),1}\beta^{(2)}\right),$$

$$\overline{C}_3^{(2),1}=\frac{3}{2}\alpha_{161}\left(-C_1^{(2),1}\gamma_3^{(2)}+C_4^{(2),1}\gamma_4^{(2)}\right)+3\alpha_{261}\left(-C_4^{(2),1}\alpha^{(2)}+C_1^{(2),1}\beta^{(2)}\right),$$

$$\overline{C}_4^{(2),1} = \frac{3}{2}\alpha_{161}\left(C_3^{(2),1}\gamma_4^{(2)} - C_2^{(2),1}\gamma_3^{(2)}\right) + 3\alpha_{261}\left(-C_3^{(2),1}\alpha^{(2)} + C_2^{(2),1}\beta^{(2)}\right). \tag{4.7.50}$$

Thus, substituting (4.7.46) in (4.7.36) and grouping by the $b_{ij}^{(2)}\left(x_2^{(2)}\right)$ with accounting (4.7.47) we obtain the following equations to determination of the unknown constants $F_i^{(2)} (i = 1,2,3,4)$.

$$F_1^{(2)}\left(\alpha_{110}\gamma_3^{(2)} - 4(2\alpha_{120} + \alpha_{660})\gamma_1^{(2)} + 16\alpha_{220}\right) +$$

$$F_4^{(2)}\left(\alpha_{110}\gamma_6^{(2)} - 4(2\alpha_{120} + \alpha_{660})\gamma_2^{(2)}\right) = \overline{C}_1^{(2),1},$$

$$F_1^{(2)}\left(-\alpha_{110}\gamma_6^{(2)} + 4(2\alpha_{120} + \alpha_{660})\gamma_2^{(2)} + 16\alpha_{220}\right) + F_4^{(2)}\left(\alpha_{110}\gamma_5^{(2)} -\right.$$

$$\left.4(2\alpha_{120} + \alpha_{660})\gamma_1^{(2)} + 16\alpha_{220}\right) = \overline{C}_4^{(2),1},$$

$$F_2^{(2)}\left(-\alpha_{110}\gamma_5^{(2)} - 4(2\alpha_{120} + \alpha_{660})\gamma_1^{(2)} + 16\alpha_{220}\right) + F_3^{(2)}\left(\alpha_{110}\gamma_6^{(2)} -\right.$$

$$\left.4(2\alpha_{120} + \alpha_{660})\gamma_2^{(2)} + 16\alpha_{220}\right) = \overline{C}_2^{(2),1},$$

$$F_2^{(2)}\left(-\alpha_{110}\gamma_6^{(2)} + 4(2\alpha_{120} + \alpha_{660})\gamma_2^{(2)}\right) + F_3^{(2)}\left(\alpha_{110}\gamma_3^{(2)} -\right.$$

$$\left.4(2\alpha_{120} + \alpha_{660})\gamma_1^{(2)} + 16\alpha_{220}\right) = \overline{C}_3^{(2),1}. \tag{4.7.51}$$

So, after determination of the unknown constants $F_i^{(2)} (i = 1,2,3,4)$ from equation (4.7.51) we find completely the function $\Phi^{(2),2}$.

Case II (4.7.12). Accounting (4.7.33) the expression (4.7.45) can be written as

$$\frac{3}{2}\alpha_{161}\frac{ds_{11}^{(2),1}\left(x_2^{(2)}\right)}{dx_2^{(2)}} + 3\alpha_{261}s_{12}^{(2),1}\left(x_2^{(2)}\right) = \overline{C}_1^{(2),1}\cosh\left(\kappa_1^{(2)}\alpha x_2^{(2)}\right) +$$

$$\overline{C}_2^{(2),1}\sinh\left(\kappa_1^{(2)}\alpha x_2^{(2)}\right) + \overline{C}_3^{(2),1}\cosh\left(\kappa_1^{(2)}\alpha x_2^{(2)}\right) + \overline{C}_4^{(2),1}\sinh\left(\kappa_2^{(2)}\alpha x_2^{(2)}\right), \tag{4.7.52}$$

where

$$\overline{C}_{1,2}^{(2),1} = \left(\frac{3}{2}\alpha_{161}\left(\kappa_1^{(2)}\right)^3 - 3\alpha_{261}\kappa_1^{(2)}\right)C_{1,2}^{(2),1},$$

$$\overline{C}_{3,4}^{(2),1} = \left(\frac{3}{2}\alpha_{161}\left(\kappa_2^{(2)}\right)^3 - 3\alpha_{261}\kappa_2^{(2)}\right)C_{3,4}^{(2),1}. \qquad (4.7.53)$$

As in the case I (4.7.10), substituting (4.7.46) and (4.7.52) in equation (4.7.36), taking account of the expression (4.7.33), and performing some known operations we obtain the following expressions for the unknown constants entering (4.7.45).

$$F_{1,2}^{(2)} = \left[\alpha_{110}\left(\kappa_1^{(2)}\right)^4 - 4\left(2\alpha_{120} + \alpha_{660}\right)\left(\kappa_1^{(2)}\right)^2 + 16\alpha_{220}\right]\overline{C}_{1,2}^{(2),1},$$

$$F_{3,4}^{(2)} = \left[\alpha_{110}\left(\kappa_2^{(2)}\right)^4 - 4\left(2\alpha_{120} + \alpha_{660}\right)\left(\kappa_2^{(2)}\right)^2 + 16\alpha_{220}\right]\overline{C}_{3,4}^{(2),1}. \qquad (4.7.54)$$

Thus we determine completely the function $\Phi^{(2),2}$.

Thus, using the relations (4.7.3), (4.7.32)-(4.7.34), (4.7.37)-(4.7.54) from contact conditions (4.7.35) we derive the algebraic equation for the unknown constants $C_{i2}^{(2),2}$ (i=1,2,3,4) which enter (4.7.38) and (4.7.39). After finding these constants we determine all the values of the second approximation related to the curvilinear anisotropic filler layer. Continuing this procedure we can determine also the values of each subsequent approximation.

4.8 . Numerical Results: Rectilinear Anisotropy

We consider some numerical results on the stress distributions in the composites with anisotropic periodically curved alternating layers. Co-phase (Fig.4.2.1) and anti-phase (Fig.4.2.3) location of the layers will be treated separately. As in section 4.3 it will be assumed that the composites are loaded by $\sigma_{11}(\infty) = p$. The matrix layers will be taken as isotropic, and the filler layers transversally isotropic with isotropy axis Ox_1. We will suppose the relations (4.7.29) and (4.7.30) hold, and that

$$\nu_1^{(2)} = \nu^{(2)} = \nu^{(1)} = 0.3, \quad \frac{E_1^{(2)}}{E^{(1)}} = 50. \qquad (4.8.1)$$

We investigate the influence of the ratios $E^{(2)}/E_1^{(2)}$, $E_1^{(2)}/G_{12}^{(2)}$ on the distributions of the self-balanced parts of the stresses which arise as a result of the curving of the layers.

4.8.1. CO-PHASE CURVING OF THE LAYERS

Low filler concentration. First we consider the case $H^{(1)} >> H^{(2)}$, i.e., we consider a low concentration of filler layers in the composite. Using the periodicity of the curving, we investigate the stress distribution on CA'C' on the surface S^+ (Fig.4.2.1) which corresponds to the parameter αt_1 in $[\pi/2, 3\pi/2]$.

Fig.4.8.1 gives the stresses for fixed values of the problem parameters χ, ε, $E^{(2)}/E_1^{(2)}$, $E_1^{(2)}/G_{12}^{(2)}$; the self-balanced stress σ_{nn} and self-balanced part of the stresses $\sigma_{\tau\tau}^{(k)} (k=1,2)$ obtain their absolute maximal values at C and C', i.e., at $\alpha t_1 = \pi/2$ and $\alpha t_1 = 3\pi/2$; the tangential stress $\sigma_{n\tau}$ obtains its absolute maximal value at A', i.e. at $\alpha t_1 = \pi$. These results agree with the corresponding ones obtained in section 4.3. However, as a result of the anisotropy of the filler layer the self-balanced parts of the stresses along CA'C' are less than those obtained for the isotropic case. Similar results are obtained on the dependence of the stresses on the parameter χ, as shown in Fig.4.8.2 for fixed problem parameters.

Table 4.24 shows the values of the relations $\sigma_{nn}/\sigma_{11}^{(1),0}$, $\sigma_{n\tau}/\sigma_{11}^{(1),0}$, $\sigma_{\tau\tau}^{(k)}/\sigma_{11}^{(k),0}$ (k=1,2) for various $E^{(2)}/E_1^{(2)}$, $E_1^{(2)}/G_{12}^{(2)}$ and ε. Analogous results for isotropic filler layers were given in Table 4.1. Table 4.24 shows that with increasing

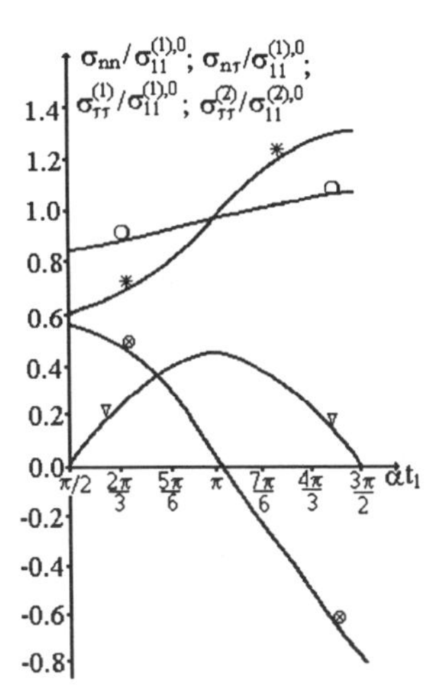

Fig.4.8.1. The graphs of the relations between $\sigma_{nn}/\sigma_{11}^{(1),0}$, $\sigma_{n\tau}/\sigma_{11}^{(1),0}$, $\sigma_{\tau\tau}^{(k)}/\sigma_{11}^{(k),0}$ and αt_1 for $\chi = 0.3$, $\varepsilon = 0.015$, $E_1^{(2)}/G_{12}^{(2)} = 2$, $E^{(2)}/E_1^{(2)} = 0.01$ ⊗-- $\sigma_{nn}/\sigma_{11}^{(1),0}$, ∇ -- $\sigma_{n\tau}/\sigma_{11}^{(1),0}$, ○ -- $\sigma_{\tau\tau}^{(1)}/\sigma_{11}^{(1),0}$, ✱ -- $\sigma_{\tau\tau}^{(2)}/\sigma_{11}^{(2),0}$

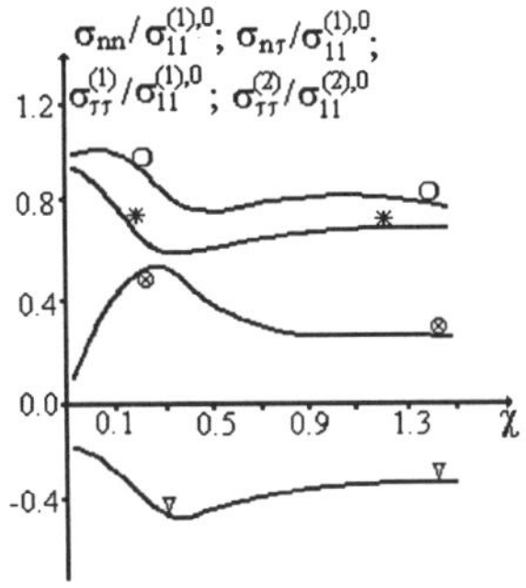

Fig.4.8.2. The graphs of the relations between $\sigma_{nn}/\sigma_{11}^{(1),0}$, $\sigma_{n\tau}/\sigma_{11}^{(1),0}$, $\sigma_{\tau\tau}^{(k)}/\sigma_{11}^{(k),0}$ and χ for $\varepsilon = 0.015$, $E_1^{(2)}/G_{12}^{(2)} = 2$, $E^{(2)}/E_1^{(2)} = 0.01$

⊗-- $\sigma_{nn}/\sigma_{11}^{(1),0}$ at point C,

∇ -- $\sigma_{n\tau}/\sigma_{11}^{(1),0}$ at point A',

○ -- $\sigma_{\tau\tau}^{(1)}/\sigma_{11}^{(1),0}$ at point C,

✱ -- $\sigma_{\tau\tau}^{(2)}/\sigma_{11}^{(2),0}$ at point C

Table 4.24. The influence of the filler layer material properties on the stress values at particular points on the inter medium surface for $\chi = 0.3$.

$\frac{E_1^{(2)}}{G_{12}^{(2)}}$	$\frac{E^{(2)}}{E_1^{(2)}}$	ε	$\frac{\sigma_{nn}(C)}{\sigma_{11}^{(1),0}}$	$\frac{\sigma_{n\tau}(A')}{\sigma_{11}^{(1),0}}$	$\frac{\sigma_{\tau\tau}^{(1)}(C)}{\sigma_{11}^{(1),0}}$	$\frac{\sigma_{\tau\tau}^{(2)}(C)}{\sigma_{11}^{(2),0}}$
	0.9	0.015	0.6599	-0.4595	0.8710	0.5938
		0.020	0.8388	-0.5856	0.8160	0.4637
2.00	0.05	0.015	0.6139	-0.4439	0.8702	0.6124
		0.020	0.7761	-0.5686	0.8148	0.4889
	0.01	0.015	0.5701	-0.4458	0.8481	0.6086
		0.020	0.7097	-0.5745	0.7855	0.4874
	0.9	0.015	0.5196	-0.4826	0.7273	0.5091
		0.020	0.6457	-0.6102	0.6556	0.3844
20.00	0.05	0.015	0.4899	-0.4667	0.7322	0.5265
		0.020	0.6083	-0.5924	0.6584	0.4029
	0.01	0.015	0.4641	-0.4666	0.7171	0.5222
		0.020	0.5687	-0.5930	0.6352	0.3964
	0.9	0.015	0.3862	-0.5042	0.5809	0.4187
		0.020	0.4515	-0.6110	0.5283	0.3387
50.00	0.05	0.015	0.3681	-0.4900	0.5896	0.4350
		0.020	0.4321	-0.5990	0.5275	0.3461
	0.01	0.015	0.3527	-0.4896	0.5893	0.4322
		0.020	0.4055	-0.5988	0.5110	0.3407
	0.9	0.015	0.2603	-0.5168	0.4835	0.3742
		0.020	0.2503	-0.5652	0.5760	0.4707
	0.05	0.015	0.2520	-0.5068	0.4861	0.3802
100.0		0.020	0.2487	-0.5667	0.5424	0.4379
	0.01	0.015	0.2422	-0.5066	0.4799	0.3782
		0.020	0.2282	-0.5671	0.5285	0.4327

(decreasing) $E_1^{(2)}/G_{12}^{(2)}$ ($E^{(2)}/E_1^{(2)}$) the ratio $\sigma_{nn}/\sigma_{11}^{(1),0}$ decreases monotonically. In this case the influence of $E_1^{(2)}/G_{12}^{(2)}$ on the values of $\sigma_{nn}/\sigma_{11}^{(1),0}$ is more significant than that of $E^{(2)}/E_1^{(2)}$. Moreover, this table shows that when $E_1^{(2)}/G_{12}^{(2)}=2$, $E^{(2)}/E_1^{(2)}$ tending to unity the values of $\sigma_{nn}/\sigma_{11}^{(1),0}$ approach those given in Table 4.1. This results holds for the ratios $\sigma_{n\tau}/\sigma_{11}^{(1),0}$, $\sigma_{\tau\tau}^{(k)}/\sigma_{11}^{(k),0}$ (k=1,2) also. However, decreasing (increasing) $E^{(2)}/E_1^{(2)}$ ($E_1^{(2)}/G_{12}^{(2)}$) causes the ratio $\sigma_{n\tau}/\sigma_{11}^{(1),0}$ to diminish. Moreover, the Table 4.24 shows that with growing $E_1^{(2)}/G_{12}^{(2)}$

the self-balanced part of the ratios $\sigma_{\tau\tau}^{(k)}/\sigma_{11}^{(k),0}$ increases significantly; but the influence of the ratio $E^{(2)}/E_1^{(2)}$ on $\sigma_{\tau\tau}^{(k)}/\sigma_{11}^{(k),0}$ is insignificant.

High filler concentration. Now we consider the numerical results when interaction between the filler layers affects the stress distribution. Fig.4.8.3 and 4.8.4 show the dependence of $\sigma_{nn}/\sigma_{11}^{(1),0}$, $\sigma_{n\tau}/\sigma_{11}^{(1),0}$, $\sigma_{\tau\tau}^{(k)}/\sigma_{11}^{(k),0}$ (k=1,2) on αt_1 with various filler concentrations, i.e. with various $\eta^{(2)}$, under fixed problem parameters. As a result of the interaction between the filler layers the distributions of the stresses σ_{nn} and $\sigma_{\tau\tau}^{(1)}$ on the inter-medium surface changes not only quantitatively but also qualitatively. In particular, for $\alpha t_1 \in [\pi/2, 5\pi/6]$ and $\alpha t_1 \in [7\pi/6, 3\pi/2]$, with increasing $\eta^{(2)}$, σ_{nn} decreases but the self-balanced part of $\sigma_{\tau\tau}^{(1)}$ increases. However, outside these intervals, i.e., for $\alpha t_1 \in [5\pi/6, 7\pi/6]$, σ_{nn} increases with $\eta^{(2)}$, but the self-balanced part of $\sigma_{\tau\tau}^{(1)}$ decreases. The figures show that the distribution of the stresses $\sigma_{n\tau}$ and $\sigma_{\tau\tau}^{(2)}$ changes only quantitatively with $\eta^{(2)}$: the absolute values of $\sigma_{n\tau}$ and self-balanced part of the $\sigma_{\tau\tau}^{(2)}$ increase with $\eta^{(2)}$.

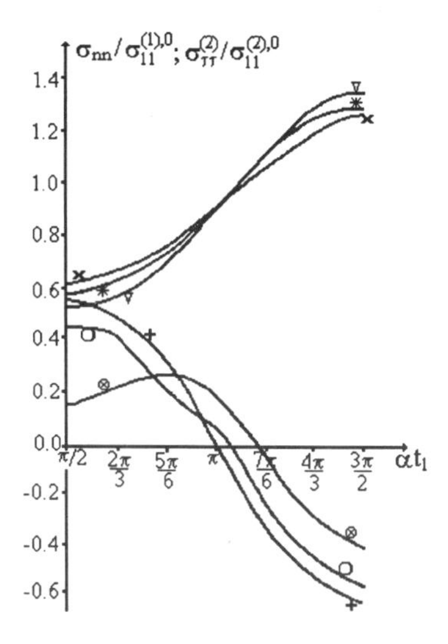

Fig.4.8.3. The graphs of the relations $\sigma_{nn}/\sigma_{11}^{(1),0}$, $\sigma_{\tau\tau}^{(2)}/\sigma_{11}^{(2),0}$ and αt_1 with $\eta^{(2)}$ =0.5, 0.2, 0.1 for $\varepsilon = 0.015$, $\chi = 0.3$, $E_1^{(2)}/G_{12}^{(2)} = 2$, $E^{(2)}/E_1^{(2)} = 0.01$, ⊗, ○, ✚ -- $\sigma_{nn}/\sigma_{11}^{(1),0}$, ∇, ✳, ✖ -- $\sigma_{\tau\tau}^{(2)}/\sigma_{11}^{(2),0}$

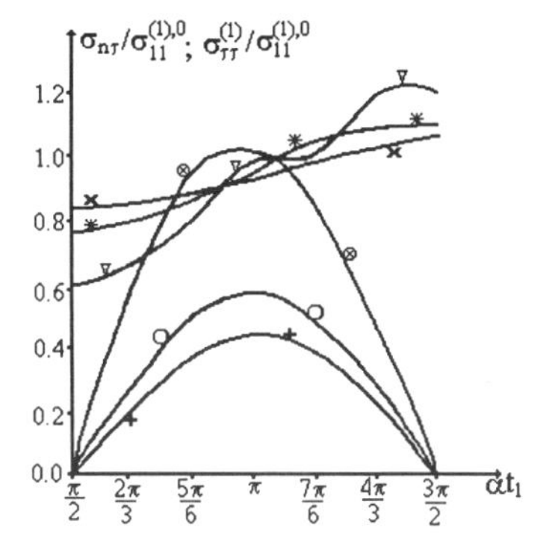

Fig.4.8.4. The graphs of the relations $\sigma_{n\tau}/\sigma_{11}^{(1),0}$, $\sigma_{\tau\tau}^{(1)}/\sigma_{11}^{(1),0}$ and αt_1 with $\eta^{(2)}$ =0.5, 0.2, 0.1 for $\varepsilon = 0.015$, $\chi = 0.3$, $E_1^{(2)}/G_{12}^{(2)} = 2$, $E^{(2)}/E_1^{(2)} = 0.01$, ⊗, ○, ✚ -- $\sigma_{n\tau}/\sigma_{11}^{(1),0}$, ∇, ✳, ✖ -- $\sigma_{\tau\tau}^{(1)}/\sigma_{11}^{(1),0}$

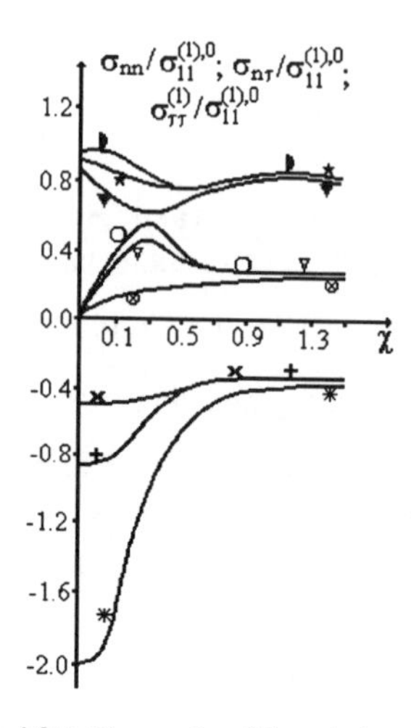

Fig.4.8.5. The graphs of the relations between $\sigma_{nn}/\sigma_{11}^{(1),0}$, $\sigma_{n\tau}/\sigma_{11}^{(1),0}$, $\sigma_{\tau\tau}^{(1)}/\sigma_{11}^{(1),0}$ and χ with $\eta^{(2)}$=0.5, 0.2, 0.1 for $\varepsilon = 0.015$, $E_1^{(2)}/G_{12}^{(2)} = 2$, $E^{(2)}/E_1^{(2)} = 0.01$,
⊗, ∇, ○ -- $\sigma_{nn}/\sigma_{11}^{(1),0}$ at point C (Fig.4.2.1)
✱, ✚, ✖ -- $\sigma_{n\tau}/\sigma_{11}^{(1),0}$ at point A' (Fig.4.2.1)
▼, ★, ◗ -- $\sigma_{\tau\tau}^{(1)}/\sigma_{11}^{(1),0}$ at point C (Fig.4.2.1)

Fig.4.8.6. The graphs of the relations between $\sigma_{\tau\tau}^{(2)}/\sigma_{11}^{(2),0}$ and χ with $\eta^{(2)}$=0.5, 0.2, 0.1 for $\varepsilon = 0.015$, $E_1^{(2)}/G_{12}^{(2)} = 2$, $E^{(2)}/E_1^{(2)} = 0.01$,
⊗, ∇, ○ -- $\sigma_{\tau\tau}^{(2)}/\sigma_{11}^{(2),0}$ at point C (Fig.4.2.1)

Figs.4.8.5 and 4.8.6 show the dependence of $\sigma_{nn}/\sigma_{11}^{(1),0}$, $\sigma_{n\tau}/\sigma_{11}^{(1),0}$, $\sigma_{\tau\tau}^{(k)}/\sigma_{11}^{(k),0}$ (k=1,2) on χ with various $\eta^{(2)}$ under fixed problem parameters. Analysis of these graphs shows that the non-monotonic character of the dependence of $\sigma_{nn}/\sigma_{11}^{(1),0}$, $\sigma_{n\tau}/\sigma_{11}^{(1),0}$ on χ disappears with growing $\eta^{(2)}$. However, the non-monotonic character of the dependence between $\sigma_{\tau\tau}^{(k)}/\sigma_{11}^{(k),0}$ and χ occurs for all $\eta^{(2)}$.

Table 4.25. The influence of the filler layer material properties on the stress values at particular points of the inter medium surface S^+ (Fig.4.2.1) for $\chi = 0.3$, $\varepsilon = 0.015$

$\eta^{(2)}$	$\frac{E_1^{(2)}}{G_{12}^{(2)}}$	$\frac{E^{(2)}}{E_1^{(2)}}$	$\frac{\sigma_{nn}(C)}{\sigma_{11}^{(1),0}}$	$\frac{\sigma_{n\tau}(A')}{\sigma_{11}^{(1),0}}$	$\frac{\sigma_{\tau\tau}^{(1)}(C)}{\sigma_{11}^{(1),0}}$	$\frac{\sigma_{\tau\tau}^{(2)}(C)}{\sigma_{11}^{(2),0}}$
	2.6	1	0.2258	-1.1377	0.5854	0.4905
		0.90	0.2209	-1.1211	0.5896	0.4969
	2.00	0.05	0.1909	-1.0840	0.6007	0.5206
		0.01	0.1709	-1.0853	0.6020	0.5303
		0.90	0.2263	-1.1124	0.5870	0.4919
0.5	20	0.05	0.2119	-1.0710	0.5996	0.5105
		0.01	0.2061	-1.0660	0.5978	0.5113
1	2	3	4	5	6	7

Table 4.25 (Continuation)

1	2	3	4	5	6	7
		0.90	0.2204	-1.0584	0.5551	0.4626
	50	0.05	0.2133	-1.0211	0.5692	0.4796
		0.01	0.2148	-1.0157	0.5683	0.4781
		0.90	0.2038	-0.9536	0.5305	0.4449
	100	0.05	0.1975	-0.9270	0.5374	0.4544
		0.01	0.1986	-0.9234	0.5359	0.4525
	2.6	1	0.5450	-0.6217	0.7782	0.5493
		0.90	0.5448	-0.6151	0.7820	0.5538
	2	0.05	0.5024	-0.5928	0.7886	0.5776
		0.01	0.4664	-0.5880	0.7761	0.5802
		0.90	0.4582	-0.5933	0.6903	0.4979
0.2	20	0.05	0.4322	-0.5722	0.6976	0.5158
		0.01	0.4106	-0.5675	0.6859	0.5135
		0.90	0.3663	-0.5620	0.5747	0.4209
	50	0.05	0.3492	-0.5450	0.5841	0.4374
		0.01	0.3378	-0.5418	0.5731	0.4049
		0.90	0.2657	-0.5223	0.4897	0.3781
	100	0.05	0.2569	-0.5116	0.4923	0.3844
		0.01	0.2474	-0.5090	0.4862	0.3824

Table 4.25 shows the stresses for various $E_1^{(2)}/G_{12}^{(2)}$, $E^{(2)}/E_1^{(2)}$ and $\eta^{(2)}$ at particular points of the surface S^+ (Fig.4.2.1). These data show that with decreasing $E^{(2)}/E_1^{(2)}$ the values of $\sigma_{nn}/\sigma_{11}^{(1),0}$ decrease. However, the influence of the $E^{(2)}/E_1^{(2)}$ on $\sigma_{n\tau}/\sigma_{11}^{(1),0}$, $\sigma_{\tau\tau}^{(1),0}/\sigma_{11}^{(1),0}$, $\sigma_{\tau\tau}^{(2),0}/\sigma_{11}^{(2),0}$ is insignificant. As $E_1^{(2)}/G_{12}^{(2)}$ increase, the absolute values of $\sigma_{nn}/\sigma_{11}^{(1),0}$ and $\sigma_{n\tau}/\sigma_{11}^{(1),0}$ decrease.

Note that all numerical results presented in this subsection had been obtained in the framework of the first four approximations (4.1.15); the convergence of these results was adequate.

4.8.2. ANTI-PHASE CURVING OF THE LAYERS

As in the previous sub-section we investigate the stress distribution in the composite material schematically shown in Fig.4.2.3 in the case where the material of the filler layers of this composite is rectilinearly transversally-isotropic about the Ox_1 axis, and the loading is $\sigma_{11}(\infty) = p$. Using the periodicity of the filler layer curving we will consider the distribution of the stresses along AB'C' of the surface S_1^+ (Fig.4.2.3) which corresponds to the change of αt_1 from $\pi/2$ to $3\pi/2$.

Figs. 4.8.7 and 4.8.8 show the distribution of the stresses with respect to αt_1, for various $\eta^{(2)}$ and fixed $E^{(2)}/E_1^{(2)}$ and $E_1^{(2)}/G_{12}^{(2)}$. These numerical results show that the anisotropy of the filler layers material does not change the character of the

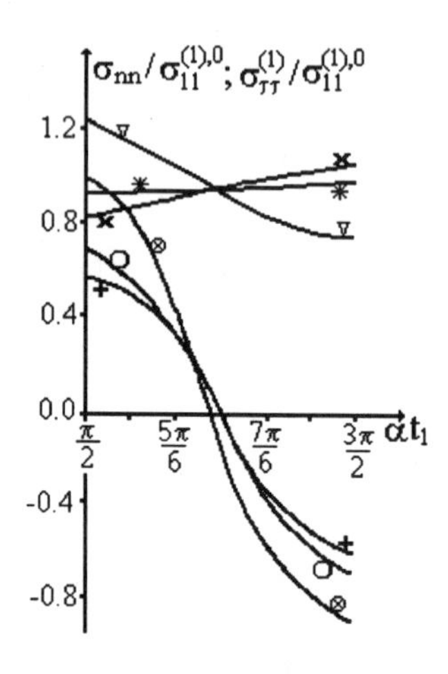

Fig.4.8.7. The graphs of the relations $\sigma_{nn}/\sigma_{11}^{(1),0}$, $\sigma_{\tau\tau}^{(1)}/\sigma_{11}^{(1),0}$ and αt_1 with $\eta^{(2)}$ =0.5, 0.2, 0.1 for $\varepsilon = 0.015$, $\chi = 0.3$, $E_1^{(2)}/G_{12}^{(2)} = 2$, $E^{(2)}/E_1^{(2)} = 0.01$, ⊗, ❍, ✚ -- $\sigma_{nn}/\sigma_{11}^{(1),0}$, ∇, ✳, ✖ -- $\sigma_{\tau\tau}^{(1)}/\sigma_{11}^{(1),0}$

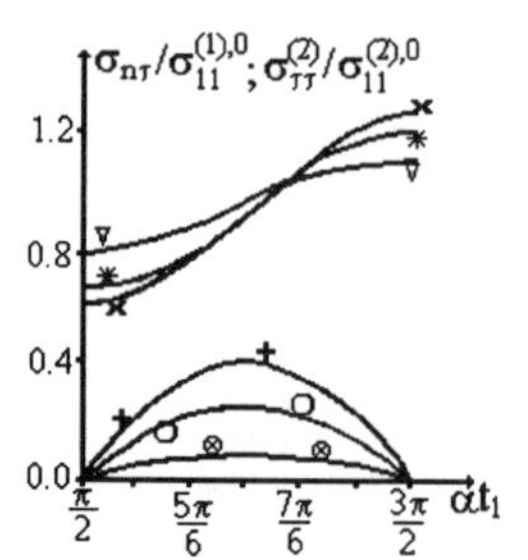

Fig.4.8.8. The graphs of the relations $\sigma_{n\tau}/\sigma_{11}^{(1),0}$, $\sigma_{\tau\tau}^{(2)}/\sigma_{11}^{(2),0}$ and αt_1 with $\eta^{(2)}$ =0.5, 0.2, 0.1 for $\varepsilon = 0.015$, $\chi = 0.3$, $E_1^{(2)}/G_{12}^{(2)} = 2$, $E^{(2)}/E_1^{(2)} = 0.01$, ⊗, ❍, ✚ -- $\sigma_{n\tau}/\sigma_{11}^{(1),0}$, ∇, ✳, ✖ -- $\sigma_{\tau\tau}^{(2)}/\sigma_{11}^{(2),0}$

relations. In other words, the influence of the parameter $\eta^{(2)}$ on these distributions is similar to that for the isotropic case. Similar conclusions apply for the distribution of these stresses with respect to the parameter χ, shown in Figs. 4.8.9 and 4.8.10.

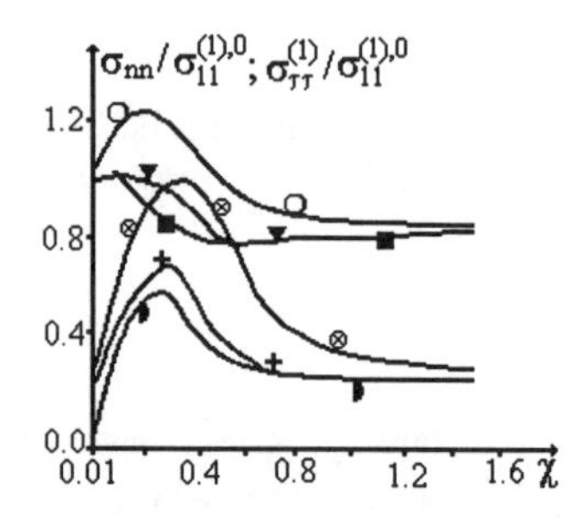

Fig.4.8.9. The graphs of the relations Between $\sigma_{nn}/\sigma_{11}^{(1),0}$, $\sigma_{\tau\tau}^{(1)}/\sigma_{11}^{(1),0}$ and χ with $\eta^{(2)}$ =0.5, 0.2, 0.1 for $\varepsilon = 0.015$, $E_1^{(2)}/G_{12}^{(2)} = 2$, $E^{(2)}/E_1^{(2)} = 0.01$,
⊗, ✚, ◗ -- $\sigma_{nn}/\sigma_{11}^{(1),0}$ at point C (Fig.4.2.3)
❍, ▼, ■ -- $\sigma_{\tau\tau}^{(1)}/\sigma_{11}^{(1),0}$ at point C (Fig.4.2.3)

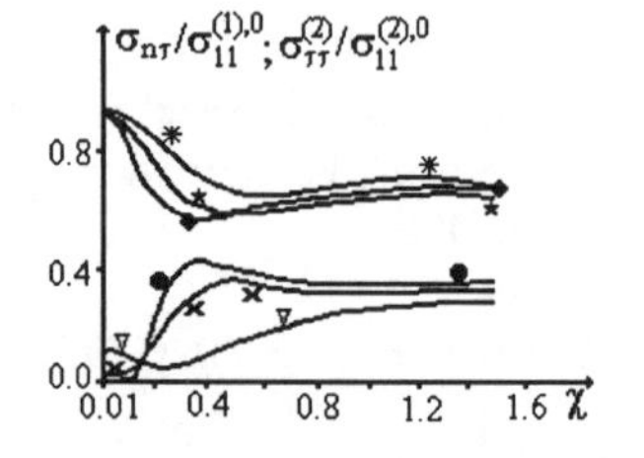

Fig.4.8.10. The graphs of the relations between $\sigma_{n\tau}/\sigma_{11}^{(1),0}$, $\sigma_{\tau\tau}^{(2)}/\sigma_{11}^{(2),0}$ and χ with $\eta^{(2)}$ =0.5, 0.2, 0.1 for $\varepsilon = 0.015$, $E_1^{(2)}/G_{12}^{(2)} = 2$, $E^{(2)}/E_1^{(2)} = 0.01$,
✳, ★, ◆ -- $\sigma_{\tau\tau}^{(2)}/\sigma_{11}^{(2),0}$ at point C (Fig.4.2.3)
∇, ✖, ● -- $\sigma_{n\tau}/\sigma_{11}^{(1),0}$ at point A' (Fig.4.2.3)

Table 4.26. The influence of the filler layer material properties on the stress values at particular points of the inter medium surface S_1^+ (Fig.4.2.3) for $\chi = 0.3$, $\varepsilon = 0.015$

$\eta^{(2)}$	$\frac{E_1^{(2)}}{G_{12}^{(2)}}$	$\frac{E^{(2)}}{E_1^{(2)}}$	$\frac{\sigma_{nn}(A)}{\sigma_{11}^{(1),0}}$	$\frac{\sigma_{n\tau}(B')}{\sigma_{11}^{(1),0}}$	$\frac{\sigma_{\tau\tau}^{(1)}(A)}{\sigma_{11}^{(1),0}}$	$\frac{\sigma_{\tau\tau}^{(2)}(A)}{\sigma_{11}^{(2),0}}$
	2.6	1	1.3824	0.0361	1.4137	0.8331
		0.90	1.2965	0.0308	1.4349	0.8484
	2.00	0.05	1.2405	0.0369	1.3661	0.8451
		0.01	1.0710	0.0489	1.2460	0.7962
		0.90	0.8929	0.1459	0.9641	0.5891
0.5	20	0.05	0.8034	0.1410	0.9415	0.6041
		0.01	0.7137	0.1423	0.8884	0.5886
		0.90	0.5904	0.2267	0.6653	0.4174
	50	0.05	0.5320	0.2188	0.6579	0.4344
		0.01	0.4732	0.2163	0.6286	0.4298
		0.90	0.3941	0.2806	0.5053	0.3398
	100	0.05	0.3499	0.2735	0.4921	0.3451
		0.01	0.3010	0.2701	0.4707	0.3442
	2.6	1	0.8190	0.2735	0.9820	0.6380
		0.90	0.8205	0.2672	0.9918	0.6472
	2	0.05	0.7611	0.2624	0.9808	0.6611
		0.01	0.7010	0.2757	0.9419	0.6475
		0.90	0.5926	0.3626	0.7728	0.5239
0.2	20	0.05	0.5575	0.3528	0.7747	0.5406
		0.01	0.5247	0.3585	0.7542	0.5338
		0.90	0.4094	0.4451	0.5891	0.4171
	50	0.05	0.3895	0.4337	0.5970	0.4334
		0.01	0.3715	0.4359	0.5861	0.4302
		0.90	0.2598	0.5048	0.4790	0.3699
	100	0.05	0.2509	0.4956	0.4811	0.3757
		0.01	0.2400	0.4965	0.4747	0.3739

Table 4.26 shows the influence of the relations $E^{(2)}/E_1^{(2)}$, $E_1^{(2)}/G_{12}^{(2)}$ on $\sigma_{nn}/\sigma_{11}^{(1),0}$, $\sigma_{n\tau}/\sigma_{11}^{(1),0}$, $\sigma_{\tau\tau}^{(k)}/\sigma_{11}^{(k),0}$ (k =1,2) at characteristic points of the surface S_1^+ (Fig. 4.2.3). It follows from this table that, in general, anisotropy reduces the stresses; also, the influence of $E_1^{(2)}/G_{12}^{(2)}$ on the stresses is greater than that of $E^{(2)}/E_1^{(2)}$.

As before the numerical results have been obtained in the framework of the first four approximations. Table 4.27 shows the convergence of the $\sigma_{nn}/\sigma_{11}^{(1),0}$ for various $E_1^{(2)}/G_{12}^{(2)}$, $E^{(2)}/E_1^{(2)}$; the convergence is adequate.

Table 4.27. The values $\sigma_{nn}(A)/\sigma_{11}^{(1),0}$ (Fig.4.2.3) in various approximations (4.1.15) and $E^{(2)}/E_1^{(2)}$, $E_1^{(2)}/G_{12}^{(2)}$ for $\eta^{(2)}=0.5$, $\chi=0.3$.

$E^{(2)}/E_1^{(2)}$	Numb. of appr.	$E_1^{(2)}/G_{12}^{(2)}$			
		2	20	50	100
0.9	1	1.1911	0.7979	0.5525	0.3943
	2	1.3857	0.9341	0.6583	0.4827
	3	1.3916	0.8969	0.5972	0.3966
	4	1.3965	0.8929	0.5904	0.3941
0.01	1	1.0166	0.7194	0.5132	0.3716
	2	1.1104	0.7663	0.5438	0.3970
	3	1.0744	0.7177	0.4827	0.3178
	4	1.0710	0.7137	0.4731	0.3010

4.9. Numerical Results: Curvilinear Anisotropy

We investigate the influence of the curvilinear anisotropy of the filler layers material on the distributions of the self-balanced normal σ_{nn} and tangential $\sigma_{n\tau}$ stresses in composites with co-phase (Fig.4.2.1) and anti-phase curved layers under the loading $\sigma_{11}(\infty)=p$.

4.9.1. CO-PHASE CURVING OF THE LAYERS

First we consider the case where the filler concentration is very small, so that interaction between the filler layers can be neglected. Fig.4.9.1 shows the relations between $\sigma_{nn}/\sigma_{11}^{(1),0}$, $\sigma_{n\tau}/\sigma_{11}^{(1),0}$ and αt_1 for various $E_1^{(2)}/G_{12}^{(2)}$ and fixed problem parameters; the curvilinearity of the filler layers significantly changes the distribution of the stresses on the inter-layer surface; the absolute maximal values of $\sigma_{n\tau}/\sigma_{11}^{(1),0}$ ($\sigma_{nn}/\sigma_{11}^{(1),0}$) increase (decrease) with growing $E_1^{(2)}/G_{12}^{(2)}$.

Fig.4.9.2 shows that the relation between $\sigma_{nn}/\sigma_{11}^{(1),0}$ and χ for various $E_1^{(2)}/G_{12}^{(2)}$; for $E_1^{(2)}/G_{12}^{(2)}=50$ it shows relation between $\sigma_{n\tau}/\sigma_{11}^{(1),0}$ and χ. These graphs show that the non-monotonic character of these relations holds for curvilinear anisotropy of the filler layers. Note that the graph of $\sigma_{nn}/\sigma_{11}^{(1),0}$ on χ is displaced wholly “down”.

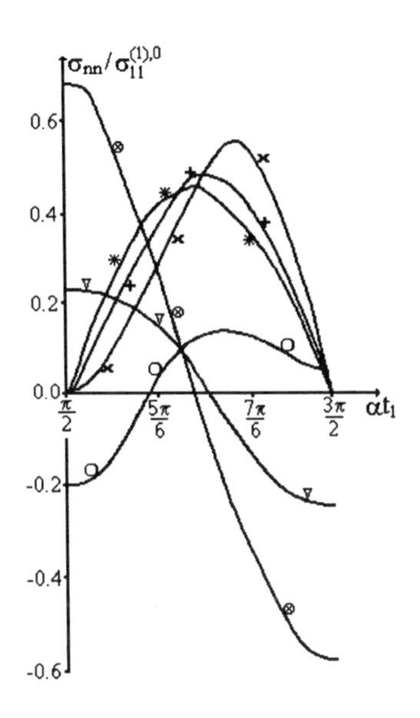

Fig.4.9.1. The graphs of the relations between $\sigma_{nn}/\sigma_{11}^{(1),0}$, $\sigma_{n\tau}/\sigma_{11}^{(1),0}$ and αt_1 with $E_1^{(2)}/G_{12}^{(2)} = 2; 20; 50$ for $\varepsilon = 0.015$, $\chi = 0.3$, $E^{(2)}/E_1^{(2)} = 0.05$, ⊗, ∇, ○ -- $\sigma_{nn}/\sigma_{11}^{(1),0}$; ✳, ✚, ✖ -- $\sigma_{n\tau}/\sigma_{11}^{(1),0}$

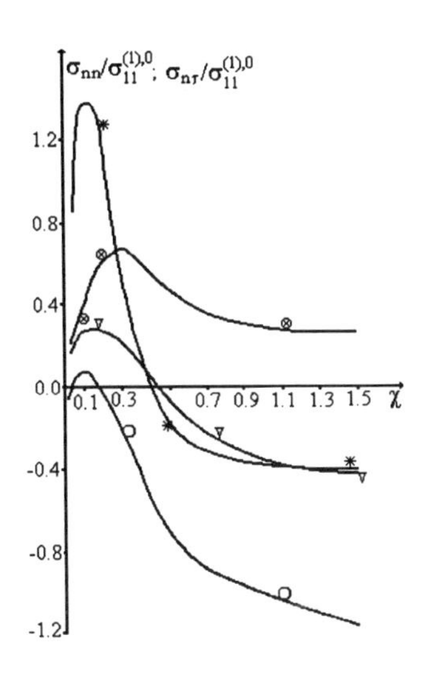

Fig.4.9.2. The graphs of the relations between $\sigma_{nn}/\sigma_{11}^{(1),0}$ and χ with $E_1^{(2)}/G_{12}^{(2)} = 2; 20; 50$ for $\varepsilon = 0.015, 2$, $E^{(2)}/E_1^{(2)} = 0.05$, ⊗, ∇, ○ -- $\sigma_{nn}/\sigma_{11}^{(1),0}$ at point C (Fig.4.2.1) ✳ -- $\sigma_{n\tau}/\sigma_{11}^{(1),0}$ at point A' (Fig.4.2.1) for $E_1^{(2)}/G_{12}^{(2)} = 50$.

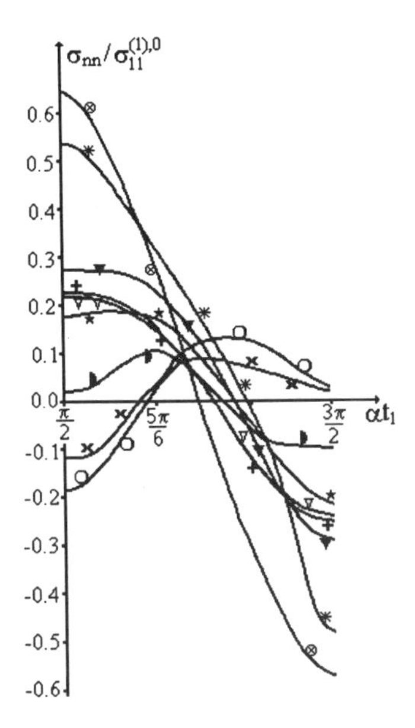

Fig.4.9.3. The graphs of the dependence of $\sigma_{nn}/\sigma_{11}^{(1),0}$ on αt_1 with $E_1^{(2)}/G_{12}^{(2)} = 2, 20, 50$ and various $\eta^{(2)}$ for $\varepsilon = 0.015$, $\chi = 0.3$, $E^{(2)}/E_1^{(2)} = 0.05$: ⊗, ∇, ○ -- $\eta^{(2)} = 0.1$; ✳, ✚, ✖ -- $\eta^{(2)} = 0.2$; ▼, ★, ◗ -- $\eta^{(2)} = 0.5$.

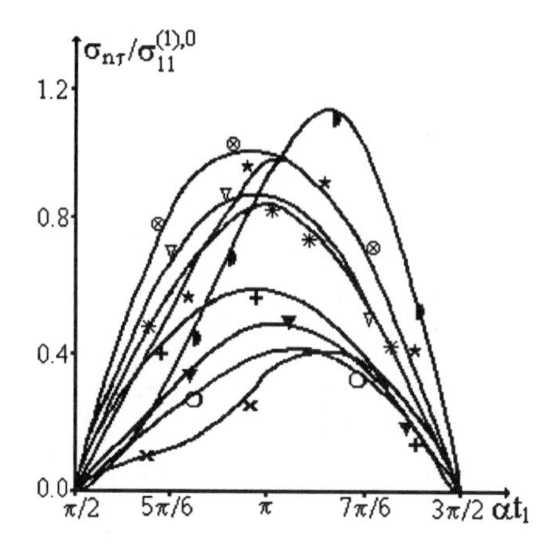

Fig.4.9.4. The graphics of the dependence of $\sigma_{n\tau}/\sigma_{11}^{(1),0}$ on αt_1 with $E_1^{(2)}/G_{12}^{(2)}$ =2, 20, 50 and various $\eta^{(2)}$ for $\varepsilon = 0.015$, $\chi = 0.3$, $E^{(2)}/E_1^{(2)} = 0.05$: ⊗, ∇, ○ -- $\eta^{(2)}$ =0.5; ✱, ✚, ✖ -- $\eta^{(2)}$ =0.2; ▼, ★, ◗ -- $\eta^{(2)}$ =0.1.

Table 4.28 shows that the influence of the $E_1^{(2)}/G_{12}^{(2)}$ on the stresses is more significant than that of $E^{(2)}/E_1^{(2)}$. Moreover with increasing $E_1^{(2)}/G_{12}^{(2)}$, of $\sigma_{n\tau}/\sigma_{11}^{(1),0}$ increases but $\sigma_{nn}/\sigma_{11}^{(1),0}$ decreases.

Table 4.28. The values of self-balanced normal and tangential stresses in the composite with curvilinear anisotropic layer shown in Fig.4.2.1 with various $E_1^{(2)}/G_{12}^{(2)}$, $E^{(2)}/E_1^{(2)}$ for $\eta^{(2)}$ =0.00, $\chi = 0.3$, $\varepsilon = 0.015$.

$E_1^{(2)}/G_{12}^{(2)}$	$E^{(2)}/E_1^{(2)}$	$\frac{\sigma_{nn}(C)}{\sigma_{11}^{(1),0}}$	$\frac{\sigma_{n\tau}(A')}{\sigma_{11}^{(1),0}}$
	0.90	0.6781	0.4584
2	0.05	0.6720	0.4453
	0.01	0.6893	0.4659
	0.90	0.1771	0.4950
20	0.05	0.2254	0.4888
	0.01	0.3062	0.5048
	0.90	-0.3004	0.4806
50	0.05	-0.1988	0.4904
	0.01	-0.0749	0.5061

Now consider the case when the concentration of filler layers is very high and the interaction between them must be taken into account. Fig.4.9.3 shows the dependence of $\sigma_{nn}/\sigma_{11}^{(1),0}$ on αt_1, for various $E_1^{(2)}/G_{12}^{(2)}$ and $\eta^{(2)}$ and fixed ε, χ and $E^{(2)}/E_1^{(2)}$. Analogous graphs of the dependence of $\sigma_{n\tau}/\sigma_{11}^{(1),0}$ on αt_1 are given in Fig.4.9.4 .

These graphs show that for all selected $\eta^{(2)}$, $\sigma_{nn}/\sigma_{11}^{(1),0}$ decreases as $E_1^{(2)}/G_{12}^{(2)}$ increases. However the dependence between $\sigma_{n\tau}/\sigma_{11}^{(1),0}$ and $E_1^{(2)}/G_{12}^{(2)}$ is complex. For instance, for $\eta^{(2)}$=0.5, 0.2, $\sigma_{n\tau}/\sigma_{11}^{(1),0}$ decreases as $E_1^{(2)}/G_{12}^{(2)}$ increases, but increase for $\eta^{(2)}$=0.1.

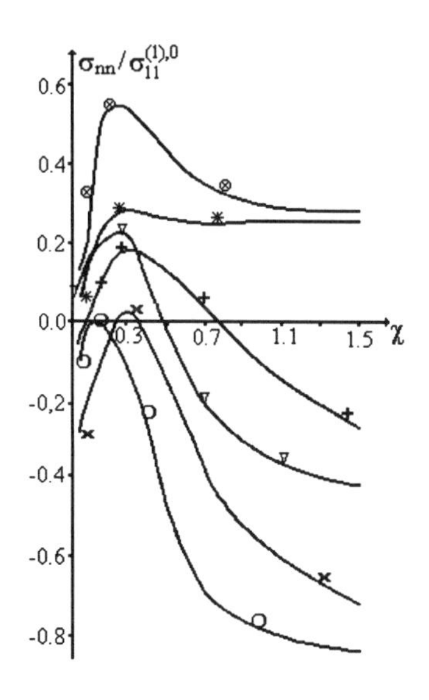

Fig.4.9.5. The graphs of the dependence of $\sigma_{nn}(C)/\sigma_{11}^{(1),0}$ (Fig.4.2.1) on χ with $E_1^{(2)}/G_{12}^{(2)}$ =2, 20, 50 and various $\eta^{(2)}$ for ε = 0.015, 0.3, $E^{(2)}/E_1^{(2)}$ = 0.05
⊗, ∇, ○ -- $\eta^{(2)}$=0.2; ✱, ✚, ✖ -- $\eta^{(2)}$=0.5;

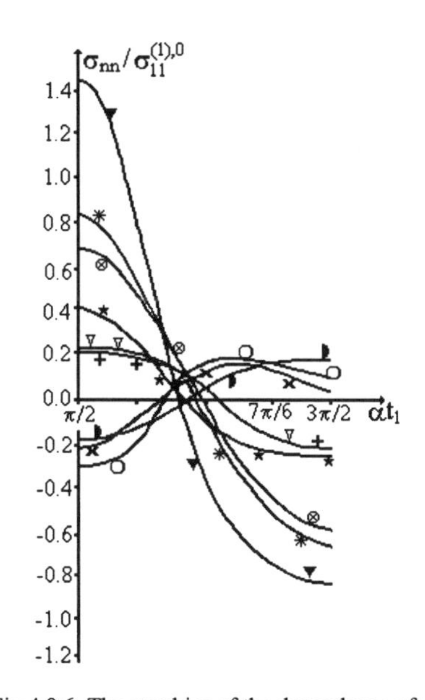

Fig.4.9.6. The graphics of the dependence of $\sigma_{nn}/\sigma_{11}^{(1),0}$ on αt_1 with $E_1^{(2)}/G_{12}^{(2)}$ =2, 20, 50 and various $\eta^{(2)}$ for ε = 0.015, χ = 0.3, $E^{(2)}/E_1^{(2)}$ = 0.05 ⊗, ∇, ○ -- $\eta^{(2)}$=0.1; ✱, ✚, ✖ -- $\eta^{(2)}$=0.2; ▼, ★, ◗ -- $\eta^{(2)}$=0.5 (anti-phase curving: Fig.4.2.3).

Fig.4.9.5 shows the relations between $\sigma_{nn}/\sigma_{11}^{(1),0}$ and parameter χ for various $E_1^{(2)}/G_{12}^{(2)}$ and $\eta^{(2)}$. The comparison of the various curves shows that, as in the case $\eta^{(2)}$=0.00, they are displaced wholly "down" with increasing $E_1^{(2)}/G_{12}^{(2)}$; moreover as $\eta^{(2)}$ increases the absolute values of $\sigma_{nn}/\sigma_{11}^{(1),0}$ decrease.

All the numerical results have been obtained using the first three approximations of the presentation (4.1.15). Table 4.29 shows the convergence of $\sigma_{nn}/\sigma_{11}^{(1),0}$ for different approximations; again the convergence is acceptable.

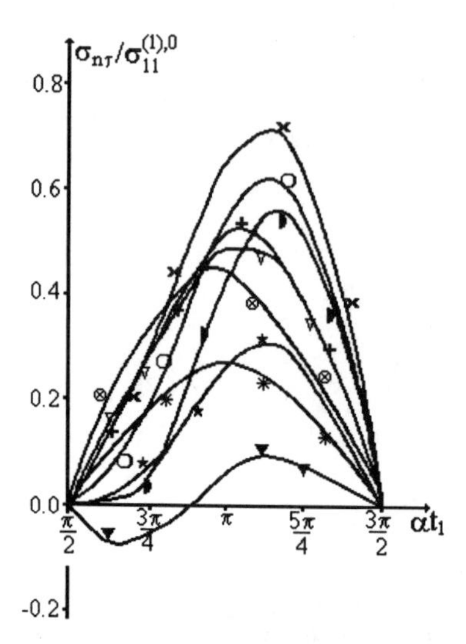

Fig.4.9.7. The graphs of the dependence of $\sigma_{n\tau}/\sigma_{11}^{(1),0}$ on αt_1 with $E_1^{(2)}/G_{12}^{(2)}$ =2, 20, 50 and various $\eta^{(2)}$ for $\varepsilon = 0.015$, $\chi = 0.3$, $E^{(2)}/E_1^{(2)}$ =0.05 ⊗, ∇, O -- $\eta^{(2)}$ =0.1; ✳, ✚, ✖ -- $\eta^{(2)}$ =0.2; ▼, ★, ◗ -- $\eta^{(2)}$ =0.5. (anti-phase curving: Fig.4.2.3)

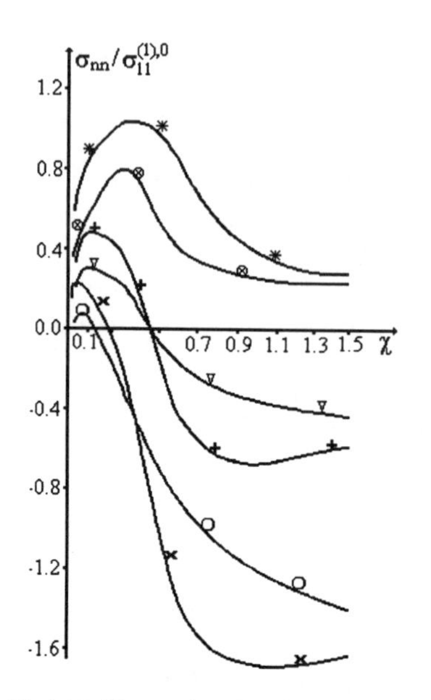

Fig.4.9.8. The graphs of the dependence of $\sigma_{nn}(A)/\sigma_{11}^{(1),0}$ (Fig.4.2.3) on χ with $E_1^{(2)}/G_{12}^{(2)}$ =2, 20, 50 and various $\eta^{(2)}$ for $\varepsilon = 0.015, 0.3$, $E^{(2)}/E_1^{(2)} = 0.05$. ⊗, ∇, O -- $\eta^{(2)}$ =0.2; ✳, ✚, ✖ -- $\eta^{(2)}$ =0.5;

4.9.2. ANTI-PHASE CURVING OF THE LAYERS

Consider the numerical results obtained for the composite with alternating anti-phase curving curvilinear anisotropic layers (Fig.4.2.3) under loading $\sigma_{11}(\infty) = p$. Note that these results are obtained in the framework of the assumptions and notation accepted in the previous subsection.

Fig.4.9.6 which shows the dependence of $\sigma_{nn}/\sigma_{11}^{(1),0}$ on αt_1 for various $\eta^{(2)}$ and $E_1^{(2)}/G_{12}^{(2)}$ for fixed problem parameters. Similar graphs for the distribution of the $\sigma_{n\tau}/\sigma_{11}^{(1),0}$ are given in Fig.4.9.7. According to these graphs, we may conclude that the curvilinearity of the filler material anisotropy significantly changes the distribution of the self-balanced stresses on the inter-layer surface.

Fig.4.9.8 shows the dependence of $\sigma_{nn}/\sigma_{11}^{(1),0}$ on χ for different $\eta^{(2)}$ and $E_1^{(2)}/G_{12}^{(2)}$; the graphs of these dependences displaced wholly “down” by increasing

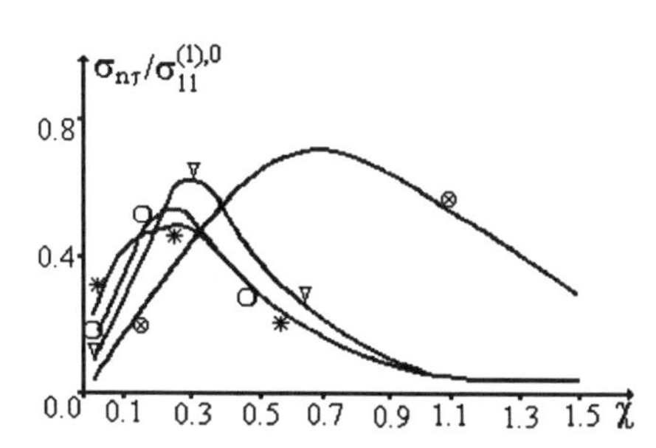

Fig.4.9.9. The graphs of the dependence of $\sigma_{n\tau}(B')/\sigma_{11}^{(1),0}$ (Fig.4.2.3) on χ with various $\eta^{(2)}$ for $\varepsilon = 0.015, 0.3$, $E^{(2)}/E_1^{(2)} = 0.05$, $E_1^{(2)}/G_{12}^{(2)} = 50$.

⊗, ∇, ○, ✳ -- $\eta^{(2)} = 0.02, 0.1, 0.2, 0.5$

$E_1^{(2)}/G_{12}^{(2)}$. Moreover, these graphs show that the absolute values of $\sigma_{nn}/\sigma_{11}^{(1),0}$ grow with increasing $\eta^{(2)}$. Note also that the graphs of the relations between $\sigma_{n\tau}/\sigma_{11}^{(1),0}$ and χ for various $\eta^{(2)}$ are illustrated in Fig.4.9.9. The results show that for all selected $\eta^{(2)}$ these dependences are non-monotonic and $\sigma_{n\tau}/\sigma_{11}^{(1),0}$ decrease with $\eta^{(2)}$.

Table 4.29. The values $\sigma_{nn}(C)/\sigma_{11}^{(1),0}$ (Fig.4.2.1) in various approximations (4.1.15) and $E^{(2)}/E_1^{(2)}$, $E_1^{(2)}/G_{12}^{(2)}$ for $\eta^{(2)} = 0.5$, $\chi = 0.3$, $\varepsilon = 0.015$.

$E^{(2)}/E_1^{(2)}$	Numb. of appr.	$E_1^{(2)}/G_{12}^{(2)}$	
		20	50
0.9	1	0.234	0.113
	2	0.105	-0.101
	3	0.069	-0.166
0.01	1	0.231	0.121
	2	0.356	0.279
	3	0.324	0.231

4.10. Bibliographical Notes

The solution method presented in sections 4.1 and 4.2 was proposed by S.D. Akbarov and A.N. Guz [29]. The study of the stress state in the curved composite under low filler concentration was also made by S.D. Akbarov and A.N. Guz [31]. However the remaining numerical results discussed in section 4.3 were obtained by S.D. Akbarov [5, 10, 11] and by S.D.Akbarov and A. N.Guz [33, 37]. The stress state in the composites considered in section 4.3 under action at $|x_2| \to \infty$ of uniformly distributed tangential forces with intensity τ was investigated by S.D. Akbarov [8]; that these results have not been included here.

The development of the solution method for composites with partially curved layers was made by S.D. Akbarov and S.A. Aliev [21, 22, 23]. The solution method for viscoelastic composites was proposed by S.D. Akbarov [3], and the influence of the rheological parameters of the matrix material to the stress distribution has been also investigated by S.D. Akbarov [2, 7].

The solution methods and numerical results related to the composites with anisotropic layers and given in sections 4.7-4.9 were presented by S.D. Akbarov, A.N.Guz and S.M. Mustafaev [43], for the rectilinear anisotropy, by S.D. Akbarov [15] and by S.D.Akbarov and M.S. Mustafaev [47] for the curvilinear anisotropy.

CHAPTER 5

COMPOSITES WITH SPATIALLY PERIODIC CURVED LAYERS

In this chapter the problem formulation and solution method presented in the previous chapter are developed for three-dimensional problems, namely, for composites with spatially curved layers. The solution method is detailed for periodic curving. Many numerical results are given for the stress distribution in such composites; all investigations are carried out for a piecewise-homogeneous body model.

As before, we use tensor notation and sum repeated indices over their ranges; underlined repeated indices are not summed.

5.1. Formulation

Consider an infinite elastic body, reinforced by an arbitrary number of non-intersecting spatially curved layers. As before, values related to matrix will be denoted by upper indices (1), to the filler by upper indices (2). The rectangular Cartesian system of coordinates $O_m^{(k)}x_{1m}^{(k)}x_{2m}^{(k)}x_{3m}^{(k)}$ which are obtained from the system of coordinates $Ox_1x_2x_3$ by parallel transfer along the Ox_2, will refer to each layer. We isolate from the composite the part at $x_{1m}^{(k)} = \text{const}$ and $x_{3m}^{(k)} = \text{const}$ shown in Fig.5.1.1. The reinforcing layers will be assumed to be located in planes $O_m^{(2)}x_{1m}^{(2)}x_{3m}^{(2)}$ and the thickness of each filler layer will be assumed constant.

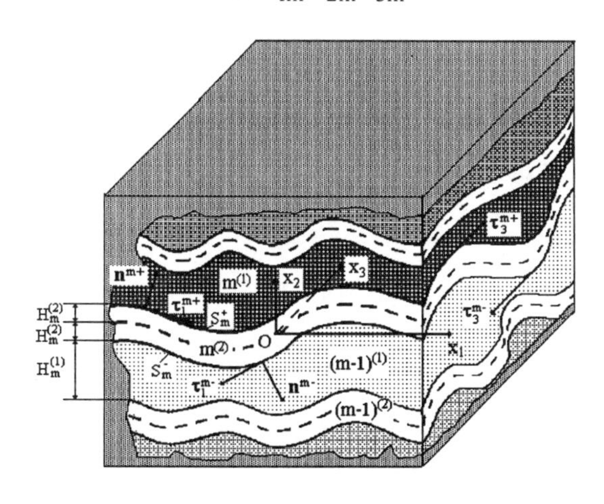

Fig.5.1.1.A part of the composite material with spatially curved structures at $x_1^{(k)}$ =const, $x_3^{(k)}$ =const.

Furthermore, we assume that the reinforcing layers are curved both in the direction of the Ox_1 axis and in the direction of the Ox_3 axis. It will also be assumed that the matrix and filler materials are homogeneous, isotropic and linearly elastic. We investigate the stress distribution in this composite under the action at infinity of uniformly distributed forces. Within each layer we write the equilibrium equations, Hooke's law, and Cauchy's geometrical relations.

$$\frac{\partial \sigma_{ij}^{(k)\underline{m}}}{\partial x_{j\underline{m}}^{(k)}}=0,\quad \sigma_{ij}^{(k)\underline{m}}=\lambda^{(k)\underline{m}}\theta^{(k)\underline{m}}\delta_i^j+2\mu^{(k)\underline{m}}\varepsilon_{ij}^{(k)\underline{m}},$$

$$i;j=1,2,3,\ k=1,2,\ q=1,2,\ldots,$$

$$\varepsilon_{ij}^{(k)\underline{m}}=\frac{1}{2}\left(\frac{\partial u_i^{(k)\underline{m}}}{\partial x_{j\underline{m}}^{(k)}}+\frac{\partial u_j^{(k)\underline{m}}}{\partial x_{i\underline{m}}^{(k)}}\right),\quad \theta^{(k)\underline{m}}=\frac{\partial u_i^{(k)\underline{m}}}{\partial x_{i\underline{m}}^{(k)}}. \tag{5.1.1}$$

In (5.1.1) the well-known notation is used.

As in subsection 4.1.1, we assume that there is complete cohesion between layers. Denoting the upper and lower surfaces of the $m^{(2)}$th filler layer by S_m^+ and S_m^- respectively, and introducing the notation $m_1=m-1$, we may write these cohesion conditions as follows:

$$\sigma_{ij}^{(1)\underline{m}}\Big|_{S_{\underline{m}}^+} n_j^{\underline{m},+}=\sigma_{ij}^{(2)\underline{m}}\Big|_{S_{\underline{m}}^+} n_j^{\underline{m},+};\quad \sigma_{ij}^{(1)m_1}\Big|_{S_{\underline{m}}^-} n_j^{\underline{m},-}=\sigma_{ij}^{(2)\underline{m}}\Big|_{S_{\underline{m}}^-} n_j^{\underline{m},-},$$

$$u_i^{(1)\underline{m}}\Big|_{S_{\underline{m}}^+}=u_i^{(2)\underline{m}}\Big|_{S_{\underline{m}}^+};\quad u_i^{(1)m_1}\Big|_{S_{\underline{m}}^-}=u_i^{(2)\underline{m}}\Big|_{S_{\underline{m}}^-}, \tag{5.1.2}$$

where $n_j^{m,\pm}$ are the components of the unit normal vector to the surface $S_m^{\pm}$.

We assume that the equation of the middle surface of the $m^{(2)}$th filler layer is given in the form

$$x_{2\underline{m}}^{(2)}=F_{\underline{m}}\left(x_{1\underline{m}}^{(2)},x_{3\underline{m}}^{(2)}\right)=\varepsilon f_{\underline{m}}\left(x_{1\underline{m}}^{(2)},x_{3\underline{m}}^{(2)}\right) \tag{5.1.3}$$

Here ε is a dimensionless small parameter ($0\le\varepsilon<1$) whose geometrical meaning will become apparent in a specific form of the function (5.1.3).

This gives the general formulation of the problems; we now consider some preparatory procedures for the exposition of the solution method.

5.2. The Equation of Contact Surfaces

As in subsection 4.1.2 using the condition of uniform filler layer thickness, and equation (5.1.3) we attempt to obtain the equation for the interface surfaces $S_m^{\pm}$. We introduce the parameters t_{1m} and t_{3m}, where $t_{1m}, t_{3m} \in (-\infty, +\infty)$ and rewrite the equation (5.1.3) in the following form

$$x_{1m}^{(2)} = t_{1m}, \quad x_{3m}^{(2)} = t_{3m}, \quad x_{2\underline{m}}^{(2)} = F_{\underline{m}}(t_{1\underline{m}}, t_{3\underline{m}}) = \varepsilon f_{\underline{m}}(t_{1\underline{m}}, t_{3\underline{m}}). \quad (5.2.1)$$

We assume that the function $F_m(t_{1m}, t_{3m})$ and its first order partial derivatives are continuous, and from (5.2.1) we obtain the expressions

$$\frac{\partial x_{2\underline{m}}^{(2)}}{\partial x_{1\underline{m}}^{(2)}} = \frac{\partial F_{\underline{m}}(t_{1\underline{m}}, t_{3\underline{m}})}{\partial t_{1\underline{m}}}, \quad \frac{\partial x_{2\underline{m}}^{(2)}}{\partial x_{3\underline{m}}^{(2)}} = \frac{\partial F_{\underline{m}}(t_{1\underline{m}}, t_{3\underline{m}})}{\partial t_{3\underline{m}}}, \quad (5.2.2)$$

for the slopes of the tangent line to the surface (5.2.1) at $x_{3m}^{(2)} = \mathrm{const}$ and at $x_{1m}^{(2)} = \mathrm{const}$, respectively. Taking (5.2.2) into account, we can write the equation of the straight line passing through the point $(t_{1\underline{m}}, t_{3\underline{m}}, F_{\underline{m}}(t_{1\underline{m}}, t_{3\underline{m}}))$ and perpendicular to these tangent lines as follows:

$$x_{1\underline{m}}^{(2)} - t_{1\underline{m}} = -\frac{\partial F_{\underline{m}}(t_{1\underline{m}}, t_{3\underline{m}})}{\partial t_{1\underline{m}}}\left(x_{2\underline{m}}^{(2)} - F_{\underline{m}}(t_{1\underline{m}}, t_{3\underline{m}})\right),$$

$$x_{3\underline{m}}^{(2)} - t_{3\underline{m}} = -\frac{\partial F_{\underline{m}}(t_{1\underline{m}}, t_{3\underline{m}})}{\partial t_{3\underline{m}}}\left(x_{2\underline{m}}^{(2)} - F_{\underline{m}}(t_{1\underline{m}}, t_{3\underline{m}})\right). \quad (5.2.3)$$

As the surfaces $S_m^{\pm}$ and the line (5.2.3) have common points, the coordinates of the intersections with surfaces must satisfy equation (5.2.3). Moreover, taking into account the uniformity of the filler layer thickness, we obtain the following system of equations for the coordinates of the points of the surfaces $S_m^{\pm}$:

$$\left(x_{1\underline{m}}^{(2)\pm} - t_{1\underline{m}}\right) = -\frac{\partial F_{\underline{m}}(t_{1\underline{m}}, t_{3\underline{m}})}{\partial t_{1\underline{m}}}\left(x_{2\underline{m}}^{(2)\pm} - F_{\underline{m}}(t_{1\underline{m}}, t_{3\underline{m}})\right),$$

$$\left(x_{3\underline{m}}^{(2)\pm} - t_{3\underline{m}}\right) = -\frac{\partial F_{\underline{m}}(t_{1\underline{m}}, t_{3\underline{m}})}{\partial t_{3\underline{m}}}\left(x_{2\underline{m}}^{(2)\pm} - F_{\underline{m}}(t_{1\underline{m}}, t_{3\underline{m}})\right),$$

$$\left(x_{1\underline{m}}^{(2)\pm} - t_{1\underline{m}}\right)^2 + \left(x_{2\underline{m}}^{(2)\pm} - F_{\underline{m}}(t_{1\underline{m}}, t_{3\underline{m}})\right)^2 + \left(x_{3\underline{m}}^{(2)\pm} - t_{3\underline{m}}\right)^2 = \left(H_{\underline{m}}^{(2)}\right)^2. \quad (5.2.4)$$

In (5.2.4) $x_{im}^{(2)\pm}$ (i=1,2,3) are the coordinates of the points of the surfaces $S_m^{\pm}$; $H_m^{(2)}$ is the half-thickness of the $m^{(2)}$th filler layer.

Thus, from (5.2.4) we have the following equation for the surfaces $S_m^{\pm}$:

$$x_{1\underline{m}}^{(2)\pm} = t_{1\underline{m}} \mp \frac{H_{\underline{m}}^{(2)}}{V(t_{1\underline{m}}, t_{3\underline{m}})} \frac{\partial F_{\underline{m}}\left(t_{1\underline{m}}, t_{3\underline{m}}\right)}{\partial t_{1\underline{m}}},$$

$$x_{2\underline{m}}^{(2)\pm} = F_{\underline{m}}\left(t_{1\underline{m}}, t_{3\underline{m}}\right) \pm \frac{H_{\underline{m}}^{(2)}}{V(t_{1\underline{m}}, t_{3\underline{m}})},$$

$$x_{3\underline{m}}^{(2)\pm} = t_{3\underline{m}} \mp \frac{H_{\underline{m}}^{(2)}}{V(t_{1\underline{m}}, t_{3\underline{m}})} \frac{\partial F\left(t_{1\underline{m}}, t_{3\underline{m}}\right)}{\partial t_{3\underline{m}}}, \tag{5.2.5}$$

where

$$V\left(t_{1\underline{m}}, t_{3\underline{m}}\right) = \left[1 + \left(\frac{\partial F_{\underline{m}}\left(t_{1\underline{m}}, t_{3\underline{m}}\right)}{\partial t_{1\underline{m}}}\right)^2 + \left(\frac{\partial F_{\underline{m}}\left(t_{1\underline{m}}, t_{3\underline{m}}\right)}{\partial t_{3\underline{m}}}\right)^2\right]^{\frac{1}{2}}. \tag{5.2.6}$$

In all further investigations we will assume that

$$\varepsilon^2 \left[\left(\frac{\partial f_{\underline{m}}\left(t_{1\underline{m}}, t_{3\underline{m}}\right)}{\partial t_{1\underline{m}}}\right)^2 + \left(\frac{\partial f_{\underline{m}}\left(t_{1\underline{m}}, t_{3\underline{m}}\right)}{\partial t_{3\underline{m}}}\right)^2\right] < 1. \tag{5.2.7}$$

Introducing the orthonormal base vectors $\mathbf{i}_{km}$ (k=1,2,3) in Cartesian coordinate system $O_m^{(2)} x_{1m}^{(2)} x_{2m}^{(2)} x_{3m}^{(2)}$, we can write the equation of the surfaces $S_m^{\pm}$ in vector form:

$$\mathbf{r}_{\underline{m}}^{\pm}\left(t_{1\underline{m}}, t_{3\underline{m}}\right) = x_{1\underline{m}}^{(2)\pm}\left(t_{1\underline{m}}, t_{3\underline{m}}\right)\mathbf{i}_{1\underline{m}} + x_{2\underline{m}}^{(2)\pm}\left(t_{1\underline{m}}, t_{3\underline{m}}\right)\mathbf{i}_{2\underline{m}} + x_{3\underline{m}}^{(2)\pm}\left(t_{1\underline{m}}, t_{3\underline{m}}\right)\mathbf{i}_{3\underline{m}}. \tag{5.2.8}$$

It is known that the first order partial derivatives of the vector (5.2.8), i.e., the vectors

$$\mathbf{r}_{j\underline{m}}^{\pm}\left(t_{1\underline{m}}, t_{3\underline{m}}\right) = \frac{\partial x_{k\underline{m}}^{(2)\pm}}{\partial t_{j\underline{m}}} \mathbf{i}_{k\underline{m}}, \quad k=1,2,3\,;\, j=1,3 \tag{5.2.9}$$

are the tangent vectors to the surfaces $S_{\underline{m}}^{\pm}$. Thus, using the formulae (5.2.9) we obtain the following expression for the normal vector $\mathbf{N}_{\underline{m}}^{\pm}$ to these surfaces:

$$\mathbf{N}_{\underline{m}}^{\pm} = \mathbf{r}_{3\underline{m}}^{\pm} \times \mathbf{r}_{1\underline{m}}^{\pm} = A_{k\underline{m}}^{\pm}\left(t_{1\underline{m}}, t_{3\underline{m}}\right)\mathbf{i}_{k\underline{m}}, \quad k=1,2,3; \tag{5.2.10}$$

where

$$A_{1\underline{m}}^{\pm} = \frac{\partial x_{3\underline{m}}^{(2)\pm}}{\partial t_{1\underline{m}}}\frac{\partial x_{2\underline{m}}^{(2)\pm}}{\partial t_{3\underline{m}}} - \frac{\partial x_{2\underline{m}}^{(2)\pm}}{\partial t_{1\underline{m}}}\frac{\partial x_{3\underline{m}}^{(2)\pm}}{\partial t_{3\underline{m}}},$$

$$A_{2\underline{m}}^{\pm} = \frac{\partial x_{1\underline{m}}^{(2)\pm}}{\partial t_{1\underline{m}}}\frac{\partial x_{3\underline{m}}^{(2)\pm}}{\partial t_{3\underline{m}}} - \frac{\partial x_{3\underline{m}}^{(2)\pm}}{\partial t_{1\underline{m}}}\frac{\partial x_{1\underline{m}}^{(2)\pm}}{\partial t_{3\underline{m}}},$$

$$A_{3\underline{m}}^{\pm} = \frac{\partial x_{1\underline{m}}^{(2)\pm}}{\partial t_{1\underline{m}}}\frac{\partial x_{2\underline{m}}^{(2)\pm}}{\partial t_{3\underline{m}}} - \frac{\partial x_{2\underline{m}}^{(2)\pm}}{\partial t_{1\underline{m}}}\frac{\partial x_{1\underline{m}}^{(2)\pm}}{\partial t_{3\underline{m}}}. \tag{5.2.11}$$

Using the equations (5.2.10) and (5.2.11) and doing some calculations we obtain the following expressions for the components of the unit normal vector of the surfaces $S_{\underline{m}}^{\pm}$:

$$n_i^{\underline{m},\pm} = \frac{A_{i\underline{m}}^{\pm}(t_{1\underline{m}}, t_{3\underline{m}})}{A_{\underline{m}}(t_{1\underline{m}}, t_{3\underline{m}})}, \quad i=1,2,3, \tag{5.2.12}$$

where

$$A_{\underline{m}}(t_{1\underline{m}}, t_{3\underline{m}}) = \left[\left(A_{1\underline{m}}^{\pm}\left(t_{1\underline{m}}, t_{3\underline{m}}\right)\right)^2 + \left(A_{2\underline{m}}^{\pm}\left(t_{1\underline{m}}, t_{3\underline{m}}\right)\right)^2 + \left(A_{3\underline{m}}^{\pm}\left(t_{1\underline{m}}, t_{3\underline{m}}\right)\right)^2\right]^{\frac{1}{2}}. \tag{5.2.13}$$

Taking (5.2.6), (5.2.7), (5.2.11), (5.2.13) into account we represent the expressions (5.2.5) and (5.2.12) as power series in the small parameter ε as follows:

$$x_{1\underline{m}}^{(2)\pm} = t_{1\underline{m}} \mp H_{\underline{m}}^{(2)}\varepsilon\frac{\partial f_{\underline{m}}}{\partial t_{1\underline{m}}} \pm \frac{1}{2}H_{\underline{m}}^{(2)}\varepsilon^3\frac{\partial f_{\underline{m}}}{\partial t_{1\underline{m}}}\Phi\left(t_{1\underline{m}}, t_{3\underline{m}}\right) \mp \frac{3}{8}H_{\underline{m}}^{(2)}\varepsilon^5\frac{\partial f_{\underline{m}}}{\partial t_{1\underline{m}}}\times$$

$$\left(\Phi\left(t_{1\underline{m}}, t_{3\underline{m}}\right)\right)^2 + \ldots = t_{1\underline{m}} + \sum_{k=1}^{\infty}(-1)^k\varepsilon^{2k-1}a_{1k}\left(\frac{\partial f_{\underline{m}}}{\partial t_{1\underline{m}}}, \frac{\partial f_{\underline{m}}}{\partial t_{3\underline{m}}}, \pm H_{\underline{m}}^{(2)}\right) \quad ;$$

$$x_{1\underline{m}}^{(2)\pm} = \pm H_{\underline{m}}^{(2)} \pm \varepsilon f_{\underline{m}} \mp \frac{1}{2} H_{\underline{m}}^{(2)} \varepsilon^2 \Phi\left(t_{1\underline{m}}, t_{3\underline{m}}\right) \pm \frac{3}{8} H_{\underline{m}}^{(2)} \varepsilon^4 \left(\Phi\left(t_{1\underline{m}}, t_{3\underline{m}}\right)\right)^2 + \ldots =$$

$$\pm H_{\underline{m}}^{(2)} + \varepsilon f_{\underline{m}} + \sum_{k=1}^{\infty} (-1)^k \varepsilon^{2k} a_{2k} \left(\frac{\partial f_{\underline{m}}}{\partial t_{1\underline{m}}}, \frac{\partial f_{\underline{m}}}{\partial t_{3\underline{m}}}, \pm H_{m}^{(2)} \right);$$

$$x_{3\underline{m}}^{(2)\pm} = t_{3\underline{m}} \mp H_{\underline{m}}^{(2)} \varepsilon \frac{\partial f_{\underline{m}}}{\partial t_{3\underline{m}}} \pm \frac{1}{2} H_{\underline{m}}^{(2)} \varepsilon^3 \frac{\partial f_{\underline{m}}}{\partial t_{3\underline{m}}} \Phi\left(t_{1\underline{m}}, t_{3\underline{m}}\right) \mp \frac{3}{8} H_{\underline{m}}^{(2)} \varepsilon^5 \frac{\partial f_{\underline{m}}}{\partial t_{3\underline{m}}} \times$$

$$\left(\Phi\left(t_{1\underline{m}}, t_{3\underline{m}}\right)\right)^2 + \ldots = t_{3m} + \sum_{k=1}^{\infty} (-1)^k \varepsilon^{2k-1} a_{3k} \left(\frac{\partial f_{\underline{m}}}{\partial t_{1\underline{m}}}, \frac{\partial f_{\underline{m}}}{\partial t_{3\underline{m}}}, \pm H_{m}^{(2)} \right);$$

$$n_{1}^{\underline{m},\pm} = -\frac{\partial f_{\underline{m}}}{\partial t_{1\underline{m}}} \varepsilon + \frac{1}{2} \varepsilon^3 \frac{\partial f_{\underline{m}}}{\partial t_{1\underline{m}}} \Phi\left(t_{1\underline{m}}, t_{3\underline{m}}\right) - \frac{3}{8} \varepsilon^5 \frac{\partial f_{\underline{m}}}{\partial t_{1\underline{m}}} \left(\Phi\left(t_{1\underline{m}}, t_{3\underline{m}}\right)\right)^2 + \ldots =$$

$$\sum_{k=1}^{\infty} (-1)^k \varepsilon^{2k-1} b_{1k} \left(\frac{\partial f_{\underline{m}}}{\partial t_{1\underline{m}}}, \frac{\partial f_{\underline{m}}}{\partial t_{3\underline{m}}} \right);$$

$$n_{2}^{\underline{m},\pm} = 1 - \frac{1}{2} \varepsilon^2 \Phi\left(t_{1\underline{m}}, t_{3\underline{m}}\right) + \frac{3}{8} \varepsilon^4 \left(\Phi\left(t_{1\underline{m}}, t_{3\underline{m}}\right)\right)^2 + \ldots =$$

$$1 + \sum_{k=1}^{\infty} (-1)^k \varepsilon^{2k} b_{2k} \left(\frac{\partial f_{\underline{m}}}{\partial t_{1\underline{m}}}, \frac{\partial f_{\underline{m}}}{\partial t_{3\underline{m}}} \right);$$

$$n_{3}^{\underline{m},\pm} = -\frac{\partial f_{\underline{m}}}{\partial t_{3\underline{m}}} \varepsilon + \frac{1}{2} \varepsilon^3 \frac{\partial f_{\underline{m}}}{\partial t_{3\underline{m}}} \Phi\left(t_{1\underline{m}}, t_{3\underline{m}}\right) - \frac{3}{8} \varepsilon^5 \frac{\partial f_{\underline{m}}}{\partial t_{3\underline{m}}} \left(\Phi\left(t_{1\underline{m}}, t_{3\underline{m}}\right)\right)^2 + \ldots =$$

$$\sum_{k=1}^{\infty} (-1)^k \varepsilon^{2k-1} b_{3k} \left(\frac{\partial f_{\underline{m}}}{\partial t_{1\underline{m}}}, \frac{\partial f_{\underline{m}}}{\partial t_{3\underline{m}}} \right), \tag{5.2.14}$$

where

$$\Phi\left(t_{1\underline{m}}, t_{3\underline{m}}\right) = \left(\frac{\partial f_{\underline{m}}}{\partial t_{1\underline{m}}} \right)^2 + \left(\frac{\partial f_{\underline{m}}}{\partial t_{3\underline{m}}} \right)^2 . \tag{5.2.15}$$

5.3. The Presentation of the Governing Relations in Series Form

It is difficult or even impossible to obtain the exact solutions to the problems formulated in section 5.1 for composite materials with spatially curved layers, because the contact conditions are given on the curved surfaces which do not coincide with the coordinate surfaces. To solve these problems, as in Chapter 4, we must use series expansions in the small parameter ε; the condition (5.2.7) limits the degree of curving of the reinforcing layers in the structure of the composite materials. Thus, we generalize the operations carried out in subsections 4.1.3 and 4.1.4 for the investigation of the stress- state in composite materials with spatially curved layers.

First, we present the sought values of the $m^{(k)}$ th layer as power series in the parameter ε as follows:

$$\left\{\sigma_{ij}^{(k)m}, \varepsilon_{ij}^{(k)m}, u_i^{(k)m}\right\}\left(x_{1m}^{(k)}, x_{2m}^{(k)}, x_{3m}^{(k)}\right) = \sum_{q=1}^{\infty} \varepsilon^q \left\{\sigma_{ij}^{(k)m,q}, \varepsilon_{ij}^{(k)m,q},\right.$$

$$\left. u_i^{(k)m,q}\right\}\left(x_{1m}^{(k)}, x_{2m}^{(k)}, x_{3m}^{(k)}\right) \tag{5.3.1}$$

Due to linearity, the equations (5.1.1) will be satisfied for each approximation separately, i.e. substituting (5.3.1) in (5.1.1) and grouping by equal powers of ε we obtain

$$\frac{\partial \sigma_{ij}^{(\underline{k})\underline{m},q}}{\partial x_{j\underline{m}}^{(\underline{k})}} = 0, \quad \sigma_{ij}^{(\underline{k})\underline{m},q} = \lambda^{(\underline{k})\underline{m}} \theta^{(\underline{k})\underline{m},q} \delta_i^j + 2\mu^{(\underline{k})\underline{m}} \varepsilon_{ij}^{(\underline{k})\underline{m},q},$$

$$i;j = 1,2,3, \quad k=1,2, \quad q=1,2,\ldots,$$

$$\varepsilon_{ij}^{(\underline{k})\underline{m},q} = \frac{1}{2}\left(\frac{\partial u_i^{(\underline{k})\underline{m},q}}{\partial x_{j\underline{m}}^{(\underline{k})}} + \frac{\partial u_j^{(\underline{k})\underline{m},q}}{\partial x_{i\underline{m}}^{(\underline{k})}}\right), \quad \theta^{(\underline{k})\underline{m},q} = \frac{\partial u_i^{(\underline{k})\underline{m},q}}{\partial x_{i\underline{m}}^{(\underline{k})}}. \tag{5.3.2}$$

Now we consider obtaining the contact conditions for each approximation. For this purpose we rewrite the equation of the surfaces S_m^+ and S_m^- in the system of coordinates $O_m^{(1)} x_{1m}^{(1)} x_{2m}^{(1)} x_{3m}^{(1)}$ and $O_{m_1}^{(1)} x_{1m_1}^{(1)} x_{2m_1}^{(1)} x_{3m_1}^{(1)}$, respectively. Taking into account the relations $x_{1m}^{(1)} = x_{1m_1}^{(1)} = x_{1m}^{(2)}$, $x_{3m}^{(1)} = x_{3m_1}^{(1)} = x_{3m}^{(2)}$, $x_{2m}^{(2)} = x_{2m}^{(1)} - H_m^{(1)} - H_m^{(2)}$, $x_{2m}^{(2)} = x_{2m_1}^{(1)} + H_{m_1}^{(1)} + H_m^{(2)}$ in equation (5.2.14) we obtain the following:

the equation of the surface S_m^+ in the system of coordinates $O_m^{(1)} x_{1m}^{(1)} x_{2m}^{(1)} x_{3m}^{(1)}$:

$$x_{1\underline{m}}^{(1)+} = t_{1\underline{m}} + \sum_{k=1}^{\infty} \varepsilon^{2k-1}(-1)^k a_{1k}\left(\frac{\partial f_{\underline{m}}}{\partial t_{1\underline{m}}}, \frac{\partial f_{\underline{m}}}{\partial t_{3\underline{m}}}, +H_{\underline{m}}^{(2)}\right),$$

$$x_{3\underline{m}}^{(1)+} = t_{3\underline{m}} + \sum_{k=1}^{\infty} \varepsilon^{2k-1}(-1)^k a_{3k}\left(\frac{\partial f_{\underline{m}}}{\partial t_{1\underline{m}}}, \frac{\partial f_{\underline{m}}}{\partial t_{3\underline{m}}}, +H_{\underline{m}}^{(2)}\right),$$

$$x_{2\underline{m}}^{(1)+} = -H_{\underline{m}}^{(1)} + \varepsilon f_{\underline{m}} + \sum_{k=1}^{\infty} \varepsilon^{2k}(-1)^k a_{2k}\left(\frac{\partial f_{\underline{m}}}{\partial t_{1\underline{m}}}, \frac{\partial f_{\underline{m}}}{\partial t_{3\underline{m}}}, +H_{\underline{m}}^{(2)}\right); \quad (5.3.3)$$

the equation of the surface S_m^- in the system of coordinates $O_{m_1}^{(1)} x_{1m_1}^{(1)} x_{2m_1}^{(1)} x_{3m_1}^{(1)}$:

$$x_{1m_1}^{(1)+} = t_{1\underline{m}} + \sum_{k=1}^{\infty} \varepsilon^{2k-1}(-1)^k a_{1k}\left(\frac{\partial f_{\underline{m}}}{\partial t_{1\underline{m}}}, \frac{\partial f_{\underline{m}}}{\partial t_{3\underline{m}}}, -H_{\underline{m}}^{(2)}\right),$$

$$x_{1m_1}^{(1)+} = t_{3\underline{m}} + \sum_{k=1}^{\infty} \varepsilon^{2k-1}(-1)^k a_{3k}\left(\frac{\partial f_{\underline{m}}}{\partial t_{1\underline{m}}}, \frac{\partial f_{\underline{m}}}{\partial t_{3\underline{m}}}, -H_{\underline{m}}^{(2)}\right)$$

$$x_{2m_1}^{(1)+} = -H_{m_1}^{(1)} + \varepsilon f_{\underline{m}} + \sum_{k=1}^{\infty} \varepsilon^{2k}(-1)^k a_{2k}\left(\frac{\partial f_{\underline{m}}}{\partial t_{1\underline{m}}}, \frac{\partial f_{\underline{m}}}{\partial t_{3\underline{m}}}, -H_{\underline{m}}^{(2)}\right). \quad (5.3.4)$$

Note that, the expressions for the components $n_i^{m,+}$ and $n_i^{m,-}$ in the system of coordinates $O_m^{(1)} x_{1m}^{(1)} x_{2m}^{(1)} x_{3m}^{(1)}$ and $O_{m_1}^{(1)} x_{1m_1}^{(1)} x_{2m_1}^{(1)} x_{3m_1}^{(1)}$ respectively, remain as in (5.2.14). Thus, taking the equations (5.2.14), (5.3.3) and (5.3.4) into account we rewrite the contact conditions (5.1.2) in the following form:

$$\sigma_{ij}^{(1)\underline{m}}\left(x_{1\underline{m}}^{(1)+}, x_{2\underline{m}}^{(1)+}, x_{3\underline{m}}^{(1)+}\right) n_{\underline{j}}^{m,+} = \sigma_{ij}^{(2)\underline{m}}\left(x_{1\underline{m}}^{(2)+}, x_{2\underline{m}}^{(2)+}, x_{3\underline{m}}^{(2)+}\right) n_{\underline{j}}^{m,+},$$

$$\sigma_{ij}^{(1)\underline{m_1}}\left(x_{1\underline{m_1}}^{(1)-}, x_{2\underline{m_1}}^{(1)-}, x_{3\underline{m_1}}^{(1)-}\right) n_{\underline{j}}^{m,-} = \sigma_{ij}^{(2)\underline{m}}\left(x_{1\underline{m}}^{(2)-}, x_{2\underline{m}}^{(2)-}, x_{3\underline{m}}^{(2)-}\right) n_{\underline{j}}^{m,-},$$

$$u_i^{(1)\underline{m}}\left(x_{1\underline{m}}^{(1)+}, x_{2\underline{m}}^{(1)+}, x_{3\underline{m}}^{(1)+}\right) = u_i^{(2)\underline{m}}\left(x_{1\underline{m}}^{(2)+}, x_{2\underline{m}}^{(2)+}, x_{3\underline{m}}^{(2)+}\right),$$

$$u_i^{(1)\underline{m_1}}\left(x_{1\underline{m_1}}^{(1)-}, x_{2\underline{m_1}}^{(1)-}, x_{3\underline{m_1}}^{(1)-}\right) = u_i^{(2)\underline{m}}\left(x_{1\underline{m}}^{(2)-}, x_{2\underline{m}}^{(2)-}, x_{3\underline{m}}^{(2)-}\right). \quad (5.3.5)$$

Substituting (5.3.1) in (5.3.5) we obtain

$$\sum_{q=1}^{\infty}\varepsilon^q\sigma_{ij}^{(1)\underline{m},q}\left(x_{1\underline{m}}^{(1)+},x_{2\underline{m}}^{(1)+},x_{3\underline{m}}^{(1)+}\right)n_j^{\underline{m},+}=\sum_{q=1}^{\infty}\varepsilon^q\sigma_{ij}^{(2)\underline{m},q}\left(x_{1\underline{m}}^{(2)+},x_{2\underline{m}}^{(2)+},x_{3\underline{m}}^{(2)+}\right)n_j^{\underline{m},+}$$

$$\sum_{q=1}^{\infty}\varepsilon^q\sigma_{ij}^{(1)\underline{m_1},q}\left(x_{1\underline{m_1}}^{(1)-},x_{2\underline{m_1}}^{(1)-},x_{3\underline{m_1}}^{(1)-}\right)n_j^{\underline{m},-}=\sum_{q=1}^{\infty}\varepsilon^q\sigma_{ij}^{(2)\underline{m},q}\left(x_{1\underline{m}}^{(2)-},x_{2\underline{m}}^{(2)-},x_{3\underline{m}}^{(2)-}\right)n_j^{\underline{m},-},$$

$$\sum_{q=1}^{\infty}\varepsilon^q u_i^{(1)\underline{m},q}\left(x_{1\underline{m}}^{(1)+},x_{2\underline{m}}^{(1)+},x_{3\underline{m}}^{(1)+}\right)=\sum_{q=1}^{\infty}\varepsilon^q u_i^{(2)\underline{m},q}\left(x_{1\underline{m}}^{(2)+},x_{2\underline{m}}^{(2)+},x_{3\underline{m}}^{(2)+}\right),$$

$$\sum_{q=1}^{\infty}\varepsilon^q u_i^{(1)\underline{m_1},q}\left(x_{1\underline{m_1}}^{(1)-},x_{2\underline{m_1}}^{(1)-},x_{3\underline{m_1}}^{(1)-}\right)=\sum_{q=1}^{\infty}\varepsilon^q u_i^{(2)\underline{m},q}\left(x_{1\underline{m}}^{(2)-},x_{2\underline{m}}^{(2)-},x_{3\underline{m}}^{(2)-}\right). \qquad (5.3.6)$$

Expanding the values $\sigma_{ij}^{(1)\underline{m},q}\left(x_{1\underline{m}}^{(1)+},x_{2\underline{m}}^{(1)+},x_{3\underline{m}}^{(1)+}\right)$, $u_i^{(1)\underline{m},q}\left(x_{1\underline{m}}^{(1)+},x_{2\underline{m}}^{(1)+},x_{3\underline{m}}^{(1)+}\right)$ in a Taylor's series in the vicinity of $\left(t_{1\underline{m}},-H_{\underline{m}}^{(1)},t_{3\underline{m}}\right)$, the values $\sigma_{ij}^{(2)\underline{m},q}\left(x_{1\underline{m}}^{(2)\pm},x_{2\underline{m}}^{(2)\pm},x_{3\underline{m}}^{(2)\pm}\right)$, $u_i^{(2)\underline{m},q}\left(x_{1\underline{m}}^{(2)\pm},x_{2\underline{m}}^{(2)\pm},x_{3\underline{m}}^{(2)\pm}\right)$ in the vicinity of $\left(t_{1\underline{m}},\pm H_{\underline{m}}^{(2)},t_{3\underline{m}}\right)$, and the values $\sigma_{ij}^{(1)\underline{m_1},q}\left(x_{1\underline{m_1}}^{(1)-},x_{2\underline{m_1}}^{(1)-},x_{3\underline{m_1}}^{(1)-}\right)$, $u_i^{(1)\underline{m_1},q}\left(x_{1\underline{m_1}}^{(1)-},x_{2\underline{m_1}}^{(1)-},x_{3\underline{m_1}}^{(1)-}\right)$ in the vicinity of $\left(t_{1\underline{m}},+H_{\underline{m_1}}^{(1)},t_{3\underline{m}}\right)$, and grouping by identical powers of ε we obtain the following representations:

$$\left\{\sigma_{ij}^{(2)\underline{m}};u_i^{(2)\underline{m}}\right\}\left(x_{1\underline{m}}^{(2)\pm},x_{2\underline{m}}^{(2)\pm},x_{3\underline{m}}^{(2)\pm}\right)=\sum_{n=0}^{\infty}\varepsilon^n\left\{P_{ij}^{(2)\underline{m},n};U_i^{(2)\underline{m},n}\right\}\left(t_{1\underline{m}},\pm H_{\underline{m}}^{(2)},t_{3\underline{m}}\right),$$

$$\left\{\sigma_{ij}^{(1)\underline{m}};u_i^{(1)\underline{m}}\right\}\left(x_{1\underline{m}}^{(1)+},x_{2\underline{m}}^{(1)+},x_{3\underline{m}}^{(1)+}\right)=\sum_{n=0}^{\infty}\varepsilon^n\left\{P_{ij}^{(1)\underline{m},n};U_i^{(1)\underline{m},n}\right\}\left(t_{1\underline{m}},-H_{\underline{m}}^{(1)},t_{3\underline{m}}\right),$$

$$\left\{\sigma_{ij}^{(1)\underline{m_1}};u_i^{(1)\underline{m_1}}\right\}\left(x_{1\underline{m_1}}^{(1)-},x_{2\underline{m_1}}^{(1)-},x_{3\underline{m_1}}^{(1)-}\right)=\sum_{n=0}^{\infty}\varepsilon^n\left\{P_{ij}^{(2)\underline{m_1},n};U_i^{(2)\underline{m_1},n}\right\}\left(t_{1\underline{m_1}},+H_{\underline{m_1}}^{(1)},t_{3\underline{m_1}}\right)$$

(5.3.7)

We write the formulae for the $P_{ij}^{(k)m,q}$ and the $U_i^{(k)m,q}$ at $\left(t_{1m},\pm H_m^{(2)},t_{3m}\right)$ in the case where the zeroth approximation (5.3.1) corresponds to the homogeneous stress state:

the formulae for $P_{ij}^{(k)m,q}$:

$$P_{ij}^{(k)m,0} = \sigma_{ij}^{(k)m,0}, \quad P_{ij}^{(k)m,1} = \sigma_{ij}^{(k)m,1},$$

$$P_{ij}^{(k)m,q} = \sigma_{ij}^{(k)m,q} + \sum_{s=1}^{q-1} R_s \sigma_{ij}^{(k)m,q-s} \quad \text{for } q \geq 2; \tag{5.3.8}$$

the formulae for $U_i^{(k)m,q}$:

$$U_i^{(k)m,0} = u_i^{(k)m,0}, \quad U_i^{(k)m,1} = u_i^{(k)m,1} + R_1 u_i^{(k)m,0},$$

$$U_i^{(\underline{k})\underline{m},2} = u_i^{(\underline{k})\underline{m},2} + R_1 u_i^{(\underline{k})\underline{m},1} \mp b_1 \frac{\partial u_i^{(\underline{k})\underline{m},0}}{\partial x_{2\underline{m}}^{(\underline{k})}},$$

$$U_i^{(k)m,q} = u_i^{(k)m,q} + \sum_{s=1}^{q-1} R_s u_i^{(k)m,q-s} \quad \text{for } q \geq 3. \tag{5.3.9}$$

In (5.3.8) and (5.3.9) R_s the linear differential operators. Here we present the expressions of these operators for the first five approximations. Note that $H_m^{(2)}$, $x_{1m}^{(k)}$, $x_{2m}^{(k)}$, $x_{3m}^{(k)}$, f_m are denoted as H, x_1, x_2, x_3 and f, respectively, in these expressions.

$$R_1 = \mp a_1 \frac{\partial}{\partial x_1} + f \frac{\partial}{\partial x_2} \mp c_1 \frac{\partial}{\partial x_3},$$

$$R_2 = \mp b_1 \frac{\partial}{\partial x_2} + \frac{a_1^2}{2} \frac{\partial^2}{\partial x_1^2} + \frac{1}{2} f^2 \frac{\partial^2}{\partial x_2^2} + \frac{1}{2} c_1^2 \frac{\partial^2}{\partial x_3^2} \mp$$

$$a_1 f \frac{\partial^2}{\partial x_1 \partial x_2} + a_1 c_1 \frac{\partial^2}{\partial x_1 \partial x_3} \mp c_1 f \frac{\partial^2}{\partial x_2 \partial x_3},$$

$$R_3 = R_{31} + R_{32} + R_{33}, \quad R_4 = R_{41} + R_{42} + R_{43} + R_{44} + R_{45} + R_{46}, \tag{5.3.10}$$

where

$$R_{31} = \pm a_2 \frac{\partial}{\partial x_1} \pm c_2 \frac{\partial}{\partial x_3} \mp f b_1 \frac{\partial^2}{\partial x_2^2} + a_1 b_1 \frac{\partial^2}{\partial x_1 \partial x_2} + b_1 c_1 \frac{\partial^2}{\partial x_2 \partial x_3} ,$$

$$R_{32} = \mp \frac{a_1^3}{6} \frac{\partial^3}{\partial x_1^3} + \frac{f^3}{6} \frac{\partial^3}{\partial x_2^3} \mp \frac{c_1^3}{6} \frac{\partial^3}{\partial x_3^3} + \frac{a_1^2}{2} f \frac{\partial^3}{\partial x_1^2 \partial x_2} \mp \frac{a_1^2}{2} c_1 \frac{\partial^3}{\partial x_1^2 \partial x_3} ,$$

$$R_{33} = \mp \frac{a_1}{2} f^2 \frac{\partial^3}{\partial x_1 \partial x_2^2} \mp \frac{a_1}{2} c_1^2 \frac{\partial^3}{\partial x_1 \partial x_3^2} +$$

$$\frac{c_1^2}{2} f \frac{\partial^2}{\partial x_2 \partial x_3^2} \mp \frac{c_1}{2} f^2 \frac{\partial^3}{\partial x_2^2 \partial x_3} + a_1 c_1 f \frac{\partial^3}{\partial x_1 \partial x_2 \partial x_3} ,$$

$$R_{41} = \pm b_2 \frac{\partial}{\partial x_2} - a_1 a_2 \frac{\partial^2}{\partial x_1^2} + \frac{1}{2} b_1^2 \frac{\partial^2}{\partial x_2^2} - c_1 c_2 \frac{\partial^2}{\partial x_3^2} \pm a_2 f \frac{\partial^2}{\partial x_1 \partial x_2} ,$$

$$R_{42} = -(a_1 c_2 + a_2 c_1) \frac{\partial^2}{\partial x_1 \partial x_3} \pm f c_2 \frac{\partial^2}{\partial x_2 \partial x_3} \mp$$

$$\frac{b_1}{2} f^2 \frac{\partial^3}{\partial x_2^3} \mp \frac{a_1^2 b_1}{2} \frac{\partial^3}{\partial x_1^2 \partial x_2} + a_1 b_1 f \frac{\partial^3}{\partial x_1 \partial x_2^2} ,$$

$$R_{43} = \mp \frac{b_1 c_1^2}{2} \frac{\partial^3}{\partial x_2 \partial x_3^2} + a_1 b_1 c_1 \frac{\partial^3}{\partial x_1 \partial x_2 \partial x_3} + \frac{a_1^4}{24} \frac{\partial^4}{\partial x_1^4} + \frac{f^4}{24} \frac{\partial^4}{\partial x_2^4} + \frac{c_1^4}{24} \frac{\partial^4}{\partial x_3^4} ,$$

$$R_{44} = \mp \frac{a_1^3 f}{6} \frac{\partial^4}{\partial x_1^3 \partial x_2} + \frac{a_1^3 c_1}{6} \frac{\partial^4}{\partial x_1^3 \partial x_3} \mp \frac{a_1}{6} f^3 \frac{\partial^4}{\partial x_1 \partial x_2^3} + \frac{a_1 c_1^3}{6} \frac{\partial^4}{\partial x_1 \partial x_3^3} ,$$

$$R_{45} = \mp \frac{c_1^3}{6} f \frac{\partial^4}{\partial x_2 \partial x_3^3} \mp \frac{c_1}{6} f^3 \frac{\partial^4}{\partial x_2^3 \partial x_3} +$$

$$\frac{a_1^2 f^2}{4} \frac{\partial^4}{\partial x_1^2 \partial x_2^2} + \frac{a_1^2 c_1^2}{4} \frac{\partial^4}{\partial x_1^2 \partial x_3^2} + \frac{c_1^2 f^2}{4} \frac{\partial^4}{\partial x_2^2 \partial x_3^2} ,$$

$$R_{46} = \mp \frac{a_1}{2} c_1^2 f \frac{\partial^4}{\partial x_1 \partial x_2 \partial x_3^2} + \frac{a_1}{2} c_1 f^2 \frac{\partial^4}{\partial x_1 \partial x_2^2 \partial x_3} \mp \frac{a_1^2}{2} c_1 f \frac{\partial^4}{\partial x_1^2 \partial x_2 \partial x_3} . \quad (5.3.11)$$

In (5.3.10) and (5.3.11) the following notation is used.

$$a_1 = H\frac{\partial f}{\partial x_1}\ ,\ b_1 = \frac{H}{2}\Phi(x_1,x_3)\ ,\ c_1 = H\frac{\partial f}{\partial x_3}\ ,\quad a_2 = \frac{H}{2}\frac{\partial f}{\partial x_1}\Phi(x_1,x_3)$$

$$b_2 = \frac{3}{8}\Phi(x_1,x_3),\ c_2 = \frac{H}{2}\frac{\partial f}{\partial x_3}\Phi(x_1,x_3)\ ,\ \Phi(x_1,x_3) = \left(\frac{\partial f}{\partial x_1}\right)^2 + \left(\frac{\partial f}{\partial x_3}\right)^2 . \tag{5.3.12}$$

It should be noted that the formulae for calculating $P_{ij}^{(1)m,q}$, $U_i^{(1)m,q}$ at $\left(t_{1m}, -H_m^{(1)}, t_{3m}\right)$ and $P_{ij}^{(1)m_1,q}$, $U_i^{(1)m_1,q}$ at $\left(t_{1m}, H_{m_1}^{(1)}, t_{3m}\right)$ are obtained from the formulae (5.3.8)-(5.3.12) by replacing $\pm H_m^{(2)}$, $\mp H_m^{(2)}$, $\pm H_m^{(k)}$ with $-H_m^{(2)}$, $H_m^{(2)}$, $-H_m^{(1)}$ and with $H_m^{(2)}$, $-H_m^{(2)}$, $H_{m_1}^{(1)}$ respectively.

Thus, taking (5.2.14) and (5.3.7)-(5.3.12) into account from the (5.3.6) we obtain the following contact conditions for the q-th approximation of the series representation (5.3.1):

$$\left.\left(-\frac{\partial f_{\underline{m}}}{\partial x_{1\underline{m}}^{(2)}} P_{i1}^{(1)\underline{m},n-1} - \frac{\partial f_{\underline{m}}}{\partial x_{3\underline{m}}^{(2)}} P_{i3}^{(1)\underline{m},n-1} + P_{i2}^{(1)\underline{m},n}\right)\right|_{\left(t_{1\underline{m}}, -H_{\underline{m}}^{(1)}, t_{3\underline{m}}\right)} =$$

$$\left.\left(-\frac{\partial f_{\underline{m}}}{\partial x_{1\underline{m}}^{(2)}} P_{i1}^{(2)\underline{m},n-1} - \frac{\partial f_{\underline{m}}}{\partial x_{3\underline{m}}^{(2)}} P_{i3}^{(2)\underline{m},n-1} + P_{i2}^{(2)\underline{m},n}\right)\right|_{\left(t_{1\underline{m}}, +H_{\underline{m}}^{(2)}, t_{3\underline{m}}\right)} ,$$

$$\left.\left(-\frac{\partial f_{\underline{m}}}{\partial x_{1\underline{m}}^{(2)}} P_{i1}^{(2)\underline{m},n-1} - \frac{\partial f_{\underline{m}}}{\partial x_{3\underline{m}}^{(2)}} P_{i3}^{(2)\underline{m},n-1} + P_{i2}^{(2)\underline{m},n}\right)\right|_{\left(t_{1\underline{m}}, -H_{\underline{m}}^{(2)}, t_{3\underline{m}}\right)} =$$

$$\left.\left(-\frac{\partial f_{\underline{m}}}{\partial x_{1\underline{m}}^{(2)}} P_{i1}^{(2)m_1,n-1} - \frac{\partial f_{\underline{m}}}{\partial x_{3\underline{m}}^{(2)}} P_{i3}^{(2)m_1,n-1} + P_{i2}^{(2)m_1,n}\right)\right|_{\left(t_{1\underline{m}}, +H_{m_1}^{(1)}, t_{3\underline{m}}\right)} ,$$

$$U_i^{(1)\underline{m},q}\Big|_{\left(t_{1\underline{m}}, -H_{\underline{m}}^{(1)}, t_{3\underline{m}}\right)} = U_i^{(2)\underline{m},q}\Big|_{\left(t_{1\underline{m}}, H_{\underline{m}}^{(2)}, t_{3\underline{m}}\right)},$$

$$U_i^{(2)\underline{m},q}\Big|_{\left(t_{1\underline{m}},-H_{\underline{m}}^{(2)},t_{3\underline{m}}\right)}=U_i^{(1)\underline{m_1},q}\Big|_{\left(t_{1\underline{m}},H_{\underline{m_1}}^{(1)},t_{3\underline{m}}\right)}. \tag{5.3.13}$$

It follows from (5.3.13) that the contact conditions related to the q-th approximation involve all previous approximations and are independent of the character of the mechanical relations of the components of composite materials. Note that when we neglect the terms involving $\partial/\partial x_3$ in the relations (5.3.3)-(5.3.4), (5.3.10), (5.3.11) and (5.3.12) we recover the relations obtained in section 4.1.

Hence we have a closed system of equations and corresponding contact conditions for each approximation for the stress state in a composite with arbitrary spatial curving of the reinforcing layers. In the next section we will develop a method for determining the values of each approximation for periodic curving of the reinforcing layers; we will then investigate the stress distribution in the material.

5.4. Method of Solution

In this section, we consider the determination of the values of each approximation (5.3.1) in a concrete problem: the stress-state in a composite material with co-phase periodically spatially curved layers. Thus, in this section we will develop the method proposed in section 4.2 for three-dimensional problems.

We consider a layered composite material which has an infinite number of spatially curved layers alternating in the direction of the Ox_2 axis. We suppose that these layers are periodically curved in the directions of the Ox_1 and Ox_3 axes, and that the material is subjected to loading at infinity by uniformly distributed normal forces of intensity p_1 and p_3 in the direction of the Ox_1 and Ox_3 axes, respectively. The structure of this material is shown schematically in Fig.5.4.1.

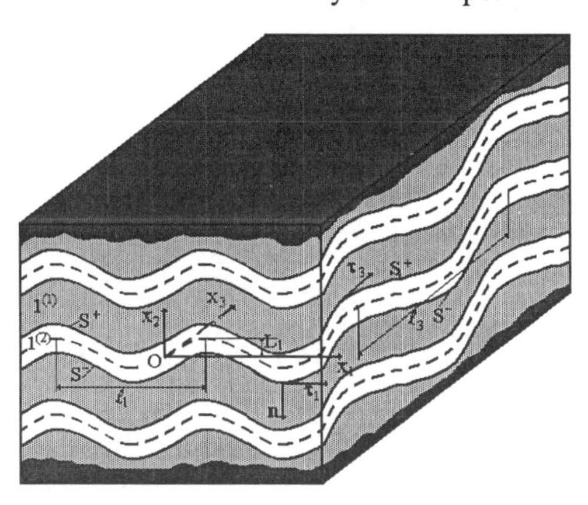

Fig.5.4.1. The structure of the composite material with co-phasically periodically spatial-curved layers.

The periodicity of the composite structure in the direction of the Ox_2 axis is $2\left(H^{(2)}+H^{(1)}\right)$, where $2H^{(1)}$ is the thickness of a matrix layer and $2H^{(2)}$ is the thickness of a filler layer. Among the layers we single out two of them, i.e. $1^{(1)}$ and $1^{(2)}$ (Fig.5.4.1), and discuss them below. The equation (5.1.3) for the middle surface of the $1^{(2)}$-th layer we take as follows:

$$x_2^{(2)} = L_1 \sin\left(\frac{2\pi}{\ell_1} x_1^{(2)}\right) \cos\left(\frac{2\pi}{\ell_3} x_3^{(2)}\right), \tag{5.4.1}$$

where L_1 is the amplitude of the curving and ℓ_1, ℓ_3 are the wavelengths of the curving in the directions Ox_1 and Ox_3, respectively. With the assumption $L_1 < \ell_1$, the value of the small parameter is selected as

$$\varepsilon = \frac{L_1}{\ell_1}. \tag{5.4.2}$$

Moreover we introduce the following notation:

$$\alpha = \frac{2\pi}{\ell_1}, \quad \beta = \frac{2\pi}{\ell_3}, \quad \gamma = \frac{\ell_1}{\ell_3}. \tag{5.4.3}$$

We will omit the index m which enters the expressions given in the previous section. Taking account the periodicity of the material structure in the direction of the Ox_2 axis, we can rewrite the contact relations (5.3.13) for this problem as follows:

$$\left(-\frac{\partial f}{\partial x_1^{(2)}} P_{i1}^{(1),n-1} - \frac{\partial f}{\partial x_3^{(2)}} P_{i3}^{(1),n-1} + P_{i2}^{(1),n}\right)\Bigg|_{(t_1,\mp H^{(1)},t_3)} =$$

$$\left(-\frac{\partial f}{\partial x_1^{(2)}} P_{i1}^{(2),n-1} - \frac{\partial f}{\partial x_3^{(2)}} P_{i3}^{(2),n-1} + P_{i2}^{(2),n}\right)\Bigg|_{(t_1,\pm H^{(2)},t_3)},$$

$$U_i^{(1),n}\Big|_{(t_1,\mp H^{(1)},t_3)} = U_i^{(2),n}\Big|_{(t_1,\pm H^{(2)},t_3)}, \quad i=1,2,3. \tag{5.4.4}$$

We consider the determination of each approximation separately.
The zeroth approximation. As noted in the subsection 4.2.1, this approximation correspond to the stress-strain state in the composite material with ideal (uncurved) layout of the layers under the action of the same external forces. Taking into account that the body is infinite in the directions of the Ox_1 and Ox_3 axes, we do not consider the effect of the edges on which the external forces act on the stress distribution.

Thus for this approximation we obtain the following contact conditions from (5.3.8), (5.3.9), (5.3.13) and (5.4.4):

$$\sigma_{i2}^{(1),0}\Big|_{(t_1,\mp H^{(1)},t_3)} = \sigma_{i2}^{(2),0}\Big|_{(t_1,\pm H^{(2)},t_3)}; \quad u_i^{(1),0}\Big|_{(t_1,\mp H^{(1)},t_3)} = u_i^{(2),0}\Big|_{(t_1,\pm H^{(2)},t_3)}. \tag{5.4.5}$$

According to the statement of the problem in the zeroth approximation we can write

$$\sigma_{12}^{(k),0}=\sigma_{22}^{(k),0}=\sigma_{13}^{(k),0}=\sigma_{32}^{(k),0}=0, \qquad k=1,2. \tag{5.4.6}$$

Further, taking (5.4.5), (5.4.6) and the relations

$$\sigma_{11}^{(1),0}\eta^{(1)}+\sigma_{11}^{(2),0}\eta^{(2)}=p_1,\ \sigma_{33}^{(1),0}\eta^{(1)}+\sigma_{33}^{(2),0}\eta^{(2)}=p_3 \tag{5.4.7}$$

into account and using Hooke's law, we find the values of the zeroth approximation as follows:

$$\sigma_{11}^{(2),0}=\left(\frac{1}{e_1}\left(p_1-\nu^{(1)}p_3\right)+\frac{e_2}{(e_1)^2}\left(p_3-\nu^{(1)}p_1\right)\right)\Bigg/\left(1-\frac{(e_2)^2}{(e_1)^2}\right),$$

$$\sigma_{11}^{(1),0}=\left(p_1-\sigma_{11}^{(2),0}\eta^{(2)}\right)\Big/\eta^{(1)},\quad \sigma_{33}^{(2),0}=\left(p_3-\nu^{(1)}p_1+e_2\sigma_{11}^{(1),0}\right)\Big/e_1,$$

$$\sigma_{33}^{(1),0}=\left(p_3-\sigma_{33}^{(2),0}\eta^{(2)}\right)\Big/\eta^{(1)},\quad u_1^{(k)}=\frac{1}{E^{(k)}}\left(\sigma_{11}^{(k),0}-\nu^{(k)}\sigma_{33}^{(k),0}\right)x_1^{(k)},$$

$$u_2^{(\underline{k}),0}=-\frac{\nu^{(\underline{k})}}{E^{(\underline{k})}}\left(\sigma_{11}^{(\underline{k}),0}+\sigma_{33}^{(\underline{k}),0}\right)x_2^{(\underline{k})}+C^{(\underline{k})},\quad C^{(k)}=\text{const},$$

$$u_3^{(\underline{k})}=\frac{1}{E^{(\underline{k})}}\left(\sigma_{33}^{(\underline{k}),0}-\nu^{(\underline{k})}\sigma_{11}^{(\underline{k}),0}\right)x_3^{(\underline{k})},\quad \eta^{(k)}=\frac{H^{(k)}}{H^{(1)}+H^{(2)}}, \tag{5.4.8}$$

where

$$e_1=\eta^{(2)}+\eta^{(1)}\frac{E^{(1)}}{E^{(2)}},\quad e_2=\nu^{(1)}\eta^{(2)}+\nu^{(2)}\eta^{(1)}\frac{E^{(1)}}{E^{(2)}}. \tag{5.4.9}$$

(5.4.6), (5.4.8) and (5.4.9) give the zeroth approximation. It follows from these expressions that the stress state corresponding to the zeroth approximation balances the external forces p_1 and p_3. Therefore the values of the first and subsequent approximations will correspond to self-balanced stresses arising as a result of the curving of the reinforcing layers.

The first approximation. For this approximation we obtain the following contact conditions from (5.3.8), (5.3.9), (5.3.13), (5.4.4), (5.4.6) and (5.4.8).

$$\sigma_{i2}^{(1),1}\Big|_{\left(t_1,\pm H^{(1)},t_3\right)}-\sigma_{i2}^{(2),1}\Big|_{\left(t_1,\pm H^{(2)},t_3\right)}=2\pi\delta_i^1\left(\sigma_{11}^{(1),0}-\sigma_{11}^{(2),0}\right)\cos(\alpha t_1)\cos(\beta t_3)+$$

$$2\pi\delta_i^3\left(\sigma_{33}^{(2),0}-\sigma_{33}^{(1),0}\right)\sin(\alpha t_1)\sin(\beta t_3),$$

$$U_i^{(1),1}\Big|_{\left(t_1,\pm H^{(1)},t_3\right)}-U_i^{(2),1}\Big|_{\left(t_1,\mp H^{(2)},t_3\right)}=\ell_1\delta_i^2\left(\frac{\nu^{(1)}}{E^{(1)}}\left(\sigma_{11}^{(1),0}+\sigma_{33}^{(1),0}\right)-\right.$$

$$\left.\frac{\nu^{(2)}}{E^{(2)}}\left(\sigma_{11}^{(2),0}+\sigma_{33}^{(2),0}\right)\right)\sin(\alpha t_1)\cos(\beta t_3). \qquad (5.4.10)$$

Since the equations (5.3.2) are satisfied for each approximation separately, the values of the first and subsequent approximations can be sought with the use of *Galerkin* representations, for which the displacements can be written as follows:

$$u_i^{(k),1}=\frac{2\nu^{(k)}-2}{2\nu^{(k)}-1}\Delta F_i^{(k),1}+\frac{1}{2\nu^{(k)}-1}\frac{\partial}{\partial x_i^{(k)}}\left(\frac{\partial F_1^{(k),1}}{\partial x_1^{(k)}}+\frac{\partial F_2^{(k),1}}{\partial x_2^{(k)}}+\frac{\partial F_3^{(k),1}}{\partial x_3^{(k)}}\right), \qquad (5.4.11)$$

where Δ is the Laplace operator, i.e.

$$\Delta=\frac{\partial^2}{\partial x_1^2}+\frac{\partial^2}{\partial x_2^2}+\frac{\partial^2}{\partial x_3^2}, \qquad (5.4.12)$$

and $F_i^{(k),1}$ (i=1,2,3) are unknown biharmonic functions, i.e.

$$\Delta\Delta F_i^{(k),1}=0 \qquad (5.4.13)$$

For the first approximation we choose $F_i^{(k),1}$ to have the form

$$F_3^{(k),1}=0;\quad F_2^{(k),1}=x_2^{(k)}\varphi_2^{(k)};\quad F_1^{(k)}=\varphi_1^{(k)}+x_2^{(k)}\psi^{(k)}. \qquad (5.4.14)$$

In (5.4.14) $\varphi_1^{(k)}$, $\varphi_2^{(k)}$ and $\psi^{(k)}$ are harmonic functions, which with the help of the contact relations (5.4.10) are found as follows:

$$\varphi_1^{(k)}=\left[A_1^{(k),1}\sinh\theta+B_1^{(k),1}\cosh\theta\right]\cos\alpha x_1^{(k)}\cos\beta x_3^{(k)},$$

$$\varphi_2^{(k)}=\left[D_1^{(k),1}\sinh\theta+C_1^{(k),1}\cosh\theta\right]\sin\alpha x_1^{(k)}\cos\beta x_3^{(k)},$$

$$\psi^{(k)}=\left[A_2^{(k),1}\sinh\theta+B_2^{(k),1}\cosh\theta\right]\cos\alpha x_1^{(k)}\cos\beta x_3^{(k)}, \qquad (5.4.15)$$

where

$$\theta = \alpha\left(1+\gamma^2\right)^{\frac{1}{2}} x_2^{(k)}. \tag{5.4.16}$$

As in the previous chapter, using the relations (5.4.11)-(5.4.15) we determine the unknown constants $A_1^{(k),1}$, $B_1^{(k),1}$, $D_1^{(k),1}$, $C_1^{(k),1}$, $A_2^{(k),1}$, $B_2^{(k),1}$ from the contact conditions (5.4.10). In this way we find the values of the first approximation which can be presented as follows:

$$u_i^{(k),1} = \varphi_{i1}^{(k),1}(x_2^{(k)})\left(\delta_i^1 \cos\alpha x_1 \cos\beta x_3 + \delta_i^2 \sin\alpha x_1 \cos\beta x_3 + \delta_i^3 \sin\alpha x_1 \sin\beta x_3\right),$$

$$\sigma_{ij}^{(k),1} = \psi_{ij}^{(k),1}(x_2^{(k)})\left(\delta_i^j \sin\alpha x_1 \cos\beta x_3 + \delta_i^1\delta_j^2 \cos\alpha x_1 \cos\beta x_3 + \right.$$

$$\left.\delta_i^1\delta_j^3 \cos\alpha x_1 \sin\beta x_3 + \delta_i^2\delta_j^3 \sin\alpha x_1 \sin\beta x_3\right). \tag{5.4.17}$$

In (5.4.17) δ_i^j are Kronecker symbols and in writing (5.4.16) we have taken into account that $x_1^{(1)} = x_1^{(2)} = x_1$, $x_3^{(1)} = x_3^{(2)} = x_3$. Note that the expressions for the functions $\varphi_{i1}^{(k),1}\left(x_2^{(k)}\right)$ and $\psi_{ij}^{(k),1}\left(x_2^{(k)}\right)$ can easily be determined from relations (5.4.16), (5.4.11) and (5.3.2).

The second approximation. Taking (5.4.17), (5.4.16) and (5.4.8) we obtain from (5.4.4) the following contact conditions for this approximation.

$$\sigma_{12}^{(1),2}\Big|_{\left(t_1,\mp H^{(1)},t_3\right)} - \sigma_{12}^{(2),2}\Big|_{\left(t_1,\pm H^{(2)},t_3\right)} = \left(Q_{112}^{(1)} - Q_{112}^{(2)}\right)\sin 2\alpha t_1 +$$

$$\left(Q_{122}^{(1)} - Q_{122}^{(2)}\right)\sin 2\alpha t_1 \cos 2\beta t_3,$$

$$\sigma_{22}^{(1),2}\Big|_{\left(t_1,\mp H^{(1)},t_3\right)} - \sigma_{22}^{(2),2}\Big|_{\left(t_1,\pm H^{(2)},t_3\right)} = \left(Q_{212}^{(1)} - Q_{212}^{(2)}\right)\cos 2\alpha t_1 +$$

$$\left(Q_{222}^{(1)} - Q_{222}^{(2)}\right)\cos 2\beta t_3 + \left(Q_{224}^{(1)} - Q_{224}^{(2)}\right)\cos 2\alpha t_1 \cos 2\beta t_3,$$

$$\sigma_{23}^{(1),2}\Big|_{\left(t_1,\mp H^{(1)},t_3\right)} - \sigma_{23}^{(2),2}\Big|_{\left(t_1,\pm H^{(2)},t_3\right)} = \left(Q_{312}^{(1)} - Q_{312}^{(2)}\right)\sin 2\beta t_3 +$$

$$\left(Q_{322}^{(1)} - Q_{322}^{(2)}\right) \cos 2\alpha t_1 \sin 2\beta t_3 ,$$

$$u_1^{(1),2}\Big|_{\left(t_1,\mp H^{(1)},t_3\right)} - u_1^{(2),2}\Big|_{\left(t_1,\pm H^{(2)},t_3\right)} = \left(R_{112}^{(1)} - R_{112}^{(2)}\right) \sin 2\alpha t_1 +$$

$$\left(R_{122}^{(1)} - R_{122}^{(2)}\right) \sin 2\alpha t_1 \cos 2\beta t_3 ,$$

$$u_2^{(1),2}\Big|_{\left(t_1,\mp H^{(1)},t_3\right)} - u_2^{(2),2}\Big|_{\left(t_1,\pm H^{(2)},t_3\right)} = \left(R_{210}^{(1)} - R_{210}^{(2)}\right) + \left(R_{212}^{(1)} - R_{212}^{(2)}\right) \cos 2\alpha t_1$$

$$+ \left(R_{222}^{(1)} - R_{222}^{(2)}\right) \cos 2\beta t_3 + \left(R_{224}^{(1)} - R_{224}^{(2)}\right) \cos 2\alpha t_1 \cos 2\beta t_3 ,$$

$$u_3^{(1),2}\Big|_{\left(t_1,\mp H^{(1)},t_3\right)} - u_3^{(2),2}\Big|_{\left(t_1,\pm H^{(2)},t_3\right)} = \left(R_{312}^{(1)} - R_{312}^{(2)}\right) \sin 2\beta t_3 +$$

$$\left(R_{322}^{(1)} - R_{322}^{(2)}\right) \cos 2\alpha t_1 \sin 2\beta t_3 . \qquad (5.4.18)$$

In (5.4.18) the following notation has been used.

$$Q_{112}^{(k)} = A\left(1-\gamma^2\right) \psi_{12} + B\psi_{12}' + \frac{\pi}{2}\psi_{11}, \quad Q_{122}^{(k)} = A\left(1+\gamma^2\right)\psi_{12} + B\psi_{12}' +$$

$$\frac{\pi}{2}\left(\psi_{11} - \psi_{13}\right), \quad Q_{212}^{(k)} = A\left(1-\gamma^2\right) \psi_{22} - B\psi_{22}' + \frac{\pi}{2}\left(\psi_{12} + \gamma\psi_{23}\right),$$

$$Q_{222}^{(k)} = A\left(1-\gamma^2\right)\psi_{22} + B\psi_{22}' + \frac{\pi}{2}\left(\psi_{12} + \gamma\psi_{23}\right),$$

$$Q_{224}^{(k)} = A\left(1+\gamma^2\right)\psi_{22} - B\psi_{22}' + \frac{\pi}{2}\left(\psi_{12} - \gamma\psi_{23}\right),$$

$$Q_{312}^{(k)} = A\left(1-\gamma^2\right) \psi_{23} + B\psi_{23}' + \frac{\pi}{2}\left(\psi_{13} - \gamma\psi_{33}\right), \quad Q_{322}^{(k)} = A\left(1+\gamma^2\right)\psi_{23} - B\psi_{23}' +$$

$$\frac{\pi}{2}\left(\psi_{13} + \gamma\psi_{33}\right), \quad R_{112}^{(k)} = A\left(1-\gamma^2\right) \varphi_{11} + B\varphi_{11}', \quad R_{122}^{(k)} = A\left(1+\gamma^2\right) \varphi_{11} + B\varphi_{11}',$$

$$R_{210}^{(k)} = A\left(1+\gamma^2\right) \varphi_{21} + B\varphi_{21}', \quad R_{212}^{(k)} = A\left(1-\gamma^2\right) \varphi_{21} - B\varphi_{21}',$$

$$R_{222}^{(k)} = A\left(1-\gamma^2\right)\varphi_{21} + B\varphi'_{21}, \quad R_{224}^{(k)} = A\left(1+\gamma^2\right)\varphi_{21} - B\varphi'_{21},$$

$$R_{312}^{(k)} = A\left(1-\gamma^2\right)\varphi_{31} + B\varphi'_{31}, \quad R_{322}^{(k)} = A\left(1+\gamma^2\right)\varphi_{31} - B\varphi'_{31}, \tag{5.4.19}$$

where

$$A = (-1)^{k+1}\alpha H^{(2)}\,\pi/2, \quad B = \left(1+\gamma^2\right)^{\frac{1}{2}}\pi/2, \quad \psi'_{ij}(x) = d\psi_{ij}/dx, \quad \varphi'_{ij}(x) = d\varphi_{ij}/dx$$

$$h^{(k)} = (-1)^k H^{(k)}, \quad \psi_{ij} \equiv \psi_{ij}^{(k),1}\left(h^{(k)}\right), \quad \varphi_{ij} \equiv \varphi_{ij}^{(k),1}\left(h^{(k)}\right) \tag{5.4.20}$$

As in the first approximation, we use the Galerkin representation and select the functions $F_i^{(k),2}$ as follows:

$$F_1^{(k),2} = x_2^{(k)}\psi_1^{(k)}\left(x_1^{(k)}, x_2^{(k)}\right) + \varphi_1^{(k)}\left(x_1^{(k)}, x_2^{(k)}, x_3^{(k)}\right) + x_2^{(k)}\psi^{(k)}\left(x_1^{(k)}, x_2^{(k)}, x_3^{(k)}\right),$$

$$F_2^{(k),2} = x_2^{(k)}\varphi_{21}^{(k)}\left(x_1^{(k)}, x_2^{(k)}\right) + x_2^{(k)}\varphi_2^{(k)}\left(x_2^{(k)}, x_3^{(k)}\right) + C^{(k)}\left(x_2^{(k)}\right)^2 +$$

$$x_2^{(k)}\varphi_{22}^{(k)}\left(x_1^{(k)}, x_2^{(k)}, x_3^{(k)}\right), \quad F_3^{(k),2} = x_2^{(k)}\varphi_3^{(k)}\left(x_2^{(k)}, x_3^{(k)}\right), \tag{5.4.21}$$

where the functions $\psi_1^{(k)}, \varphi_1^{(k)}, \psi^{(k)}, \varphi_{21}^{(k)}, \varphi_2^{(k)}, \varphi_{22}^{(k)}, \varphi_3^{(k)}$ are unknown harmonic functions, which by considering the contact conditions (5.4.18) are determined in the following form.

$$\varphi_1^{(k)} = \left[A_1^{(k),2}\sinh\vartheta + B_1^{(k),2}\cosh\vartheta\right]\sin 2\alpha x_1 \cos 2\beta x_3,$$

$$\varphi_{22}^{(k)} = \left[D_1^{(k),2}\sinh\vartheta + C_1^{(k),2}\cosh\vartheta\right]\cos 2\alpha x_1 \cos 2\beta x_3,$$

$$\psi^{(k)} = \left[A_2^{(k),2}\sinh\vartheta + B_2^{(k),2}\cosh\vartheta\right]\sin 2\alpha x_1 \cos 2\beta x_3,$$

$$\varphi_2^{(k)} = \left[A_{21}^{(k),2}\sinh\vartheta_1 + B_{21}^{(k),2}\cosh\vartheta_1\right]\cos 2\beta x_3,$$

$$\varphi_3^{(k)} = \left[A_{31}^{(k),2}\sinh\vartheta_1 + B_{31}^{(k),2}\cosh\vartheta_1\right]\sin 2\beta x_3,$$

$$\varphi_{21}^{(k)} = \left[A_{41}^{(k),2}\sinh\vartheta_2 + B_{41}^{(k),2}\cosh\vartheta_2\right]\cos 2\alpha x_1,$$

$$\psi_1^{(k)} = \left[A_{51}^{(k),2} \sinh \vartheta_2 + B_{51}^{(k),2} \cosh \vartheta_2 \right] \sin 2\alpha x_1, \tag{5.4.22}$$

where

$$\vartheta = 2\alpha\left(1+\gamma^2\right)^{\frac{1}{2}} x_2^{(k)}, \quad \vartheta_1 = 2\alpha\gamma\, x_2^{(k)}, \quad \vartheta_2 = 2\alpha x_2^{(k)}. \tag{5.4.23}$$

Using the relations (5.4.21), (5.4.11) from (5.4.18) we find the unknown constants entering (5.4.22). Proceeding in this way we can determine the values for each subsequent approximation (5.3.1). We can investigate the stress state in a composite with more complicated spatial periodical curving of the reinforcing layers in a similar way.

5.5. Stress Distribution

In the present section we will analyse the numerical results. Uniaxial and biaxial loading will be considered separately. All analyses will be carried out for composites with co-phase and anti-phase curved layers. In these cases we will consider only the self-balanced stresses.

5.5.1. UNIAXIAL LOADING

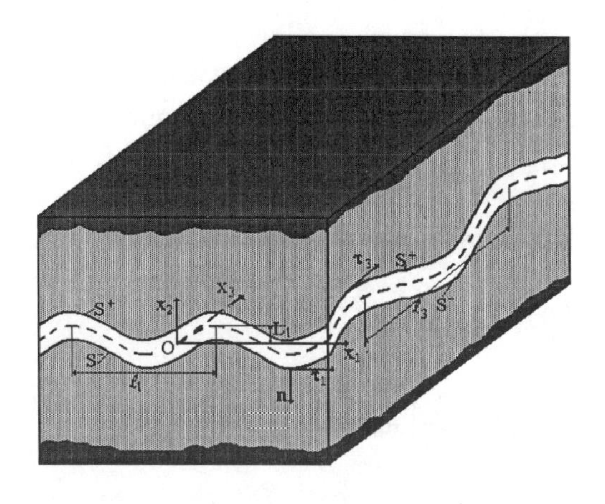

Fig.5.5.1. An infinite elastic body reinforced by a single periodically spatial curved layer of a filler.

Co-phase curving.
We investigate the distribution of the self-balanced stresses σ_{nn}, $\sigma_{n\tau_1}$, $\sigma_{n\tau_3}$ on the inter-layer surface S^+ (Fig.5.4.1) when $p_1 > 0$, $p_3 = 0$. Here σ_{nn} is the normal stress on the surface S^+ acting along the normal vector $\mathbf{n}$, while $\sigma_{n\tau_1}$ and $\sigma_{n\tau_3}$ are the tangential stresses on this surface acting along the tangent vectors $\boldsymbol{\tau}_1$ and $\boldsymbol{\tau}_3$ respectively (Fig.5.4.1). Because the curving is periodic, we can limit ourselves to the region $0 \le \alpha t_1 \le \pi/2$; $0 \le \beta t_3 \le \pi/2$. As in the previous chapter we introduce the parameter $\chi = 2\pi H^{(2)}/\ell_1$ and call it the wave generation parameter of the curving. Here $2H^{(2)}$ is the thickness of the filler layer, ℓ_1 is the wavelength of the curvature of the middle surface of the filler layer in the direction of the Ox_1 axis. We assume Poisson's ratio $\nu^{(1)} = \nu^{(2)} = 0.3$.

We first consider the case where $H^{(1)} >> H^{(2)}$ and therefore the interactions curved filler layers do not have to be taken into account, i.e. we consider the stress distribution in an infinite elastic body reinforced by a single curved layer of filler (Fig.5.5.1).

Thus we consider the graphs given in Figs.5.5.2 and 5.5.3 which show the distributions of the $\sigma_{nn}/\sigma_{11}^{(1),0}$, $\sigma_{n\tau_1}/\sigma_{11}^{(1),0}$ and $\sigma_{n\tau_3}/\sigma_{11}^{(1),0}$ with respect to αt_1 and βt_3 for various values of the $\gamma = \ell_1/\ell_3$. In these figures the graphs of the dependence of $\sigma_{n\tau_1}/\sigma_{11}^{(1),0}$, $\sigma_{n\tau_3}/\sigma_{11}^{(1),0}$ on αt_1 also accurately reflect the dependence between

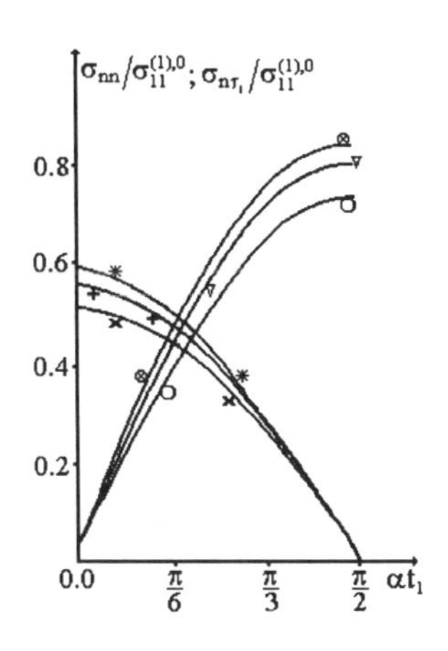

Fig.5.5.2. The distribution of the stresses σ_{nn}, $\sigma_{n\tau_1}$ with respect to αt_1 for $\beta t_3 = 0$, $E^{(2)}/E^{(1)} = 50$, $\chi = 0.3$, $\varepsilon = 0.02$, $\gamma = 0.1; 0.3; 0.5$: ⊗,∇,O -- $\sigma_{nn}/\sigma_{11}^{(1),0}$; ✳,✚,✖ -- $\sigma_{n\tau_1}/\sigma_{11}^{(1),0}$.

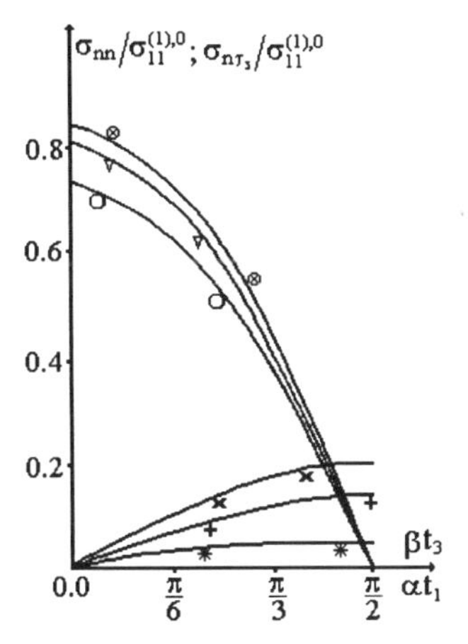

Fig.5.5.3. The distribution of the stresses σ_{nn} and $\sigma_{n\tau_3}$ with respect to βt_3 and αt_1, respectively, for $E^{(2)}/E^{(1)} = 50$, $\chi = 0.3$, $\varepsilon = 0.02$, $\gamma = 0.1; 0.3; 0.5$: ⊗,∇,O -- $\sigma_{nn}/\sigma_{11}^{(1),0}$ for $\alpha t_1 = \pi/2$;✳,✚,✖ -- $\sigma_{n\tau_3}/\sigma_{11}^{(1),0}$ for $\beta t_3 = \pi/2$.

$\sigma_{n\tau_1}/\sigma_{11}^{(1),0}$ (under $\alpha t_1 = 0$), $\sigma_{n\tau_3}/\sigma_{11}^{(1),0}$ (under $\alpha t_1 = \pi/2$) and βt_3, for the selected values of the problem parameters. These graphs show that the normal stress σ_{nn} reaches its maximum at points of the surface S^+ corresponding to $\{\alpha t_1 = \pi/2; \beta t_3 = 0\}$, while the tangential stress $\sigma_{n\tau_1}$ ($\sigma_{n\tau_3}$) reaches its maximum values at points of the surface S^+ corresponding to $\{\alpha t_1 = 0; \beta t_3 = 0\}$ ($\{\alpha t_1 = \pi/2; \beta t_3 = \pi/2\}$). Moreover these graphs show that, as γ increases σ_{nn} and $\sigma_{n\tau_1}$ decrease while $\sigma_{n\tau_3}$ increases. With decreasing γ the values of the stresses σ_{nn} and $\sigma_{n\tau_1}$ approach the corresponding values obtained in the previous chapter for plain-strain.

We now consider the distribution of these stresses when the interaction between the filler layers is taken into account. Graphs corresponding to these distributions are shown in Figs.5.5.4.-5.5.7.

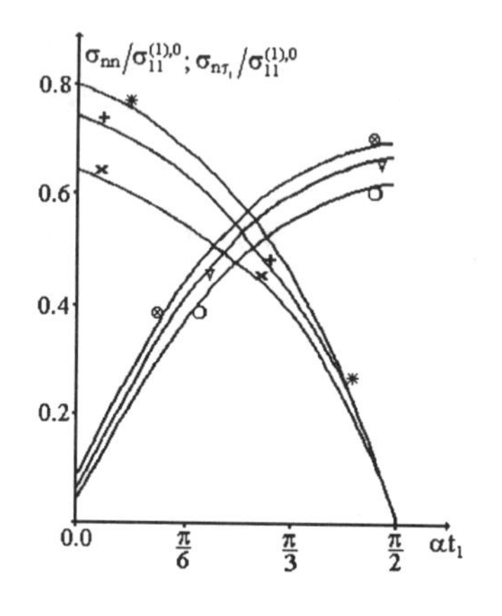

Fig.5.5.4. The distribution of the stresses σ_{nn}, $\sigma_{n\tau_1}$ with respect to αt_1 for $\beta t_3 = 0$, $\eta^{(2)} = 0.2$, $E^{(2)}/E^{(1)} = 50$, $\chi = 0.3$, $\varepsilon = 0.02$, $\gamma = 0.1; 0.3; 0.5$:
$\otimes, \nabla, \bigcirc$ -- $\sigma_{nn}/\sigma_{11}^{(1),0}$;
$*, +, \times$ -- $\sigma_{n\tau_1}/\sigma_{11}^{(1),0}$.

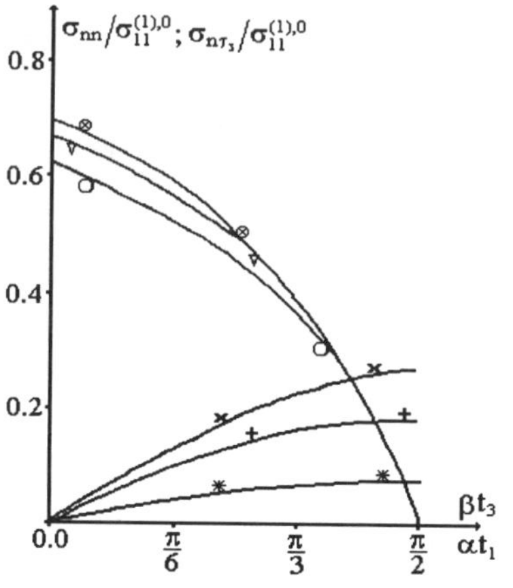

Fig.5.5.5. The distribution of the stress σ_{nn} and $\sigma_{n\tau_3}$ with respect to βt_3 and αt_1, respectively, for $\eta^{(2)} = 0.2$, $E^{(2)}/E^{(1)} = 50$, $\chi = 0.3$, $\varepsilon = 0.02$, $\gamma = 0.1; 0.3; 0.5$:
$\otimes, \nabla, \bigcirc$ -- $\sigma_{nn}/\sigma_{11}^{(1),0}$ for $\alpha t_1 = \pi/2$;
$*, +, \times$ -- $\sigma_{n\tau_3}/\sigma_{11}^{(1),0}$ for $\beta t_3 = \pi/2$.

These graphs show that the tangential self-balanced stresses $\sigma_{n\tau_1}/\sigma_{11}^{(1),0}$ and $\sigma_{n\tau_3}/\sigma_{11}^{(1),0}$ increase significantly as $\eta^{(2)}$ increases, while the self-balanced normal stress $\sigma_{nn}/\sigma_{11}^{(1),0}$ decreases. The inclusion of the interaction between the curved layers significantly changes the normal stress σ_{nn} distribution on the layers interface. Within the interval of γ considered here, the stresses σ_{nn}, $\sigma_{n\tau_1}$ decrease with increasing γ, while the stress $\sigma_{n\tau_3}$, which appears because of the curving along the Ox_3 axis, increases.

Consider the effect of varying the parameter χ on the stresses. The results show that when $\eta^{(2)} = 0$ the dependence of the stresses σ_{nn}, $\sigma_{n\tau_1}$, $\sigma_{n\tau_3}$ on χ is non-monotonic as in subsection 4.3.1. Hence there is a value of χ for which these stresses reach maximum values. However, the dependence between $\sigma_{n\tau_1}/\sigma_{11}^{(1),0}$, $\sigma_{n\tau_3}/\sigma_{11}^{(1),0}$ and χ becomes monotonic as the concentration of the filler layers $\eta^{(2)}$

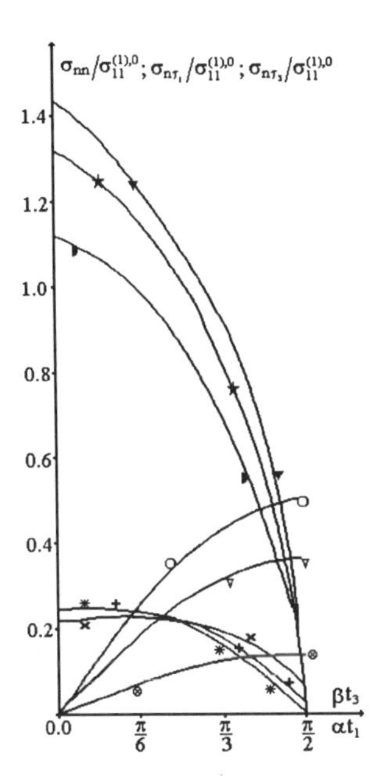

Fig.5.5.6. The relations between the $\sigma_{nn}/\sigma_{11}^{(1),0}$ and βt_3 at $\alpha t_1 = \pi/2$; $\sigma_{n\tau_1}/\sigma_{11}^{(1),0}$ and αt_1 at $\beta t_3 = 0$; $\sigma_{n\tau_3}/\sigma_{11}^{(1),0}$ and αt_1 at $\beta t_3 = \pi/2$; for $E^{(2)}/E^{(1)} = 50$, $\chi = 0.3$, $\varepsilon = 0.02$, $\eta^{(2)} = 0.5$ with $\gamma = 0.1; 0.3; 0.5$: ✳, ✚, ✖ -- $\sigma_{nn}/\sigma_{11}^{(1),0}$; ▼, ★, ◗ -- $\sigma_{n\tau_1}/\sigma_{11}^{(1),0}$; ⊗, ∇, ○ -- $\sigma_{n\tau_3}/\sigma_{11}^{(1),0}$.

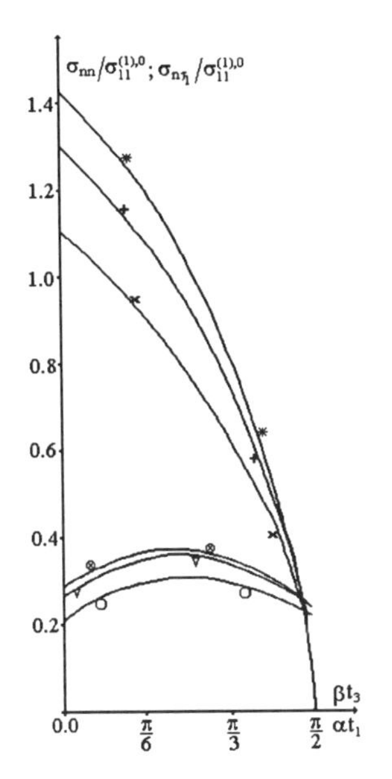

Fig.5.5.7. The relations between the $\sigma_{nn}/\sigma_{11}^{(1),0}$ and βt_1 at $\alpha t_3 = 0$; $\sigma_{n\tau_1}/\sigma_{11}^{(1),0}$ and αt_3 at $\beta t_1 = 0$; for $E^{(2)}/E^{(1)} = 50$, $\chi = 0.3$, $\varepsilon = 0.02$, $\eta^{(2)} = 0.5$ with $\gamma = 0.1; 0.3; 0.5$: ⊗, ∇, ○ -- $\sigma_{nn}/\sigma_{11}^{(1),0}$; ✳, ✚, ✖ -- $\sigma_{n\tau_1}/\sigma_{11}^{(1),0}$;

increases. In these cases the values of $\sigma_{n\tau_1}/\sigma_{11}^{(1),0}$ and $\sigma_{n\tau_3}/\sigma_{11}^{(1),0}$ increase monotonically as the parameter χ decreases and the dependence of $\sigma_{nn}/\sigma_{11}^{(1),0}$ on χ does not change with increasing $\eta^{(2)}$. We observe that when $\eta^{(2)} \geq 0.2$ the stresses $\sigma_{n\tau_1}$ and $\sigma_{n\tau_3}$ are much larger than the normal stress σ_{nn}. All of the stresses attain their most significant values when $\chi \leq 0.4$. Note that part of these conclusions are also proven by the data given in Table 5.1 in which the values $\sigma_{n\tau_1}/\sigma_{11}^{(1),0}$ and $\sigma_{nn}/\sigma_{11}^{(1),0}$ are given for different values of $\eta^{(2)}$, χ and γ.

The data of Table 5.1 show that for all values of χ and $\eta^{(2)}$ an increase in

Table 5.1. The values of $\sigma_{n\tau_1}/\sigma_{11}^{(1),0}$ (the upper rows) and $\sigma_{nn}/\sigma_{11}^{(1),0}$ (the lower rows) for $E^{(2)}/E^{(1)} = 50$, $\varepsilon = 0.02$

$\eta^{(2)}$	γ	χ			
		0.1	0.2	0.3	0.4
0.2	0.1	−1.119 / 0.310	−0.955 / 0.606	−0.794 / 0.694	−0.669 / 0.651
	0.3	−1.052 / 0.309	−0.891 / 0.595	−0.738 / 0.669	−0.621 / 0.615
	0.5	−0.936 / 0.307	−0.783 / 0.572	−0.643 / 0.619	−0.542 / 0.550
0.5	0.1	−2.504 / −0.092	−1.928 / 0.125	−1.449 / 0.243	−1.105 / 0.285
	0.3	−2.330 / −0.093	−1.782 / 0.123	−1.329 / 0.236	−1.008 / 0.274
	0.5	−2.046 / −0.088	−1.535 / 0.122	−1.128 / 0.223	−0.852 / 0.252

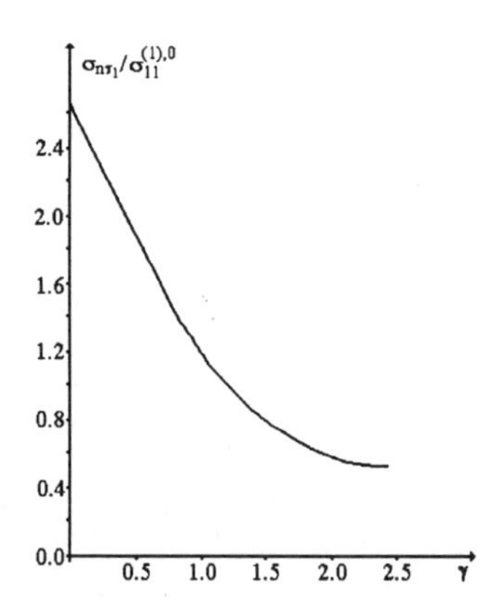

Fig.5.5.8. The relations between the $\sigma_{n\tau_1}/\sigma_{11}^{(1),0}$ (at $\{\alpha t_1 = 0; \beta t_3 = 0\}$) and γ for $E^{(2)}/E^{(1)} = 50$, $\chi = 0.1$, $\varepsilon = 0.02$, $\eta^{(2)} = 0.5$.

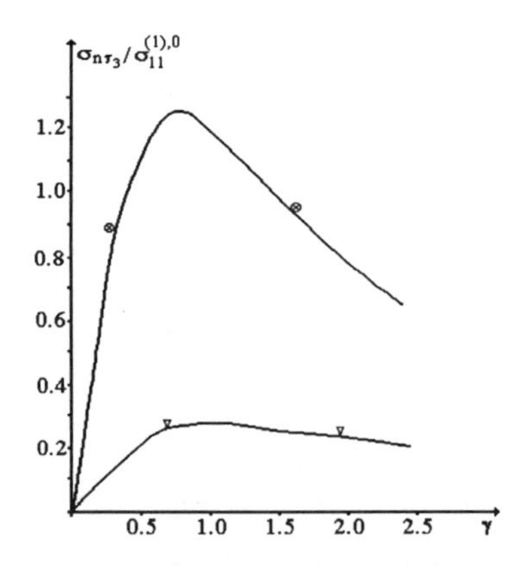

Fig.5.5.9. The relations between the $\sigma_{n\tau_3}/\sigma_{11}^{(1),0}$ (at $\{\alpha t_1 = \pi/2; \beta t_3 = \pi/2\}$) and γ for $E^{(2)}/E^{(1)} = 50$, $\chi = 0.1$, $\varepsilon = 0.02$: ∇ -- $\eta^{(2)} = 0.1$; $\otimes$ -- $\eta^{(2)} = 0.5$.

γ leads to a decrease in the self-balanced stresses σ_{nn} and $\sigma_{n\tau_1}$. However, at higher concentrations of the filler (for example, $\eta^{(2)} = 0.5$) and $\chi \leq 0.3$, the self-balanced

Table 5.2. The values of $\sigma_{n\tau_1}/\sigma_{11}^{(1),0}$ (upper rows) and $\sigma_{n\tau_3}/\sigma_{11}^{(1),0}$ (lower rows) for χ =0.1.

ε	$\eta^{(2)}$	$E^{(2)}/E^{(1)}$	γ = 0.5	1.0	1.5	2.0	2.5
0.010	0.2	20	−0.247	−0.182	−0.141	−0.119	−0.107
			0.086	0.105	0.094	0.078	0.064
		50	−0.500	−0.330	−0.222	−0.164	−0.134
			0.219	0.260	0.211	0.161	0.134
		100	−0.854	−0.522	−0.315	−0.210	−0.158
			0.413	0.470	0.378	0.278	0.201
	0.5	20	−0.514	−0.357	−0.257	−0.202	−0.172
			0.202	0.244	0.212	0.171	0.135
		50	−1.098	−0.688	−0.430	−0.295	−0.224
			0.513	0.593	0.487	0.366	0.270
		100	−1.855	−1.079	−0.608	−0.377	−0.265
			0.940	1.031	0.790	0.555	0.385
0.015	0.2	20	−0.364	−0.269	−0.209	−0.177	−0.160
			0.129	0.159	0.142	0.119	0.099
		50	−0.725	−0.480	−0.324	−0.242	−0.200
			0.330	0.393	0.336	0.267	0.211
		100	−1.219	−0.749	−0.454	−0.307	−0.235
			0.627	0.716	0.580	0.435	0.326
	0.5	20	−0.753	−0.526	−0.380	−0.300	−0.257
			0.304	0.367	0.321	0.259	0.208
		50	−1.580	−0.995	−0.623	−0.431	−0.333
			0.776	0.899	0.742	0.566	0.429
		100	−1.640	−1.539	−0.869	−0.547	−0.393
			1.447	1.589	1.229	0.883	0.634
0.025	0.2	20	−0.582	−0.434	−0.340	−0.293	−0.271
			0.219	0.270	0.245	0.214	0.195
		50	−1.155	−0.768	−0.527	−0.408	−0.354
			0.572	0.687	0.607	0.524	0.480
		100	−2.077	−1.250	−0.780	−0.561	−0.456
			1.158	1.347	1.153	0.982	0.878
	0.5	20	−1.204	−0.845	−0.614	−0.493	−0.434
			0.517	0.628	0.558	0.475	0.423
		50	−2.598	−1.614	−1.023	−0.738	−0.601
			1.387	1.628	1.398	1.181	1.047
		100	−5.083	−2.762	−1.593	−1.063	−0.799
			2.915	3.238	2.659	2.178	1.844

tangential stress $\sigma_{n\tau_1}$ is several times greater than the normal stress $\sigma_{11}^{(1),0}$, balanced by the external forces p_1, even at $\gamma = 0.5$, which is the largest value of γ considered here.

Moreover, the results show that in all cases where $0.1 \le \gamma \le 0.5$, an increase in γ leads to a decrease in σ_{nn} and $\sigma_{n\tau_1}$, and to an increase in $\sigma_{n\tau_3}$. It follows from the other numerical results obtained for $\gamma > 0.5$, with further increasing γ the values of the σ_{nn} and $\sigma_{n\tau_1}$ always decrease. However, $\sigma_{n\tau_3}$ increases with increasing γ up to a certain value $\gamma = \gamma'$ and then decreases with further increase of γ. We see from graphs shown in Figs. 5.5.8 and 5.5.9 confirming the given results that $\gamma' \approx 1.0$. Hence the self-balanced tangential stress $\sigma_{n\tau_3}$ reaches its maxima when the wavelength of the distortion along the Ox_3 axis (ℓ_3) is equal to that along the Ox_1 axis (ℓ_1).

In Table 5.2 with various problem parameters and for $\chi = 0.1$, the values of $\sigma_{n\tau_1} / \sigma_{11}^{(1),0}$ (at $\alpha t_1 = 0; \beta t_3 = 0$) and $\sigma_{n\tau_3} / \sigma_{11}^{(1),0}$ (at $\alpha t_1 = \pi/2; \beta t_3 = \pi/2$) are given. These results show that $\sigma_{n\tau_1} / \sigma_{11}^{(1),0}$ and $\sigma_{n\tau_3} / \sigma_{11}^{(1),0}$ increase monotonically, as $\eta^{(2)}$, $E^{(2)} / E^{(1)}$ and ε increase. Further, these results show that in many cases $\sigma_{n\tau_3}$ is much larger than $\sigma_{n\tau_1}$. Note that all numerical results analysed here have been obtained in the framework of the first five approximations (5.3.1). For illustration of the convergence of these results in Table 5.3 the values of $\sigma_{n\tau_3} / \sigma_{11}^{(1),0}$ (at $\alpha t_1 = \pi/2; \beta t_3 = \pi/2$) obtained for various approximations are given; the convergence is acceptable.

Table 5.3. The values of $\sigma_{n\tau_3} / \sigma_{11}^{(1),0}$ in various approximations

for $\eta^{(2)} = 0.5$, $\gamma = 1.0$, $\chi = 0.1$, $E^{(2)} / E^{(1)} = 50$

ε	Number of approximations (5.3.1)		
	2	4	5
0.010	0.592	0.569	0.593
0.020	1.184	1.001	1.233
0.025	1.480	1.122	1.628

Anti-phase curving.

We now analyse the stress distribution in a composite with anti-phase spatially periodic curved layers in the case when $p_3 = 0$. The structure of this composite is shown in Fig. 5.5.10. The period in the direction of the Ox_2 axis is $4\left(H^{(2)} + H^{(1)}\right)$; among the layers, we single out four of them, i.e. $1^{(1)}, 1^{(2)}$, $2^{(1)}$, $2^{(2)}$ (Fig.5.5.10), and discuss them below. The equation (5.1.3) for the middle surface of the $1^{(2)}$-th layer we take as

$$x_{21}^{(2)} = L_1 \sin\left(\frac{2\pi}{\ell_1} x_1\right) \cos\left(\frac{2\pi}{\ell_3} x_3\right), \tag{5.5.1}$$

but the equation for the middle surface of the $2^{(2)}$-th layer we take as

$$x_{22}^{(2)} = -L_1 \sin\left(\frac{2\pi}{\ell_1} x_1\right)\cos\left(\frac{2\pi}{\ell_3} x_3\right). \tag{5.5.2}$$

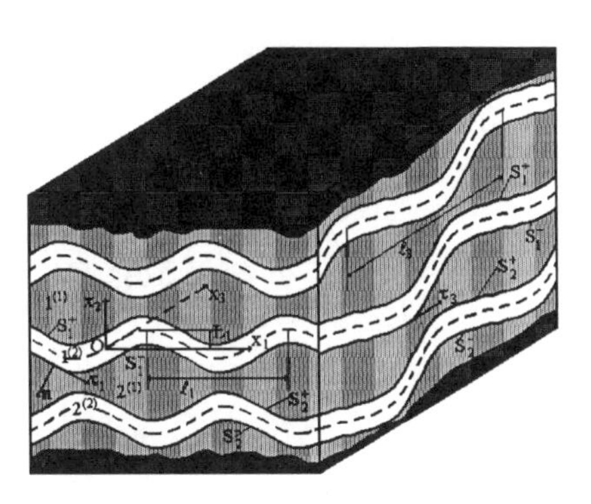

Fig.5.5.10. The structure of the composite material with anti-phase spatial periodically-curved layers.

In (5.5.1) and (5.5.2) we have taken into account that $x_{11}^{(2)} = x_{12}^{(2)} = x_1$, $x_{31}^{(2)} = x_{32}^{(2)} = x_3$. As in subsection 4.3.3 we suppose that neighbouring reinforcing layers do not make contact with each other, i.e. $L_1 < H^{(1)}$. We also assume that $L_1 < \ell_1$ and take $\varepsilon = L_1/\ell_1$ as small parameter.

Note that the values of each approximation (5.3.1) for this problem are determined as in section 5.4 for the composite with co-phase periodically curved layers, with appropriate changes. We merely analyse the numerical results. In this case we will consider only the distribution of the self-balanced normal stress σ_{nn} at the particular points of the surface S_1^+, because the values of the stresses $\sigma_{n\tau_1}$, $\sigma_{n\tau_3}$ are very small

Table 5.4. The values of $\sigma_{nn}/\sigma_{11}^{(1),0}$ for $\eta^{(2)} = 0.5$ and large values of γ.

χ	βt_3	αt_1					
		0	$\pi/4$	$\pi/2$	0	$\pi/4$	$\pi/2$
		$\gamma = 1.5$			$\gamma = 1.1$		
0.1	0	-0.048	0.404	0.646	-0.044	0.412	0.649
	$\pi/4$	-0.019	0.290	0.419	-0.018	0.297	0.437
	$\pi/2$	0.010	-0.014	-0.039	0.006	-0.007	-0.022
0.2	0	-0.091	0.603	0.995	-0.100	0.697	1.136
	$\pi/4$	-0.036	0.411	0.603	-0.043	0.491	0.733
	$\pi/2$	0.018	-0.036	-0.092	0.013	-0.024	-0.062
0.3	0	-0.067	0.450	0.736	-0.107	0.676	1.108
	$\pi/4$	-0.027	0.303	0.445	-0.047	0.469	0.705
	$\pi/2$	-0.012	-0.030	-0.072	0.012	-0.029	-0.071
0.4	0	-0.026	0.255	0.401	-0.067	0.463	0.747
	$\pi/4$	-0.010	0.174	0.251	-0.030	0.321	0.480
	$\pi/2$	0.004	-0.013	-0.030	0.006	-0.019	-0.045

Table 5.5. The values of $\sigma_{nn}/\sigma_{11}^{(1),0}$ for $\eta^{(2)}=0.5$ and small values of γ.

χ	βt_3	αt_1					
		0	$\pi/4$	$\pi/2$	0	$\pi/4$	$\pi/2$
		$\gamma=0.5$			$\gamma=0.1$		
0.1	0	-0.038	0.418	0.649	-0.036	0.420	0.648
	$\pi/4$	-0.018	0.301	0.449	-0.018	0.301	0.451
	$\pi/2$	0.001	-0.001	-0.004	0.000	0.000	0.000
0.2	0	-0.103	0.766	2.227	-0.102	0.781	1.242
	$\pi/4$	-0.049	0.556	0.845	-0.051	0.569	0.867
	$\pi/2$	0.003	-0.005	-0.015	0.000	0.000	0.000
0.3	0	-0.153	0.947	1.541	-0.163	1.010	1.636
	$\pi/4$	-0.074	0.686	1.053	-0.081	0.744	1.148
	$\pi/2$	0.004	-0.010	-0.024	0.000	0.000	-0.001
0.4	0	-0.153	0.882	1.438	-0.185	1.023	1.670
	$\pi/4$	-0.074	0.636	0.979	-0.092	0.759	1.176
	$\pi/2$	0.003	-0.010	-0.025	0.000	0.000	-0.001

compared to σ_{nn}. Moreover, we will assume that $E^{(2)}/E^{(1)}=50$, $\nu^{(1)}=\nu^{(2)}=0.3$, $\varepsilon=0.02$

Table 5.6. The values of $\sigma_{nn}/\sigma_{11}^{(1),0}$ for $\eta^{(2)}=0.2$ and large values of γ.

χ	βt_3	αt_1					
		0	$\pi/4$	$\pi/2$	0	$\pi/4$	$\pi/2$
		$\gamma=1.5$			$\gamma=1.1$		
0.1	0	-0.040	0.371	0.570	-0.042	0.398	0.618
	$\pi/4$	-0.014	0.267	0.381	-0.017	0.286	0.419
	$\pi/2$	0.011	-0.013	-0.037	0.006	-0.010	-0.027
0.2	0	-0.022	0.337	0.491	-0.036	0.480	0.718
	$\pi/4$	-0.002	0.251	0.355	-0.012	0.349	0.505
	$\pi/2$	0.016	0.012	0.009	0.012	0.003	-0.005
0.3	0	0.000	0.215	0.297	-0.005	0.340	0.485
	$\pi/4$	0.005	0.162	0.226	0.002	0.251	0.355
	$\pi/2$	0.010	0.016	0.023	0.010	0.013	0.017
0.4	0	0.010	0.148	0.196	0.007	0.236	0.325
	$\pi/4$	0.007	0.110	0.150	0.006	0.174	0.243
	$\pi/2$	0.004	0.011	0.018	0.005	0.011	0.018

Note that the numerical results obtained for $\sigma_{nn}/\sigma_{11}^{(1),0}$ are given in Tables 5.4-5.7 and they show that the stress σ_{nn} has its maximum at the point $\{\alpha t_1 = \pi/2; \beta t_3 = 0\}$; the dependence of the values of σ_{nn} on χ is non-monotonic for all $\eta^{(2)}$ and γ. However, the values $\chi = \chi'$ for which $\sigma_{nn}/\sigma_{11}^{(1),0}$ becomes maximal, decrease with growing γ. Moreover, as a result of increasing γ, the values of $\sigma_{nn}/\sigma_{11}^{(1),0}$ decrease monotonically and for all γ the ratio $\sigma_{nn}/\sigma_{11}^{(1),0}$ is significant for $\chi \le 0.4$.

The comparison of the results given here and obtained for small values of γ with corresponding ones obtained for the composite with plane-curved structure and given in subsection 4.3.3 shows that they approach each other with decreasing γ. This situation can be taken as confirmation of the numerical results.

Table 5.7. The values of $\sigma_{nn}/\sigma_{11}^{(1),0}$ for $\eta^{(2)} = 0.2$ and small values of γ.

χ	βt_3	αt_1: 0	$\pi/4$	$\pi/2$	0	$\pi/4$	$\pi/2$
		$\gamma = 0.5$			$\gamma = 0.1$		
0.1	0	-0.043	0.413	0.640	-0.043	0.415	0.642
	$\pi/4$	-0.020	0.298	0.445	-0.021	0.300	0.450
	$\pi/2$	0.001	-0.002	-0.007	0.000	0.000	0.000
0.2	0	-0.065	0.658	1.010	-0.077	0.696	1.076
	$\pi/4$	-0.031	0.479	0.712	-0.038	0.515	0.771
	$\pi/2$	0.003	-0.003	-0.008	0.000	0.000	0.000
0.3	0	-0.025	0.605	0.889	-0.039	0.685	1.046
	$\pi/4$	-0.010	0.439	0.635	-0.019	0.523	0.764
	$\pi/2$	0.003	0.002	0.001	0.000	0.000	0.000
0.4	0	0.004	0.462	0.655	0.001	0.592	0.918
	$\pi/4$	0.003	0.333	0.472	0.000	0.420	0.599
	$\pi/2$	0.002	0.005	0.007	0.000	0.000	0.000

5.5.2. TWO-AXIAL LOADING

Consider the stress distribution in the composite with co-phase curved structures (Fig.5.4.1) under biaxial loading. We assume that the composite is loaded at infinity by normal forces with intensity p_1 and p_3 in the direction of the Ox_1 and Ox_3 axes, respectively.

We introduce the parameter $\rho_1 = p_3/p_1$ and assume that $\nu^{(1)} = \nu^{(2)} = 0.3$, $E^{(2)}/E^{(1)} = 50$ and $\varepsilon = 0.02$. Taking into account the results obtained in the previous subsection we consider the influence of the parameters ρ_1, $\eta^{(2)}$ and γ on the values of σ_{nn}, $\sigma_{n\tau_1}$, $\sigma_{n\tau_3}$ obtained at points $(\alpha t_1 = \pi/2, \beta t_3 = 0)$, $(\alpha t_1 = 0, \beta t_3 = 0)$, $(\alpha t_1 = \pi/2, \beta t_3 = \pi/2)$, respectively, and for $0 \le \chi \le 0.3$.

The numerical results related to the case where $H^{(1)} >> H^{(2)}$, are given in Tables 5.8-5.10. Note that in obtaining of these results, as before, we neglected the interaction between curved layers. The analogical results for the higher filler concentration case, given in Tables 5.11-5.13, show that under $\rho_1 > 0$ $(\rho_1 < 0)$ additional loading in the direction of the Ox_3 axis leads to increases (decreases) of the absolute values of $\sigma_{nn}/\sigma_{11}^{(1),0}$, $\sigma_{n\tau_1}/\sigma_{11}^{(1),0}$, $\sigma_{n\tau_3}/\sigma_{11}^{(1),0}$.

Table 5.8. The values of $\sigma_{nn}/\sigma_{11}^{(1),0}$ for lower filler concentration.

ρ_1	χ	γ				
		0.1	0.3	0.5	1.0	1.5
0.5	0.1	0.574	0.624	0.667	0.723	0.692
	0.2	0.878	0.940	0.971	0.900	0.848
	0.3	0.893	0.943	0.928	0.827	0.782
1.0	0.1	0.601	0.706	0.801	0.961	0.982
	0.2	0.920	1.063	1.165	1.197	1.200
	0.3	0.936	1.056	1.107	1.168	1.194
1.5	0.1	0.629	0.787	0.934	1.200	1.272
	0.2	0.962	1.186	1.359	1.495	1.580
	0.3	0.979	1.178	1.291	1.341	1.401
-0.5	0.1	0.519	0.461	0.401	0.246	0.111
	0.2	0.794	0.694	0.583	0.306	0.212
	0.3	0.808	0.690	0.555	0.289	0.192
-1.0	0.1	0.492	0.380	0.268	0.008	-0.178
	0.2	0.752	0.571	0.389	0.008	-0.201
	0.3	0.765	0.568	0.370	0.007	-0.169
-1.5	0.1	0.464	0.298	0.134	-0.329	-0.463
	0.2	0.710	0.448	0.195	-0.288	-0.521
	0.3	0.723	0.446	0.186	-0.267	-0.492

Table 5.9. The values of $\sigma_{n\tau_1}/\sigma_{11}^{(1),0}$ for lower filler concentration.

ρ_1	χ	γ				
		0.1	0.3	0.5	1.0	1.5
0.5	0.1	-0.347	-0.361	-0.367	-0.360	-0.341
	0.2	-0.541	-0.561	-0.560	-0.494	-0.435
	0.3	-0.622	-0.635	-0.614	-0.547	-0.482
1	2	3	4	5	6	7

Table 5.9 (Continuation)

1	2	3	4	5	6	7
	0.1	-0.358	-0.390	-0.413	-0.453	-0.424
1.0	0.2	-0.560	-0.615	-0.647	-0.607	-0.450
	0.3	-0.644	-0.697	-0.705	-0.661	-0.483
	0.1	-0.368	-0.420	-0.459	-0.506	-0.507
1.5	0.2	-0.579	-0.669	-0.725	-0.719	-0.702
	0.3	-0.666	-0.759	-0.796	-0.772	-0.749
	0.1	-0.327	-0.302	-0.274	-0.214	-0.174
-0.5	0.2	-0.503	-0.452	-0.395	-0.269	-0.189
	0.3	-0.577	-0.510	-0.433	-0.295	-0.203
	0.1	-0.317	-0.272	-0.228	-0.141	-0.091
-1.0	0.2	-0.485	-0.398	-0.313	-0.157	-0.098
	0.3	-0.555	-0.448	-0.338	-0.176	-0.101
	0.1	-0.307	-0.243	-0.182	-0.068	-0.008
-1.5	0.2	-0.466	-0.344	-0.230	-0.044	-0.002
	0.3	-0.533	-0.385	-0.251	-0.029	-0.000

Table 5.10. The values of $\sigma_{n\tau_3} / \sigma_{11}^{(1),0}$ for lower filler concentration.

		γ				
ρ_1	χ	0.1	0.3	0.5	1.0	1.5
	0.1	0.035	0.097	0.161	0.307	0.424
0.5	0.2	0.062	0.167	0.269	0.449	0.521
	0.3	0.075	0.193	0.285	0.501	0.602
	0.1	0.049	0.132	0.224	0.457	0.667
1.0	0.2	0.082	0.217	0.359	0.602	0.812
	0.3	0.101	0.251	0.397	0.658	0.874
	0.1	0.062	0.167	0.287	0.607	0.909
1.5	0.2	0.103	0.266	0.449	0.955	1.156
	0.3	0.128	0.309	0.498	1.003	1.384
	0.1	0.008	0.028	0.035	0.007	-0.059
-0.5	0.2	0.021	0.069	0.090	0.043	0.022
	0.3	0.022	0.077	0.096	0.087	0.038
	0.1	-0.050	-0.006	0.027	-0.142	-0.302
-1.0	0.2	0.000	0.020	0.000	-0.159	-0.366
	0.3	-0.004	0.020	-0.003	-0.171	-0.457
	0.1	-0.018	-0.040	-0.089	-0.292	-0.544
-1.5	0.2	-0.013	-0.029	-0.089	-0.362	-0.661
	0.3	-0.031	-0.037	-0.104	-0.423	-0.810

Table 5.11. The values of $\sigma_{nn}/\sigma_{11}^{(1),0}$ for higher filler concentration.

ρ_1	χ	γ 0.1	0.3	0.5	1.0	1.5
0.5	0.1	-0.096	-0.107	-0.111	-0.109	-0.127
	0.2	0.131	0.141	0.153	0.162	0.177
	0.3	0.255	0.271	0.279	0.283	0.300
1.0	0.1	-0.101	-0.121	-0.133	-0.146	-0.180
	0.2	0.138	0.160	0.183	0.215	0.265
	0.3	0.267	0.306	0.334	0.371	0.402
1.5	0.1	-0.106	-0.136	-0.155	-0.182	-0.233
	0.2	0.144	0.178	0.213	0.269	0.298
	0.3	0.279	0.341	0.389	0.438	0.441
-0.5	0.1	-0.087	-0.079	-0.066	-0.037	-0.022
	0.2	0.119	0.104	0.092	0.054	0.012
	0.3	0.231	0.200	0.168	0.105	0.043
-1.0	0.1	-0.082	-0.064	-0.044	-0.001	0.030
	0.2	0.113	0.086	0.061	0.333	-0.006
	0.3	0.218	0.165	0.112	0.007	0.000
-1.5	0.1	-0.078	-0.050	-0.021	0.034	0.083
	0.2	0.107	0.068	0.031	-0.052	-0.143
	0.3	0.206	0.130	0.057	-0.004	-0.061

Table 5.12. The values of $\sigma_{n\tau_1}/\sigma_{11}^{(1),0}$ for higher filler concentration.

ρ_1	χ	γ 0.1	0.3	0.5	1.0	1.5
0.5	0.1	-2.617	-2.645	-2.498	-1.807	-1.228
	0.2	-2.013	-2.014	-1.859	-1.137	-0.625
	0.3	-1.510	-1.498	-1.350	-0.757	-0.243
1.0	0.1	-2.731	-2.960	-2.950	-2.407	-1.646
	0.2	-2.088	-2.247	-2.184	-1.698	-1.001
	0.3	-1.571	-1.657	-1.578	-0.989	-0.502
1.5	0.1	-2.845	-3.274	-3.408	-2.847	-2.065
	0.2	-2.182	-2.479	-2.508	-1.898	-1.096
	0.3	-1.632	-1.821	-1.795	-1.002	-0.243
-0.5	0.1	-2.390	-2.016	-1.594	-0.768	-0.392
	0.2	-1.843	-1.550	-1.211	-0.577	-0.300
	0.3	-1.388	-1.165	-0.906	-0.403	-0.198
-1.0	0.1	-2.276	-1.701	-1.142	-0.248	-0.009
	0.2	-1.758	-1.317	-0.886	-0.246	-0.006
	0.3	-1.327	-1.001	-0.684	-0.221	0.000
-1.5	0.1	-2.163	-1.387	-0.690	0.271	0.444
	0.2	-1.163	-1.085	-0.562	0.083	0.055
	0.3	-1.267	-0.837	-0.461	0.047	0.029

Table 5.13. The values of $\sigma_{n\tau_3}/\sigma_{11}^{(1),0}$ for higher filler concentration.

ρ_1	χ	γ				
		0.1	0.3	0.5	1.0	1.5
	0.1	0.338	0.920	1.415	1.971	1.962
0.5	0.2	0.269	0.692	1.019	1.251	1.201
	0.3	0.209	0.499	0.707	0.909	0.897
	0.1	0.417	1.108	1.771	2.410	2.894
1.0	0.2	0.346	0.851	1.296	1.750	1.945
	0.3	0.282	0.633	0.922	0.990	1.123
	0.1	0.497	1.297	2.127	3.450	3.825
1.5	0.2	0.422	1.010	1.574	2.249	2.677
	0.3	0.356	0.767	1.136	1.661	2.001
	0.1	0.179	0.542	0.703	0.493	0.100
-0.5	0.2	0.117	0.374	0.465	0.252	0.044
	0.3	0.062	0.231	0.277	0.106	0.018
	0.1	0.099	0.353	0.347	-0.245	-0.830
-1.0	0.2	0.040	0.215	0.188	-0.246	-0.867
	0.3	-0.011	0.097	0.063	-0.273	-0.922
	0.1	0.020	0.165	-0.008	-0.984	-1.762
-1.5	0.2	-0.035	0.055	-0.088	-0.745	-1.323
	0.3	-0.084	-0.036	-0.151	-0.586	-1.062

5.6. Bibliographical Notes

The solution method presented in sections 5.1-5.4 was proposed by S.D.Akbarov and A.N.Guz [29] and by S.D.Akbarov, M.D.Verdiev and A.N.Guz [49]. The study of the stress distribution in the spatially curved composite was also made by S.D.Akbarov, M.D.Verdiev and A.N.Guz [49].

CHAPTER 6

LOCALLY-CURVED COMPOSITES

In this chapter we investigate the stress state in composite material with locally plane-curved layers. All investigations are carried out for plane-strain state with the use of the relations and assumptions given in Chapter 4. For various local curving form the distribution of the self-balanced stresses on the inter-layer surfaces are studied in detail.

6.1. Formulation

As in Chapter 4, we consider an infinite elastic body, reinforced by an arbitrary number of non-intersecting locally curved filler layers. We will use the notation introduced in Chapter 4 and the rectangular Cartesian system of coordinates $O_m^{(k)}x_{1m}^{(k)}x_{2m}^{(k)}x_{3m}^{(k)}$ which are obtained from the system of coordinates $Ox_1x_2x_3$ by parallel transfer along the Ox_2 axis, will refer to each layer. The reinforcing layers will be assumed to be located in planes $O_m^{(k)}x_{1m}^{(k)}x_{3m}^{(k)}$ and the thickness of each filler layer will be assumed constant. Furthermore, we assume that the reinforcing layers are locally curved only in the direction of the Ox_1 axis, and that the matrix and filler materials are homogeneous, isotropic and linearly elastic. Taking into account that within each layer the system of equations (4.1.1) are satisfied and in the framework of the contact condition (4.1.2) and to give the relation (4.1.3) we investigate the stress distribution in this composite material under the action at infinity of uniformly distributed normal forces with intensity p in the direction of the Ox_1 axis. In this case we will assume that the function (4.1.3) and its first order derivative are continuous and satisfy the following decay conditions.

$$F_m \to 0, \quad \frac{dF_m}{dx_{1m}^{(2)}} \to 0 \quad \text{for} \quad \left|x_{1m}^{(2)}\right| \to \infty \tag{6.1.1}$$

As before, we will seek the values characterizing the stress-deformed state in each layer in the series (4.1.15) in the small parameter ε. In this case the equations (4.1.16) and the relations (4.1.26) arise. However, the method for the determination of the values of each approximation described in the preceding two chapters for periodical curving cannot be applied for the local curving case. Therefore, we now consider how to determine each approximation in the local curving case.

6.2. Method of Solution

For simplicity we consider the composite material with an infinite number of co-phase plane-curved layers alternating in the direction of the Ox_2 axis, and investigate the

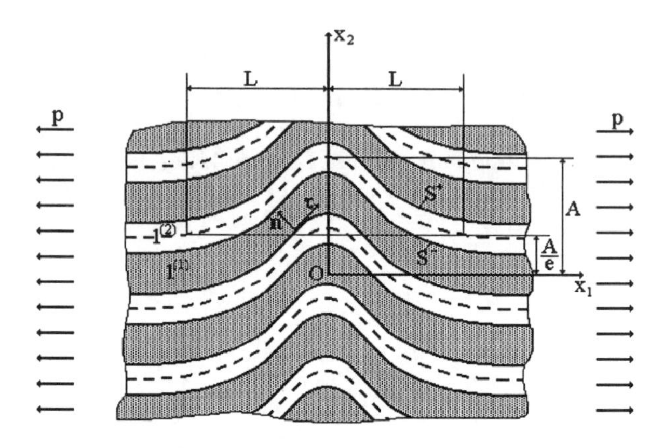

Fig.6.2.1. The structure of the composite material with cophasically locally plane-curved layers

plane-strain state under the action, at infinity, of uniformly distributed normal forces of intensity p in the direction of the Ox_1 axis. The structure of this material is shown schematically in Fig.6.2.1. The composite structure in the direction of the Ox_2 axis has period $2\left(H^{(2)}+H^{(1)}\right)$, where $2H^{(1)}$ is a thickness of the matrix layer and $2H^{(2)}$ is a thickness of the filler layer. Among the layers we single out two of them, i.e. $1^{(1)}$, $1^{(2)}$ (Fig.6.2.1); the equation of the middle surface of the $1^{(2)}$-th layer we take as $x_2^{(2)} = \varepsilon f\left(x_1^{(2)}\right)$.

Under local curving of the reinforcing layers, as in the periodic curving case, the zeroth approximation is determined through the expressions (4.2.5), (4.2.7). Therefore we consider the determination of the values of the first approximation, for which we obtain the following contact condition from (4.2.3).

$$\sigma_{i2}^{(1),1}\Big|_{\left(t_1,\mp H^{(1)}\right)} - \sigma_{i2}^{(2),1}\Big|_{\left(t_1,\pm H^{(2)}\right)} = \frac{df(x_1)}{dx_1}\Big|_{(t_1)}\left(\sigma_{11}^{(1),0}-\sigma_{11}^{(2),0}\right)\delta_i^1,$$

$$u_i^{(1),1}\Big|_{\left(t_1,\mp H^{(1)}\right)} - u_i^{(2),1}\Big|_{\left(t_1,\pm H^{(2)}\right)} = f(x_1)\big|_{(t_1)}\left(\frac{\partial u_2^{(2),0}}{\partial x_2^{(2)}}-\frac{\partial u_2^{(1),0}}{\partial x_2^{(1)}}\right)\delta_i^2, \qquad (6.2.1)$$

where δ_i^1, δ_i^2 are Kronecker symbols. As the first approximation satisfies the equations (4.1.16), it can be sought with the use of Papkovich-Neuber representations (4.2.9). After this procedure we apply the exponential Fourier transform with respect to x_1, i.e.

$$\overline{V}(\lambda, x_2) = \int_{-\infty}^{+\infty} V(x_1, x_2) e^{-ix_1\lambda} dx_1 \tag{6.2.2}$$

to equations (4.1.16), (6.2.1) and (4.2.9) and the Fourier transform of the harmonic functions $\Phi_0^{(k),1}$ and $\Phi_2^{(k),1}$ in (4.2.9) we find as follows:

$$\overline{\Phi}_0^{(k),1}\left(\lambda x_2^{(k)}\right) = \overline{C}_{01}^{(k)}(\lambda)\cosh\left(\lambda x_2^{(k)}\right) + \overline{C}_{02}^{(k)}(\lambda)\sinh\left(\lambda x_2^{(k)}\right),$$

$$\overline{\Phi}_2^{(k),1}\left(\lambda x_2^{(k)}\right) = \overline{C}_{21}^{(k)}(\lambda)\cosh\left(\lambda x_2^{(k)}\right) + \overline{C}_{22}^{(k)}(\lambda)\sinh\left(\lambda x_2^{(k)}\right). \tag{6.2.3}$$

The Fourier transform of the relations (4.2.9), (4.1.16) and the contact conditions (6.2.1) which have the form

$$\overline{\sigma}_{i2}^{(1),1}\Big|_{\left(\mp H^{(1)}\right)} - \overline{\sigma}_{i2}^{(2),1}\Big|_{\left(\pm H^{(2)}\right)} = -i\lambda\overline{f}(\lambda)\left(\sigma_{11}^{(1),0} - \sigma_{11}^{(2),0}\right)\delta_i^1,$$

$$\overline{u}_i^{(1),1}\Big|_{\left(\mp H^{(1)}\right)} - \overline{u}_i^{(2),1}\Big|_{\left(\pm H^{(2)}\right)} = \overline{f}(\lambda)\left(\frac{\partial u_2^{(2),0}}{\partial x_2^{(2)}} - \frac{\partial u_2^{(1),0}}{\partial x_2^{(1)}}\right)\delta_i^2, \tag{6.2.4}$$

give a system of non-homogeneous algebraic equations for the constants $\overline{C}_{ni}^{(k)}(\lambda)$ (i;n=0,2) which enter (6.2.3); these constants give the Fourier transform of the first approximation. The Fourier transform of the stresses and displacements can be presented as follows:

$$\overline{\sigma}_{ij}^{(k),1}\left(\lambda x_2^{(k)}\right) = \overline{\Psi}_{ij}\left(\lambda x_2^{(k)}\right), \quad \overline{u}_i^{(k),1}\left(\lambda x_2^{(k)}\right) = \overline{\varphi}_i^{(k)}\left(\lambda x_2^{(k)}\right), \tag{6.2.5}$$

where the expressions of the functions $\overline{\Psi}_{ij}\left(\lambda x_2^{(k)}\right)$, $\varphi_i^{(k)}\left(\lambda x_2^{(k)}\right)$ can be easily determined from the relations(6.2.3), (4.2.9) and (4.1.16).

We now invert of the Fourier transforms. For this purpose we use the algorithm proposed in [122]; we calculate $\overline{C}_{0i}^{(k)}(\lambda)$, $\overline{C}_{2i}^{(k)}(\lambda)$ at $\lambda = \lambda_n$ (n=1,2,3,…,N) from the contact conditions (6.2.4). In this way we determine also the values of $\overline{\sigma}_{ij}^{(k),1}\left(\lambda x_2^{(k)}\right)$, $\overline{u}_i^{(k),1}\left(\lambda x_2^{(k)}\right)$ at $\lambda = \lambda_n$ and using the summation formulae [122] we find the values of

$\sigma_{ij}^{(k),1}\left(x_1^{(k)}, x_2^{(k)}\right)$, $u_i^{(k),1}\left(x_1^{(k)}, x_2^{(k)}\right)$ for fixed $\left(x_1^{(k)}, x_2^{(k)}\right)$. The selection rule of λ_n is also given in [122]. We determine the subsequent approximations similarly.

All numerical results which will be analysed have been obtained by using the described algorithm in the framework of the first five approximation (4.1.15). The programs were written in FORTRAN-IV and realized on a PC. The function $f(x_1)$ is selected as follows:

$$F(x_1) = \varepsilon f(x_1) = A \exp\left(-\left((x_1/L)^2\right)^{\delta}\right)\cos\left(\frac{mx_1^{(2)}}{L}\right) = \varepsilon L \exp\left(-\left((x_1/L)^2\right)^{\delta}\right)\cos\left(\frac{mx_1^{(2)}}{L}\right) \tag{6.2.6}$$

and composites with both alternating co-phase and alternating anti-phase locally curved layers are considered. In (6.2.6) it is assumed that $L > A$, and the small parameter is $\varepsilon = A/L$. Here A is the maximum value of the lift of the local curving, and L is the introduced geometrical parameter shown in Fig.6.2.1. In (6.2.6) the parameter m shows the oscillatory character of the local mode of curvature, while the parameter δ characterizes the smoothness of the matrix-filler interface in the neighbourhood of the maximum rise and the change in the slope of the curved layer. We will analyse the influence of these parameters on the stress distribution.

6.3. Composite with Alternating Layers

We consider both co-phase and anti-phase locally curved layers. The local curving form of the layers is taken as

$$F(x_1) = A \exp\left(-\left(\frac{x_1}{L}\right)^2\right). \tag{6.3.1}$$

We introduce the parameter $\chi = H^{(2)}/L$, and assume that $\nu^{(1)} = \nu^{(2)} = 0.3$.

6.3.1. LOWER FILLER CONCENTRATION

As before, under lower filler concentration we understand the case $H^{(1)} >> H^{(2)}$ and the interaction between the filler layers is not taken into account. We investigate the distribution of the self-balanced normal and tangential stress on the inter-layer surfaces S^+ and S^- (Fig.6.2.1). Taking into account the symmetry of the composite structure with respect to the Ox_2 axis, we consider only the part where $x_1 \geq 0$; values related to the surfaces S^+ and S^- are denoted by the upper symbols + and - , respectively.

Fig.6.3.1 shows the dependence of $\sigma_{nn}^{+}/\sigma_{11}^{(1),0}$, $\sigma_{nn}^{-}/\sigma_{11}^{(1),0}$, $\sigma_{n\tau}^{+}/\sigma_{11}^{(1),0}$ on t_1/L for fixed problem parameters. These graphs show that the self-balanced normal stresses σ_{nn}^{+} and σ_{nn}^{-} have their absolute maximal value at $t_1/L = 0$. For the

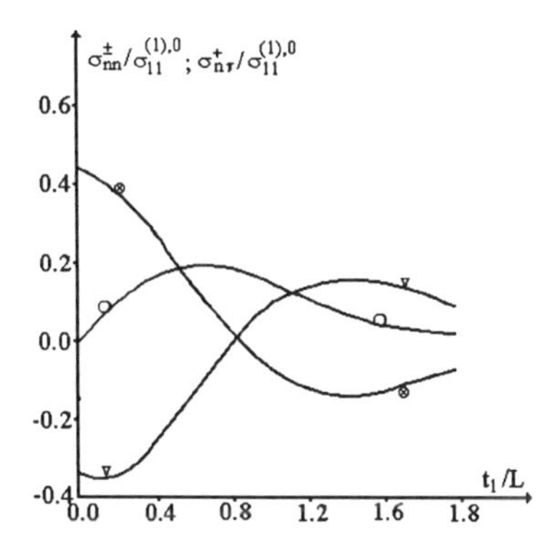

Fig.6.3.1. The graphs of the relations between $\sigma_{nn}^{\pm}$, $\sigma_{n\tau}^{+}$ and t_1/L for $E^{(2)}/E^{(1)}$ =50, $\chi = 0.1$, $\varepsilon = 0.05$. ⊗ -- σ_{nn}^{+}; ∇ -- σ_{nn}^{-}; O -- $\sigma_{n\tau}^{+}$

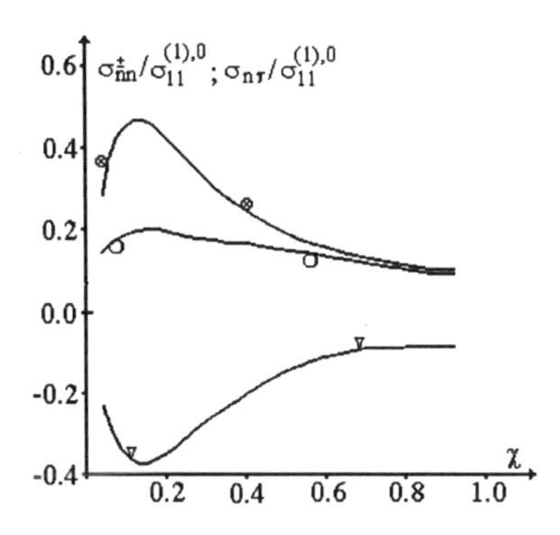

Fig.6.3.2. The graphs of the relations between $\sigma_{nn}^{\pm}$, $\sigma_{n\tau}^{+}$ and χ for $E^{(2)}/E^{(1)}$ =50, $\chi = 0.1$, $\varepsilon = 0.05$. ⊗ -- σ_{nn}^{+} at $t_1/L = 0.0$; ∇ -- σ_{nn}^{-} at $t_1/L = 0.0$; O -- $\sigma_{n\tau}^{+}$ at $t_1/L = 0.6$.

region $0 \le t_1/L \le 0.85$, σ_{nn}^{+} is tensile, but σ_{nn}^{-} is compressive. However, for $t_1/L > 0.85$ the σ_{nn}^{+} and σ_{nn}^{-} become compressive and tensile, respectively. When $t_1/L \to \infty$ the stresses $\sigma_{nn}^{\pm} \to 0$. The results show that $\sigma_{n\tau}^{+} \ge 0$ for all t_1/L and dependence between them is non-monotonic. For fixed problem parameters, $\sigma_{n\tau}^{+}$ has its maximum at $t_1/L \approx 0.7$; with increasing t_1/L, the $\sigma_{n\tau}^{+} \to 0$. Since the stresses $\sigma_{n\tau}^{+}$ and $\sigma_{n\tau}^{-}$ show similar behaviours we do not consider the distribution of the $\sigma_{n\tau}^{-}$ on S^{-}.

Fig.6.3.2 shows the dependence of $\sigma_{nn}^{+}/\sigma_{11}^{(1),0}$, $\sigma_{nn}^{-}/\sigma_{11}^{(1),0}$ and $\sigma_{n\tau}^{+}/\sigma_{11}^{(1),0}$ on the parameter χ; these dependencies are non-monotonic: i.e. there is a value of the parameter χ for which the self-balanced stresses σ_{nn}^{+}, σ_{nn}^{-} and $\sigma_{n\tau}^{+}$ have their maxima; this non-monotonic behaviour is observed for all points of the surfaces S^{+} and S^{-}.

Table 6.1. The values of $\sigma_{nn}^{+}/\sigma_{11}^{(1),0}$ at $t_1/L=0$, $\sigma_{nn}^{-}/\sigma_{11}^{(1),0}$ at $t_1/L=0$, $\sigma_{n\tau}^{+}/\sigma_{11}^{(1),0}$ at $t_1/L=0.8$, $\chi=0.2$ for $E^{(2)}/E^{(1)}=20$; $\chi=0.1$ for $E^{(2)}/E^{(1)}=50;100$.

$E^{(2)}/E^{(1)}$	ε	$\sigma_{nn}^{+}/\sigma_{11}^{(1),0}$	$\sigma_{nn}^{-}/\sigma_{11}^{(1),0}$	$\sigma_{n\tau}^{+}/\sigma_{11}^{(1),0}$
	0.01	0.035	-0.033	0.021
	0.04	0.165	-0.133	0.089
20	0.05	0.218	-0.168	0.109
	0.07	0.349	-0.246	0.170
	0.10	0.615	-0.386	0.275
	0.01	0.067	-0.064	0.030
	0.04	0.329	-0.272	0.135
50	0.05	0.450	-0.357	0.190
	0.07	0.770	-0.564	0.300
	0.10	1.546	-1.007	0.594
	0.01	0.117	-0.112	0.049
	0.04	0.606	-0.508	0.237
100	0.05	0.856	-0.690	0.328
	0.07	1.571	-1.179	0.590
	0.10	3.473	-2.380	1.300

Consider the influence of $E^{(2)}/E^{(1)}$ and ε on the ratios $\sigma_{nn}^{+}/\sigma_{11}^{(1),0}$, $\sigma_{nn}^{-}/\sigma_{11}^{(1),0}$ and $\sigma_{n\tau}^{+}/\sigma_{11}^{(1),0}$ which are given in Table 6.1. This table shows that the values of $\sigma_{nn}^{+}/\sigma_{11}^{(1),0}$, $\sigma_{nn}^{-}/\sigma_{11}^{(1),0}$ and $\sigma_{n\tau}^{+}/\sigma_{11}^{(1),0}$ increase monotonically with increasing $E^{(2)}/E^{(1)}$ and ε, and also that the normal self-balanced stresses are greater than the self-balanced tangential stress. For $E^{(2)}/E^{(1)} \geq 50$, $\varepsilon \geq 0.07$ the values of the σ_{nn}^{+}, σ_{nn}^{-}, $\sigma_{n\tau}^{+}$ can be several times greater than that of the $\sigma_{11}^{(1),0}$.

In the Table 6.2 gives the values of the $\sigma_{nn}^{+}/\sigma_{11}^{(1),0}$ for various approximations (4.1.15) and various $E^{(2)}/E^{(1)}$, ε. The comparison of these results shows that the convergence is acceptable.

Table 6.2. The values of $\sigma_{nn}^{+}/\sigma_{11}^{(1),0}$ in various approximations at $t_1/L=0$, $\chi=0.2$ for $E^{(2)}/E^{(1)}=20$; $\chi=0.1$ for $E^{(2)}/E^{(1)}=50;100$.

$\frac{E^{(2)}}{E^{(1)}}$	Numb.of approxm.	ε 0.01	0.04	0.07	0.10
	1	0.034	0.138	0.242	0.346
	2	0.035	0.153	0.289	0.442
20	3	0.035	0.164	0.347	0.612
	4	0.035	0.165	0.352	0.631
	5	0.035	0.165	0.349	0. 615
1	2	3	4	5	6

Table 6.2 (Continuation)

1	2	3	4	5	6
	1	0.065	0.262	0.458	0.655
	2	0.067	0.287	0.537	0.815
50	3	0.067	0.326	0.745	1.422
	4	0.067	0.329	0.771	1.532
	5	0.067	0.329	0.774	1.546
	1	0.113	0.452	0.791	1.131
	2	0.115	0.494	0.918	1.390
100	3	0.117	0.597	1.471	3.001
	4	0.117	0.604	1.540	3.288
	5	0.117	0.606	1.570	3.473

6. 3. 2. COMPOSITE WITH CO-PHASE CURVED LAYERS

We consider the case where the concentration of the filler layers is higher and the local curving of these layers is co-phase, and investigate the self-balanced stresses σ_{nn}^{+}, σ_{nn}^{-}, $\sigma_{n\tau}^{+}$ taking the interaction between layers into account. First, we analyse the distribution of these stresses on the surfaces S^{+} and S^{-} (Fig.6.2.1). The graphs given in Fig.6.3.3 and Fig.6.3.4 show this distribution for various values of $\eta^{(2)}$. These graphs show that at all points of S^{+} the values of $\sigma_{n\tau}^{+}/\sigma_{11}^{(1),0}$ increase monotonically with $\eta^{(2)}$ and interaction between the filler layers does not vary the character of the

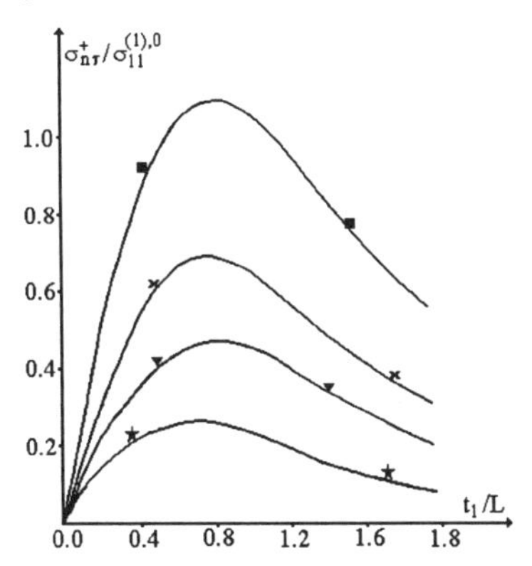

Fig.6.3.3. The relation between $\sigma_{n\tau}^{+}/\sigma_{11}^{(1),0}$ and t_1/L for $\chi = 0.1$, $\varepsilon = 0.05$, $E^{(2)}/E^{(1)} = 50$, ★-- $\eta^{(2)}=0.1$; ▼ -- $\eta^{(2)}=0.2$; ✖ -- $\eta^{(2)}=0.3$; ■ -- $\eta^{(2)}=0.5$.

Fig.6.3.4. The relation between $\sigma_{nn}^{+}/\sigma_{11}^{(1),0}$, $\sigma_{nn}^{-}/\sigma_{11}^{(1),0}$ and t_1/L for $\chi = 0.1$, $\varepsilon = 0.05$, $E^{(2)}/E^{(1)} = 50$, ✱-- $\sigma_{nn}^{+}/\sigma_{11}^{(1),0}$, $\eta^{(2)}=0.3$; ◗ -- $\sigma_{nn}^{+}/\sigma_{11}^{(1),0}$, $\eta^{(2)}=0.5$; ✚ -- $\sigma_{nn}^{-}/\sigma_{11}^{(1),0}$, $\eta^{(2)}=0.3$; ● -- $\sigma_{nn}^{-}/\sigma_{11}^{(1),0}$, $\eta^{(2)}=0.5$.

dependence of $\sigma_{n\tau}^{+}/\sigma_{11}^{(1),0}$ on t_1/L. Moreover the absolute values of $\sigma_{nn}^{+}/\sigma_{11}^{(1),0}$ and $\sigma_{nn}^{-}/\sigma_{11}^{(1),0}$ decrease with $\eta^{(2)}$; the stress σ_{nn}^{+} has its maximum at $t_1/L=0$, but the point at which σ_{nn}^{-} has its maximum moves to the right from the point $t_1/L=0$. Always σ_{nn}^{+}, σ_{nn}^{-}, $\sigma_{n\tau}^{+} \to 0$ as $t_1/L \to \infty$.

Figs.6.3.5 and 6.3.6. show the dependencies of $\sigma_{n\tau}^{+}/\sigma_{11}^{(1),0}$, $\sigma_{nn}^{+}/\sigma_{11}^{(1),0}$ and $\sigma_{nn}^{-}/\sigma_{11}^{(1),0}$ on χ for various $\eta^{(2)}$. These graphs show that $\sigma_{n\tau}^{+}/\sigma_{11}^{(1),0}$ monotonically decreases with increasing χ. However the non monotonic character of the dependencies of $\sigma_{nn}^{+}/\sigma_{11}^{(1),0}$ and $\sigma_{nn}^{-}/\sigma_{11}^{(1),0}$ on χ remains. In this case the values of $\sigma_{nn}^{+}/\sigma_{11}^{(1),0}$, $\sigma_{nn}^{-}/\sigma_{11}^{(1),0}$ decrease as $\eta^{(2)}$ increases.

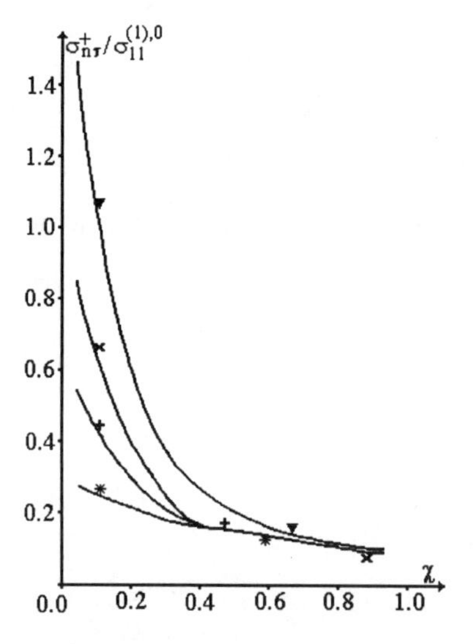

Fig.6.3.5. The relation between $\sigma_{n\tau}^{+}/\sigma_{11}^{(1),0}$ and χ for $t_1/L=0.8$, $\varepsilon=0.05$, $E^{(2)}/E^{(1)}=50$, ✳-- $\eta^{(2)}=0.1$; ✚-- $\eta^{(2)}=0.2$; ✖ -- $\eta^{(2)}=0.3$; ▼ -- $\eta^{(2)}=0.5$.

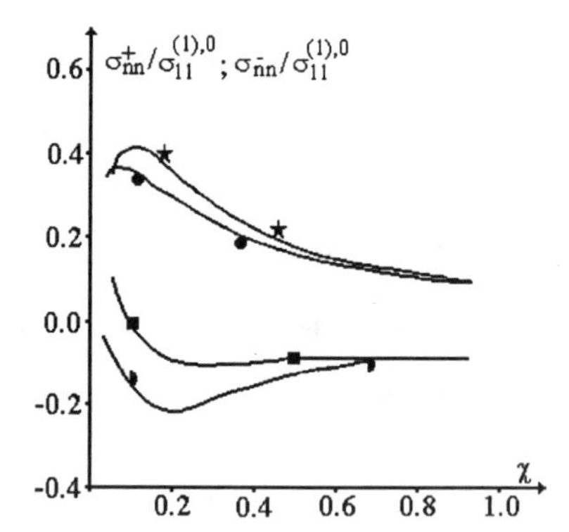

Fig.6.3.6. The relation between $\sigma_{nn}^{+}/\sigma_{11}^{(1),0}$, $\sigma_{nn}^{-}/\sigma_{11}^{(1),0}$ and χ for $t_1/L=0$, $\varepsilon=0.05$, $E^{(2)}/E^{(1)}=50$, ★ -- $\sigma_{nn}^{+}/\sigma_{11}^{(1),0}$, $\eta^{(2)}=0.3$; ●-- $\sigma_{nn}^{+}/\sigma_{11}^{(1),0}$, $\eta^{(2)}=0.5$; ◗ -- $\sigma_{nn}^{-}/\sigma_{11}^{(1),0}$, $\eta^{(2)}=0.3$; ■-- $\sigma_{nn}^{-}/\sigma_{11}^{(1),0}$, $\eta^{(2)}=0.5$.

The values of $\sigma_{n\tau}^{+}/\sigma_{11}^{(1),0}$ given in Table 6.3 for various $E^{(2)}/E^{(1)}$ and ε, show that the ratio $\sigma_{n\tau}^{+}/\sigma_{11}^{(1),0}$ increases significantly with both $E^{(2)}/E^{(1)}$ and ε. In these cases the values of $\sigma_{n\tau}^{+}$ are found to be several times greater than those of σ_{nn}^{+}, σ_{nn}^{-} and $\sigma_{11}^{(1),0}$.

Table 6.3. The values of $\sigma_{n\tau}^{+}/\sigma_{11}^{(1),0}$ for various ε, $\eta^{(2)}$, $E^{(2)}/E^{(1)}$ under $\chi = 0.05$, t_1/L.

ε	$\eta^{(2)}$	$E^{(2)}/E^{(1)}$		
		20	50	100
0.01	0.3	0.050	0.120	0.226
	0.5	0.081	0.195	0.367
0.04	0.3	0.225	0.617	1.380
	0.5	0.368	1.032	2.339
0.07	0.3	0.495	1.663	4.591
	0.5	0.820	2.877	8.129
0.10	0.3	0.943	3.861	12.63
	0.5	1.582	6.860	23.13

Table 6.4 shows the values of $\sigma_{n\tau}^{+}/\sigma_{11}^{(1),0}$ obtained in various approximations (4.1.15): the convergence is acceptable.

Table 6.4. The values of $\sigma_{n\tau}^{+}/\sigma_{11}^{(1),0}$ in various approximations for $E^{(2)}/E^{(1)} = 50$, $\chi = 0.05$, $t_1/L = 0.8$.

Numb. of approx.	$\eta^{(2)}$					
	0.3			0.5		
	ε					
	0.01	0.07	0.10	0.01	0.07	0.10
1	0.110	0.826	1.180	0.191	1.340	1.914
2	0.117	0.822	1.173	0.191	1.338	1.910
3	0.120	1.562	3.330	0.195	2.708	5.905
4	0.120	1.602	3.496	0.195	2.737	6.025
5	0.120	1.663	3.861	0.195	2.877	6.860

6. 3.3. ANTI-PHASE CURVED LAYERS

The structure of this composite is shown schematically in Fig.6.3.7. Taking into account the periodicity along the Ox_2 axis with period $4\left(H^{(2)} + H^{(1)}\right)$ we single out four layers: $1^{(1)}, 1^{(2)}, 2^{(1)}, 2^{(2)}$. The equation of the middle surface of the $1^{(2)}$-th and $2^{(2)}$-th layers we take as $x_{21}^{(2)} = A\exp(-(x_1/L)^2)$ and $x_{22}^{(2)} = -A\exp(-(x_1/L)^2)$ respectively. The parameters A and L are taken as before; the small parameter is $\varepsilon = A/L$. Because of the symmetry we consider only $x_1 \geq 0$.

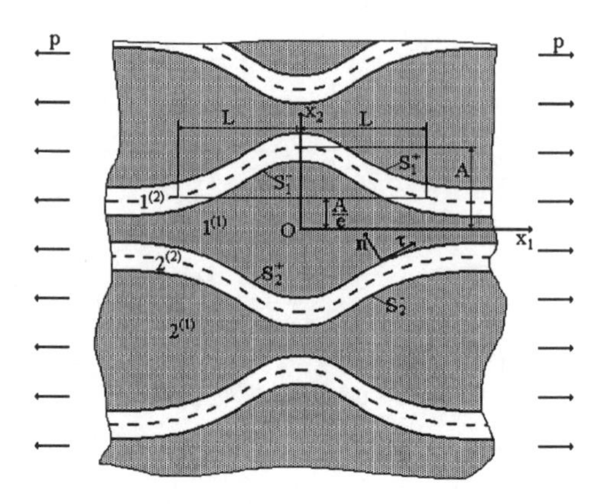

Fig.6.3.7. A composite with anti-phase locally plane-curved layers

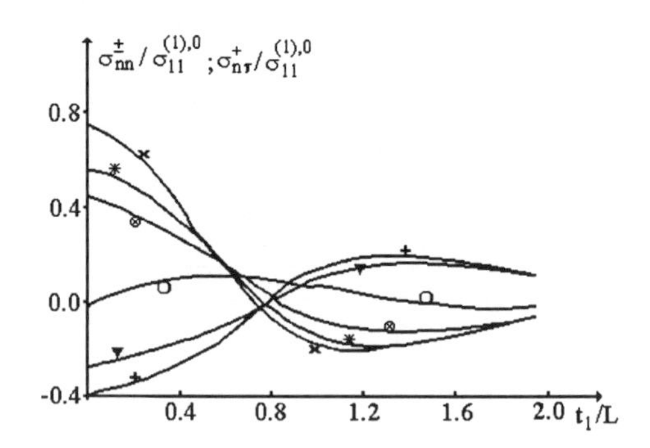

Fig.6.3.8. The relations between $\sigma_{nn}^{\pm}/\sigma_{11}^{(1),0}$, $\sigma_{n\tau}^{+}/\sigma_{11}^{(1),0}$ and t_1/L for various $\eta^{(2)}$ under $E^{(2)}/E^{(1)}=50$, $\varepsilon=0.05$, $\chi=0.1$. ⊗ -- $\sigma_{nn}^{+}/\sigma_{11}^{(1),0}$, $\eta^{(2)}=0.1$; ✳ -- $\sigma_{nn}^{+}/\sigma_{11}^{(1),0}$, $\eta^{(2)}=0.3$; ✖ -- $\sigma_{nn}^{+}/\sigma_{11}^{(1),0}$, $\eta^{(2)}=0.5$; ○ -- $\sigma_{n\tau}^{+}/\sigma_{11}^{(1),0}$, $\eta^{(2)}=0.1$; ▼ -- $\sigma_{nn}^{-}/\sigma_{11}^{(1),0}$, $\eta^{(2)}=0.5$; ✚ -- $\sigma_{nn}^{-}/\sigma_{11}^{(1),0}$, $\eta^{(2)}=0.3$

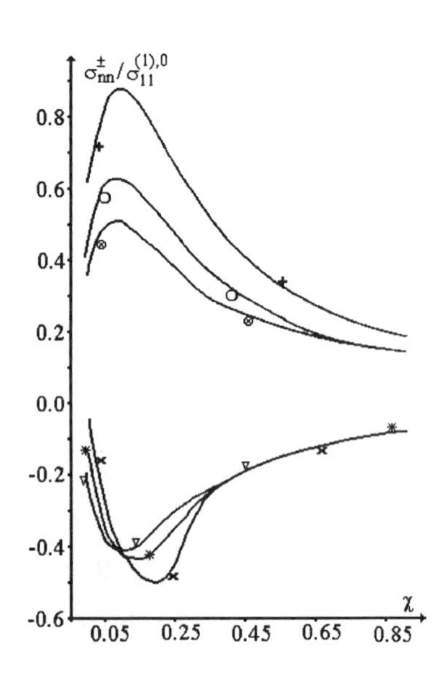

Fig. 6.3.9. The relations between $\sigma_{nn}^{\pm}/\sigma_{11}^{(1),0}$ and χ for various $\eta^{(2)}$ under $E^{(2)}/E^{(1)}=50$, $\varepsilon=0.05$, $t_1/L=0.0$.
⊗ -- $\sigma_{nn}^{+}/\sigma_{11}^{(1),0}$, $\eta^{(2)}=0.1$;
○ -- $\sigma_{nn}^{+}/\sigma_{11}^{(1),0}$, $\eta^{(2)}=0.3$;
✚ -- $\sigma_{nn}^{+}/\sigma_{11}^{(1),0}$, $\eta^{(2)}=0.5$;
∇ -- $\sigma_{nn}^{-}/\sigma_{11}^{(1),0}$, $\eta^{(2)}=0.1$;
✳ -- $\sigma_{nn}^{-}/\sigma_{11}^{(1),0}$, $\eta^{(2)}=0.3$;
✖ -- $\sigma_{nn}^{-}/\sigma_{11}^{(1),0}$, $\eta^{(2)}=0.5$

As in the previous subsection the quantities relating to the surfaces S_1^+ and S_1^- will be denoted by superscripts + and - respectively. Moreover, we assume that $\nu^{(1)} = \nu^{(2)} = 0.3$ and $A < H^{(1)}$; this assumption precludes contact between neighbouring filler layers.

Table 6.5. The values of $\sigma_{nn}^{+}/\sigma_{11}^{(1),0}$ (upper row) and $\sigma_{nn}^{-}/\sigma_{11}^{(1),0}$ (lower row) at $t_1/L = 0$, $\chi = 0.20$ for $E^{(2)}/E^{(1)} = 20;50$, $\chi = 0.15$ for $E^{(2)}/E^{(1)} = 100$.

$E^{(2)}/E^{(1)}$	$\eta^{(2)}$	ε			
		0.01	0.04	0.07	0.10
20	0.2	0.039	0.181	0.382	0.664
		−0.036	−0.145	−0.268	−0.419
	0.3	0.043	0.203	0.430	0.752
		−0.041	−0.159	−0.291	−0.444
	0.4	0.048	0.231	0.500	0.895
		−0.045	−0.170	−0.301	−0.443
	0.5	0.053	0.265	0.596	1.111
		−0.048	−0.174	−0.289	−0.398
50	0.2	0.077	0.367	0.818	1.507
		−0.073	−0.306	−0.612	−1.027
	0.3	0.089	0.426	0.944	1.733
		−0.084	−0.345	−0.668	−1.085
	0.4	0.103	0.501	1.125	2.096
		−0.096	−0.376	−0.696	−1.076
	0.5	0.118	0.593	1.378	2.671
		−0.106	−0.393	−0.679	−0.972
100	0.2	0.137	0.681	1.600	3.060
		−0.131	−0.572	−1.126	−2.078
	0.3	0.162	0.804	1.880	3.607
		−0.152	−0.633	−1.262	−2.051
	0.4	0.187	0.951	2.269	4.487
		−0.170	−0.664	−1.227	−1.851
	0.5	0.211	1.128	2.818	5.879
		−0.184	−0.656	−1.090	−1.412

Fig.6.3.8 shows the dependencies of $\sigma_{nn}^{\pm}/\sigma_{11}^{(1),0}$ and $\sigma_{n\tau}^{+}/\sigma_{11}^{(1),0}$ on t_1/L for various $\eta^{(2)}$. Detailed analysis of the results showed that with the increase of the concentration of the filler $\eta^{(2)}$, (and any χ, $E^{(2)}/E^{(1)}$, ε) the values of the tangential stress $\sigma_{n\tau}^{+}/\sigma_{11}^{(1),0}$ became insignificant in comparison with $\sigma_{nn}^{\pm}/\sigma_{11}^{(1),0}$. Since the

relation $\sigma_{n\tau}^{+} \approx \sigma_{n\tau}^{-}$ holds in this case, we will consider only the distribution of the stresses $\sigma_{nn}^{\pm}$.

The graphs presented in Fig.6.3.8 show that the stresses $\sigma_{nn}^{+}\left(\sigma_{nn}^{-}\right)$ in the case in which $0 \le t_1/L < t_{1*}^{\pm}/L$ (where t_{1*}^{+} (t_{1*}^{-}) correspond to the point of the surface S_1^{+} (S_1^{-}) at which $\sigma_{nn}^{+} = 0$ $(\sigma_{nn}^{-} = 0)$) are tensile (compressive) and $t_1 < t_{1*}^{\pm}$ are compressive (tensile). With increasing $\eta^{(2)}$, the values of $t_{1*}^{\pm}$ decrease. In addition to this, the graphs show that as $t_1/L \to \infty$ the stresses $\sigma_{nn}^{\pm}$ tend to zero, and with increasing $\eta^{(2)}$ for fixed χ (i.e. at $\chi = 0.1$) the absolute values of $\sigma_{nn}^{+}\left(\sigma_{nn}^{-}\right)$ increase (basically decrease).

Fig. 6.3.9 shows the graphs of the dependence of $\sigma_{nn}^{\pm}/\sigma_{11}^{(1),0}$ on χ for various $\eta^{(2)}$. The graphs show that these dependencies are non-monotonic for all examined $\eta^{(2)}$, which means that for any selected L there is a thickness $H^{(2)}$ for which the absolute values of σ_{nn}^{+} and σ_{nn}^{-} become maximum. In this case, an increase of $\eta^{(2)}$ increases the values $\chi_*^{\pm}$ at which $\left|\sigma_{nn}^{\pm}\right|$ becomes maximum.

The values of $\sigma_{nn}^{\pm}/\sigma_{11}^{(1),0}$ given in Table 6.5 for various $E^{(2)}/E^{(1)}$, $\eta^{(2)}$, and ε, show that the values of $\sigma_{nn}^{\pm}/\sigma_{11}^{(1),0}$ increase monotonically as $E^{(2)}/E^{(1)}$ and ε increase. In the cases where $E^{(2)}/E^{(1)} \ge 50$, $\eta^{(2)} \ge 0.3$ and $\varepsilon \ge 0.07$, the normal stresses σ_{nn}^{+}, σ_{nn}^{-} are several times higher than the stress $\sigma_{11}^{(1),0}$. It should also be noted that the values of the stress σ_{nn}^{+} greatly exceed the absolute values of the stress σ_{nn}^{-}.

Table 6.6. The values of $\sigma_{nn}^{+}/\sigma_{11}^{(1),0}$ in various approximations for the case where $E^{(2)}/E^{(1)} = 100$, $\chi = 0.15$, $t_1/L = 0$.

Number of Approxim.	$\eta^{(2)}$ = 0.3			$\eta^{(2)}$ = 0.5		
ε	0.04	0.07	0.10	0.04	0.07	0.10
1	0.624	1.093	1.561	0.784	1.373	1.962
2	0.703	1.333	2.052	0.998	2.027	3.296
3	0.800	1.853	3.567	1.105	2.603	4.975
4	0.807	1.922	3.854	1.128	2.812	5.848
5	0.804	1.880	3.607	1.128	2.818	5.879

Table 6.6 gives, for fixed $E^{(2)}/E^{(1)}$, χ, t_1/L and for different ε, $\eta^{(2)}$, the values of $\sigma_{nn}^{+}/\sigma_{11}^{(1),0}$ in the first five approximations within which all the results presented previously were obtained. Comparison of the results obtained in different approximations shows that the solution method used is highly efficient.

6.4. The Influence of Local Curving Form

In this section we investigate the influence of the curving parameters δ, m, which enter the expression of the function (6.2.6) and characterize the local curving form of the reinforcing layers. We consider co-phase and anti-phase curving separately and assume that all assumptions and notations accepted in the previous section hold.

6.4.1. CO-PHASE CURVING

The structure of the composite material is shown schematically in Fig.6.4.1. We analyse the influence of the parameters δ and m (6.2.6) on the distribution of the

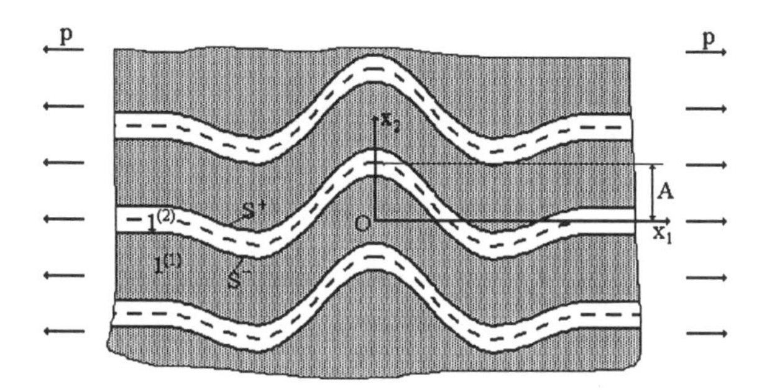

Fig.6.4.1. The structure of the composite material with co-phase complicate local curved structure.

stresses σ_{nn}^{+}, σ_{nn}^{-} and $\sigma_{n\tau}^{+}$. As before, first we consider the case where $H^{(1)} >> H^{(2)}$ and interaction between the filler layers is not taken into account.

Fig.6.4.2 shows the dependencies of $\sigma_{nn}^{\pm}/\sigma_{11}^{(1),0}$ on t_1/L for fixed $E^{(2)}/E^{(1)}$, χ, ε when m=0 and $\delta \le 1$; the corresponding graphs for the case $\delta > 1$ are given in Fig.6.4.3. The results related to the influence of the parameter m on the distribution of $\sigma_{nn}^{\pm}/\sigma_{11}^{(1),0}$ with respect to t_1/L are given in Fig.6.4.4. Similar results are obtained for the stress $\sigma_{n\tau}^{+}$: Fig.6.4.5 shows the dependence of $\sigma_{n\tau}^{+}/\sigma_{11}^{(1),0}$ on t_1/L for various δ when m=0 and $\delta \le 1$; Fig.6.4.6 shows this dependence for m=0 and $\delta > 1$; Fig.6.4.7 shows also this dependence for various values of m.

The graphs given in Fig.6.4.2 show that for $\delta \leq 1$ the stresses σ_{nn}^{+} , σ_{nn}^{-} obtain their maxima at $t_1/L=0$, which corresponds to the maximal value of the curving rise; the dependence of σ_{nn}^{+} and σ_{nn}^{-} on δ is non-monotonic. Similar results are obtained for the distribution of the stress $\sigma_{n\tau}^{+}$ when $\delta \leq 1$ and those are shown in Fig.6.4.5. However in these cases the point at which $\sigma_{n\tau}^{+}$ has its maximal value moves towards the left with decreasing δ. Fig.6.4.3 and Fig.6.4.6 show that when $\delta > 1$, the absolute maximal values of these stresses increase with increasing δ. Moreover, when $\delta > 1$, the σ_{nn}^{+} and σ_{nn}^{-} have their maxima not at $t_1/L=0$ corresponding to the maximum rise, but at points near which there is a sharp change in the slope of the curved layer, this point moves towards the right with increasing δ.

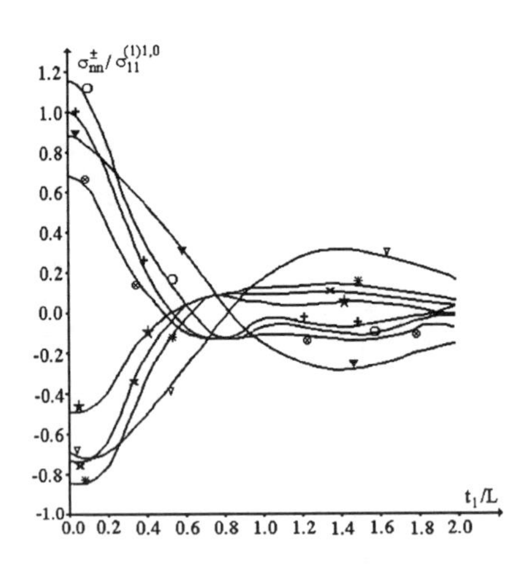

Fig.6.4.2. The distribution of the stresses σ_{nn}^{+}, σ_{nn}^{-} with respect to t_1/L when $\delta \leq 1$ for m=0, $\varepsilon = 0.05$, $E^{(2)}/E^{(1)} = 50$, $\chi = 0.1$: ⊗ -- $\sigma_{nn}^{+}/\sigma_{11}^{(1),0}$, $\delta=1$; ○ -- $\sigma_{nn}^{+}/\sigma_{11}^{(1),0}$, $\delta=1/2$; ✚ -- $\sigma_{nn}^{+}/\sigma_{11}^{(1),0}$, $\delta=1/3$; ▼ -- $\sigma_{nn}^{+}/\sigma_{11}^{(1),0}$, $\delta=1/5$; ∇ -- $\sigma_{nn}^{-}/\sigma_{11}^{(1),0}$, $\delta=1$; ✳ -- $\sigma_{nn}^{-}/\sigma_{11}^{(1),0}$, $\delta=1/2$; ✖ -- $\sigma_{nn}^{-}/\sigma_{11}^{(1),0}$, $\delta=1/3$; ★-- $\sigma_{nn}^{-}/\sigma_{11}^{(1),0}$, $\delta=1/5$.

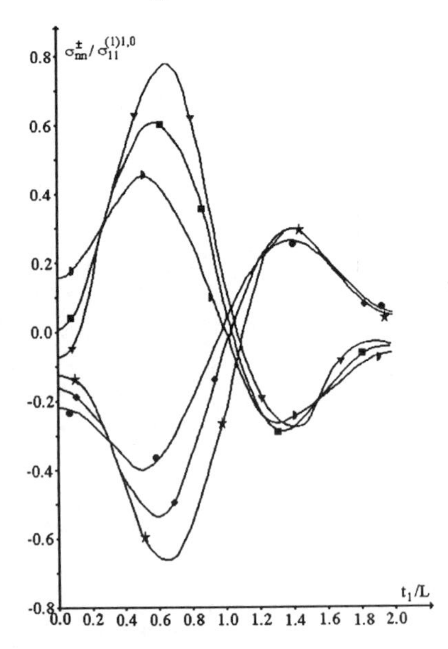

Fig.6.4.3. The distribution of the stresses σ_{nn}^{+}, σ_{nn}^{-} with respect to t_1/L when $\delta > 1$ for m=0, $\varepsilon = 0.05$, $E^{(2)}/E^{(1)} = 50$, $\chi = 0.1$: ◗ -- $\sigma_{nn}^{+}/\sigma_{11}^{(1),0}$, $\delta=2$; ■ -- $\sigma_{nn}^{+}/\sigma_{11}^{(1),0}$, $\delta=3$; ▼ -- $\sigma_{nn}^{+}/\sigma_{11}^{(1),0}$, $\delta=5$; ● -- $\sigma_{nn}^{-}/\sigma_{11}^{(1),0}$, $\delta=2$; ◆ -- $\sigma_{nn}^{-}/\sigma_{11}^{(1),0}$, $\delta=3$; ★-- $\sigma_{nn}^{-}/\sigma_{11}^{(1),0}$, $\delta=5$.

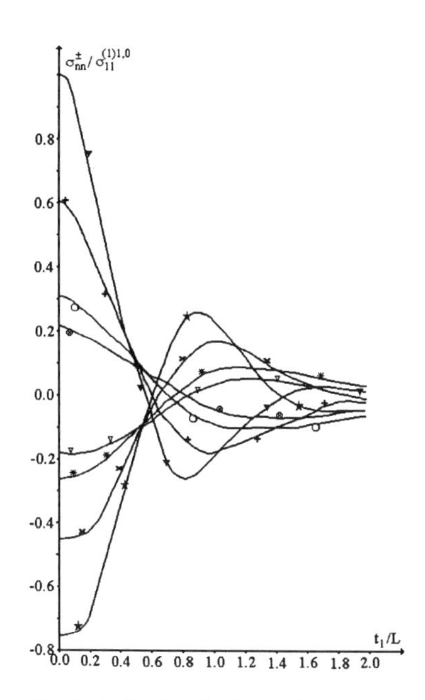

Fig.6.4.4. The distribution of the stresses σ^{+}_{nn} , σ^{-}_{nn} with respect to t_1/L for various m, and $\delta = 1$, $\varepsilon = 0.05$, $E^{(2)}/E^{(1)} = 50$, $\chi = 0.1$: ⊗ -- $\sigma^{+}_{nn}/\sigma^{(1),0}_{11}$, m=0; ○ -- $\sigma^{+}_{nn}/\sigma^{(1),0}_{11}$, m=1; ✚ -- $\sigma^{+}_{nn}/\sigma^{(1),0}_{11}$, m=2; ▼ -- $\sigma^{+}_{nn}/\sigma^{(1),0}_{11}$, m=3; ∇-- $\sigma^{-}_{nn}/\sigma^{(1),0}_{11}$, m=0; ✳ -- $\sigma^{-}_{nn}/\sigma^{(1),0}_{11}$, m=1; ✖ -- $\sigma^{-}_{nn}/\sigma^{(1),0}_{11}$, m=2; ★ -- $\sigma^{-}_{nn}/\sigma^{(1),0}_{11}$, m=3.

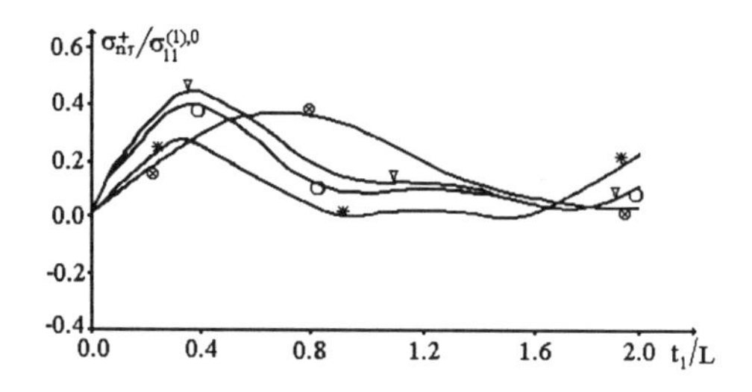

Fig.6.4.5. The distribution of the $\sigma^{+}_{n\tau}/\sigma^{(1),0}_{11}$ with respect to t_1/L when $\delta \le 1$ for m=0, $\varepsilon = 0.05$, $E^{(2)}/E^{(1)} = 50$, $\chi = 0.1$: ⊗ -- $\delta = 1$; ∇ -- $\delta = 1/2$; ○ -- $\delta = 1/3$; ✳ -- $\delta = 1/5$.

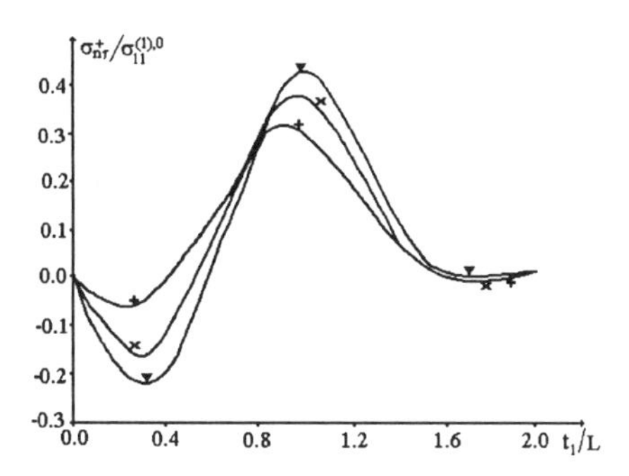

Fig.6.4.6. The distribution of the $\sigma^{+}_{n\tau}/\sigma^{(1),0}_{11}$ with respect to t_1/L when $\delta > 1$ for m=0, $\varepsilon = 0.05$, $E^{(2)}/E^{(1)} = 50$, $\chi = 0.1$: ✚ -- δ =2; ✖ -- δ =3; ▼ -- δ =5.

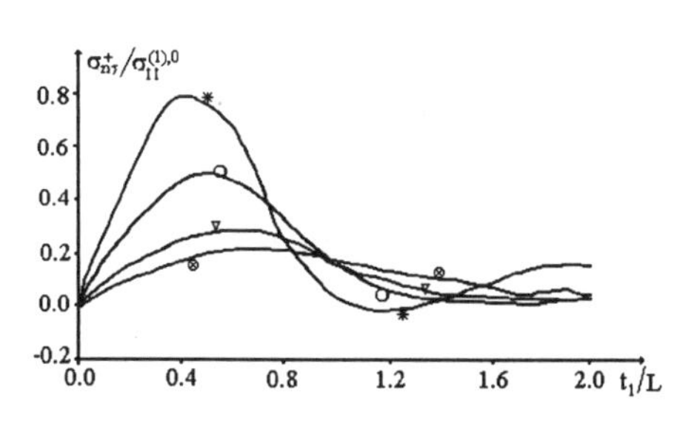

Fig.6.4.7. The distribution of the $\sigma^{+}_{n\tau}/\sigma^{(1),0}_{11}$ with respect to t_1/L for various m under $\delta = 1$, $\varepsilon = 0.05$, $E^{(2)}/E^{(1)} = 50$, $\chi = 0.1$: ⊗-- m=0, ∇-- m=1; ○ -- m=2; ✳ -- m=3.

Consider the influence of the parameter m on the distribution of these stresses shown in Figs.6.4.4 and 6.4.7. These graphs show that the absolute maximal values of the stresses σ_{nn}^{+}, σ_{nn}^{-} and $\sigma_{n\tau}^{+}$ increase monotonically with increasing m. In these cases the stresses σ_{nn}^{+} and σ_{nn}^{-} have their absolute maximal values at $t_1/L = 0$. However the point at which the stress $\sigma_{n\tau}^{+}$ has its maxima moves towards the left.

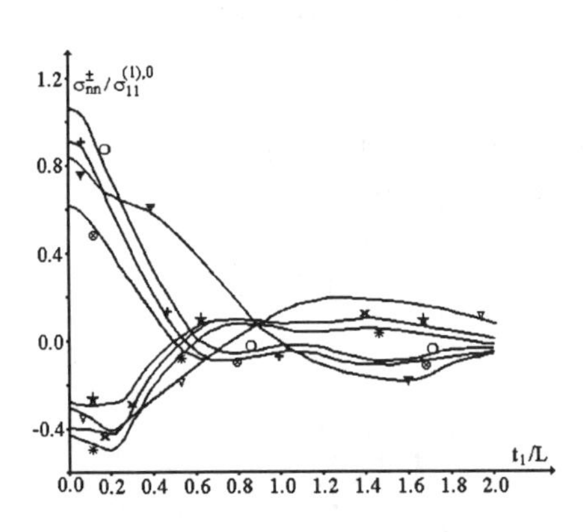

Fig.6.4.8. The distribution of the stresses σ_{nn}^{+}, σ_{nn}^{-} with respect to t_1/L when $\delta \le 1$ for m=0, $\varepsilon = 0.05$, $E^{(2)}/E^{(1)} = 50$, $\chi = 0.1$, $\eta^{(2)} = 0.3$: ⊗ -- $\sigma_{nn}^{+}/\sigma_{11}^{(1),0}$, $\delta = 1$; ○ -- $\sigma_{nn}^{+}/\sigma_{11}^{(1),0}$, $\delta = 1/2$; ✚ -- $\sigma_{nn}^{+}/\sigma_{11}^{(1),0}$, $\delta = 1/3$; ▼ -- $\sigma_{nn}^{+}/\sigma_{11}^{(1),0}$, $\delta = 1/5$; ∇ -- $\sigma_{nn}^{-}/\sigma_{11}^{(1),0}$, $\delta = 1$; ✱ -- $\sigma_{nn}^{-}/\sigma_{11}^{(1),0}$, $\delta = 1/2$; ✖ -- $\sigma_{nn}^{-}/\sigma_{11}^{(1),0}$, $\delta = 1/3$; ★-- $\sigma_{nn}^{-}/\sigma_{11}^{(1),0}$, $\delta = 1/5$.

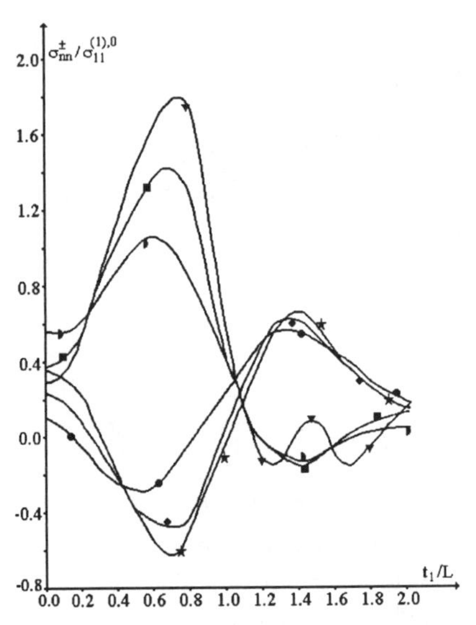

Fig.6.4.9. The distribution of the stresses σ_{nn}^{+}, σ_{nn}^{-} with respect to t_1/L when $\delta > 1$ for m=0, $\varepsilon = 0.05$, $E^{(2)}/E^{(1)} = 50$, $\chi = 0.1$, $\eta^{(2)} = 0.3$: ◗ -- $\sigma_{nn}^{+}/\sigma_{11}^{(1),0}$, $\delta = 2$; ■ -- $\sigma_{nn}^{+}/\sigma_{11}^{(1),0}$, $\delta = 3$; ▼ -- $\sigma_{nn}^{+}/\sigma_{11}^{(1),0}$, $\delta = 5$; ● -- $\sigma_{nn}^{-}/\sigma_{11}^{(1),0}$, $\delta = 2$; ◆ -- $\sigma_{nn}^{-}/\sigma_{11}^{(1),0}$, $\delta = 3$; ★-- $\sigma_{nn}^{-}/\sigma_{11}^{(1),0}$, $\delta = 5$.

Note that all the numerical results discussed here have been obtained for $\chi = 0.1$. However, other results not given here show that the distributions of the stresses σ_{nn}^{+}, σ_{nn}^{-} and $\sigma_{n\tau}^{+}$ with respect to t_1/L are qualitatively similar for other χ's. For example, these results show that for each χ, and increasing parameter m, the self-balanced stresses increase. In these cases the dependencies of $\sigma_{nn}^{\pm}/\sigma_{11}^{(1),0}$ and $\sigma_{n\tau}^{+}/\sigma_{11}^{(1),0}$ on χ are non monotonic; i.e. there is a value of χ (see $\chi = \chi'$) under which these stresses have obtained their maximal value. All numerical investigations are

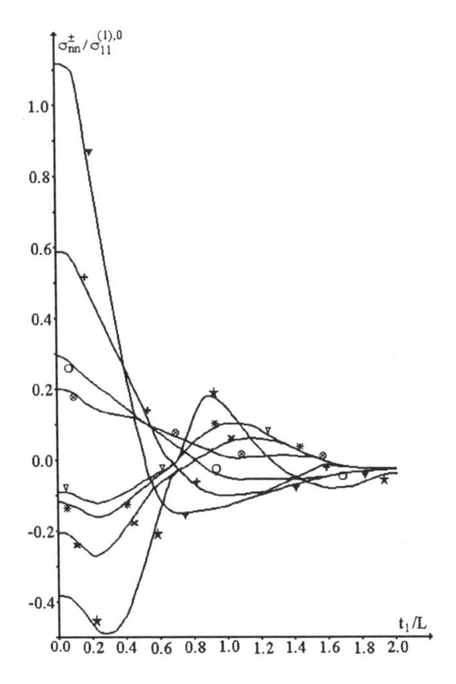

Fig.6.4.10. The distribution of the stresses σ^{+}_{nn}, σ^{-}_{nn} with respect to t_1/L for various m, and $\delta = 1$, $\varepsilon = 0.05$, $E^{(2)}/E^{(1)} = 50$, $\chi = 0.1$, $\eta^{(2)} = 0.3$: ⊗ -- $\sigma^{+}_{nn}/\sigma^{(1),0}_{11}$, m=0; ○ -- $\sigma^{+}_{nn}/\sigma^{(1),0}_{11}$, m=1; ✚ -- $\sigma^{+}_{nn}/\sigma^{(1),0}_{11}$, m=2; ▼ -- $\sigma^{+}_{nn}/\sigma^{(1),0}_{11}$, m=3; ∇-- $\sigma^{-}_{nn}/\sigma^{(1),0}_{11}$, m=0; ✳-- $\sigma^{-}_{nn}/\sigma^{(1),0}_{11}$, m=1; ✖ -- $\sigma^{-}_{nn}/\sigma^{(1),0}_{11}$, m=2; ★-- $\sigma^{-}_{nn}/\sigma^{(1),0}_{11}$, m=3.

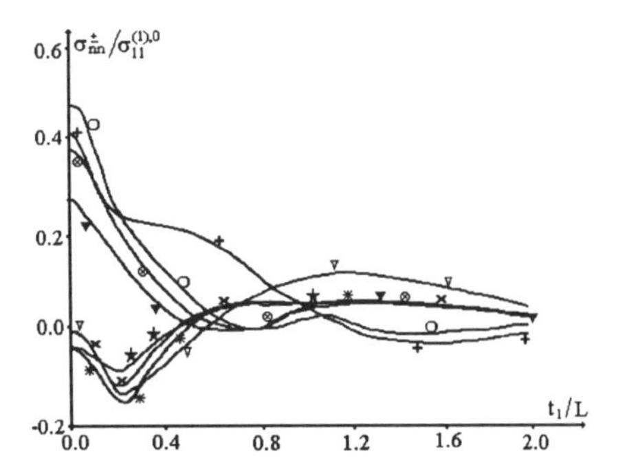

Fig.6.4.11. The distribution of the stresses σ^{+}_{nn}, σ^{-}_{nn} with respect to t_1/L when $\delta \le 1$ for m=0, $\varepsilon = 0.05$, $E^{(2)}/E^{(1)} = 50$, $\chi = 0.1$, $\eta^{(2)} = 0.5$: ⊗ -- $\sigma^{+}_{nn}/\sigma^{(1),0}_{11}$, δ =1; ○ -- $\sigma^{+}_{nn}/\sigma^{(1),0}_{11}$, δ =1/2; ✚ -- $\sigma^{+}_{nn}/\sigma^{(1),0}_{11}$, δ =1/3; ▼ -- $\sigma^{+}_{nn}/\sigma^{(1),0}_{11}$, δ =1/5; ∇ -- $\sigma^{-}_{nn}/\sigma^{(1),0}_{11}$, δ =1; ✳-- $\sigma^{-}_{nn}/\sigma^{(1),0}_{11}$, δ =1/2; ✖ -- $\sigma^{-}_{nn}/\sigma^{(1),0}_{11}$, δ =1/3; ★-- $\sigma^{-}_{nn}/\sigma^{(1),0}_{11}$, δ =1/5.

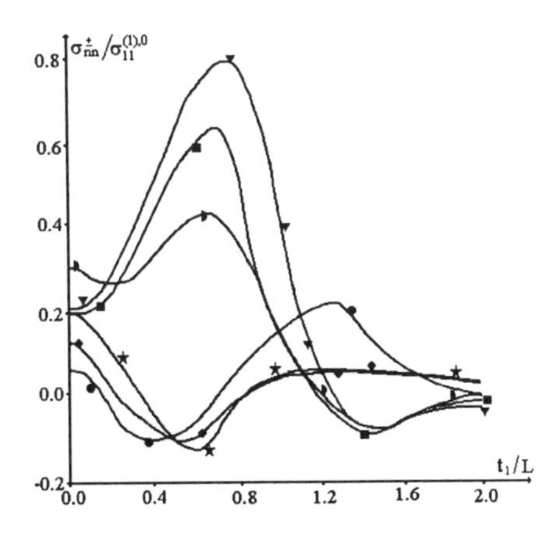

Fig.6.4.12. The distribution of the stresses σ^{+}_{nn}, σ^{-}_{nn} with respect to t_1/L when $\delta > 1$ for m=0, $\varepsilon = 0.05$, $E^{(2)}/E^{(1)} = 50$, $\chi = 0.1$, $\eta^{(2)} = 0.5$: ◗ -- $\sigma^{+}_{nn}/\sigma^{(1),0}_{11}$, δ =2; ■ -- $\sigma^{+}_{nn}/\sigma^{(1),0}_{11}$, δ =3; ▼ -- $\sigma^{+}_{nn}/\sigma^{(1),0}_{11}$, δ =5; ● -- $\sigma^{-}_{nn}/\sigma^{(1),0}_{11}$, δ =2; ◆ -- $\sigma^{-}_{nn}/\sigma^{(1),0}_{11}$, δ =3; ★-- $\sigma^{-}_{nn}/\sigma^{(1),0}_{11}$, δ =5

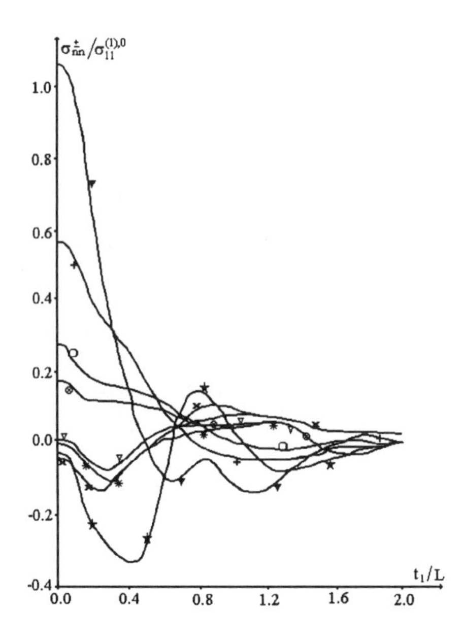

Fig.6.4.13. The distribution of the stresses σ_{nn}^{+}, σ_{nn}^{-} with respect to t_1/L for various m, and $\delta = 1$, $\varepsilon = 0.05$, $E^{(2)}/E^{(1)} = 50$, $\chi = 0.1$, $\eta^{(2)} = 0.5$:⊗ -- $\sigma_{nn}^{+}/\sigma_{11}^{(1),0}$, m=0; ❍ -- $\sigma_{nn}^{+}/\sigma_{11}^{(1),0}$, m=1; ✚ -- $\sigma_{nn}^{+}/\sigma_{11}^{(1),0}$, m=2; ▼ -- $\sigma_{nn}^{+}/\sigma_{11}^{(1),0}$, m=3; ∇-- $\sigma_{nn}^{-}/\sigma_{11}^{(1),0}$, m=0; ✳ -- $\sigma_{nn}^{-}/\sigma_{11}^{(1),0}$, m=1; ✖ -- $\sigma_{nn}^{-}/\sigma_{11}^{(1),0}$, m=2; ★ -- $\sigma_{nn}^{-}/\sigma_{11}^{(1),0}$, m=3.

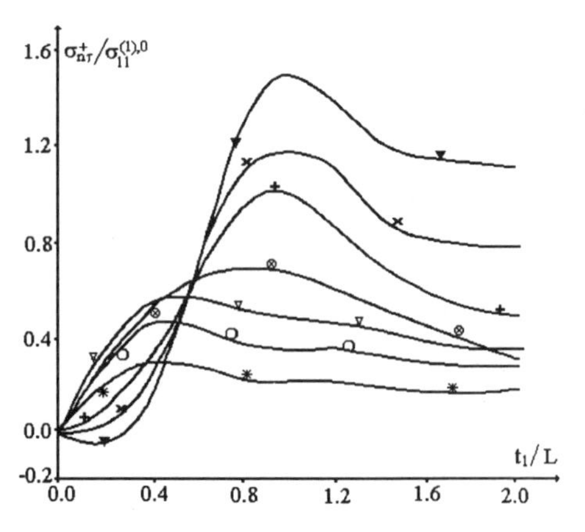

Fig.6.4.14. The distribution of the $\sigma_{n\tau}^{+}/\sigma_{11}^{(1),0}$ with respect to t_1/L in various δ for m=0, $\varepsilon = 0.05$, $E^{(2)}/E^{(1)} = 50$, $\chi = 0.1$, $\eta^{(2)} = 0.3$: ⊗ -- $\delta = 1$; ∇ -- $\delta = 1/2$; ❍ -- $\delta = 1/3$; ✳ -- $\delta = 1/5$; ✚ -- $\delta = 2$; ✖ -- $\delta = 3$; ▼ -- $\delta = 5$.

carried out for $\chi \le 1$; this inequality prevents contact between the different points of the contact surfaces $S^{\pm}$. For all selected δ and m it has been found that $\chi' \approx 0.1$ and therefore here the results corresponding to the $\chi = 0.1$ are selected for discussion.

We now consider the case where the filler layer concentration is high and it is necessary to include into account interaction between them.

The Figs. 6.4.8 - 6.4.13 reveal the influence of the parameters δ and m on the distribution of the stresses σ_{nn}^{+} and σ_{nn}^{-} with respect to t_1/L under high filler concentration, for fixed $E^{(2)}/E^{(1)}$, χ and ε. Corresponding results for the tangential stress $\sigma_{n\tau}^{+}$ are given in Figs.6.4.14-6.4.17.

In Figs.6.4.8 - 6.4.13 show that the influence of the parameters δ and m on the distribution of the stresses σ_{nn}^{+} and σ_{nn}^{-} is similar to that observed for $\eta^{(2)} = 0$. In these cases, when $\delta \le 1$, the values of $\sigma_{nn}^{\pm}/\sigma_{11}^{(1),0}$ decrease with increasing $\eta^{(2)}$. Note

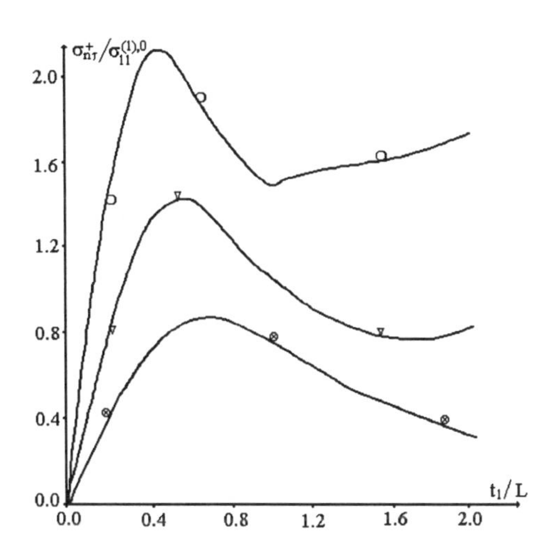

Fig.6.4.15. The distribution of the $\sigma_{n\tau}^{+}/\sigma_{11}^{(1),0}$ with respect to t_1/L for various m under $\delta=1$, $\varepsilon=0.05$, $E^{(2)}/E^{(1)}=50$, $\chi=0.1$, $\eta^{(2)}=0.3$:⊗-- m=1, ∇-- m=2; ❍ -- m=3.

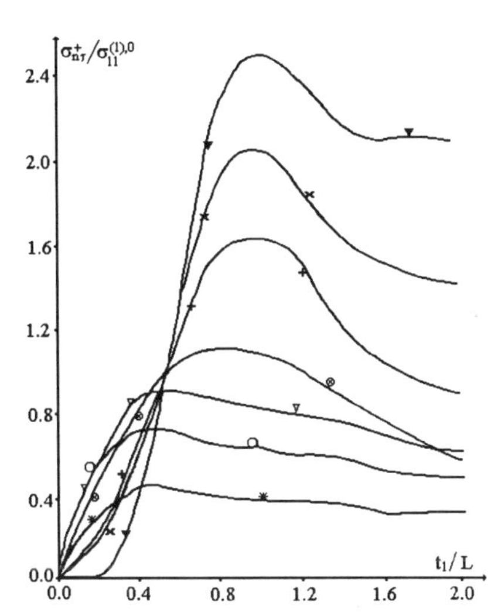

Fig.6.4.16. The distribution of the $\sigma_{n\tau}^{+}/\sigma_{11}^{(1),0}$ with respect to t_1/L in various δ for m=0, $\varepsilon=0.05$, $E^{(2)}/E^{(1)}=50$, $\chi=0.1$, $\eta^{(2)}=0.5$: ⊗ -- $\delta=1$; ∇ -- $\delta=1/2$; ❍ -- $\delta=1/3$; ✱ -- $\delta=1/5$; ✚ -- $\delta=2$; ✖ -- $\delta=3$; ▼ -- $\delta=5$.

that increasing the filler concentration does not change the character of the dependence of $\sigma_{nn}^{+}/\sigma_{11}^{(1),0}$ on t_1/L. However, with increasing $\eta^{(2)}$ the character of the dependence between $\sigma_{nn}^{-}/\sigma_{11}^{(1),0}$ and t_1/L becomes different from that obtained for $\eta^{(2)}=0$. In the cases where $\delta>1$ the values of $\sigma_{nn}^{-}/\sigma_{11}^{(1),0}$ decrease monotonically as $\eta^{(2)}$ increases. However, in the cases where $\delta>1$ the dependence between $\sigma_{nn}^{+}/\sigma_{11}^{(1),0}$ and $\eta^{(2)}$ is non-monotonic for the selected χ (i.e. for χ=0.1): at the beginning $\sigma_{nn}^{+}/\sigma_{11}^{(1),0}$ increases with $\eta^{(2)}$, but after some value of $\eta^{(2)}$ the $\sigma_{nn}^{+}/\sigma_{11}^{(1),0}$ decreases as $\eta^{(2)}$ increases. Moreover, graphs constructed at various m and $\eta^{(2)}$ show that at all selected m the values of $\sigma_{nn}^{\pm}/\sigma_{11}^{(1),0}$ decrease, basically, with growing $\eta^{(2)}$.

We analyse the results given in Figs.6.4.14-6.4.17 which show the distribution of the tangential stress $\sigma_{n\tau}^{+}$ with respect to t_1/L under various δ and m; the ratio $\sigma_{n\tau}^{+}/\sigma_{11}^{(1),0}$ increases monotonically with $\eta^{(2)}$ for all considered δ and m. When $\eta^{(2)}\geq 0.3$, the values of $\sigma_{n\tau}^{+}$ are found to be significantly greater than the

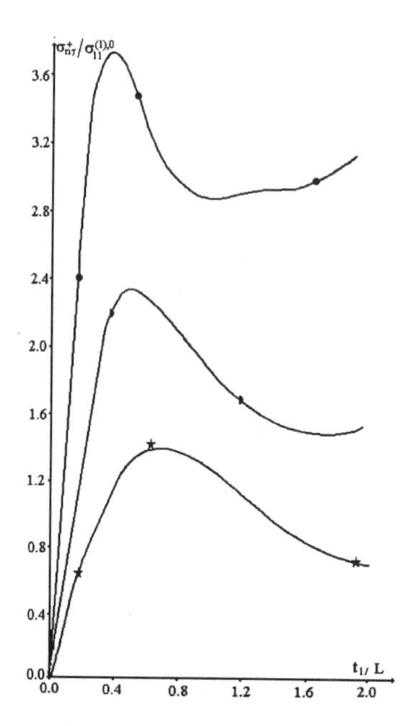

Fig.6.4.17. The distribution of the $\sigma^{+}_{n\tau}/\sigma^{(1),0}_{11}$ with respect to t_1/L for various m under $\delta=1$, $\varepsilon=0.05$, $E^{(2)}/E^{(1)}=50$, $\chi=0.1$, $\eta^{(2)}=0.5$: ★-- m=1, ◗-- m=2; ● -- m=3.

corresponding values of σ^{+}_{nn}, σ^{-}_{nn}. Moreover, when $\eta^{(2)} \geq 0.3$, and $\delta < 1$ the values of $\sigma^{+}_{n\tau}/\sigma^{(1),0}_{11}$ decrease monotonically with decreasing δ.

Numerical results which are not given here show that when $\eta^{(2)} = 0.1$, the length of the part of S^{+} (denoted by $\tilde{L}$) on which the inequality $\sigma^{+}_{n\tau}/\sigma^{(1),0}_{11} > 1$ satisfied, is insignificant with respect to L, i.e. $\tilde{L}/L \ll 1$. However, when $\eta^{(2)} = 0.3;0.5$, for m=0, $\delta > 1$ as well as for $\delta = 1$, m=1,2,3, we obtain $0.6L \leq \tilde{L} \leq 0.9L$. Moreover, when $\eta^{(2)} = 0.3;0.5$ the influence of the parameters δ and m on the vales of $\sigma^{+}_{n\tau}$ is more significant than when $\eta^{(2)} < 0.3$. Note that numerical results obtained for various χ (not given here) show that the parameters δ and m do not change the character of the dependence between $\sigma^{\pm}_{nn}/\sigma^{(1),0}_{11}$, $\sigma^{+}_{n\tau}/\sigma^{(1),0}_{11}$ and χ; they remain as in the case where $\delta = 1$, m=0.

Table 6.7. The values of $\sigma^{+}_{n\tau}/\sigma^{(1),0}_{11}$ for various δ, ε, $E^{(2)}/E^{(1)}$, $\eta^{(2)}$ under m=0, $\chi = 0.1$, $t_1/L = 0.7$.

ε	$E^{(2)}/E^{(1)}$	$\eta^{(2)}$	δ = 1	2	3	5
	20	0.3	0.045	0.067	0.081	0.094
		0.5	0.072	0.105	0.123	0.140
	50	0.3	0.099	0.137	0.158	0.176
0.01		0.5	0.157	0.214	0.244	0.267
	100	0.3	0.160	0.222	0.250	0.272
		0.5	0.253	0.344	0.385	0.414
	20	0.3	0.203	0.305	0.379	0.442
		0.5	0.323	0.483	0.586	0.675
	50	0.3	0.483	0.695	0.853	0.983
0.04		0.5	0.776	1.112	1.362	1.572
	100	0.3	0.892	1.292	1.592	1.854
		0.5	1.452	2.090	2.587	3.039
1	2	3	4	5	6	7

Table 6.7 (Continuation)

1	2	3	4	5	6	7
	20	0.3	0.431	0.691	0.906	1.061
		0.5	0.693	1.117	1.472	1.764
	50	0.3	1.174	1.931	2.617	3.224
0.07		0.5	1.911	3.232	4.461	5.652
	100	0.3	2.523	4.372	6.172	8.011
		0.5	4.235	7.482	10.79	14.35

Table 6.8. The values of $\sigma_{n\tau}^{+}/\sigma_{11}^{(1),0}$ for various m, ε, $E^{(2)}/E^{(1)}$, $\eta^{(2)}$ under $\delta=1$, $\chi=0.1$, $t_1/L=0.7$.

ε	$E^{(2)}/E^{(1)}$	$\eta^{(2)}$	m			
			0	1	2	3
	20	0.3	0.045	0.055	0.078	0.113
		0.5	0.072	0.082	0.088	0.169
	50	0.3	0.099	0.110	0.153	0.196
0.01		0.5	0.157	0.174	0.234	0.291
	100	0.3	0.160	0.182	0.230	0.263
		0.5	0.253	0.283	0.299	0.396
	20	0.3	0.203	0.255	0.393	0.594
		0.5	0.323	0.384	0.488	0.915
	50	0.3	0.483	0.584	0.930	1.327
0.04		0.5	0.776	0.943	1.489	2.211
	100	0.3	0.892	1.107	1.746	2.451
		0.5	1.452	1.785	2.784	3.993
	20	0.3	0.431	0.567	0.990	1.633
		0.5	0.693	0.902	1.447	2.649
	50	0.3	1.174	1.585	3.095	5.348
0.07		0.5	1.911	2.623	5.235	9.247
	100	0.3	2.523	3.513	7.508	13.98
		0.5	4.235	5.888	13.96	24.80

Tables 6.7, 6.8 show the values of $\sigma_{n\tau}^{+}/\sigma_{11}^{(1),0}$ for various δ, m, $\eta^{(2)}$, ε and $E^{(2)}/E^{(1)}$; for all selected δ and m, and increasing $E^{(2)}/E^{(1)}$ and ε, the values of $\sigma_{n\tau}^{+}/\sigma_{11}^{(1),0}$ increase strongly.

Table 6.9. The values of $\sigma_{n\tau}^{+}/\sigma_{11}^{(1),0}$ in various approximations for m=1, $\delta=1$, $E^{(2)}/E^{(1)}=50$, $\eta^{(2)}=0.5$, $\chi=0.1$.

ε	Numb. of approx	t_1/L		
		0.2	0.4	0.6
	1	0.082	0.144	0.173
	2	0.083	0.145	0.174
0.01	3	0.084	0.147	0.177
	4	0.084	0.147	0.177
	5	0.084	0.147	0.177
1	2	3	4	5

Table 6.9 (Continuation)

1	2	3	4	5
	1	0.328	0.576	0.695
	2	0.345	0.599	0.701
0.04	3	0.426	0.755	0.926
	4	0.431	0.763	0.932
	5	0.428	0.759	0.931
	1	0.574	1.008	1.217
	2	0.626	1.066	1.234
0.07	3	1.063	1.925	2.439
	4	1.109	1.996	2.502
	5	1.062	1.941	2.475

Table 6.9 gives the values of $\sigma_{n\tau}^{+}/\sigma_{11}^{(1),0}$ for various approximation; convergence is acceptable.

6.4.2. ANTI-PHASE CURVING

The structure of this composite is shown schematically in Fig.6.4.18. In the framework of the assumptions and notation accepted in the previous subsection we investigate the distribution of the self-balanced normal stresses $\sigma_{nn}^{\pm}$ on the inter-layer surfaces $S_1^{\pm}$ (Fig.6.4.18). As before, taking into account the periodicity of the material structure along the Ox_2 axis with period $4\left(H^{(2)}+H^{(1)}\right)$ we subdivide the body into four layers: $1^{(1)}$, $1^{(2)}$, $2^{(1)}$, $2^{(2)}$ and obtain the entire solution with them. In this case the equation of the middle surface of the $1^{(2)}$-th and $2^{(2)}$-th layers we take as $x_{21}^{(2)} = A\exp\left(-\left((x_1/L)^2\right)^{\delta}\right)\cos(mx_1/L)$, $x_{22}^{(2)} = -A\exp\left(-\left((x_1/L)^2\right)^{\delta}\right)\cos(mx_1/L)$, respectively. We assume $A < L$, and introduce a small parameter $\varepsilon = A/L$. As before we introduce the parameter $\chi = H^{(2)}/L$ and assume that $\nu^{(1)} = \nu^{(2)} = 0.3$. By virtue of the symmetry of the problem relative to the Ox_2 axis, we will restrict ourselves to examination of the distribution of the stresses $\sigma_{nn}^{\pm}$ for $x_1 \geq 0$.

Figs. 6.4.19 - 6.4.24 show the dependencies of $\sigma_{nn}^{\pm}/\sigma_{11}^{(1),0}$ on t_1/L for various δ and $\eta^{(2)}$. For each fixed δ and increasing $\eta^{(2)}$, the absolute values of $\sigma_{nn}^{+}/\sigma_{11}^{(1),0}$ increase. However, the influence of $\eta^{(2)}$ on the $\sigma_{nn}^{-}/\sigma_{11}^{(1),0}$ depends on the parameter χ: there are values $\chi' < \chi''$ of this parameter such that when $\chi < \chi'$ $(\chi > \chi'')$ the absolute values of $\sigma_{nn}^{-}/\sigma_{11}^{(1),0}$ increase insignificantly (significantly) as $\eta^{(2)}$ increases. When $\chi' < \chi < \chi''$, the dependence of $\sigma_{nn}^{-}/\sigma_{11}^{(1),0}$ on $\eta^{(2)}$ is non-monotonic; when $\chi = 0.1$ is in this interval and all numerical results given in this section are obtained for $\chi = 0.1$.

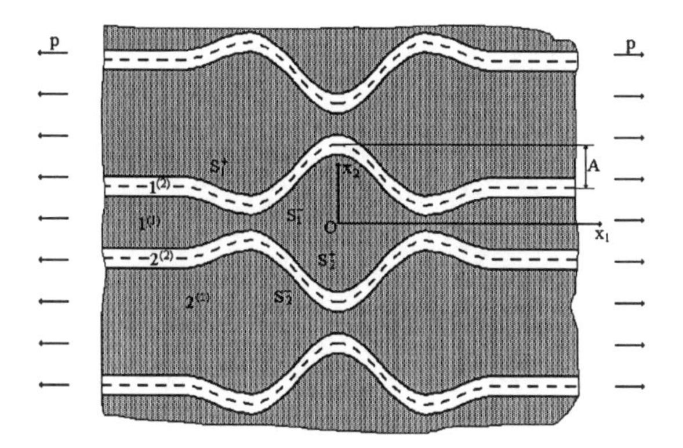

Fig.6.4.18. The structure of the composite material with anti-phase complicate local curved structure.

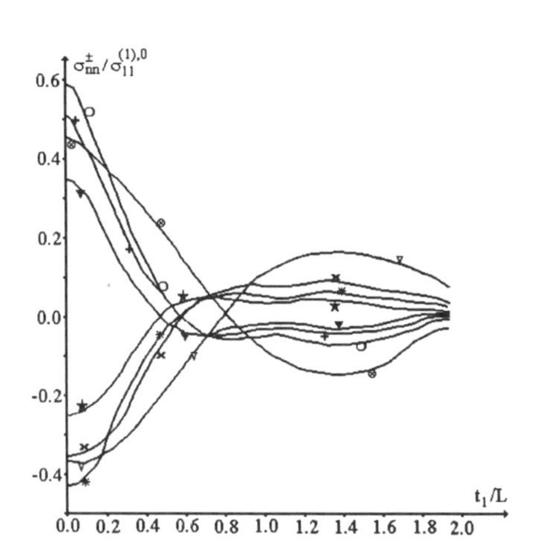

Fig.6.4.19. The distribution of the stresses σ_{nn}^{+}, σ_{nn}^{-} with respect to t_1/L when $\delta \leq 1$ for m=0, $\varepsilon = 0.05$, $E^{(2)}/E^{(1)} = 50$, $\chi = 0.1$, $\eta^{(2)} = 0.1$: ⊗ -- $\sigma_{nn}^{+}/\sigma_{11}^{(1),0}$, $\delta = 1$; O -- $\sigma_{nn}^{+}/\sigma_{11}^{(1),0}$, $\delta = 1/2$; ✚ -- $\sigma_{nn}^{+}/\sigma_{11}^{(1),0}$, $\delta = 1/3$; ▼ -- $\sigma_{nn}^{+}/\sigma_{11}^{(1),0}$, $\delta = 1/5$; ∇ -- $\sigma_{nn}^{-}/\sigma_{11}^{(1),0}$, $\delta = 1$; ✳ -- $\sigma_{nn}^{-}/\sigma_{11}^{(1),0}$, $\delta = 1/2$; ✖ -- $\sigma_{nn}^{-}/\sigma_{11}^{(1),0}$, $\delta = 1/3$; ★-- $\sigma_{nn}^{-}/\sigma_{11}^{(1),0}$, $\delta = 1/5$.

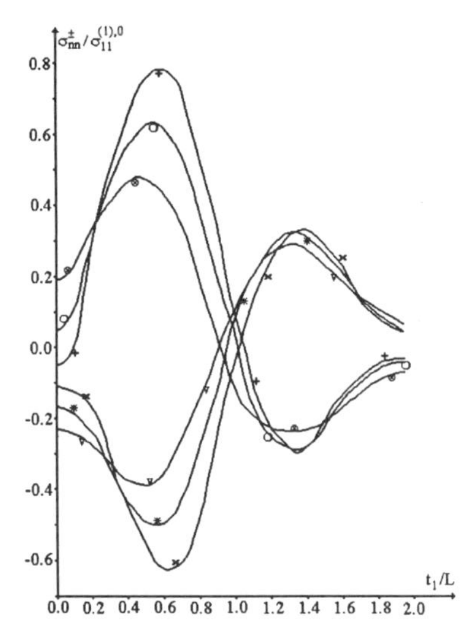

Fig.6.4.20. The distribution of the stresses σ_{nn}^{+}, σ_{nn}^{-} with respect to t_1/L when $\delta > 1$ for m=0, $\varepsilon = 0.05$, $E^{(2)}/E^{(1)} = 50$, $\chi = 0.1$, $\eta^{(2)} = 0.1$: ⊗ -- $\sigma_{nn}^{+}/\sigma_{11}^{(1),0}$, $\delta = 2$; O -- $\sigma_{nn}^{+}/\sigma_{11}^{(1),0}$, $\delta = 3$; ✚ -- $\sigma_{nn}^{+}/\sigma_{11}^{(1),0}$, $\delta = 5$; ∇ -- $\sigma_{nn}^{-}/\sigma_{11}^{(1),0}$, $\delta = 2$; ✳ -- $\sigma_{nn}^{-}/\sigma_{11}^{(1),0}$, $\delta = 3$; ✖ -- $\sigma_{nn}^{-}/\sigma_{11}^{(1),0}$, $\delta = 5$..

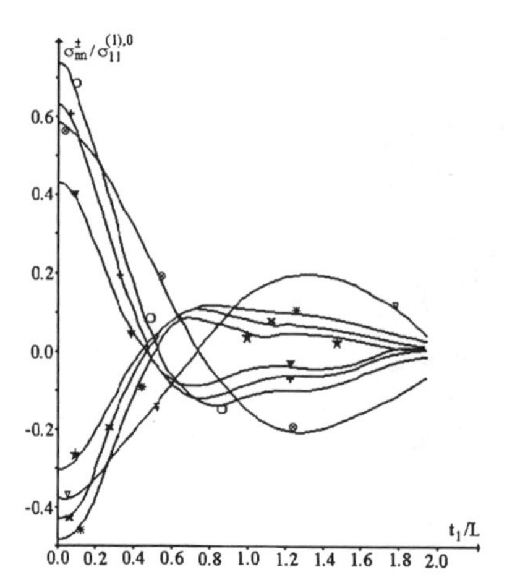

Fig.6.4.21. The distribution of the stresses σ_{nn}^{+}, σ_{nn}^{-} with respect to t_1/L when $\delta \leq 1$ for m=0, $\varepsilon = 0.05$, $E^{(2)}/E^{(1)} = 50$, $\chi = 0.1$, $\eta^{(2)} = 0.3$: ⊗ -- $\sigma_{nn}^{+}/\sigma_{11}^{(1),0}$, δ =1; ❍ -- $\sigma_{nn}^{+}/\sigma_{11}^{(1),0}$, δ =1/2; ✚ -- $\sigma_{nn}^{+}/\sigma_{11}^{(1),0}$, δ =1/3; ▼ -- $\sigma_{nn}^{+}/\sigma_{11}^{(1),0}$, δ =1/5; ∇ -- $\sigma_{nn}^{-}/\sigma_{11}^{(1),0}$, δ =1 ; ✳ -- $\sigma_{nn}^{-}/\sigma_{11}^{(1),0}$, δ =1/2; ✖ -- $\sigma_{nn}^{-}/\sigma_{11}^{(1),0}$, δ =1/3; ★-- $\sigma_{nn}^{-}/\sigma_{11}^{(1),0}$, δ =1/5

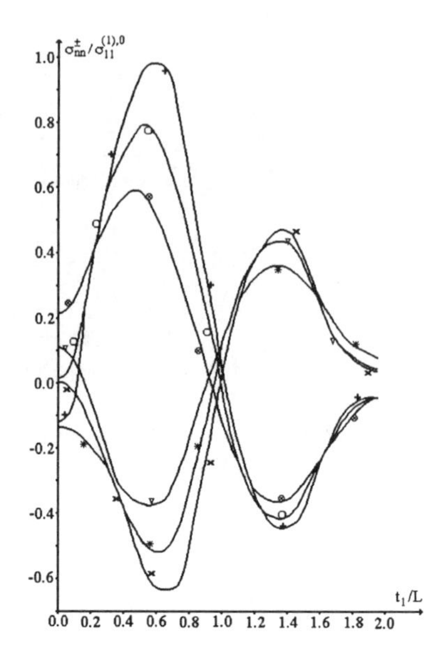

Fig.6.4.22. The distribution of the stresses σ_{nn}^{+}, σ_{nn}^{-} with respect to t_1/L when $\delta > 1$ for m=0, $\varepsilon = 0.05$, $E^{(2)}/E^{(1)} = 50$, $\chi = 0.1$, $\eta^{(2)} = 0.3$: ⊗ -- $\sigma_{nn}^{+}/\sigma_{11}^{(1),0}$, δ =2; ❍ -- $\sigma_{nn}^{+}/\sigma_{11}^{(1),0}$, δ =3; ✚ -- $\sigma_{nn}^{+}/\sigma_{11}^{(1),0}$, δ =5; ∇ -- $\sigma_{nn}^{-}/\sigma_{11}^{(1),0}$, δ =2 ; ✳ -- $\sigma_{nn}^{-}/\sigma_{11}^{(1),0}$, δ =3; ✖ -- $\sigma_{nn}^{-}/\sigma_{11}^{(1),0}$, δ =5.

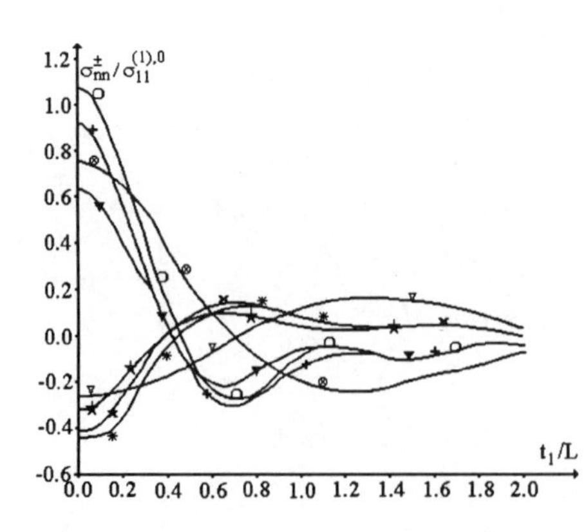

Fig.6.4.23. The distribution of the stresses σ_{nn}^{+}, σ_{nn}^{-} with respect to t_1/L when $\delta \leq 1$ for m=0, $\varepsilon = 0.05$, $E^{(2)}/E^{(1)} = 50$, $\chi = 0.1$, $\eta^{(2)} = 0.5$: ⊗ -- $\sigma_{nn}^{+}/\sigma_{11}^{(1),0}$, δ =1; ❍ -- $\sigma_{nn}^{+}/\sigma_{11}^{(1),0}$, δ =1/2; ✚ -- $\sigma_{nn}^{+}/\sigma_{11}^{(1),0}$, δ =1/3; ▼ -- $\sigma_{nn}^{+}/\sigma_{11}^{(1),0}$, δ =1/5; ∇ -- $\sigma_{nn}^{-}/\sigma_{11}^{(1),0}$, δ =1 ; ✳ -- $\sigma_{nn}^{-}/\sigma_{11}^{(1),0}$, δ =1/2; ✖ -- $\sigma_{nn}^{-}/\sigma_{11}^{(1),0}$, δ =1/3; ★-- $\sigma_{nn}^{-}/\sigma_{11}^{(1),0}$, δ =1/5

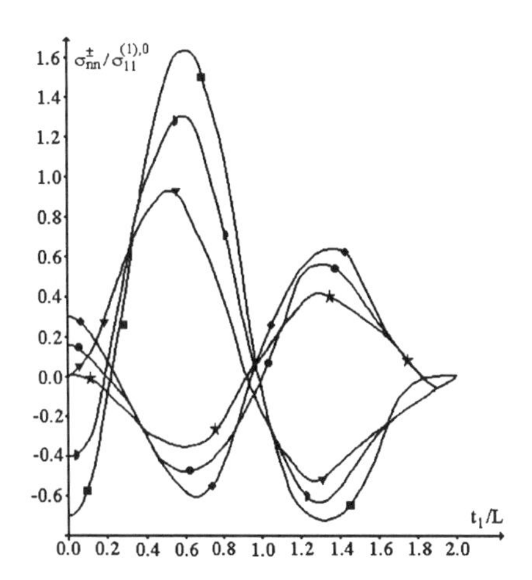

Fig.6.4.24. The distribution of the stresses σ_{nn}^{+}, σ_{nn}^{-} with respect to t_1/L when $\delta > 1$ for m=0, $\varepsilon = 0.05$, $E^{(2)}/E^{(1)} = 50$, $\chi = 0.1$, $\eta^{(2)} = 0.5$: ▼ -- $\sigma_{nn}^{+}/\sigma_{11}^{(1),0}$, $\delta = 2$; ◗ -- $\sigma_{nn}^{+}/\sigma_{11}^{(1),0}$, $\delta = 3$; ■ -- $\sigma_{nn}^{+}/\sigma_{11}^{(1),0}$, $\delta = 5$; ★ -- $\sigma_{nn}^{-}/\sigma_{11}^{(1),0}$, $\delta = 2$; ●-- $\sigma_{nn}^{-}/\sigma_{11}^{(1),0}$, $\delta = 3$; ◆-- $\sigma_{nn}^{-}/\sigma_{11}^{(1),0}$, $\delta = 5$.

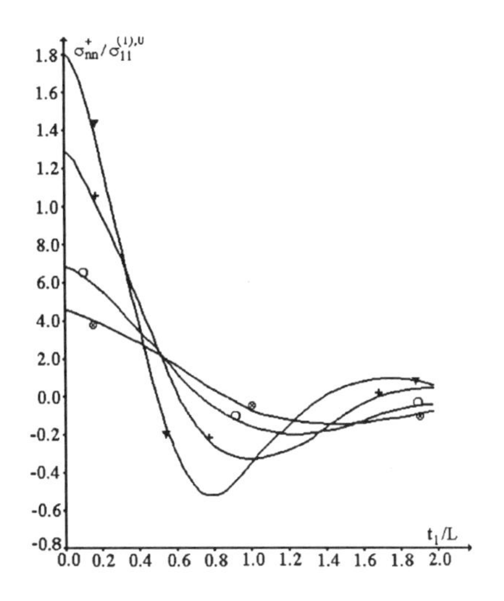

Fig.6.4.25. The distribution of the stress σ_{nn}^{+} with respect to t_1/L for various m, under $\delta = 1$, $\varepsilon = 0.05$, $E^{(2)}/E^{(1)} = 50$, $\chi = 0.1$, $\eta^{(2)} = 0.1$: ⊗ -- m=0; ❍ -- m=1, ✚ -- m=2, ▼ -- m=3.

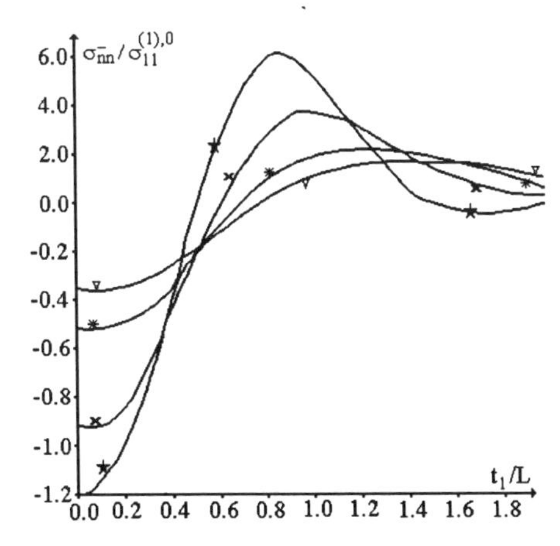

Fig.6.4.26. The distribution of the stress σ_{nn}^{-} with respect to t_1/L for various m, under $\delta = 1$, $\varepsilon = 0.05$, $E^{(2)}/E^{(1)} = 50$, $\chi = 0.1$, $\eta^{(2)} = 0.1$: ∇ -- m=0; ✳ -- m=1; ✖ -- m=2; ★-- m=3

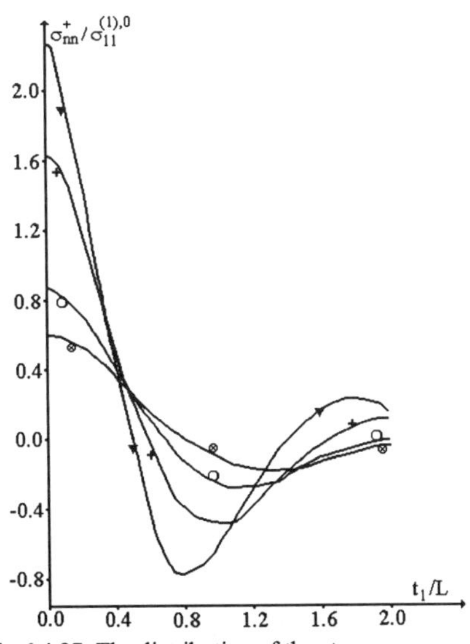

Fig.6.4.27. The distribution of the stress σ_{nn}^{+} with respect to t_1/L for various m, under $\delta = 1$, $\varepsilon = 0.05$, $E^{(2)}/E^{(1)} = 50$, $\chi = 0.1$, $\eta^{(2)} = 0.3$: ⊗ -- m=0; ❍ -- m=1, ✚ -- m=2, ▼ -- m=3.

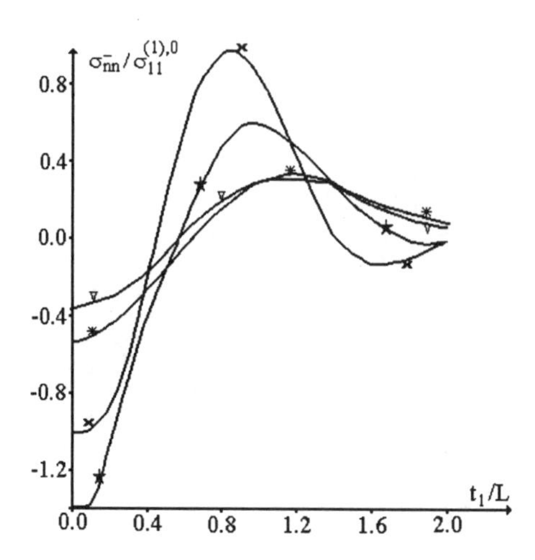

Fig.6.4.28. The distribution of the stress σ_{nn}^{-} with respect to t_1/L for various m, under $\delta=1, \varepsilon=0.05$, $E^{(2)}/E^{(1)}=50$, $\chi=0.1$, $\eta^{(2)}=0.3$: ∇ -- m=0; ✱ -- m=1; ✖ -- m=2; ★-- m=3

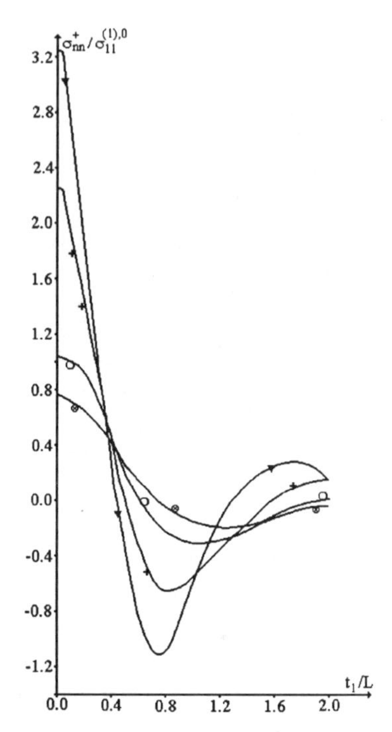

Fig.6.4.29. The distribution of the stress σ_{nn}^{+} with respect to t_1/L for the various m under $\delta=1$, $\varepsilon=0.05$, $E^{(2)}/E^{(1)}=50$, $\chi=0.1$, $\eta^{(2)}=0.5$: ⊗ -- m=0; ❍ -- m=1, ✚ -- m=2, ▼ -- m=3.

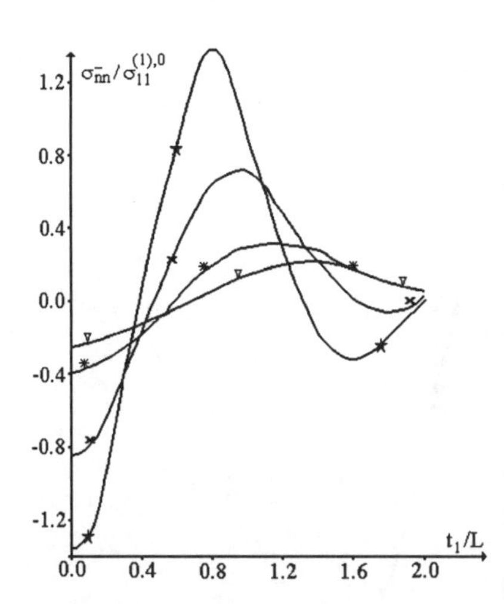

Fig.6.4.30. The distribution of the stress σ_{nn}^{-} with respect to t_1/L for the various m under $\delta=1, \varepsilon=0.05$, $E^{(2)}/E^{(1)}=50$, $\chi=0.1$, $\eta^{(2)}=0.5$: ∇ -- m=0; ✱ -- m=1; ✖ -- m=2; ★-- m=3

Figs.6.4.19, 6.4.21 and 6.4.23 show that when $1/5 \le \delta \le 1$, the normal stresses $\sigma_{nn}^{\pm}/\sigma_{11}^{(1),0}$ reach their absolute maximum value at the point $t_1/L=0$, which corresponds to the maximum rise. Here, the maximum values of $\sigma_{nn}^{+}/\sigma_{11}^{(1),0}$ and $\sigma_{nn}^{-}/\sigma_{11}^{(1),0}$ depend non-monotonically on the parameter δ, i.e., the maximum values of these stresses initially increase with a decrease in δ. After a certain $\delta=\delta^{*}$, a further reduction in δ leads to a reduction in the ratio $\sigma_{nn}^{\pm}/\sigma_{11}^{(1),0}$.

Figs. 6.4.20, 6.4.22 and 6.4.24 show that under $\delta>1$, the absolute maximum values of these stresses increase with an increase in the parameter δ. Here, the absolute maximum values of the normal stresses $\sigma_{nn}^{\pm}$ are reached not at the point corresponding to the maximum rise, but at points near which there is a sharp change in the slope of the curved layer. It should also be noted that, with an increase in δ, the point at which these stresses become maximal is shifted to the right.

The effect of the parameter m on the values of the stresses $\sigma_{nn}^{\pm}$ is in Figs.6.4.25 - 6.4.30; as m increases, the absolute maximum values of $\sigma_{nn}^{\pm}$ increase significantly.

Table 6.10. The values of $\sigma_{nn}^{+}/\sigma_{11}^{(1),0}$ at $t_1/L = 0$ for various m and χ under $E^{(2)}/E^{(1)} = 50$, $\varepsilon = 0.05$, $\delta = 1$

$\eta^{(2)}$	χ	m			
		0	1	2	3
0.3	0.05	0.386	0.608	1.278	2.115
	0.1	0.582	0.858	1.608	2.258
	0.2	0.574	0.722	1.066	1.188
	0.3	0.445	0.518	0.679	0.686
0.5	0.05	0.650	1.015	2.071	3.282
	0.1	0.774	1.170	2.256	3.292
	0.2	0.811	1.044	1.565	1.746
	0.3	0.642	0.745	0.911	0.921

Table 6.11. The values of $\sigma_{nn}^{+}/\sigma_{11}^{(1),0}$ at $t_1/L = 0.6$, m=0 for different δ, $E^{(2)}/E^{(1)}$, $\eta^{(2)}$, ε.

ε	$E^{(2)}/E^{(1)}$	$\eta^{(2)}$	δ			
			1	2	3	5
0.04	20	0.3	0.172	0.178	0.250	0.302
		0.5	0.214	0.265	0.384	0.468
	50	0.3	0.428	0.381	0.530	0.645
		0.5	0.549	0.595	0.865	1.066
	100	0.3	0.819	0.648	0.882	1.072
		0.5	1.086	1.013	1.462	1.813
0.07	20	0.3	0.378	0.449	0.665	0.839
		0.5	0.522	0.731	1.115	1.404
	50	0.3	0.948	1.031	1.547	2.001
		0.5	1.392	1.750	2.724	3.513
	100	0.3	1.950	1.851	2.765	3.632
		0.5	2.869	3.107	4.894	6.441

All these results have been obtained for $\chi = 0.1$. Many other numerical results not given here show that the qualitative aspects of the distributions of the stresses $\sigma_{nn}^{\pm}$

with respect to t_1/L remain valid for other values of the parameter χ as well. Table 6.10 shows the values of $\sigma_{nn}^{+}/\sigma_{11}^{(1),0}$ calculated for various m and χ; there is no change in the character of the relation between $\sigma_{nn}^{+}/\sigma_{11}^{(1),0}$ and χ with a change in the values of the parameter m. Here $\sigma_{nn}^{+}/\sigma_{11}^{(1),0}$ depends always non-monotonically on χ, and for any χ the values of $\sigma_{nn}^{+}/\sigma_{11}^{(1),0}$ increase with increasing m.

Tables 6.11 and 6.12 show that any chosen δ and m, $\sigma_{nn}^{+}/\sigma_{11}^{(1),0}$ increases with $E^{(2)}/E^{(1)}$, as well as with ε. Table 6.13 shows values $\sigma_{nn}^{+}/\sigma_{11}^{(1),0}$ for different approximations at a certain $E^{(2)}/E^{(1)}$ and ε; convergence is acceptable

Table 6.12. The values of $\sigma_{nn}^{+}/\sigma_{11}^{(1),0}$ at $t_1/L=0$, $\delta=1$ for different m, $E^{(2)}/E^{(1)}$, $\eta^{(2)}$, ε.

ε	$E^{(2)}/E^{(1)}$	$\eta^{(2)}$	m			
			0	1	2	3
0.04	20	0.3	0.172	0.252	0.472	0.717
		0.5	0.214	0.320	0.615	0.957
	50	0.3	0.428	0.615	1.114	1.569
		0.5	0.549	0.813	1.527	2.254
	100	0.3	0.819	1.156	2.019	2.650
		0.5	1.086	1.586	2.899	4.046
0.07	20	0.3	0.378	0.586	1.186	1.763
		0.5	0.522	0.821	1.667	2.492
	50	0.3	0.978	1.511	2.998	4.145
		0.5	1.392	2.199	4.449	6.314
	100	0.3	1.950	2.983	5.780	7.638
		0.5	2.869	4.520	9.015	12.148

Table 6.13. The values of $\sigma_{nn}^{+}/\sigma_{11}^{(1),0}$ in different approximations under $t_1/L=0$, $\delta=1$, m=3, $\eta^{(2)}=0.5$.

$E^{(2)}/E^{(1)}$	ε	Number of approxim.				
		1	2	3	4	5
20	0.01	0.153	0.166	0.168	0.168	0.168
	0.04	0.614	0.827	0.935	0.959	0.957
	0.07	1.075	1.728	2.304	2.529	2.492
100	0.01	0.570	0.620	0.632	0.633	0.633
	0.04	2.282	3.087	3.858	4.074	4.046
	0.07	3.993	6.459	10.59	12.62	12.14

6.5. Bibliographical Notes

The solution method presented in section 6.2 was proposed by S.D.Akbarov [9]. The study given in section 6.3 was also made by S.D. Akbarov [12,13]. The investigations presented in section 6.4 were carried out by A.N.Guz and S.D.Akbarov [103] and by S.D. Akbarov [14].

CHAPTER 7

FIBROUS COMPOSITES

In this chapter the problem formulation and solution methods presented in the previous chapters are developed for fibrous composites with curved structures, and various concrete problems are investigated.

7.1. Formulation

We consider an infinite body containing any number of non-intersecting curved fibers. The values related to the fibers are denoted by upper index (2)m, where m shows the number of a fiber, and the values related to the matrix by upper index (1). We associate a Cartesian system of coordinates $O_m x_{1m} x_{2m} x_{3m}$ and cylindrical system of coordinates $O_m r_m \theta_m z_m$ with m-th fiber (Fig.7.1.1). We will assume that the fibers lie along the $O_m x_{3m}$ axis and $x_{31} = x_{32} = \ldots = x_{3m} = \ldots = x_3 = z_1 = z_2 = \ldots \; z_m = z$.

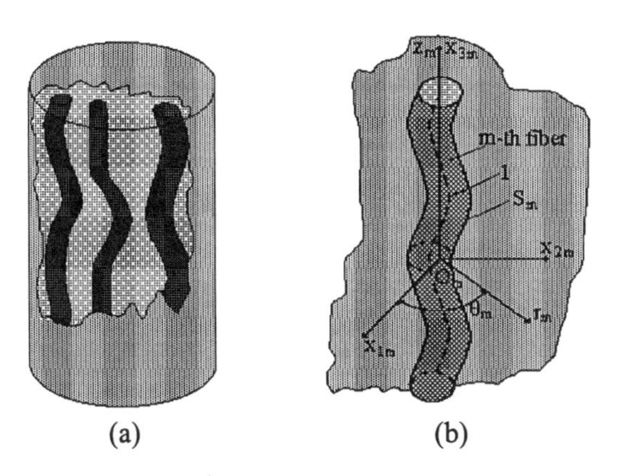

Fig.7.1.1. The schematic illustration of the structure of the composite with curved fibers (a); isolated m-th fiber (b).

We examine the case where the middle line of the fiber (shown by the number 1 in Fig.7.1.1) lies on the $O_m x_{2m} x_{3m}$ plane. Here we suppose that the cross sections of the fiber normal to the middle line of the fiber are circles, with constant radius R_m along the entire length of the fiber. We suppose that the materials of the fibers and matrix are homogeneous and isotropic. Moreover we assume that this body is loaded at infinity by uniformly distributed normal forces with intensity p acting in the direction of the $O_m x_{3m}$ axis.

Thus, we write the equilibrium equation, Hooke's law, and the Cauchy relations within the limits of the fibers and matrix.

$$\frac{\partial \sigma_{rr}}{\partial r} + \frac{1}{r}\frac{\partial \sigma_{r\theta}}{\partial \theta} + \frac{\partial \sigma_{rz}}{\partial z} + \frac{\sigma_{rr} - \sigma_{\theta\theta}}{r} = 0 ,$$

$$\frac{\partial \sigma_{r\theta}}{\partial r} + \frac{1}{r}\frac{\partial \sigma_{\theta\theta}}{\partial \theta} + \frac{\partial \sigma_{z\theta}}{\partial z} + \frac{2}{r}\sigma_{r\theta} = 0 ,$$

$$\frac{\partial \sigma_{zr}}{\partial r} + \frac{1}{r}\frac{\partial \sigma_{z\theta}}{\partial \theta} + \frac{\partial \sigma_{zz}}{\partial z} + \frac{\sigma_{zr}}{r} = 0 . \tag{7.1.1}$$

$$\sigma_{ij} = \lambda e \delta_i^j + 2\mu\varepsilon_{ij} ;\; ij = rr, \theta\theta, zz, r\theta, rz, \theta z ,\quad e = \varepsilon_{rr} + \varepsilon_{\theta\theta} + \varepsilon_{zz} , \tag{7.1.2}$$

$$\varepsilon_{rr} = \frac{\partial u_r}{\partial r},\; \varepsilon_{\theta\theta} = \frac{\partial u_\theta}{r\partial\theta} + \frac{u_r}{r},\; \varepsilon_{zz} = \frac{\partial u_z}{\partial z},\; \varepsilon_{r\theta} = \frac{1}{2}\left(\frac{\partial u_r}{r\partial\theta} + \frac{\partial u_\theta}{\partial r} - \frac{u_\theta}{r}\right),$$

$$\varepsilon_{\theta z} = \frac{1}{2}\left(\frac{\partial u_\theta}{\partial z} + \frac{\partial u_z}{r\partial\theta}\right) ;\; \varepsilon_{zr} = \frac{1}{2}\left(\frac{\partial u_z}{\partial r} + \frac{\partial u_r}{\partial z}\right). \tag{7.1.3}$$

In (7.1.1)-(7.1.3) the conventional notation is used. Moreover in (7.1.1)-(7.1.3) the upper indices which are omitted will be taken into account when applying these equations.

We assume that there is complete cohesion conditions on the fiber-matrix interfaces S_m:

$$\left.\left(\sigma_{rr}^{(2)m} n_r^m + \sigma_{r\theta}^{(2)m} n_\theta^m + \sigma_{rz}^{(2)m} n_z^m\right)\right|_{S_m} = \left.\left(\sigma_{rr}^{(1)m} n_r^m + \sigma_{r\theta}^{(1)m} n_\theta^m + \sigma_{rz}^{(1)m} n_z^m\right)\right|_{S_m} ,$$

$$\left.\left(\sigma_{r\theta}^{(2)m} n_r^m + \sigma_{\theta\theta}^{(2)m} n_\theta^m + \sigma_{\theta z}^{(2)m} n_z^m\right)\right|_{S_m} = \left.\left(\sigma_{r\theta}^{(1)m} n_r^m + \sigma_{\theta\theta}^{(1)m} n_\theta^m + \sigma_{\theta z}^{(1)m} n_z^m\right)\right|_{S_m} ,$$

$$\left.\left(\sigma_{rz}^{(2)m} n_r^m + \sigma_{\theta z}^{(2)m} n_\theta^m + \sigma_{zz}^{(2)m} n_z^m\right)\right|_{S_m} = \left.\left(\sigma_{rz}^{(1)m} n_r^m + \sigma_{\theta z}^{(1)m} n_\theta^m + \sigma_{zz}^{(1)m} n_z^m\right)\right|_{S_m} ,$$

$$\left. u_r^{(2)m}\right|_{S_m} = \left. u_r^{(1)m}\right|_{S_m} ,\quad \left. u_\theta^{(2)m}\right|_{S_m} = \left. u_\theta^{(1)m}\right|_{S_m} ,\quad \left. u_z^{(2)m}\right|_{S_m} = \left. u_z^{(1)m}\right|_{S_m} , \tag{7.1.4}$$

where n_r^m, n_θ^m, n_z^m are the components of the unit normal vector to the surface S_m in the cylindrical system of coordinates $O_m r_m \theta_m z_m$.

The equation of the middle line of the m-th fiber is written as follows:

$$x_{2m} = F_m(x_{3m}) = \varepsilon\delta_m(x_{3m}), \quad x_{1m} = 0, \tag{7.1.5}$$

where ε is a dimensionless small parameter. The geometric significance of this parameter will be indicated with a specifically prescribed form of the function (7.1.5).

Thus the investigation of the stress distribution in the fibrous composites with curved structures under the action of the given external forces is reduced to the solution to the system of equation (7.1.1) - (7.1.3), the contact conditions (7.1.4), and equation (7.1.5).

7.2. Method of Solution for Lower Fiber Concentration

Consider the case where the filler concentration is low and interaction between them can be neglected. We develop the method of solution formulated in the previous section. We will omit the index m, and investigate the stress state in an infinite elastic body containing a single curved fiber (Fig.7.2.1).

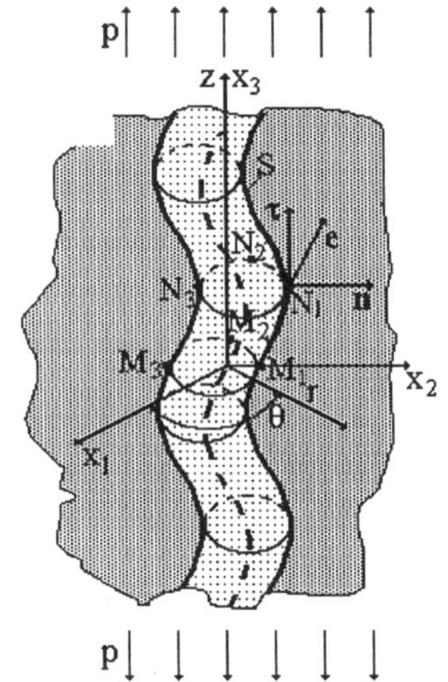

Fig.7.2.1. An infinite elastic body containing a single periodically curved fiber

7.2.1. GOVERNING EQUATIONS AND RELATIONS

First, using (7.1.5) and the condition on the cross section of the fiber, we derive the equation of the surface S in the cylindrical system of coordinates $Or\theta z$. For this purpose we introduce the parameter $t \in (-\infty, +\infty)$, and equation (7.1.5) rewrite through this parameter.

$$x_1 = 0, \quad x_3 = t, \quad x_2 = \varepsilon\delta(t). \tag{7.2.1}$$

Moreover we write the equation of the plane which is perpendicular to the tangent of the line (7.2.1)

$$\delta'(t_3)x_2 + x_3 - t = 0, \; \delta'(t) = \frac{d\delta(t)}{dt}. \tag{7.2.2}$$

We add the condition on the cross section of the fiber, which can be written as

$$x_1^2 + (x_2 - \delta(t))^2 + (x_3 - t)^2 = R^2. \tag{7.2.3}$$

For a point on the surface S, the coordinates x_1, x_2 and x_3 satisfy the equations (7.2.2) and (7.2.3). Thus, making a transformation to the cylindrical system of coordinates $Or\theta z$ and using (7.2.2), (7.2.3), we obtain the following equation for the surface S:

$$r = \left(1+\varepsilon^2(\delta'(t))^2 \sin^2\theta\right)^{-1}\Big\{\left(\varepsilon\delta(t)+\varepsilon^3\delta(t)(\delta'(t))^2\right)\sin\theta +$$

$$\left[R^2-\varepsilon^2(\delta(t))^2-\varepsilon^4(\delta'(t))^2(\delta(t))^2\left(1+\varepsilon^2(\delta'(t))^2\right)\sin^2\theta\right]^{\frac{1}{2}}\Big\},$$

$$z = t-\varepsilon\delta'(t)\,r(t)\sin\theta+\varepsilon^2\delta(t)\,\delta'(t). \tag{7.2.4}$$

After certain transformations, we obtain the following expressions from (7.2.4) for the components of the unit normal vector to the surface S:

$$n_r = r(\theta,t)\frac{\partial z(\theta,t)}{\partial t}[A(\theta,t)]^{-1},\quad n_\theta = \left[\frac{\partial z(\theta,t)}{\partial\theta}\frac{\partial r(\theta,t)}{\partial t}-\right.$$

$$\left.\frac{\partial r(\theta,t)}{\partial\theta}\frac{\partial z(\theta,t)}{\partial t}\right][A(\theta,t)]^{-1},\quad n_z = -r(\theta,t)\frac{\partial r(\theta,t)}{\partial t}[A(\theta,t)]^{-1}, \tag{7.2.5}$$

where

$$A(\theta,t) = \left[\left(r(\theta,t)\frac{\partial z(\theta,t)}{\partial t}\right)^2+\left(\frac{\partial z(\theta,t)}{\partial\theta}\frac{\partial r(\theta,t)}{\partial t}-\frac{\partial z(\theta,t)}{\partial t}\frac{\partial r(\theta,t)}{\partial\theta}\right)^2+\right.$$

$$\left.\left(r(\theta,t)\frac{\partial r(\theta,t)}{\partial t}\right)^2\right]^{\frac{1}{2}}. \tag{7.2.6}$$

As in the previous chapters we will seek quantities characterising the stress-strain state of the matrix and fiber in the form of series in positive powers of the small parameter ε:

$$\sigma_{rr}^{(k)} = \sum_{q=0}^{\infty}\varepsilon^q\sigma_{rr}^{(k),q},\ \ldots,\ \varepsilon_{rr}^{(k)} = \sum_{q=0}^{\infty}\varepsilon^q\varepsilon_{rr}^{(k),q},\ \ldots,\ u_r^{(k)} = \sum_{q=0}^{\infty}\varepsilon^q u_r^{(k),q},\ \ldots. \tag{7.2.7}$$

We expand the expressions (7.2.4) - (7.2.6) as series in ε:

$$r = R + \sum_{q=1}^{\infty} \varepsilon^q a_{rq}(R,\theta,t), \quad z = t + \sum_{q=1}^{\infty} \varepsilon^q a_{zq}(R,\theta,t),$$

$$n_r = 1 + \sum_{q=1}^{\infty} \varepsilon^q b_{rq}(R,\theta,t), \quad n_\theta = \sum_{q=1}^{\infty} \varepsilon^q b_{\theta q}(R,\theta,t), \quad n_z = \sum_{q=1}^{\infty} \varepsilon^q b_{zq}(R,\theta,t). \quad (7.2.8)$$

The expressions for the coefficients of the ε^q in (7.2.8) can be obtained from (7.2.4) - (7.2.6) by employing routine operations; we omit details.

Substituting the expressions (7.2.7), (7.2.8) in (7.1.4) and expanding the components of each approximation in Taylor's series in the vicinity (R,θ,t), we obtain contact conditions for each approximation. We record them for the zeroth, first and second approximations.

The zeroth approximation.

$$[\sigma_{rr}]_{1,0}^{2,0} = [\sigma_{r\theta}]_{1,0}^{2,0} = [\sigma_{rz}]_{1,0}^{2,0} = 0, \quad [u_r]_{1,0}^{2,0} = [u_\theta]_{1,0}^{2,0} = [u_z]_{1,0}^{2,0} = 0, \quad (7.2.9)$$

The first approximation.

$$[\sigma_{rr}]_{1,1}^{2,1} + f_1\left[\frac{\partial\sigma_{rr}}{\partial r}\right]_{1,0}^{2,0} + \varphi_1\left[\frac{\partial\sigma_{rr}}{\partial z}\right]_{1,0}^{2,0} + \gamma_r[\sigma_{rr}]_{1,0}^{2,0} + \gamma_\theta[\sigma_{r\theta}]_{1,0}^{2,0} + \gamma_z[\sigma_{rz}]_{1,0}^{2,0} = 0$$

$$[u_r]_{1,1}^{2,1} + f_1\left[\frac{\partial u_r}{\partial r}\right]_{1,0}^{2,0} + \varphi_1\left[\frac{\partial u_r}{\partial z}\right]_{1,0}^{2,0} = 0. \quad (7.2.10)$$

The second approximation.

$$[\sigma_{rr}]_{1,2}^{2,2} + f_1\left[\frac{\partial\sigma_{rr}}{\partial r}\right]_{1,1}^{2,1} + \varphi_1\left[\frac{\partial\sigma_{rr}}{\partial z}\right]_{1,1}^{2,1} + f_2\left[\frac{\partial\sigma_{rr}}{\partial r}\right]_{1,0}^{2,0} + \varphi_2\left[\frac{\partial\sigma_{rr}}{\partial z}\right]_{1,0}^{2,0} +$$

$$(f_1)^2\left[\frac{\partial^2\sigma_{rr}}{\partial r^2}\right]_{1,0}^{2,0} + f_1\varphi_1\left[\frac{\partial^2\sigma_{rr}}{\partial r\partial z}\right]_{1,0}^{2,0} + \frac{1}{2}(\varphi_1)^2\left[\frac{\partial^2\sigma_{rr}}{\partial z^2}\right]_{1,0}^{2,0} +$$

$$\gamma_r[\sigma_{rr}]_{1,1}^{2,1} + f_1\gamma_r\left[\frac{\partial\sigma_{rr}}{\partial r}\right]_{1,0}^{2,0} + \varphi_1\gamma_r\left[\frac{\partial\sigma_{rr}}{\partial z}\right]_{1,0}^{2,0} + \gamma_\theta[\sigma_{r\theta}]_{1,1}^{2,1} +$$

$$f_1\gamma_\theta\left[\frac{\partial\sigma_{r\theta}}{\partial r}\right]_{1,0}^{2,0}+\varphi_1\gamma_\theta\left[\frac{\partial\sigma_{r\theta}}{\partial z}\right]_{1,0}^{2,0}+\gamma_z[\sigma_{rz}]_{1,0}^{2,0}+f_1\gamma_z\left[\frac{\partial\sigma_{rz}}{\partial r}\right]_{1,0}^{2,0}+$$

$$\beta_r[\sigma_{rr}]_{1,0}^{2,0}+\beta_\theta[\sigma_{r\theta}]_{1,0}^{2,0}+\beta_z[\sigma_{rz}]_{1,0}^{2,0}=0$$

$$[u_r]_{1,2}^{2,2}+f_1\left[\frac{\partial u_r}{\partial r}\right]_{1,1}^{2,1}+\varphi_1\left[\frac{\partial u_r}{\partial z}\right]_{1,1}^{2,1}+f_2\left[\frac{\partial u_r}{\partial r}\right]_{1,0}^{2,0}+\varphi_2\left[\frac{\partial u_r}{\partial z}\right]_{1,0}^{2,0}+$$

$$(f_1)^2\left[\frac{\partial^2 u_r}{\partial r^2}\right]_{1,0}^{2,0}+f_1\varphi_1\left[\frac{\partial^2 u_r}{\partial r\partial z}\right]_{1,0}^{2,0}+\frac{1}{2}(\varphi_1)^2\left[\frac{\partial^2 u_r}{\partial z^2}\right]_{1,0}^{2,0}=0. \qquad (7.2.11)$$

In (7.2.10) and (7.2.11) the following notation is used.

$$[F]_{1,k}^{2,k}=F^{(1),k}(R,\theta,t)-F^{(2),k}(R,\theta,t),\quad k=0,1,2,$$

$$f_1=\delta(t)\sin\theta,\quad f_2=\frac{R}{2}\left[\frac{(\delta(t))^2}{R^2}\cos^2\theta+(\delta'(t)\sin\theta)^2\right],\quad \varphi_1=-R\delta'(t)\sin\theta,$$

$$\varphi_2=-\delta'(t)\delta(t)\sin^2\theta,\quad \gamma_r=\left(\frac{\delta(t)}{R}-\delta''(t)R\right)\sin\theta,\quad \gamma_\theta=-\frac{\delta(t)}{R}\cos\theta,$$

$$\beta_r=-\left[2\delta(t)\delta''(t)+(\delta'(t))^2\right]\sin^2\theta-\frac{1}{2}\left[\frac{(\delta(t)^2}{R^2}\cos^2\theta+(\delta'(t)\sin\theta)^2\right],$$

$$\beta_\theta=\frac{1}{2}\sin 2\theta\left[\delta(t)\delta''(t)-\frac{(\delta(t))^2}{R^2}\right],\quad \gamma_z=-\delta'(t)\sin\theta,$$

$$\beta_z=\frac{\delta(t)}{R}\delta(t)\cos 2\theta+R\delta'(t)\delta''(t)\sin^2\theta. \qquad (7.2.12)$$

Moreover, in (7.2.10), (7.2.11) we have given the contact relations for radial force $(\sigma_{rr}n_r+\sigma_{r\theta}n_\theta+\sigma_{rz}n_z)$ and radial displacement u_r. The rest of the contact relations for the first and second approximations are obtained from (7.2.10) and (7.2.11), respectively, by means of cyclic permutation of the indices r, θ and z only in the components stress tensor (first index is permuted) and displacement vector. These contact conditions remain valid for any model of a deformable solid body under small deformation.

Similarly, from (7.1.4), (7.1.5), (7.2.4) and (7.2.8) we can write the contact conditions for the subsequent approximations of (7.2.7). Since the equations (7.1.1)-(7.1.3) are satisfied for each approximation separately by virtue of linearity, to find each approximation we have a closed system of equations (7.1.1) - (7.1.3), and corresponding boundary and contact conditions. The zeroth approximation corresponds to the stress-strain state of the body with a rectilinear fiber balanced by external forces. The first, second and subsequent approximations correspond to self-balanced stresses caused by the curvatures of the fiber in the matrix. Consider the determination of each approximation and first assume that the materials of fiber and matrix are isotropic. Then we will consider the case in which these materials are rectilinear transversal-isotropic, with symmetry axis Ox_3 (Fig.7.2.1).

7.2.2. PERIODICAL CURVING FORM

The fiber and matrix material is isotropic.
In this case the elasticity relations both for the fiber and matrix are taken as (7.1.2) and instead of $E^{(k)}, \nu^{(k)}$ the notation E_k, ν_k $(k = 1,2)$ is used for material constants.

The zeroth approximation. We can write

$$\sigma_{zz}^{(1),0} = p, \quad \varepsilon_{zz}^{(1),0} = \varepsilon_{zz}^{(2),0} = \frac{p}{E_1}, \quad u_z^{(2),0} = u_z^{(1),0} = \varepsilon_{zz}^{(1),0} z,$$

$$u_\theta^{(1),0} = u_\theta^{(2),0} = 0, \; \sigma_{r\theta}^{(1),0} = \sigma_{r\theta}^{(2),0} = \sigma_{rz}^{(1),0} = \sigma_{rz}^{(2),0} = \sigma_{\theta z}^{(1),0} = \sigma_{\theta z}^{(2),0} = 0 \qquad (7.2.13)$$

Taking into account (7.2.13) and solving the equilibrium equation for the displacements, we obtain

$$u_r^{(1),0} = \frac{C_1}{r} + Cr; \quad u_r^{(2),0} = C_2 r, \qquad (7.2.14)$$

where C, C_1 and C_2 are unknown constants.

Using (7.2.9) and conditions $\sigma_{rr}^{(1),0} = \sigma_{\theta\theta}^{(1),0} \to 0$ as $r \to \infty$, we find these unknown constants in the form

$$C = -\nu_1 \varepsilon_{zz}^{(1),0}, \; C_1 = \frac{R^2(\nu_1 - \nu_2)}{\Lambda} \varepsilon_{zz}^{(1),0}, \; C_2 = \left(\frac{\nu_1 - \nu_2}{\Lambda} - \nu_1 \right) \varepsilon_{zz}^{(1),0}, \qquad (7.2.15)$$

where

$$\Lambda = \frac{1+\nu_1 + \frac{E_1}{E_2}(1+\nu_2)(1-2\nu_2)}{1+\nu_1}. \tag{7.2.16}$$

Thus, using (7.2.14)-(7.2.16) we write the expressions for the remaining non trivial components of the stress tensor:

$$\sigma_{rr}^{(2),0} = \sigma_{\theta\theta}^{(2),0} = \frac{(\nu_2 - \nu_1)}{(1+\nu_1)\Lambda} p, \quad \sigma_{rr}^{(1),0} = -\sigma_{\theta\theta}^{(1),0} = \frac{R^2}{r^2} \frac{(\nu_2 - \nu_1)}{1+\nu_1} \frac{p}{\Lambda},$$

$$\sigma_{zz}^{(2),0} = \frac{E_2}{E_1(1+\nu_1)} \left(\frac{\nu_1}{1+\nu_1} \frac{1+\nu_1 + (1-2\nu_1)\frac{E_2}{E_1}(1+\nu_2)}{\Lambda} + 1 \right) p. \tag{7.2.17}$$

We now consider the determination of the values of the first and second approximations. In this case we consider in detail the periodical curving of the fiber and make some remarks related to the local curving form.

The equation (7.1.5) of the middle line of the fiber we select as follows

$$x_2 = L \sin \frac{2\pi}{\ell} x_3 = \varepsilon \ell \sin \alpha x_3, \; x_1 = 0, \tag{7.2.18}$$

where $\alpha = 2\pi/\ell$. Assume that $L < \ell$ and as the small parameter we take $\varepsilon = L/\ell$.

The first approximation. For this approximation we obtain the following contact relation from (7.2.10), (7.2.12) - (7.2.17) and (7.2.18).

$$[\sigma_{rr}]_{1,1}^{2,1} = -\Pi \sin \alpha t \sin \theta, \quad [\sigma_{r\theta}]_{1,1}^{2,1} = \Pi \sin \alpha t \cos \theta,$$

$$[\sigma_{rz}]_{1,1}^{2,1} = 2\pi\left(\sigma_{zz}^{(1),0} - \sigma_{zz}^{(2),0}\right) \cos \alpha t \sin \theta, \quad [u_r]_{1,1}^{2,1} = \frac{R}{E_1} \Pi \sin \alpha t \sin \theta,$$

$$[u_\theta]_{1,1}^{2,1} = [u_z]_{1,1}^{2,1} = 0,. \tag{7.2.19}$$

where

$$\Pi = 4\pi \frac{p}{\Lambda} \frac{(\nu_1 - \nu_2)}{\alpha R} \tag{7.2.20}$$

We use a Papkovich-Neuber representation which can be written for displacements in the following form.

$$u_r^{(k),1} = 4(1-\nu_k)B_r^{(k),1} - \frac{\partial\left(rB_r^{(k),1}\right)}{\partial r} - \frac{\partial B_0^{(k),1}}{\partial r},$$

$$u_\theta^{(k),1} = 4(1-\nu_k)B_\theta^{(k),1} - \frac{\partial\left(B_r^{(k),1}\right)}{\partial \theta} - \frac{\partial B_0^{(k),1}}{r\partial \theta},$$

$$u_3^{(k),1} = -\frac{\partial\left(rB_r^{(k),1}\right)}{\partial x_3} - \frac{\partial B_0^{(k),1}}{\partial x_3}. \qquad (7.2.21)$$

The functions $B_r^{(k),1}$, $B_\theta^{(k),1}$ and $B_0^{(k),1}$ in (7.2.21) satisfy the equations

$$\Delta B_0^{(k),1} = 0, \ \Delta B_r^{(k),1} - \frac{B_r^{(k),1}}{r^2} - \frac{2}{r^2}\frac{\partial B_\theta^{(k),1}}{\partial \theta} = 0, \ \Delta B_\theta^{(k),1} - \frac{B_\theta^{(k),1}}{r^2} + \frac{2}{r^2}\frac{\partial B_r^{(k),1}}{\partial \theta} = 0, \qquad (7.2.22)$$

where

$$\Delta = \frac{\partial^2}{\partial r^2} + \frac{1}{r}\frac{\partial}{\partial r} + \frac{1}{r^2}\frac{\partial^2}{\partial \theta^2} + \frac{\partial^2}{\partial z^2}. \qquad (7.2.23)$$

Note that the presentation (7.2.21) and the equations (7.2.22), (7.2.23) remain valid for any subsequent approximation as well.

Noting the form of the expressions of the right hand side of the contact relations (7.2.19), we select the unknown functions $B_0^{(k),1}$, $B_\theta^{(k),1}$, $B_r^{(k),1}$ in the following form:

$$B_r^{(k),1} = g_r^{(k),1}(r)\sin\alpha z\sin\theta, \ B_\theta^{(k),1} = g_\theta^{(k),1}(r)\sin\alpha z\cos\theta,$$

$$B_0^{(k),1} = g_0^{(k),1}(r)\sin\alpha z\sin\theta. \qquad (7.2.24)$$

Substituting (7.2.24) in (7.2.22) we obtain ordinary differential equations for the functions $g_0^{(k),1}(r)$, $g_r^{(k),1}(r)$ and $g_\theta^{(k),1}(r)$. After solving these equations we obtain the following expressions for the functions (7.2.24).

$$B_r^{(1),1} = A_0^{(1)} K_1(\alpha r) \sin \alpha z \sin \theta , \qquad B_r^{(1),1} = \left[A_1^{(1)} K_0(\alpha r) + A_2^{(1)} K_2(\alpha r)\right] \sin \alpha z \sin \theta ,$$

$$B_\theta^{(1),1} = \left[A_1^{(1)} K_0(\alpha r) - A_2^{(1)} K_2(\alpha r)\right] \sin \alpha z \cos \theta$$

$$B_r^{(2),1} = A_0^{(2)} I_1(\alpha r) \sin \alpha z \sin \theta , \; B_r^{(2),1} = \left[A_1^{(2)} I_0(\alpha r) + A_2^{(2)} I_2(\alpha r)\right] \sin \alpha z \sin \theta ,$$

$$B_\theta^{(2),1} = \left[A_1^{(2)} I_0(\alpha r) - A_2^{(2)} I_2(\alpha r)\right] \sin \alpha z \cos \theta , \qquad (7.2.25)$$

Here $K_n(\alpha r)$ is a Macdonald function, $I_n(\alpha r)$ is a Bessel function of a purely imaginary argument.

From (7.2.25), (7.2.21), (7.2.19), (7.1.2) and (7.1.3) we find the values of the first approximation.

The second approximation. In this case, using the expressions of the previous approximations we derive the following contact relations from (7.2.11).

$$\left[\sigma_{rr}\right]_{1,2}^{2,2} = \Pi_{10} + \Pi_{11} \cos 2\theta + \Pi_{12} \cos 2\alpha t + \Pi_{13} \cos 2\alpha t \cos 2\theta ,$$

$$\left[\sigma_{r\theta}\right]_{1,2}^{2,2} = \Pi_{21} \sin 2\theta + \Pi_{23} \sin 2\theta \cos 2\alpha t_3 ,$$

$$\left[\sigma_{rz}\right]_{1,2}^{2,2} = \Pi_{31} \sin 2\alpha t_3 + \Pi_{33} \cos 2\theta \sin 2\alpha t_3 ,$$

$$\left[u_r\right]_{1,2}^{2,2} = \Pi_{40} + \Pi_{41} \cos 2\theta + \Pi_{42} \cos 2\alpha t_3 + \Pi_{43} \cos 2\alpha t_3 \cos 2\theta ,$$

$$\left[u_\theta\right]_{1,2}^{2,2} = \Pi_{51} \sin 2\theta + \Pi_{53} \sin 2\theta \cos 2\alpha t_3 ,$$

$$\left[u_z\right]_{1,2}^{2,2} = \Pi_{61} \sin 2\alpha t_3 + \Pi_{63} \cos 2\theta \sin 2\alpha t_3 . \qquad (7.2.26)$$

The constants Π_{ij} in (7.2.26) are expressed through the value of the preceding approximations; we omit details. Noting the form of the expressions entering the right hand side of (7.2.26) and using the presentation (7.2.21) - (7.2.23) the solutions to the equations (7.2.22) for the second approximation we find as follows:

$$B_r^{(1),2} = C_0^{(1)} \frac{1}{r} + \left(C_1^{(1)} \frac{1}{r^3} + C_2^{(1)} \frac{1}{r} \right) \sin 2\theta + C_3^{(1)} K_1(2\alpha r) \cos 2\alpha z +$$

$$\left[C_4^{(1)} K_3(2\alpha r) + C_5^{(1)} K_1(2\alpha r)\right] \cos 2\theta \cos 2\alpha z ,$$

$$B_r^{(2),2} = C_0^{(2)} r + \left(C_1^{(2)} r^3 + C_2^{(2)} r\right)\sin 2\theta + C_3^{(2)} I_1(2\alpha r)\cos 2\alpha z +$$

$$\left[C_4^{(2)} I_3(2\alpha r) + C_5^{(2)} I_1(2\alpha r)\right]\cos 2\theta \cos 2\alpha z,$$

$$B_\theta^{(1),2} = \left(C_1^{(1)} \frac{1}{r^3} - C_2^{(1)} \frac{1}{r}\right)\sin 2\theta + \left[C_4^{(1)} K_3(2\alpha r) - C_5^{(1)} K_1(2\alpha r)\right]\sin 2\theta \cos 2\alpha z,$$

$$B_\theta^{(2),2} = \left(C_1^{(2)} r^3 - C_2^{(2)} r\right)\sin 2\theta + \left[C_4^{(2)} I_3(2\alpha r) - C_5^{(2)} I_1(2\alpha r)\right]\sin 2\theta \cos 2\alpha z,$$

$$B_0^{(1),2} = C_6^{(1)} K_0(2\alpha r)\cos(2\alpha z) + C_7^{(1)} K_2(2\alpha r)\cos 2\theta \cos 2\alpha z,$$

$$B_0^{(2),2} = C_6^{(2)} I_0(2\alpha r)\cos(2\alpha z) + C_7^{(2)} I_2(2\alpha r)\cos 2\theta \cos 2\alpha z. \qquad (7.2.27)$$

Using the known procedure from the contact relations (7.2.26) we find the unknown constants entering (7.2.27) and in this way we determine the values of the second approximation and subsequent approximations.

The fiber and matrix material is transversely-isotropic.
We assume that the symmetry axis is the Ox_3 axis (Fig.7.2.1); the elasticity relations (7.1.2) are replaced by the following ones:

$$\varepsilon_{rr}^{(k)} = \frac{1}{E_1^{(k)}} \sigma_{rr}^{(k)} - \frac{\nu^{(k)}}{E_1^{(k)}} \sigma_{\theta\theta}^{(k)} - \frac{\nu_1^{(k)}}{E_3^{(k)}} \sigma_{zz}^{(k)}, \quad \varepsilon_{r\theta}^{(k)} = \frac{1+\nu^{(k)}}{E_1^{(k)}} \sigma_{r\theta}^{(k)}$$

$$\varepsilon_{\theta\theta}^{(k)} = -\frac{\nu^{(k)}}{E_1^{(k)}} \sigma_{rr}^{(k)} + \frac{1}{E_1^{(k)}} \sigma_{\theta\theta}^{(k)} - \frac{\nu_1^{(k)}}{E_3^{(k)}} \sigma_{zz}^{(k)}, \quad \varepsilon_{rz}^{(k)} = \frac{1}{2G_1^{(k)}} \sigma_{rz}^{(k)}$$

$$\varepsilon_{zz}^{(k)} = -\frac{\nu_1^{(k)}}{E_3^{(k)}} \left(\sigma_{rr}^{(k)} + \sigma_{\theta\theta}^{(k)}\right) + \frac{1}{E_3^{(k)}} \sigma_{zz}^{(k)}, \quad \varepsilon_{\theta z}^{(k)} = \frac{1}{2G_1^{(k)}} \sigma_{\theta z}^{(k)}. \qquad (7.2.28)$$

Consider the determination of each approximation. We restrict ourselves to the determination of the zeroth and first approximations.

The zeroth approximation. For this approximation the relations (7.2.13) also hold with E_1 replaced by $E_3^{(1)}$, i.e.

$$\varepsilon_{zz}^{(1),0} = \varepsilon_{zz}^{(2),0} = \frac{p}{E_3^{(1)}} \tag{7.2.29}$$

As in the isotropic case we obtain the following expressions for the displacements and non-zero stresses:

$$u_r^{(1),0} = \frac{C_1}{r} + Cr, \quad u_r^{(2),0} = C_2 r, \quad u_z^{(1),0} = u_z^{(2),0} = \varepsilon_{zz}^{(1),0} z,$$

$$\sigma_{rr}^{(1),0} = -\sigma_{\theta\theta}^{(1),0} = -\frac{E_1^{(1)}}{(1+\nu^{(1)})} \frac{C_1 r^2}{R^2},$$

$$\sigma_{rr}^{(2),0} = \sigma_{\theta\theta}^{(2),0} = \frac{A^{(2)}}{\nu_1^{(2)}} \frac{C_1}{R^2} + \frac{A^{(2)}}{\nu_1^{(2)}} \left(\nu_1^{(2)} - \nu_1^{(1)}\right) \varepsilon_{zz}^{(1),0},$$

$$\sigma_{zz}^{(1),0} = p, \quad \sigma_{zz}^{(2),0} = \frac{E_3^{(2)}}{E_3^{(1)}} p + 2A^{(2)} \left(\frac{C_1}{R^2} + \left(\nu_1^{(2)} - \nu_1^{(1)}\right) \varepsilon_{zz}^{(1),0} \right), \tag{7.2.30}$$

where

$$C = -\nu_1^{(1)} \varepsilon_{zz}^{(1),0}, \quad C_1 = \left(\nu_1^{(1)} - \nu_1^{(2)}\right) \frac{R^2 p}{E_3^{(1)}} \frac{\left(1+\nu^{(1)}\right)}{\Lambda},$$

$$C_2 = \frac{C_1}{R^2} - \nu_1^{(1)} \varepsilon_{zz}^{(1),0}, \quad A^{(2)} = \frac{\nu_1^{(2)} E_1^{(2)} E_3^{(2)}}{\left(1-\nu^{(2)}\right) E_3^{(2)} - 2\left(\nu_1^{(2)}\right)^2 E_1^{(2)}}. \tag{7.2.31}$$

In (7.2.31) the following notation is used.

$$\Lambda = 1 + \nu^{(1)} + \nu_1^{(2)} E_1^{(1)} \left(\frac{1-\nu^{(2)}}{\nu_1^{(2)} E_1^{(2)}} - \frac{2\nu_1^{(2)}}{E_3^{(2)}} \right). \tag{7.2.32}$$

The first approximation. For this approximation the contact conditions (7.2.19) hold , but in this case the coefficient $R/E^{(1)}$ of Π in the contact condition for displacement u_r must be replaced by $\left(1+\nu^{(1)}\right)/\left(RE_1^{(1)}\right)$. Moreover the expression for Π in (7.2.20) must be replaced by the following one.

$$\Pi = 4\pi \frac{E_1^{(1)} C_1}{(1+\nu^{(1)})\alpha R}. \tag{7.2.33}$$

We use the following representation for transversal - isotropic body in the cylindrical system of coordinates [94, 95, 101]:

$$u_r^{(k),1} = \frac{1}{r}\frac{\partial}{\partial\theta}\Psi^{(k)} - \frac{\partial^2}{\partial r\partial z}X^{(k)}, \quad u_\theta^{(k),1} = -\frac{\partial}{\partial r}\Psi^{(k)} - \frac{1}{r}\frac{\partial^2}{\partial\theta\partial z}X^{(k)},$$

$$u_3^{(k)} = \left(A_{13}^{(k)} + G_1^{(k)}\right)^{-1}\left(A_{11}^{(k)}\Delta_1 + G_1^{(k)}\frac{\partial^2}{\partial z^2}\right)X^{(k)}, \tag{7.2.34}$$

where

$$\Delta_1 = \frac{\partial^2}{\partial r^2} + \frac{1}{r}\frac{\partial}{\partial r} + \frac{1}{r^2}\frac{\partial^2}{\partial\theta^2}. \tag{7.2.35}$$

The functions $\Psi^{(k)}$ and $X^{(k)}$ are determined by the following equations:

$$\left(\Delta_1 + \left(\xi_1^{(k)}\right)^2\frac{\partial^2}{\partial z^2}\right)\Psi = 0; \quad \left(\Delta_1 + \left(\xi_2^{(k)}\right)^2\frac{\partial^2}{\partial z^2}\right)\left(\Delta_1 + \left(\xi_3^{(k)}\right)^2\frac{\partial^2}{\partial z^2}\right)X = 0, \tag{7.2.36}$$

In (7.2.36) $\xi_i^{(k)}$ are the constants and are determined from the following relations.

$$\xi_1^{(k)} = \left(\frac{G_1^{(k)}2\left(1+\nu^{(k)}\right)}{E_1^{(k)}}\right)^{\frac{1}{2}}, \quad \xi_{2;3}^{(k)} = \left(2A_{11}^{(k)}G_1^{(k)}\right)^{-\frac{1}{2}}\left\{\left[A_{33}^{(k)}A_{11}^{(k)} + \left(G_1^{(k)}\right)^2 - \right.\right.$$

$$\left.\left.\left(G_1^{(k)} + A_{13}^{(k)}\right)^2\right] \pm \left[\left(A_{33}^{(k)}A_{11}^{(k)} + \left(G_1^{(k)}\right)^2 - \left(G_1^{(k)} + A_{13}^{(k)}\right)^2\right)^2 - 4A_{33}^{(k)}A_{11}^{(k)}\left(G_1^{(k)}\right)^2\right]\right\}^{\frac{1}{2}} \tag{7.2.37}$$

Note that the mechanical constants $A_{ij}^{(k)}$ (i,j=1,2,3) are determined from the constans entering the elasticity relations (7.2.28) by the following formulae:

$$A_{ij}^{(k)} = \frac{\mathrm{Cofactor}\left(a_{ij}^{(k)}\right)}{\det\left\|a_{nm}^{(k)}\right\|}, \quad i;j;n;m=1,2,3, \tag{7.2.38}$$

where

$$a_{11}^{(k)} = a_{22}^{(k)} = \frac{1}{E_1^{(k)}}, \quad a_{33}^{(k)} = \frac{1}{E_3^{(k)}}, \quad a_{12}^{(k)} = -\frac{\nu^{(k)}}{E_1^{(k)}},$$

$$a_{12}^{(k)} = -\frac{\nu^{(k)}}{E_1^{(k)}}, \quad a_{13}^{(k)} = a_{23}^{(k)} = \frac{\nu_1^{(k)}}{E_3^{(k)}}, \quad a_{ij}^{(k)} = a_{ji}^{(k)}. \tag{7.2.39}$$

By direct verification we determine that in almost all practical cases the following relations hold:

$$\left(\xi_1^{(k)}\right)^2 > 0, \quad \left(\xi_2^{(k)}\right)^2 > 0, \quad \left(\xi_3^{(k)}\right)^2 > 0, \quad \left(\xi_2^{(k)}\right)^2 \neq \left(\xi_3^{(k)}\right)^2. \tag{7.2.40}$$

Noting (7.2.40), we select the solution to equations (7.2.36) as follows:

$$\Psi^{(1)} = \alpha A_0^{(1)} K_1\left(\alpha\xi_1^{(1)} r\right)\sin\alpha z\cos\theta, \quad \Psi^{(2)} = \alpha A_0^{(2)} I_1\left(\alpha\xi_1^{(2)} r\right)\sin\alpha z\cos\theta,$$

$$X^{(1)} = \left[A_1^{(1)} K_1\left(\alpha\xi_2^{(1)} r\right) + A_2^{(1)} K_1\left(\alpha\xi_3^{(1)} r\right)\right]\cos\alpha z\sin\theta,$$

$$X^{(2)} = \left[A_1^{(2)} I_1\left(\alpha\xi_2^{(2)} r\right) + A_2^{(2)} I_1\left(\alpha\xi_3^{(2)} r\right)\right]\cos\alpha z\sin\theta. \tag{7.2.41}$$

The unknown constants entering (7.2.41) are determined from the relations (7.2.34), (7.2.28) and (7.2.19). This gives the first approximation; the second and subsequent approximations are determined likewise.

7.2.3. LOCAL CURVING FORM

We assume that the material of the fiber and matrix is transversely-isotropic with symmetry axis Oz. Moreover, we suppose that all assumptions and notation accepted in the previous subsection hold, and the curving of the fiber is local, given by

$$x_2 = F(z) = \varepsilon\delta(z) = A\exp\left(-\left(\frac{z}{L}\right)^{2\beta}\right)\cos\left(m\frac{z}{L}\right), \tag{7.2.42}$$

where the parameters A, L, β and m have the same meaning as those entering (6.2.6).

Consider the determination of the values for each approximation. The zeroth approximation is found as before. To get the first approximation we apply the exponential Fourier transform

$$\overline{V}(r,\theta,\lambda) = \int_{-\infty}^{+\infty} V(r,\theta,z)e^{-iz\lambda}dz\,, \qquad (7.2.43)$$

to all the corresponding equations and relations and the functions $\Psi^{(k)}$ and $X^{(k)}$ are presented as follows:

$$\Psi^{(1)} = \frac{1}{2\pi}\int_{-\infty}^{+\infty} \overline{A}_0^{(1)}(\lambda)K_1\left(\lambda\xi_1^{(1)}r\right)e^{i\lambda z}d\lambda\cos\theta\,,$$

$$\Psi^{(2)} = \frac{1}{2\pi}\int_{-\infty}^{+\infty} \overline{A}_0^{(2)}(\lambda)I_1\left(\lambda\xi_1^{(2)}r\right)e^{i\lambda z}d\lambda\cos\theta\,,$$

$$X^{(1)} = \frac{1}{2\pi}\int_{-\infty}^{+\infty}\left[\overline{A}_1^{(1)}(\lambda)K_1\left(\lambda\xi_2^{(1)}r\right)+\overline{A}_2^{(1)}(\lambda)K_1\left(\lambda\xi_3^{(1)}r\right)\right]e^{i\lambda z}d\lambda\sin\theta\,,$$

$$X^{(2)} = \frac{1}{2\pi}\int_{-\infty}^{+\infty}\left[\overline{A}_1^{(2)}(\lambda)I_1\left(\lambda\xi_2^{(2)}r\right)+\overline{A}_2^{(2)}(\lambda)I_1\left(\lambda\xi_3^{(2)}r\right)\right]e^{i\lambda z}d\lambda\sin\theta\,. \qquad (7.2.44)$$

Applying the transformation (7.2.43) to the contact conditions (7.2.10) and taking into account the expressions (7.2.12), we obtain the contact relations for transformed quantities. Note that these relations can be written as (7.2.19) by replacing $\sigma_{rr}^{(k),1}$, $\sigma_{r\theta}^{(k),1}$, $\sigma_{rz}^{(k),1}$, $u_r^{(k),1}$, $u_\theta^{(k),1}$, $u_z^{(k),1}$, $\sin\alpha z$ and $\cos\alpha z$ by $\overline{\sigma}_{rr}^{(k),1}$, $\overline{\sigma}_{r\theta}^{(k),1}$, $\overline{\sigma}_{rz}^{(k),1}$, $\overline{u}_r^{(k),1}$, $\overline{u}_\theta^{(k),1}$, $\overline{u}_z^{(k),1}$, $\overline{\delta}(\lambda)$ and $(i\lambda)\overline{\delta}(\lambda)$, respectively. Applying the algorithm described in the previous chapter we can determine the values of the stresses arising as a result of the fiber local curving; subsequent approximations can be obtained likewise.

7.3. Method of Solution for Higher Fiber Concentrations

We suppose that the fibers are close together and interact. For simplicity we take three concrete examples: two fibers, row fibers and double periodically fibers. We assume that the curving and materials of the fibers are identical, and the fibers and matrix materials are transversely-isotropic, with symmetry axis Ox_3.

7.3.1. TWO FIBERS

Consider an infinite body (matrix) containing two neighbouring and non-intersecting periodically curved fibers. With each fiber we associate the Cartesian $O_m x_{1m} x_{2m} x_{3m}$ and cylindrical $O_m r_m \theta_m z_m$ system of coordinates (Fig.7.3.1). According to Fig.7.3.1 we can write the following relations:

$$x_{31} = x_{32} = x_3, \quad x_{11} = x_{12}, \; x_{21} = R_{12} + x_{22} \; ,$$

$$r_1 \cos\theta_1 = r_2 \cos\theta_2 , \qquad r_1 \sin\theta_1 = R_{12} + r_2 \sin\theta_2 . \tag{7.3.1}$$

We assume that the middle line of fibers lies on the plane $x_{11} = x_{12} = 0$ and consider the following cases: *1) the curving of the fibers is co-phase*, i.e. the equation of the middle lines is taken as follows:

$$x_{21} = \varepsilon\delta(x_3), \quad x_{11} = 0 \quad \text{for the 1st fiber (Fig.7.3.1)},$$

$$x_{22} = \varepsilon\delta(x_3), \; x_{12} = 0 \quad \text{for the 2nd fiber (Fig.7.3.1)}, \tag{7.3.2}$$

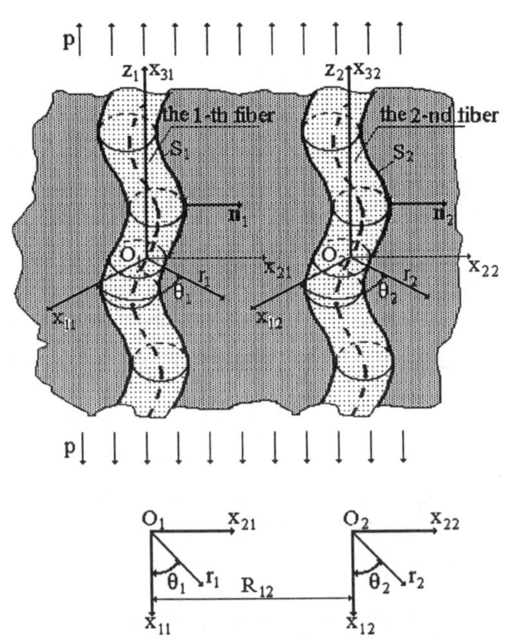

Fig. 7.3.1. A co-phase curved two neighbouring fibers and the system of coordinates associated with them.

2) the curving of the fibers is anti-phase,

$$x_{21} = \varepsilon\delta(x_3), \quad x_{11} = 0 \quad \text{for the 1st fiber (Fig.7.3.2)},$$

$$x_{22} = -\varepsilon\delta(x_3), \; x_{12} = 0 \quad \text{for the 2nd fiber (Fig.7.3.2)}. \tag{7.3.3}$$

where ε is a dimensionless small parameter. We will consider periodic curving of the fibers and therefore the expressions for the parameter ε and for the function $\delta(x_3)$ are taken as

$$\varepsilon = \frac{L}{\ell}, \qquad \delta(x_3) = \ell \sin \alpha x_3, \tag{7.3.4}$$

where $L < \ell$, $\alpha = 2\pi/\ell$; L and ℓ are respectively the amplitude and length of the periodic curving form. Repeating the operations performed in subsection 7.2.1 we obtain the equation for the surfaces S_m and for the components for orthonormal vectors to these surfaces. Since these equations can be written as (7.2.4) and (7.2.5),

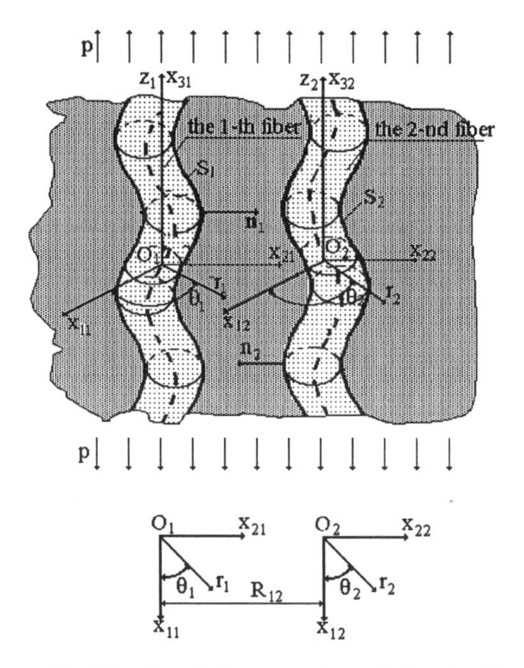

Fig. 7.3.2. An anti-phase curved two neighbouring fibers and the system of coordinates associated with them.

respectively, by adding lower index m (where m is the number of fiber) to the coordinates r and θ, and the equations (7.3.2), (7.3.3) are taken into account, they are not rewritten here.

We assume that within each fiber and within matrix the equilibrium equation (7.1.1), the elasticity relations (7.2.28) and the geometrical relations (7.1.3) hold. Further we assume that the complete cohesion conditions (7.1.4) on the fiber-matrix interfaces S_m hold as well.

Thus, we investigate the stress distribution in the body with the structure shown in Figs.7.3.1 and 7.3.2 under the action at infinity of normal forces with intensity p in the direction of the $O_m x_3$ axis. For this purpose, as in the previous section the various quantities are presented as series in ε. Note that these representations have the form (7.2.7) with obvious changes in the corresponding indices. Repeating the operations of the previous section we obtain the contact relations for each approximation of (7.2.7). Due to linearity of the governing equations we obtain a closed system of equations for each approximation separately.

We consider the determination of the values of the zeroth and first approximations.

The zeroth approximation. The determination of this approximation is reduced to the investigation of the generalized plane strain state in the infinite plane containing two neighbouring circular inclusions. This investigation must be made in the case where

$$u_z^{(1),0} = u_z^{(2)1,0} = u_z^{(2)2,0} = \varepsilon_{zz}^{(1),0} z, \quad \varepsilon_{zz}^{(1),0} = \frac{p}{E_3^{(1)}}. \tag{7.3.5}$$

It should be noted that the method for solution to such problems have been improved by many researchers; the list of contributors can be found in [24,73,74] and elsewhere. It follows from the results of these investigations that the stresses arising under (7.3.5) in the matrix (plane) and in the fibers (in each circular inclusion) can be represented as follows:

$$\left\{\sigma_{rr}^{(2)m,0}; \sigma_{\theta\theta}^{(2)m,0}; \sigma_{r\theta}^{(2)m,0}, \sigma_{rr}^{(1),0}; \sigma_{\theta\theta}^{(1),0}; \sigma_{r\theta}^{(1),0}\right\}$$

$$= \left(\nu_1^{(1)} - \nu_1^{(2)}\right)\left\{s_{rr}^{(2)m,0}; s_{\theta\theta}^{(2)m,0}; s_{r\theta}^{(2)m,0}, s_{rr}^{(1),0}; s_{\theta\theta}^{(1),0}; s_{r\theta}^{(1),0}\right\}. \tag{7.3.6}$$

The relation (7.3.6) shows that in the cases where

$$\nu_1^{(1)} = \nu_1^{(2)}. \tag{7.3.7}$$

the zeroth approximation corresponds to the homogeneous stress state and only the stresses

$$\sigma_{zz}^{(1),0} = p, \ \sigma_{zz}^{(2)1,0} = \sigma_{zz}^{(2)2,0} = \frac{E_3^{(2)}}{E_3^{(1)}} p \tag{7.3.8}$$

remain non-zero.

For simplicity of the future discussions we assume that the relation (7.3.7) holds; no great difficulties arise if it does not hold.

The first approximation. We consider only the co-phase curving case (Fig.7.3.2) and note that the anti-phase case (Fig.7.3.3) can be analysed similarly. Thus, for the first approximation we obtain the following contact conditions.

$$\left[\sigma_{rr}\right]_{1,1}^{2m,1} = 0, \quad \left[\sigma_{r\theta}\right]_{1,1}^{2m,1} = 0,$$

$$\left[\sigma_{rz}\right]_{1,1}^{2m,1} = 2\pi\left(\sigma_{zz}^{(1),0} - \sigma_{zz}^{(2)m,0}\right)\cos\alpha t \sin\theta_m,$$

$$\left[u_r\right]_{1,1}^{2m,1} = \left[u_\theta\right]_{1,1}^{2m,1} = \left[u_z\right]_{1,1}^{2m,1} = 0, \quad m=1,2, \qquad (7.3.9)$$

where all jumps are evaluated at (R, θ_m, t) and $[F]_{1,1}^{2m,1} = F^{(1),1} - F^{(2)m,1}$. In addition to the conditions (7.3.9), we impose the attenuation and limitation conditions

$$\left\{\sigma_{rr}^{(1),1};...;\sigma_{r3}^{(1),1};u_r^{(1),1};..;u_3^{(1),1}\right\}\Big|_{r_1;r_2\to\infty} \to 0.$$

$$\left\{\left|\sigma_{rr}^{(2)m,1}\right|;...;\left|\sigma_{r3}^{(2)m,1}\right|;\left|u_r^{(2)m,1}\right|;..;\left|u_3^{(2)m,1}\right|\right\}\Big|_{r_m\to 0} < M, \quad M=\text{const}. \qquad (7.3.10)$$

Thus the determination of the first approximation is reduced to the solution of the system of equations (7.1.1), (7.2.28) and (7.2.3) satisfied within each fiber and matrix separately in the framework of the conditions (7.3.9) and (7.3.10). For the solution to this problem we use the representation (7.2.34) and according to the contact conditions (7.3.9), we select the solution to the equations (7.2.36) in the following form

$$\begin{Bmatrix}\sin\alpha z\\ \cos\alpha z\end{Bmatrix}\sum_{n=-\infty}^{+\infty} f_n(\alpha r)\exp(in\theta). \qquad (7.3.11)$$

Substituting (7.3.11) into equations (7.2.36) we obtain the corresponding Bessel equation to determine the functions $f_n(\alpha r)$. The selection of the solutions of latter equations depends on the sign and character (real or complex) of the constants $\left(\xi_i^{(2)m}\right)^2$ and $\left(\xi_i^{(1)}\right)^2$. We assume that the conditions (7.2.40) are satisfied for these constants. Taking this situation and the conditions (7.3.10) into account we obtain the following solution to the equation (7.2.35).

For fibers:

$$\psi^{(2)m} = \alpha\sin\alpha z \sum_{n=-\infty}^{+\infty} C_n^{(2)m} I_n\left(\xi_1^{(2)}\alpha r_m\right)\exp(in\theta_m),$$

$$X^{(2)m} = \cos\alpha z \sum_{n=-\infty}^{+\infty} \left[A_n^{(2)m} I_n\left(\xi_2^{(2)}\alpha r_m\right) + B_n^{(2)m} I_n\left(\xi_3^{(2)}\alpha r_m\right)\right]\exp(in\theta_m), \quad (7.3.12)$$

For matrix:

$$\psi^{(1)} = \alpha \sin\alpha z \sum_{m=1}^{2} \sum_{n=-\infty}^{+\infty} C_n^{(1)m} K_n\left(\xi_1^{(1)}\alpha r_m\right)\exp(in\theta_m),$$

$$X^{(1)} = \cos\alpha z \sum_{m=1}^{2} \sum_{n=-\infty}^{+\infty} \left[A_n^{(1)m} K_n\left(\xi_2^{(1)}\alpha r_m\right) + B_n^{(1)m} K_n\left(\xi_3^{(1)}\alpha r_m\right)\right]\exp(in\theta_m). \quad (7.3.13)$$

In (7.3.12) and (7.3.13) $I_n(x)$ is a Bessel function of a purely imaginary argument, $K_n(x)$ is a MacDonald function, $A_n^{(k)m}$, $B_n^{(k)m}$ and $C_n^{(k)m}$ (k=1,2; m=1,2) are unknown constants. Note that to obtain the solution (7.3.13) for the matrices, they considered as multi-connected (two-connected) regions and the solution is constructed by summing the solutions to the corresponding singly-connected regions.

As follows from (7.3.12), (7.3.13), (7.2.34) and (7.2.28) the selected solution satisfies the conditions (7.3.10). Now we attempt to satisfy the contact condition (7.3.9). For this purpose we must represent the expressions (7.3.12) and (7.3.13) in the m-th cylindrical coordinate system to satisfy the contact conditions on the m-th fiber-matrix interface S_m. The expressions (7.3.12) are already presented in the m-th cylindrical system of coordinates. To make these operations for the expressions (7.3.13) we use the summation theorem [147] for the $K_n(x)$ function, which can be written for the case at hand as follows:

$$r_m \exp i\theta_m = R_{mn} \exp i\varphi_{mn} + r_n \exp i\theta_n,$$

$$K_\nu(cr_n)\exp i\nu\theta_n = \sum_{k=-\infty}^{+\infty}(-1)^\nu I_k(cr_m)K_{\nu-n}(cR_{mn})\exp[i(\nu-n)\varphi_{mn}]\exp ik\theta_m,$$

$$mn = 12;21, \quad m;n=1,2, \quad c = \text{const}, \quad r_m < R_{mn}, \quad R_{12} = R_{21}, \quad \varphi_{12} = \frac{\pi}{2} \quad \varphi_{21} = \frac{3\pi}{2}. \quad (7.3.14)$$

Thus applying the summation theorem (7.3.14) to the solutions (7.3.13), we obtain the following expressions for each cylindrical system of coordinates.

In the $O_1r_1\theta_1z_1$ coordinate system:

$$\Psi^{(1)} = \alpha \sin\alpha z \sum_{n=-\infty}^{+\infty} \left\{C_n^{(1)1} K_n\left(\xi_1^{(1)}\alpha r_1\right) + I_n\left(\xi_1^{(1)}\alpha r_1\right)\times\right.$$

$$\sum_{\nu=-\infty}^{+\infty}(-1)^{\nu}\exp\left[i(\nu-n)\frac{\pi}{2}\right]C_{\nu}^{(1)2}K_{\nu-n}\left(\xi_1^{(1)}\alpha R_{12}\right)\Bigg\}\exp in\theta_1,$$

$$X^{(1)}=\cos\alpha z\sum_{n=-\infty}^{+\infty}\left\{A_n^{(1)1}K_n\left(\xi_2^{(1)}\alpha r_1\right)+B_n^{(1)1}K_n\left(\xi_3^{(1)}\alpha r_1\right)+I_n\left(\xi_1^{(1)}\alpha r_1\right)\times\right.$$

$$\left.\sum_{\nu=-\infty}^{+\infty}(-1)^{\nu}\exp\left[i(\nu-n)\frac{\pi}{2}\right]\left[A_{\nu}^{(1)2}K_{\nu-n}\left(\xi_2^{(1)}\alpha R_{12}\right)+B_{\nu}^{(1)2}K_{\nu-n}\left(\xi_3^{(1)}\alpha R_{12}\right)\right]\right\}\exp in\theta_1$$

$$r_1<R_{12}, \tag{7.3.15}$$

In the $O_2r_2\theta_2z_2$ *coordinate system:*

$$\Psi^{(1)}=\alpha\sin\alpha z\sum_{n=-\infty}^{+\infty}\left\{C_n^{(1)2}K_n\left(\xi_1^{(1)}\alpha r_2\right)+I_n\left(\xi_1^{(1)}\alpha r_2\right)\times\right.$$

$$\left.\sum_{\nu=-\infty}^{+\infty}(-1)^{\nu}\exp\left[i(\nu-n)\frac{3\pi}{2}\right]C_{\nu}^{(1)1}K_{\nu-n}\left(\xi_1^{(1)}\alpha R_{12}\right)\right\}\exp in\theta_2,$$

$$X^{(1)}=\cos\alpha z\sum_{n=-\infty}^{+\infty}\left\{A_n^{(1)2}K_n\left(\xi_2^{(1)}\alpha r_2\right)+B_n^{(1)2}K_n\left(\xi_3^{(1)}\alpha r_2\right)+I_n\left(\xi_1^{(1)}\alpha r_2\right)\times\right.$$

$$\left.\sum_{\nu=-\infty}^{+\infty}(-1)^{\nu}\exp\left[i(\nu-n)\frac{3\pi}{2}\right]\left[A_{\nu}^{(1)1}K_{\nu-n}\left(\xi_2^{(1)}\alpha R_{12}\right)+B_{\nu}^{(1)1}K_{\nu-n}\left(\xi_3^{(1)}\alpha R_{12}\right)\right]\right\}\exp in\theta_2,$$

$$r_2<R_{12} \tag{7.3.16}$$

According to the form of the expressions (7.3.12) and (7.3.13) the unknown constants $A_n^{(k)m}$, $B_n^{(k)m}$ and $C_n^{(k)m}$ (k=1,2; m=1,2) must be taken as complex numbers and must satisfy the following relations.

$$A_{-n}^{(k)m}=\overline{A_n^{(k)m}},\quad B_{-n}^{(k)m}=\overline{B_n^{(k)m}},\quad C_{-n}^{(k)m}=\overline{C_n^{(k)m}},$$

$$\operatorname{Im}A_0^{(k)m}=0,\quad \operatorname{Im}B_0^{(k)m}=0,\quad \operatorname{Im}C_0^{(k)m}=0, \tag{7.3.17}$$

where the over-line indicates complex conjugate. Thus, using (7.3.15), (7.3.16), (7.3.12) and (7.2.34) we obtain an infinite system of algebraic equations with respect to the unknown constants (7.3.17). Introducing the notation

$$C_n^{(1)m} K_n\left(\xi_1^{(1)}\chi\right) = y_{n1}^{(1)m} + iz_{n1}^{(1)m}, \qquad A_n^{(1)m} K_n\left(\xi_2^{(1)}\chi\right) = z_{n2}^{(1)m} + iy_{n2}^{(1)m},$$

$$B_n^{(1)m} K_n\left(\xi_3^{(1)}\chi\right) = z_{n3}^{(1)m} + iy_{n3}^{(1)m}, \qquad C_n^{(2)m} I_n\left(\xi_1^{(2)}\chi\right) = y_{n1}^{(2)m} + iz_{n1}^{(2)m},$$

$$A_n^{(2)m} I_n\left(\xi_2^{(1)}\chi\right) = z_{n2}^{(2)m} + iy_{n2}^{(2)m}, \qquad B_n^{(2)m} I_n\left(\xi_3^{(1)}\chi\right) = z_{n3}^{(2)m} + iy_{n3}^{(2)m},$$

$$Z_n^{(k)m} = \left\| \begin{matrix} z_{n1}^{(k),m} \\ z_{n2}^{(k),m} \\ z_{n3}^{(k),m} \end{matrix} \right\|, \quad Y_n^{(k)m} = \left\| \begin{matrix} y_{n1}^{(k),m} \\ y_{n2}^{(k),m} \\ y_{n3}^{(k),m} \end{matrix} \right\|, \quad D_{n\nu}^{(1)m} = \left\| d_{rs}^{(1)m}(n,\nu) \right\|, \quad D_n^{(2)m} = \left\| d_{rs}^{(2)m}(n) \right\|,$$

$$F_{n\nu}^{(1)m} = \left\| f_{rs}^{(1)m}(n,\nu) \right\|, \quad F_n^{(2)m} = \left\| f_{rs}^{(2)m}(n) \right\|, \quad m = 1,2, \quad r;s = 1,2,3, \quad \chi = \frac{2\pi R}{\ell} \tag{7.3.18}$$

we can unite this infinite set of equations in the following form:

$$\begin{cases} Z_n^{(1)1} + \sum\limits_{\nu=0}^{\infty} D_{n\nu}^{(1)2} Z_\nu^{(1)2} + D_n^{(2)1} Z_n^{(2)1} = 0, \\ Z_n^{(1)2} + \sum\limits_{\nu=0}^{\infty} D_{n\nu}^{(1)1} Z_\nu^{(1)1} + D_n^{(2)2} Z_n^{(2)2} = 0, \end{cases} \tag{7.3.19}$$

$$\begin{cases} Y_n^{(1)1} + \sum\limits_{\nu=0}^{\infty} F_{n\nu}^{(1)2} Y_\nu^{(1)2} + F_n^{(2)1} Y_n^{(2)1} = 2\pi\delta_n^3\left(\sigma_{33}^{(1),0} - \sigma_{33}^{(2)1,0}\right), \\ Y_n^{(1)2} + \sum\limits_{\nu=0}^{\infty} F_{n\nu}^{(1)1} Y_\nu^{(1)1} + F_n^{(2)2} Y_n^{(2)2} = 2\pi\delta_n^3\left(\sigma_{33}^{(1),0} - \sigma_{33}^{(2)2,0}\right). \end{cases} \tag{7.3.20}$$

where

$$n = 0,1,2,\ldots,\infty, \qquad \delta_n^3 = \begin{cases} 1 & \text{if } n = 3 \\ 0 & \text{if } n \neq 3 \end{cases} \tag{7.3.21}$$

Note that the equations (7.3.19) and (7.3.20) are obtained for the co-phase curving of the fibers (Fig.7.3.1). For the anti-phase case (Fig.7.3.2) the equations (7.3.19) remain valid and instead of equations (7.3.20) the following ones are obtained:

$$\begin{cases} Y_n^{(1)1} + \sum\limits_{\nu=0}^{\infty} F_{n\nu}^{(1)2} Y_\nu^{(1)2} + F_n^{(2)1} Y_n^{(2)1} = 2\pi\delta_n^3\left(\sigma_{33}^{(1),0} - \sigma_{33}^{(2)1,0}\right), \\ Y_n^{(1)2} + \sum\limits_{\nu=0}^{\infty} F_{n\nu}^{(1)1} Y_\nu^{(1)1} + F_n^{(2)2} Y_n^{(2)2} = -2\pi\delta_n^3\left(\sigma_{33}^{(1),0} - \sigma_{33}^{(2)2,0}\right). \\ n = 0,1,2,\ldots,\infty \end{cases} \tag{7.3.22}$$

We omit the detailed expressions for $F_{n\nu}^{(1)m}, D_n^{(2)m}, D_{n\nu}^{(1)m}$ and $F_n^{(2)m}$.

The equation (7.3.19) shows that for both co-phase and anti-phase fiber curving

$$Z_n^{(k)m} = 0, \; k = 1,2, \quad m = 1,2. \tag{7.3.23}$$

Moreover, as $\sigma_{33}^{(2)1,0} = \sigma_{33}^{(2)2,0}$, then according to the mechanical consideration and according to the equations (7.3.20), (7.3.22) we write the following relations:

For the co-phase curving,

$$Y_n^{(k)1} = Y_n^{(k)2}. \tag{7.3.24}$$

For the anti-phase curving.

$$Y_n^{(k)1} = -Y_n^{(k)2}. \tag{7.3.25}$$

Taking (7.3.24) and (7.3.25) into account, we reduce the equation (7.3.20) to the equation

$$Y_n^{(1)1} + \sum_{\nu=0}^{\infty} F_{n\nu}^{(1)2} Y_\nu^{(1)1} + F_n^{(2)1} Y_n^{(2)1} = 2\pi\delta_n^3 \left(\sigma_{33}^{(1),0} - \sigma_{33}^{(2)1,0} \right), \tag{7.3.26}$$

and equation (7.3.22) is reduced to

$$Y_n^{(1)1} - \sum_{\nu=0}^{\infty} F_{n\nu}^{(1)2} Y_\nu^{(1)1} + F_n^{(2)1} Y_n^{(2)1} = 2\pi\delta_n^3 \left(\sigma_{33}^{(1),0} - \sigma_{33}^{(2)1,0} \right). \tag{7.3.27}$$

For numerical investigations the infinite system of algebraic equations (7.3.25) and (7.3.26) must be approximated by a finite system. To validate such a replacement it must be shown that the determinant of these infinite system of equations must be of normal type [113]; this hold if we can prove the convergence of the series

$$M = \sum_{n=0}^{\infty} \sum_{\nu=0}^{\infty} \left| F_{n\nu}^{(1)2} \right|. \tag{7.3.28}$$

For investigation of the series (7.3.28) we use the following asymptotic estimates of the functions $I_n(x)$ and $K_n(x)$:

$$I_n(x) < c_1 \frac{1}{n!} \left(\frac{|x|}{2} \right)^n, \; c_1 = \text{const}, \quad K_n(x) \approx c_2 (n-1)! \left(\frac{2}{|x|} \right)^n, \; c_2 = \text{const}. \tag{7.3.29}$$

These hold for large n and fixed x. Moreover we use the following inequality

$$\rho = \frac{R}{R_{12}} > 2 , \tag{7.3.30}$$

which means that the fibers do not touch. Thus taking (7.3.29) and (7.3.30) into account and analysing the expressions of $F_{nv}^{(1)2}$ we obtain the following estimate for the series (7.3.28):

$$M < c_3 \sum_{n=0}^{\infty} n^{c_4} (\rho - 1)^{-n}, \quad c_3; c_4 = \text{const} . \tag{7.3.31}$$

As the series on the right hand side converges, so does (7.3.38). Note that similar convergence estimates were made in [1, 32]. Thus the determinant of the infinite system of equations (7.3.26), (7.3.27) is normal and the infinite systems can be replaced by finite system for numerical purposes. The requisite number of equations in these finite systems must be determined from the convergence numerical results.

This concludes discussion of the first approximation. Subsequent approximations can be found likewise. Other configurations can be analysed similarly.

7.3.2. PERIODICALLY LOCATED ROW FIBERS

Consider an infinite body containing periodically located row fibers (Fig.7.3.3). With each fiber we associate the Cartesian $O_{k\pm}x_{1k\pm}x_{2k\pm}x_{3k\pm}$ and

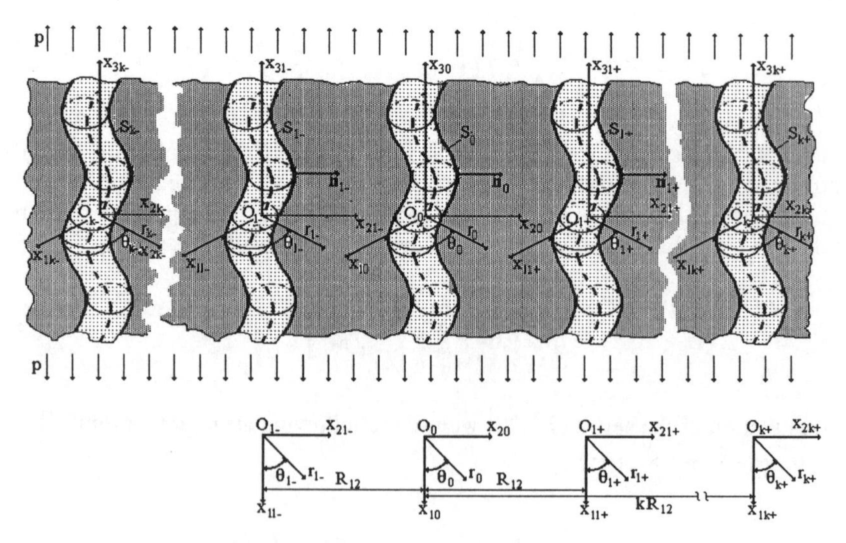

Fig.7.3.3. Periodically located co-phase curved row fibers.

cylindrical $O_{k\pm}r_{k\pm}\theta_{k\pm}z_{k\pm}$ system of coordinate which are obtained from the system of coordinates $O_0x_{10}x_{20}x_{30}$ and $O_0r_0\theta_0z_0$ respectively, by parallel transfer along the O_0x_{10} axis.

We denote by R_{12} and kR_{12} the distances between the neighbouring coordinate origins and the distance between the origins O_0 and O_{k+} or O_{k-} respectively (Fig.7.3.3). We assume that these fibers are periodically curved and the middle line lies in the plane $O_{k\pm}x_{2k\pm}x_{3k\pm}$. We consider the following two configuration: 1) co-phase curving (Fig.7.3.3); 2)anti-phase curving (Fig.7.3.4). In both cases the following relations between the system of coordinates hold.

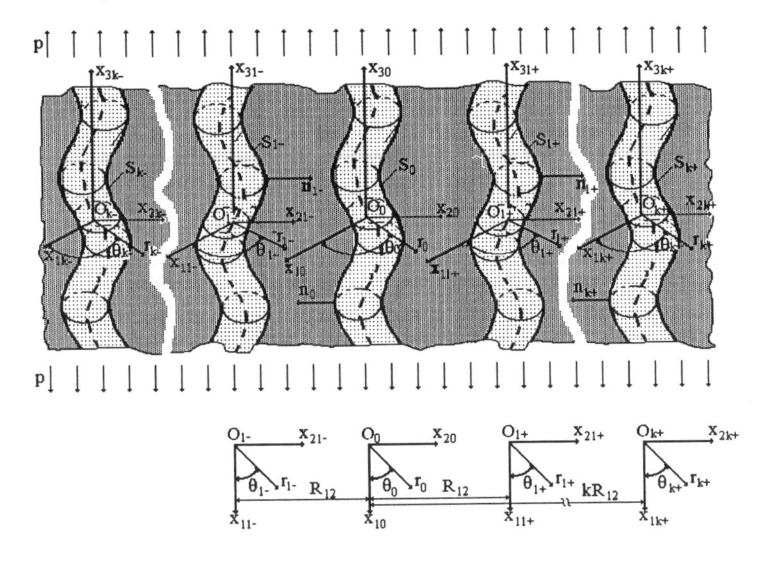

Fig.7.3.4. Periodically located anti-phase curved row fibers.

$$x_{3j\pm} = x_{30}\,;\quad x_{1j} = x_{10}\,;\quad x_{2j\pm} = x_{20} \mp jR_{12}\,;\quad j = 0,1,2,\ldots,\infty\,,$$

$$x_{2j\pm} = x_{2k\pm} + (\mp j \mp k)R_{12}$$

$$r_{j\pm}\exp(i\theta_{j\pm}) = r_0\exp(i\theta_0) \mp jR_{12}\exp\left(i\frac{\pi}{2}\right);$$

$$r_{j\pm}\exp(i\theta_{j\pm}) = r_{k\pm}\exp(i\theta_{k\pm}) + (\mp j \pm k)R_{12}\exp\left(i\frac{\pi}{2}\right). \tag{7.3.32}$$

We introduce the following notation:

$$x_{10} = x_1,\quad x_{20} = x_2,\quad x_{30} = x_3\,. \tag{7.3.33}$$

As in the previous subsection we write the full system equations (7.1.1), (7.2.28) and (7.1.3) within each fiber and matrix, and assume that on the matrix-fibers interfaces $S_{k\pm}$ the complete cohesion conditions (7.1.4) are satisfied. The equations of the middle lines we select as follows:

For co-phase curved fibers:

$$x_{2k\pm} = \varepsilon\delta(x_3), \quad x_{1k\pm} = 0. \tag{7.3.34}$$

For anti-phase curved fibers:

$$x_{2k\pm} = (-1)^k \varepsilon\delta(x_3), \quad x_{1k\pm} = 0, \tag{7.3.35}$$

where

$$\varepsilon = \frac{L}{\ell} \text{ and } \delta(x_3) = \ell \sin \alpha x_3. \tag{7.3.36}$$

The meanings of the parameters in (7.3.36) are the same as in the previous subsection.

According to Fig.7.3.3 in the co-phase curving case, periodicity conditions must be satisfied:

$$\begin{Bmatrix} u_i^{(1)} \\ \sigma_{ij}^{(1)} \end{Bmatrix}(x_1, x_2 \pm jR_{12}, x_3) = \begin{Bmatrix} u_i^{(1)} \\ \sigma_{ij}^{(1)} \end{Bmatrix}(x_1, x_2, x_3), \quad j = 0,1,2,\ldots,\infty \tag{7.3.37}$$

Similarly Fig.7.3.4 shows that in the anti-phase curving case, the periodicity conditions can be written as

$$\begin{Bmatrix} u_i^{(1)} \\ \sigma_{ij}^{(1)} \end{Bmatrix}(x_1, x_2 + (\pm 2j+1)R_{12}, x_3) = -\begin{Bmatrix} u_i^{(1)} \\ \sigma_{ij}^{(1)} \end{Bmatrix}(x_1, x_2, x_3)$$

$$\begin{Bmatrix} u_i^{(1)} \\ \sigma_{ij}^{(1)} \end{Bmatrix}(x_1, x_2 \pm 2jR_{12}, x_3) = \begin{Bmatrix} u_i^{(1)} \\ \sigma_{ij}^{(1)} \end{Bmatrix}(x_1, x_2, x_3), \quad j = 0,1,2,\ldots,\infty. \tag{7.3.38}$$

Also in both cases the following symmetry conditions with respect to $x_1 = 0$ plane can be taken into account.

$$u_1^{(1)}(x_1, x_2, x_3) = -u_1^{(1)}(-x_1, x_2, x_3), \quad u_2^{(1)}(x_1, x_2, x_3) = u_2^{(1)}(-x_1, x_2, x_3). \tag{7.3.39}$$

Thus in the co-phase (anti-phase) curving of the row fibers the investigation of the problem is reduced to the investigation of the system equations (7.1.1), (7.2.28), (7.1.3), (7.1.4) in the framework of the periodicity conditions (7.3.37) ((7.3.38)) and symmetry conditions (7.3.39). As in the previous section, for the solution to this problem, first, we derive the equations of the surfaces $S_{k\pm}$ and their normals. Note that these equations can be written as (7.2.4)-(7.2.6) by replacing r, θ and z with $r_{k\pm}$, $\theta_{k\pm}$ and $z_{k\pm}$ respectively. Carrying out the same operations as before, and representing the quantities as series in ε we obtain a closed system of equations for each approximation. We consider the determination of the zeroth and first approximations.

The zeroth approximation. We assume that the relation (7.3.7) is satisfied, therefore for this approximation we obtain only the following non-zero stresses:

$$\sigma_{zz}^{(1),0} = p, \quad \sigma_{zz}^{(2)k\pm,0} = \sigma_{zz}^{(2),0} = \frac{E_3^{(2)}}{E_3^{(1)}} p. \tag{7.3.40}$$

Also, the following relations hold:

$$u_z^{(1),0} = u_z^{(2)k\pm,0} = u_z^{(2),0} = \varepsilon_{zz}^{(1),0} z, \quad \varepsilon_{zz}^{(1),0} = \frac{p}{E_3^{(1)}}. \tag{7.3.41}$$

The first approximation. For this approximation we obtain the following contact conditions under co-phase fibers curving:

$$\left[\sigma_{rr}\right]_{1,1}^{2k\pm,1} = 0, \quad \left[\sigma_{r\theta}\right]_{1,1}^{2k\pm,1} = 0,$$

$$\left[\sigma_{rz}\right]_{1,1}^{2k\pm,1} = 2\pi\left(\sigma_{zz}^{(1),0} - \sigma_{zz}^{(2)k\pm,0}\right)\cos\alpha t \sin\theta_{k\pm},$$

$$\left[u_r\right]_{1,1}^{2k\pm,1} = \left[u_\theta\right]_{1,1}^{2k\pm,1} = \left[u_z\right]_{1,1}^{2k\pm,1} = 0, \quad k = 0,1,2,\ldots,\infty, \tag{7.3.42}$$

where all jumping are evaluated at $\left(R, \theta_{k\pm}, t\right)$ and $\left[F\right]_{1,1}^{2k\pm,1} = F^{(1),1} - F^{(2)k\pm,1}$.

As in the previous subsection, in addition to (7.3.42), we write the following decay and limitation conditions:

$$\left\{\sigma_{rr}^{(1),1};\ldots;\sigma_{r3}^{(1),1};u_r^{(1),1};\ldots;u_3^{(1),1}\right\}\Big|_{|x_1|\to\infty} \to 0$$

$$\left\{\left|\sigma_{rr}^{(2)k\pm,1}\right|;\ldots;\left|\sigma_{r3}^{(2)k\pm,1}\right|;\left|u_r^{(2)k\pm,1}\right|;\ldots;\left|u_3^{(2)k\pm,1}\right|\right\}\Big|_{r_{k\pm}\to 0} < M, \quad M=\text{const}. \tag{7.3.43}$$

Moreover according to the periodicity of the material structure in the co-phase curving case we can write the relation.

$$\left\{\sigma_{ij}^{(2)k\pm,1};u_n^{(2)k\pm,1}\right\}=\left\{\sigma_{ij}^{(2)0,1};u_n^{(2)0,1}\right\},\quad ij=rr;\theta\theta;zz;r\theta;rz,\theta z,\quad n=r;\theta;z\,,\quad(7.3.44)$$

and note that the conditions (7.3.37), (7.3.39) are satisfied for each approximation separately.

Thus the determination of the first approximation is reduced the solution of the equations (7.1.1), (7.2.28), (7.1.3) in the framework of the conditions (7.3.42), (7.3.43), (7.3.44), (7.3.37) and (7.3.39). For the solution of this problem we use the presentations (7.2.34) and assume that the relations (7.2.40) hold. In this case we obtain the following solution to the equations (7.2.36):

For the $k\pm$ -th fiber:

$$\Psi^{(2)k\pm}=\alpha\sin\alpha z\sum_{n=-\infty}^{\infty}C_n^{(2)k\pm}I_n\left(\xi_1^{(2)}\alpha r_{k\pm}\right)\exp\left(in\theta_{k\pm}\right),$$

$$X^{(2)k\pm}=\cos\alpha z\sum_{n=-\infty}^{\infty}\left[A_n^{(2)k\pm}I_n\left(\xi_2^{(2)}\alpha r_{k\pm}\right)+B_n^{(2)k\pm}I_n\left(\xi_3^{(2)}\alpha r_{k\pm}\right)\right]\exp\left(in\theta_{k\pm}\right).\quad(7.3.45)$$

For the matrix:

$$\Psi^{(1)}=\alpha\sin\alpha z\left[\sum_{k=0}^{\infty}\sum_{n=-\infty}^{\infty}C_n^{(1)k+}K_n\left(\xi_1^{(1)}\alpha r_{k+}\right)\exp\left(in\theta_{k+}\right)+\right.$$

$$\left.\sum_{k=1}^{\infty}\sum_{n=-\infty}^{\infty}C_n^{(1)k-}K_n\left(\xi_1^{(1)}\alpha r_{k-}\right)\exp\left(in\theta_{k-}\right)\right],$$

$$X^{(1)}=\cos\alpha z\left\{\sum_{k=0}^{\infty}\sum_{n=-\infty}^{\infty}\left[A_n^{(1)k+}K_n\left(\xi_2^{(1)}\alpha r_{k+}\right)+B_n^{(1)k+}K_n\left(\xi_3^{(1)}\alpha r_{k+}\right)\right]\exp\left(in\theta_{k+}\right)+\right.$$

$$\left.\sum_{k=1}^{\infty}\sum_{n=-\infty}^{\infty}\left[A_n^{(1)k-}K_n\left(\xi_2^{(1)}\alpha r_{k-}\right)+B_n^{(1)k-}K_n\left(\xi_3^{(1)}\alpha r_{k-}\right)\right]\exp\left(in\theta_{k-}\right)\right\}.\quad(7.3.46)$$

Physical considerations demand the functions $\Psi^{(2)k\pm}$, $X^{(2)k\pm}$, $\Psi^{(1)}$ and $X^{(1)}$ be real, and therefore the unknown constants entering (7.3.45), (7.3.46) must satisfy the following relations:

$$A_{-n}^{(q)k\pm}=\overline{A_n^{(q)k\pm}}\,,\ B_{-n}^{(q)k\pm}=\overline{B_n^{(q)k\pm}}\,,\ C_{-n}^{(q)k\pm}=\overline{C_n^{(q)k\pm}}\,,$$

$$\operatorname{Im} A_0^{(q)k\pm} = 0\,,\ \operatorname{Im} B_0^{(q)k\pm} = 0\,,\ \operatorname{Im} C_0^{(q)k\pm} = 0\,,\quad q = 1{,}2\,;\ k = -\infty,\ldots,0,\ldots,\infty\,. \qquad (7.3.47)$$

The periodicity of the fiber location and the symmetry of the material structure with respect to plane $x_{20} = 0$ in the anti-phase curving lead to the following relations:

For the co-phase curving case.

$$A_n^{(q)k\pm} = A_n^{(q)0}\,,\ B_n^{(q)k\pm} = B_n^{(q)0}\,,\ C_n^{(q)k\pm} = C_n^{(q)0}\,,\ q = 1{,}2\,, \qquad (7.3.48)$$

For the anti-phase curving case.

$$A_n^{(q)k\pm} = (-1)^k A_n^{(q)0}\,,\ B_n^{(q)k\pm} = (-1)^k B_n^{(q)0}\,,\ C_n^{(q)k\pm} = (-1)^k C_n^{(q)0}\,. \qquad (7.3.49)$$

Taking (7.3.48), (7.3.49) into account we can satisfy the contact conditions (7.3.42) on the S_0 surface only and find all unknowns. However in this case we must rewrite the expressions (7.3.46) in the cylindrical coordinate system $O_0 r_0 \theta_0 z_0$. For this purpose we use the following summation formulae [147] for the $K_n(x)$ functions:

$$K_n(cr_{k+})\exp(in\theta_{k+}) = \sum_{\nu=-\infty}^{\infty} (-1)^n I_\nu(cr_0) K_{n-\nu}(ckR_{12}) \exp\left(i(n-\nu)\frac{3\pi}{2}\right) \exp(i\nu\theta_0)\,,$$

$$K_n(cr_{k-})\exp(in\theta_{k-}) = \sum_{\nu=-\infty}^{\infty} (-1)^n I_\nu(cr_0) K_{n-\nu}(ckR_{12}) \exp\left(i(n-\nu)\frac{\pi}{2}\right) \exp(i\nu\theta_0)\,,$$

$$r_{k+}\exp(i\theta_{k+}) = r_0\exp(i\theta_0) + kR_{12}\exp\left(i\frac{3\pi}{2}\right),$$

$$r_{k-}\exp(i\theta_{k-}) = r_0\exp(i\theta_0) + kR_{12}\exp\left(i\frac{\pi}{2}\right),\ c = \mathrm{const}\,,\ r_0 < R_{12}\,. \qquad (7.3.50)$$

Substituting (7.3.50) in (7.3.46) and carrying some transformations we obtain

$$\Psi^{(1)} = \alpha\sin\alpha x_3 \sum_{n=-\infty}^{\infty}\Bigg[C_n^{(1)0} K_n\left(\xi_1^{(1)}\alpha r_0\right) + I_n\left(\xi_1^{(1)}\alpha r_0\right) \sum_{\nu=-\infty}^{\infty} (-1)^\nu C_\nu^{(1),0} \times$$

$$\left(\exp\left(i(\nu-n)\frac{3\pi}{2}\right) + \exp\left(i(\nu-n)\frac{\pi}{2}\right)\right) \sum_{k=1}^{\infty} K_{\nu-n}\left(\xi_1^{(1)}\alpha kR_{12}\right)\Bigg] \exp(in\theta_0),$$

$$X^{(1)} = \cos\alpha x_3 \sum_{n=-\infty}^{\infty} \left[A_n^{(1)0} K_n\left(\xi_2^{(1)}\alpha r_0\right) + I_n\left(\xi_2^{(1)}\alpha r_0\right) \sum_{\nu=-\infty}^{\infty} (-1)^\nu A_\nu^{(1),0} \times \right.$$

$$\left(\exp\left(i(\nu-n)\frac{3\pi}{2} \right) + \exp\left(i(\nu-n)\frac{\pi}{2} \right) \right) \sum_{k=1}^{\infty} K_{\nu-n}\left(\xi_2^{(1)}\alpha k R_{12}\right) +$$

$$B_n^{(1)0} K_n\left(\xi_3^{(1)}\alpha r_0\right) + I_n\left(\xi_3^{(1)}\alpha r_0\right) \sum_{\nu=-\infty}^{\infty} (-1)^\nu B_\nu^{(1),0} \times$$

$$\left. \left(\exp\left(i(\nu-n)\frac{3\pi}{2} \right) + \exp\left(i(\nu-n)\frac{\pi}{2} \right) \right) \sum_{k=1}^{\infty} K_{\nu-n}\left(\xi_3^{(1)}\alpha k R_{12}\right) \right] \exp(in\theta_0). \quad (7.3.51)$$

Note that the expressions have been written for the co-phase curving of the fibers. For anti-phase curving the corresponding expressions are obtained from (7.3.51) by replacing $K_{\nu-n}(c\alpha k R_{12})$ with $(-1)^k K_{\nu-n}(c\alpha k R_{12})$.

Thus substituting these expressions in (7.2.34) and using (7.1.3) and (7.2.28) we determine the stresses and displacement for the matrix and for the 0-th fiber and introducing the notation for unknown constants as in (7.3.18) under m=0, from the symmetry conditions (7.3.39) we obtain

$$z_{n,1}^{(q)0} = z_{n,2}^{(q)0} = z_{n,3}^{(q)0} = 0, \quad q = 1,2; \quad n = -\infty,\ldots,0,\ldots,\infty. \quad (7.3.52)$$

Moreover, from the contact conditions (7.3.42) we obtain

$$Y_n^{(1)0} + \sum_{\nu=0}^{\infty} Y_\nu^{(1)0} \sum_{k=1}^{\infty} F_{n\nu k}^{(1)0} + Y_n^{(2)0} = 2\pi\delta_n^3\left(\sigma_{zz}^{(1),0} - \sigma_{zz}^{(2),0}\right), \quad (7.3.53)$$

where $n = 0,1,2,\ldots,\infty$, and δ_n^3 is determined as in (7.3.21). In (7.3.53) by replacing $F_{n\nu k}^{(1)0}$ with $(-1)^k F_{n\nu k}^{(1)0}$ we obtain the corresponding equations for anti-phase curving of the fibers. We omit the details of the expressions for the $F_{n\nu k}^{(1)0}$.

Thus the investigation of the stress distribution of the fibrous composite material with structure shown in Figs.7.3.3 and 7.3.4 is reduced to the solution of the infinite system of equation (7.3.53). As before, we may show that the determinant of these equations is normal [113].

We can also determine the subsequent approximations similarly.

7.3.3. DOUBLY-PERIODICALLY LOCATED FIBERS

In the framework of the assumptions and notation we consider the case where periodically curved fibers are located in a doubly-periodic pattern with parameters R_{01},

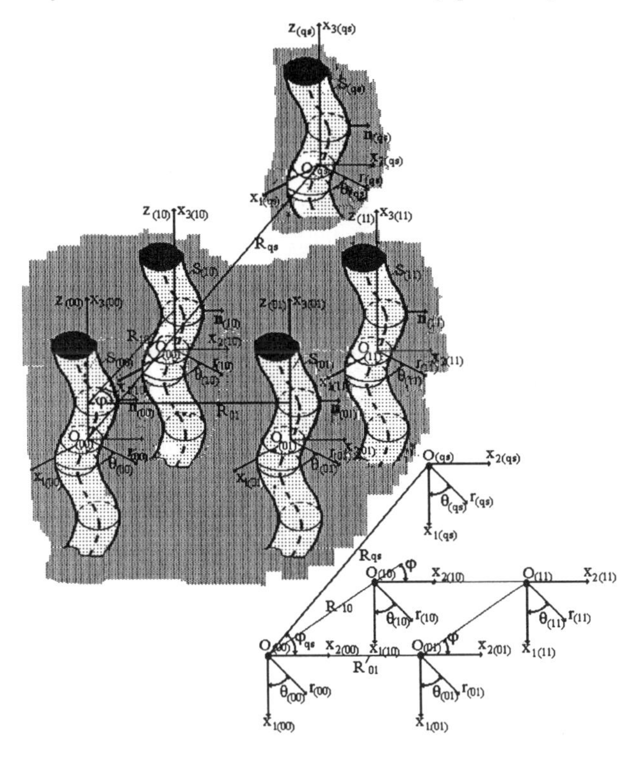

Fig.7.3.5. A structure of the composite material with double-periodically located curved fibers.

R_{10} and φ. With each fiber we associate the Cartesian $O_{(qs)}x_{1(qs)}x_{2(qs)}x_{3(qs)}$ and cylindrical $O_{(qs)}r_{(qs)}\theta_{(qs)}z_{(qs)}$ system of coordinates, where $q,s=-\infty,...,0,...,\infty$. Fig.7.3.5 shows that we can write the following relations:

$$x_{1(00)}=R_{qs}\sin\varphi_{qs}+x_{1(qs)},\ x_{2(00)}=R_{qs}\cos\varphi_{qs}+x_{2(qs)},\ x_{3(00)}=x_{3(qs)},$$

$$r_{(00)}\exp(i\theta_{(00)})=iR_{qs}\exp(-i\varphi_{qs})+r_{(qs)}\exp(i\theta_{(qs)}),\ q,s=-\infty,...,0,...\infty. \quad (7.3.54)$$

We introduce the following notation:

$$x_{1(00)}=x_1,\ x_{2(00)}=x_2,\ x_{3(00)}=x_3=z,\ r_{(00)}=r,\ \theta_{(00)}=\theta. \quad (7.3.55)$$

We assume that the curving of the fibers is co-phase and periodic, and the middle lines of these fibers lie on the $O_{(qs)}x_{2(qs)}x_{3(qs)}$ planes. The equation of these middle lines is taken as follows:

$$x_{2(qs)} = \varepsilon\delta(x_3) = L\sin\left(\frac{2\pi}{\ell}x_3\right) = \frac{L}{\ell}\ell\sin\left(\frac{2\pi}{\ell}x_3\right) = \varepsilon\ell\sin(\alpha x_3). \tag{7.3.56}$$

Symmetry implies the following relations:

$$\begin{Bmatrix} u_i^{(k)} \\ \sigma_{ij}^{(k)} \end{Bmatrix}(x_1 - qR_{10}\sin\varphi, x_2 + tR_{01}, x_3) = \begin{Bmatrix} u_i^{(k)} \\ \sigma_{ij}^{(k)} \end{Bmatrix}(x_1, x_2, x_3). \tag{7.3.57}$$

As before, we assume that the complete cohesion condition is satisfied on each $S_{(qs)}$ surface.

Thus we investigate the stress distribution under the action at infinity (i.e. at $|x_3| \to \infty$) of uniformly distributed normal forces with intensity p in the direction Ox_3 axis. The solution proceeds as before.

The zeroth approximation. We assume that the condition (7.3.7) holds and only the following non-zero stresses remain.

$$\sigma_{zz}^{(1),0} = p\left(\eta^{(1)} + \frac{E_3^{(2)}}{E_3^{(1)}}\eta^{(2)}\right)^{-1}, \ \sigma_{zz(qs)}^{(2),0} = \sigma_{zz(00)}^{(2),0} = \sigma_{zz}^{(2)} = p\frac{E_3^{(2)}}{E_3^{(1)}}\left(\eta^{(1)} + \frac{E_3^{(2)}}{E_3^{(1)}}\right)^{-1},$$

$$q, s = -\infty, \ldots, 0, \ldots, \infty. \tag{7.3.58}$$

Here $\eta^{(2)}$ is a fiber concentration and $\eta^{(1)} = 1 - \eta^{(2)}$. Moreover the following relations hold:

$$u_z^{(1),0} = u_{z(qt)}^{(2),0} = u_{z(00)}^{(2),0} = u_z^{(2),0} = \varepsilon_{zz}^{(1),0}z, \ \varepsilon_{zz}^{(1),0} = \frac{\sigma_{zz}^{(1),0}}{E_3^{(1)}}. \tag{7.3.59}$$

The first approximation. For this approximation the following contact conditions are selected.

$$\left[\sigma_{rr}\right]_{1,1}^{2(qs),1} = 0, \ \left[\sigma_{r\theta}\right]_{1,1}^{2(qs),1} = 0,$$

$$\left[\sigma_{rz}\right]_{1,1}^{2(qt),1} = 2\pi\left(\sigma_{zz}^{(1),0} - \sigma_{zz}^{(2),0}\right)\cos\alpha t\sin\theta_{(qs)},$$

$$\left[u_r\right]_{1,1}^{2(qs),1} = \left[u_\theta\right]_{1,1}^{2(qs),1} = \left[u_z\right]_{1,1}^{2(qs),1} = 0\,, \tag{7.3.60}$$

where all jumps are evaluated at $\left(R, \theta_{(qs)}, t\right)$ and $\left[F\right]_{1,1}^{2(qs),1} = F^{(1),1} - F_{(qs)}^{(2),1}$.
According to (7.3.60) the solution to the equations (7.2.36) is selected as follows:

For the fibers:

$$\Psi_{(qs)}^{(2)} = \alpha \sin \alpha z \sum_{n=-\infty}^{+\infty} C_n^{(2)(qs)} I_n\left(\xi_1^{(2)} \alpha r_{(qs)}\right) \exp\left(in\theta_{(qs)}\right),$$

$$X_{(qs)}^{(2)} = \cos \alpha z \sum_{n=-\infty}^{+\infty} \left[A_n^{(2)(qs)} I_n\left(\xi_2^{(2)} \alpha r_{(qs)}\right) + B_n^{(2)(qs)} I_n\left(\xi_3^{(2)} \alpha r_{(qs)}\right)\right] \exp\left(in\theta_{(qs)}\right). \tag{7.3.61}$$

For the matrix:

$$\Psi^{(1)} = \alpha \sin \alpha z \sum_{n=-\infty}^{\infty} \sum_{q,t=-\infty}^{\infty} C_n^{(1)(qs)} K_n\left(\xi_1^{(1)} \alpha r_{(qs)}\right) \exp\left(in\theta_{(qs)}\right),$$

$$X^{(1)} = \cos \alpha z \sum_{n=-\infty}^{\infty} \sum_{q,t=-\infty}^{\infty} \left[A_n^{(1)(qs)} K_n\left(\xi_2^{(1)} \alpha r_{(qs)}\right) + B_n^{(1)(qs)} K_n\left(\xi_3^{(1)} \alpha r_{(qs)}\right)\right] \exp\left(in\theta_{(qs)}\right), \tag{7.3.62}$$

where the unknown constants satisfy the following relations:

$$A_{-n}^{(k)(qs)} = \overline{A_n^{(k)(qs)}}\,,\ B_{-n}^{(k)(qs)} = \overline{B_n^{(k)(qs)}}\,,\ C_{-n}^{(k)(qs)} = \overline{C_n^{(k)(qs)}}\,,$$

$$\operatorname{Im} A_0^{(k)(qs)} = \operatorname{Im} B_0^{(k)(qs)} = \operatorname{Im} C_0^{(k)(qs)} = 0\,. \tag{7.3.63}$$

The periodicity conditions (7.3.57) employ

$$A_n^{(k)(qs)} = A_n^{(k)(00)} = A_n^{(k)};\ B_n^{(k)(qs)} = B_n^{(k)(00)} = B_n^{(k)},\ C_n^{(k)(qs)} = C_n^{(k)(00)} = C_n^{(k)}. \tag{7.3.64}$$

As before we apply the summation formula [147] for $K_n(x)$:

$$K_\nu\left(cr_{(qs)}\right)\exp\left(i\nu\theta_{(qs)}\right) = \sum_{n=-\infty}^{\infty} (-1)^\nu I_n(cr)\, K_{\nu-n}\left(cR_{qs}\right)\exp\left[i(\nu-n)\left(\varphi_{qs} + \frac{\pi}{2}\right)\right] \times$$

$$\exp(in\theta), \quad r < R_{qs}, \quad c = \text{const}, \quad R_{qs} = \left[q^2 R_{10}^2 + s^2 R_{01}^2 + 2qsR_{01}R_{10}\cos\varphi\right]^{\frac{1}{2}},$$

$$\varphi_{qs} = a\tan\left[\frac{qR_{10}\sin\varphi}{sR_{01} + qR_{10}\sin\varphi}\right]. \tag{7.3.65}$$

Using (7.3.64), (7.3.65) we can rewrite the expressions for the functions $\Psi^{(1)}$ and $X^{(1)}$ as follows:

$$\Psi^{(1)} = \alpha\sin\alpha z \sum_{n=-\infty}^{\infty} \left\{ C_n^{(1)} K_n\left(\xi_1^{(1)}\alpha r\right) + I_n\left(\xi_1^{(1)}\alpha r\right) \sum_{\nu=-\infty}^{\infty} (-1)^{\nu} C_{\nu}^{(1)} \times \right.$$

$$\left. \sum_{q,s=-\infty}^{\infty}{}' K_{\nu-n}\left(\xi_1^{(1)}\alpha R_{qs}\right) \exp\left[i(\nu-n)\left(\varphi_{qs} + \frac{\pi}{2}\right)\right]\right\} \exp(in\theta),$$

$$X^{(1)} = \cos\alpha z \sum_{n=-\infty}^{\infty} \left\{ A_n^{(1)} K_n\left(\xi_2^{(1)}\alpha r\right) + I_n\left(\xi_2^{(1)}\alpha r\right) \sum_{\nu=-\infty}^{\infty} (-1)^{\nu} A_{\nu}^{(1)} \times \right.$$

$$\sum_{q,s=-\infty}^{\infty}{}' K_{\nu-n}\left(\xi_2^{(1)}\alpha R_{qs}\right) \exp\left[i(\nu-n)\left(\varphi_{qs} + \frac{\pi}{2}\right)\right] +$$

$$B_n^{(1)} K_n\left(\xi_3^{(1)}\alpha r\right) + I_n\left(\xi_3^{(1)}\alpha r\right) \sum_{\nu=-\infty}^{\infty} (-1)^{\nu} B_{\nu}^{(1)} \times$$

$$\left. \sum_{q,s=-\infty}^{\infty}{}' K_{\nu-n}\left(\xi_3^{(1)}\alpha R_{qs}\right) \exp\left[i(\nu-n)\left(\varphi_{qs} + \frac{\pi}{2}\right)\right]\right\} \exp(in\theta). \tag{7.3.66}$$

The prime on the summation symbol in (7.3.66) means that the term $q = s = 0$ is omitted. Thus, satisfying the contact conditions (7.3.60) on the $S_{(00)}$ surface only, we obtain an infinite system of non-homogeneous algebraic equations for unknown constants in (7.3.61) and (7.3.66). Again, the determinant is normal.

We can also determine subsequent approximations similarly. Now we consider the simplification of the expressions (7.3.66) for the case where the cell of the periodicity is a rectangle, i.e.

$$\varphi = \frac{\pi}{2}, \quad R_{01} \neq R_{10}. \tag{7.3.67}$$

In this case (7.3.65) implies

$$R_{(qs)} = \left(q^2R_{10}^2 + s^2R_{01}^2\right)^{\frac{1}{2}}, \quad \varphi_{qs} = a\tan\left(\frac{qR_{10}}{sR_{01}}\right). \tag{7.3.68}$$

The symmetry with respect to the planes $x_1 = 0$ and $x_2 = 0$ and relations (7.3.64) lead to the following relations for the unknown constants.

$$\operatorname{Re} A_{2n}^{(k)} = 0, \ \operatorname{Re} B_{2n}^{(k)} = 0, \ \operatorname{Im} C_{2n}^{(k)} = 0, \ \operatorname{Im} A_{n}^{(k)} = 0, \ \operatorname{Im} B_{n}^{(k)} = 0,$$

$$\operatorname{Re} C_{n}^{(k)} = 0, \ k=1,2, \quad n = 0,1,2,\ldots,\infty \ . \tag{7.3.69}$$

Taking (7.3.64), (7.3.69) into consideration, we can rewrite the expressions for the functions Ψ and X as follows:

For the (00)-th fiber:

$$\Psi_{(00)}^{(2)} = \Psi^{(2)} = -2\alpha \sin\alpha z \sum_{n=0}^{\infty} \left(\operatorname{Im} C_{2n+1}\right) I_n\left(\xi_1^{(2)}\alpha r\right) \sin(2n+1)\theta,$$

$$X_{(00)}^{(2)} = X^{(2)} = 2\cos\alpha z \sum_{n=0}^{\infty} \left[\left(\operatorname{Re} A_{2n+1}^{(2)}\right) I_{2n+1}\left(\xi_2^{(2)}\alpha r\right) + \left(\operatorname{Re} B_{2n+1}^{(2)}\right) I_{2n+1}\left(\xi_3^{(2)}\alpha r\right)\right] \times$$

$$\cos(2n+1)\theta \tag{7.3.70}$$

For the matrix:

$$\Psi^{(1)} - 2\alpha \sin\alpha z \sum_{n=0}^{\infty} \left[\left(\operatorname{Im} C_{2n+1}^{(1)}\right) K_{2n+1}\left(\xi_1^{(1)}\alpha r\right) + I_n\left(\xi_1^{(1)}\alpha r\right) \sum_{\nu=0}^{\infty} \left(\operatorname{Im} C_{2n+1}^{(1)}\right) \times\right.$$

$$\left.\left(D_{2\nu-2n,1} - D_{2\nu+2n+2,1}\right)\right] \sin(2n+1)\theta,$$

$$X^{(1)} = 2\cos\alpha z \sum_{n=0}^{\infty} \left[\left(\operatorname{Re} A_{2n+1}^{(1)}\right) K_{2n+1}\left(\xi_2^{(1)}\alpha r\right) + \left(\operatorname{Re} B_{2n+1}^{(1)}\right) K_{2n+1}\left(\xi_3^{(1)}\alpha r\right) + \right.$$

$$I_{2n+1}\left(\xi_2^{(1)}\alpha r\right) \sum_{\nu=0}^{\infty} \left(\operatorname{Re} A_{2\nu+1}^{(1)}\right)\left(D_{2\nu-2n,2} + D_{2\nu+2n+2,2}\right) +$$

$$\left. I_{2n+1}\left(\xi_3^{(1)}\alpha r\right) \sum_{\nu=0}^{\infty} \left(\operatorname{Re} B_{2\nu+1}^{(1)}\right)\left(D_{2\nu-2n,3} + D_{2\nu+2n+2,3}\right)\right] \cos(2n+1)\theta. \tag{7.3.71}$$

where

$$D_{\nu-n,j} = \sum_{q,s=-\infty}^{\infty}{}' K_{\nu-n}\left(\xi_j^{(1)}\alpha R_{qt}\right)\exp\left[i(\nu-n)\left(\varphi_{qt}+\frac{\pi}{2}\right)\right]. \qquad (7.3.72)$$

In (7.3.72) $j=1,2,3$ and the prime on the summation symbol indicates that the term $q=s=0$ is omitted.

7.4. Numerical Results

We analyse some numerical results related to the stress distribution in a fibrous composite with periodically curved structure, with low concentration. We consider the stress distribution in an infinite elastic body containing periodically curved fiber under the action at infinity of uniformly distributed normal forces with intensity p in the direction of the Oz axis (Fig.7.2.1). We assume that a material of the fiber and matrix are isotropic and homogeneous, and study the distribution of the normal stress σ_{nn} and tangential stresses $\sigma_{n\tau}$ and σ_{ne} on the matrix-fiber interface. Moreover we study the distribution of the normal stresses $\sigma_{\tau\tau}^{(k)}$ and $\sigma_{ee}^{(k)}$ ($k=1,2$) where they act in the direction of the tangent vectors $\boldsymbol{\tau}$ and $\mathbf{e}$ (Fig.7.2.1) respectively. First we consider the case where the matrix and fiber materials are elastic, and later the case where the matrix is visco-elastic, but the fiber elastic.

7.4.1. ELASTIC COMPOSITE

We introduce the parameter $\chi = 2\pi R/\ell$ and call it the parameter of the wave generation of the curving, where R is a radius of the fiber cross section which is perpendicular to its middle line, ℓ is a wavelength of a form of curvature of the middle line. Assume that $\nu_1 = \nu_2 = 0.3$, where ν_1 and ν_2 is a Poisson's ratios for matrix and fiber materials respectively. We use the notation introduced in section 7.2 and the notation $\kappa = E_2/E_1$, where E_1 and E_2 is an elasticity moduli for matrix and fiber materials respectively.

Due to the periodicity of the fiber curvature and the symmetry relative to the plane Ox_2x_3, we will limit ourselves to examining those sections of the interface which correspond to a change in the parameters αt and θ in the intervals $[0,\pi/2]$ and $[\pi/2,3\pi/2]$ respectively. For example the values $\{\alpha t=0,\theta=\pi/2\}$, $\{\alpha t=0,\theta=\pi\}$, $\{\alpha t=0,\theta=3\pi/2\}$, $\{\alpha t=\pi/2,\theta=\pi/2\}$, $\{\alpha t=\pi/2,\theta=3\pi/2\}$ correspond to the points M_1, M_2, M_3, N_1 and N_3 respectively (Fig.7.2.1).

Thus we consider the graphs given in Fig.7.4.1. which show the dependencies of $\sigma_{nn}/\sigma_{zz}^{(1),0}$, $\sigma_{n\tau}/\sigma_{zz}^{(1),0}$ and $\sigma_{ne}/\sigma_{zz}^{(1),0}$ on the parameter αt for various κ. These graphs show that the stresses σ_{nn}, σ_{ne} are minimal and the stress $\sigma_{n\tau}$ maximal at

$\alpha t = 0$. The values of σ_{nn} and σ_{ne} increase monotonically and $\sigma_{n\tau}$ decreases with an increase in αt. The maximum values of the stresses σ_{nn}, σ_{ne} correspond to $\alpha t = \pi/2$. The stress $\sigma_{n\tau}$ is nearly zero at the same value of αt. These graphs also show that the stresses σ_{nn}, σ_{ne}, $\sigma_{n\tau}$ increase monotonically with an increase in the elastic moduli ratio κ.

Fig.7.4.2 shows the dependencies of $\sigma_{nn}/\sigma_{zz}^{(1),0}$, $\sigma_{n\tau}/\sigma_{zz}^{(1),0}$ and $\sigma_{ne}/\sigma_{zz}^{(1),0}$ on the parameter χ for various values of κ. To obtain these results we take $\chi \leq 1.5$, which ensures that different points of the fiber-matrix interface do not coincide.

These graphs show that the dependence of the self-balanced stresses σ_{nn}, σ_{ne}, $\sigma_{n\tau}$ on χ is non-monotonic, i.e., for each value of the κ there is a value of $\chi = \chi^*$, such that the absolute values of σ_{nn}, σ_{ne}, $\sigma_{n\tau}$ are maximal. At $\chi > \chi^*$ ($\chi < \chi^*$) these stresses monotonically decrease (increase) with χ. It should also be noted that the values of χ^* decrease with an increase in κ.

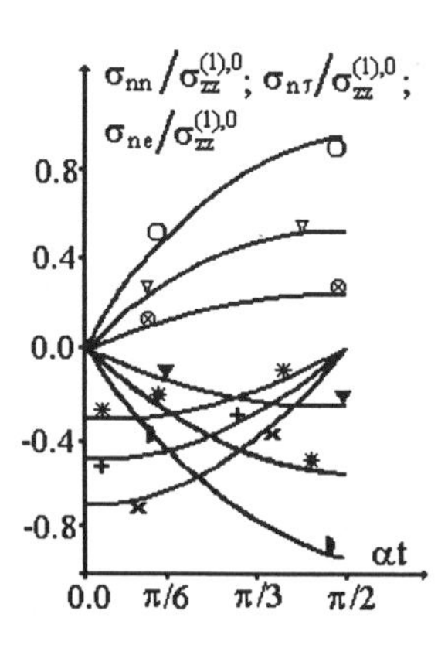

Fig.7.4.1. The distribution of the stresses $\sigma_{nn}, \sigma_{n\tau}, \sigma_{ne}$ with respect to αt in the case where $\chi = 0.4$, $\varepsilon = 0.015$, $\kappa = 20, 50, 100$:
⊗,∇,○ -- $\sigma_{nn}/\sigma_{zz}^{(1),0}$, $\theta = \pi/2$;
✱,✚,✖ -- $\sigma_{n\tau}/\sigma_{zz}^{(1),0}$, $\theta = \pi/2$;
▼,✱,◗ -- $\sigma_{ne}/\sigma_{zz}^{(1),0}$, $\theta = \pi$.

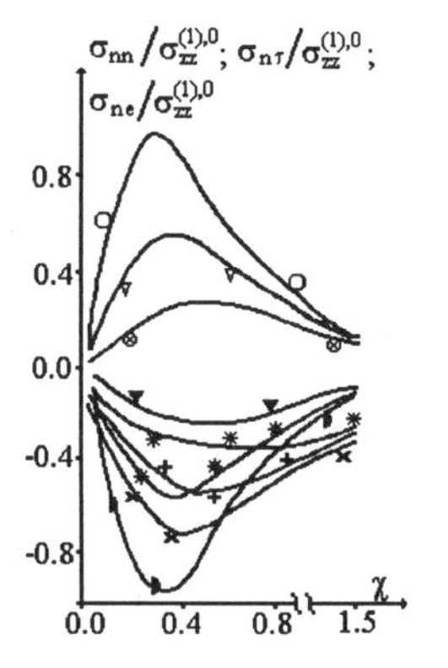

Fig.7.4.2. The graphs of the relations between $\sigma_{nn}, \sigma_{n\tau}, \sigma_{ne}$ and χ in the case where $\varepsilon = 0.015$, $\kappa = 20, 50, 100$:
⊗,∇,○ -- $\sigma_{nn}/\sigma_{zz}^{(1),0}$, $\alpha t = \pi/2$, $\theta = \pi/2$;
✱,✚,✖ -- $\sigma_{n\tau}/\sigma_{zz}^{(1),0}$, $\alpha t = 0, \theta = \pi/2$;
▼,✱,◗ -- $\sigma_{ne}/\sigma_{zz}^{(1),0}$, $\alpha t = \pi/2, \theta = \pi$.

Fig.7.4.3 shows the dependencies of $\sigma_{nn}/\sigma_{zz}^{(1),0}$, $\sigma_{n\tau}/\sigma_{zz}^{(1),0}$ and $\sigma_{ne}/\sigma_{zz}^{(1),0}$ on θ: the normal stress σ_{nn} is tensile at $\pi/2 \leq \theta < \theta^*$, where the value $\theta = \theta^*$ corresponds to the points at which $\sigma_{nn} = 0$, while it is compressive at $\theta^* < \theta \leq 3\pi/2$. The values of θ^* decrease with an increase in the ratio κ. The maximum tensile and compressive normal stresses are obtained at points $\theta = \pi/2$ and $\theta = 3\pi/2$ respectively. The dependence of σ_{ne} on θ is non-monotonic and the values of σ_{ne} at all the selected values of θ are negative, while $\sigma_{ne} = 0$ at $\theta = \pi/2$ and $\theta = 3\pi/2$.

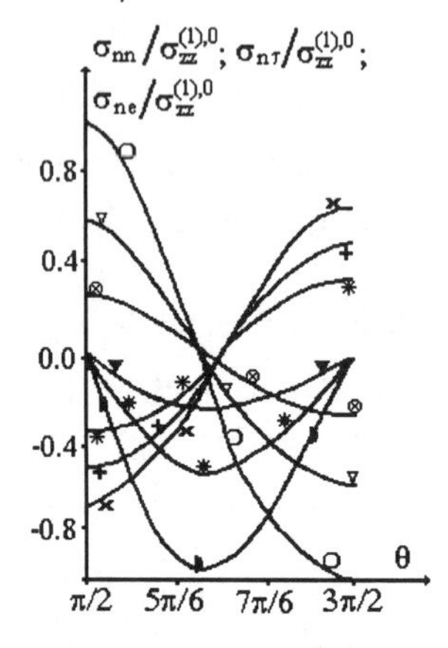

Fig.7.4.3. The graphs of the relations between $\sigma_{nn}, \sigma_{n\tau}, \sigma_{ne}$ and θ in the case where $\varepsilon = 0.015$, $\kappa = 20, 50, 100$: ⊗,∇,○ -- $\sigma_{nn}/\sigma_{zz}^{(1),0}$, $\alpha t = \pi/2$, ; $\chi = 0.4$ ✳,✚,✖ -- $\sigma_{n\tau}/\sigma_{zz}^{(1),0}$, $\alpha t = 0$, ; $\chi = 0.4$ ▼,✱,◗ -- $\sigma_{ne}/\sigma_{zz}^{(1),0}$, $\alpha t = \pi/2, \chi = 0.3$.

The graphs show that the stress $\sigma_{n\tau}$ changes its sign, i.e., $\sigma_{n\tau} < 0$ at $\pi/2 \le \theta < \pi$ and $\sigma_{n\tau} > 0$ at $\pi < \theta \le 3\pi/2$. The maximum shear stress $\sigma_{n\tau}$ is obtained at points M_1 and M_3 (Fig.7.2.1), i.e., at $\theta = \pi/2$ and $\theta = 3\pi/2$.

Table 7.1 shows values of $\sigma_{nn}/\sigma_{zz}^{(1),0}$, $\sigma_{n\tau}/\sigma_{zz}^{(1),0}$ and $\sigma_{ne}/\sigma_{zz}^{(1),0}$ for different ε and κ. The stresses increase with ε and κ; when $\kappa \ge 100$, even for very slight fiber curvature their values are several fold greater than the normal stress in the matrix $\sigma_{zz}^{(1),0}$.

Table 7.2 shows the values of $\sigma_{nn}/\sigma_{zz}^{(1),0}$, $\sigma_{n\tau}/\sigma_{zz}^{(1),0}$ and $\sigma_{ne}/\sigma_{zz}^{(1),0}$ obtained at various approximations; convergence of the numerical results is acceptable.

We now consider the distribution of the normal stresses $\sigma_{\tau\tau}^{(k)}$, $\sigma_{ee}^{(k)}$ $(k = 1,2)$. Note that when $\nu_1 = \nu_2$, which we have assumed, in the zeroth approximation $\sigma_{ee}^{(1),0} = \sigma_{\theta\theta}^{(1)} = \sigma_{ee}^{(2),0} = \sigma_{\theta\theta}^{(2),0} = 0$, while the stresses $\sigma_{\tau\tau}^{(1)}, \sigma_{\tau\tau}^{(2)}$ (the values of these stresses in the zeroth approximation will coincide with $\sigma_{zz}^{(1),0}$, $\sigma_{zz}^{(2),0}$ respectively) are non-trivial and are balanced by external forces p. This conclusion follows from the expressions (7.2.17). Hence, here we will determine the effect of curvature on the value of $\sigma_{ee}^{(k)}$, $\sigma_{\tau\tau}^{(k)}$ in relation to their self-balanced parts. In these cases the numerical results show that $\sigma_{ee}^{(2)}/\sigma_{zz}^{(2),0}$ is of the order of $(\kappa)^{-1}\sigma_{\tau\tau}^{(2)}/\sigma_{zz}^{(2),0}$. Since we assume that $\kappa >> 1$, we have $\sigma_{ee}^{(2)} << \sigma_{\tau\tau}^{(2)}$. Taking this into account we investigate only the distributions of the stresses $\sigma_{\tau\tau}^{(k)}$, $\sigma_{ee}^{(1)}$. In this case we take the values of these stresses calculated at points of the matrix-fiber interface S (Fig.7.2.1).

Table 7.1. The values of $\sigma_{nn}/\sigma_{zz}^{(1),0}$ {at $\alpha t = \pi/2$, $\theta = \pi/2$ }, $\sigma_{n\tau}/\sigma_{zz}^{(1),0}$ {at $\alpha t = 0$, $\theta = \pi/2$ }, $\sigma_{ne}/\sigma_{zz}^{(1),0}$ {at $\alpha t = \pi/2$, $\theta = \pi$ }, for $\chi = 0.4$ under $\kappa = 20,50$; $\chi = 0.3$ under $\kappa = 100,150$; $\chi = 0.2$ under $\kappa = 200$ for various ε and κ.

κ	ε	$\sigma_{nn}/\sigma_{zz}^{(1),0}$	$\sigma_{n\tau}/\sigma_{zz}^{(1),0}$	$\sigma_{ne}/\sigma_{zz}^{(1),0}$
	0.010	0.179	-0.242	-0.166
20	0.015	0.269	-0.363	-0.253
	0.020	0.359	-0.484	-0.343
	0.025	0.449	-0.605	-0.436
	0.010	0.386	-0.354	-0.373
50	0.015	0.581	-0.531	-0.555
	0.020	0.777	-0.708	-0.777
	0.025	0.973	-0.885	-0.991
	0.010	0.666	-0.478	-0.659
100	0.015	1.002	-0.717	-1.015
	0.020	1.340	-0.956	-1.388
	0.025	1.680	-1.195	-1.781
	0.010	0.911	-0.751	-0.906
150	0.015	1.371	-0.856	-1.396
	0.020	1.834	-1.142	-1.910
	0.025	2.300	-1.427	-2.449
	0.010	1.116	-0.642	-1.112
200	0.015	1.679	-0.963	-1.713
	0.020	2.246	-1.284	-2.344
	0.025	2.816	-1.605	-3.006

Fig.4.4.4 shows the distribution of the stresses $\sigma_{ee}^{(1)}$, $\sigma_{\tau\tau}^{(k)}$ $(k=1,2)$ with respect to αt for various κ: the absolute values of the self-balanced parts of the stresses $\sigma_{\tau\tau}^{(1)}$, $\sigma_{\tau\tau}^{(2)}$ and $\sigma_{ee}^{(1)}$ depend monotonically on the parameter αt. Here, $\alpha t = 0$ corresponds to the minimum values of the self-balanced parts of these stresses, while $\alpha t = \pi/2$ corresponds to their maximum values. The absolute values of the self-balanced parts of these stresses increase with an increase in κ.

Consider the change in the stresses $\sigma_{\tau\tau}^{(1)}$, $\sigma_{\tau\tau}^{(2)}$ and $\sigma_{ee}^{(1)}$ in relation to the parameter χ. The graphs of these change are shown in Fig.7.4.5; when $0.01 < \chi \le \chi'$, the self balanced part of the stress $\sigma_{\tau\tau}^{(1)}$ is positive, while it is negative when $\chi' < \chi \le 1.5$. Here, χ' is the abscissa of the point of intersection of the graphs of $\sigma_{\tau\tau}^{(1)}/\sigma_{zz}^{(1),0}$ with the straight line $\sigma_{\tau\tau}^{(1)}/\sigma_{zz}^{(1),0} = 1$. The value of χ' increases with an increase in κ. It should be noted that in each interval there is a value of χ (we designate them through $\chi_{I}^{(1)*} \in (0.01, \chi')$ and $\chi_{II}^{(1)*} \in (\chi', 1.5)$ respectively) at which the absolute value of the self-balanced part of the stress $\sigma_{\tau\tau}^{(1)}$ is maximal.

Table 7.2. The values of $\sigma_{nn}/\sigma_{zz}^{(1),0}$ {at $\alpha t = \pi/2$, $\theta = \pi/2$ }, $\sigma_{n\tau}/\sigma_{zz}^{(1),0}$ {at $\alpha t = 0$, $\theta = \pi/2$ }, $\sigma_{ne}/\sigma_{zz}^{(1),0}$ {at $\alpha t = \pi/2$, $\theta = \pi$ }, for $\chi = 0.4$ under $\kappa = 50$; $\chi = 0.3$ under $\kappa = 100$; $\chi = 0.2$ under $\kappa = 200$ with various ε and κ in various approximations.

κ	ε	Numb. of appr.	$\sigma_{nn}/\sigma_{zz}^{(1),0}$	$\sigma_{n\tau}/\sigma_{zz}^{(1),0}$	$\sigma_{ne}/\sigma_{zz}^{(1),0}$
		0	0.000	0.000	0.000
	0.010	1	0.385	-0.354	-0.370
50		2	0.386	-0.354	-0.370
		0	0.000	0.000	0.000
	0.025	1	0.962	-0.885	-0.925
		2	0.973	-0.885	-0.925
		0	0.000	0.000	0.000
	0.010	1	0.661	-0.478	-0.645
100		2	0.666	-0.478	-0.645
		0	0.000	0.000	0.000
	0.025	1	1.654	-1.195	-1.614
		2	1.680	-1.195	-1.614
		0	0.000	0.000	0.000
	0.010	1	1.109	-0.642	-1.090
200		2	1.116	-0.642	-1.090
		0	0.000	0.000	0.000
	0.025	1	2.774	-1.605	-2.725
		2	2.816	-1.605	-2.725

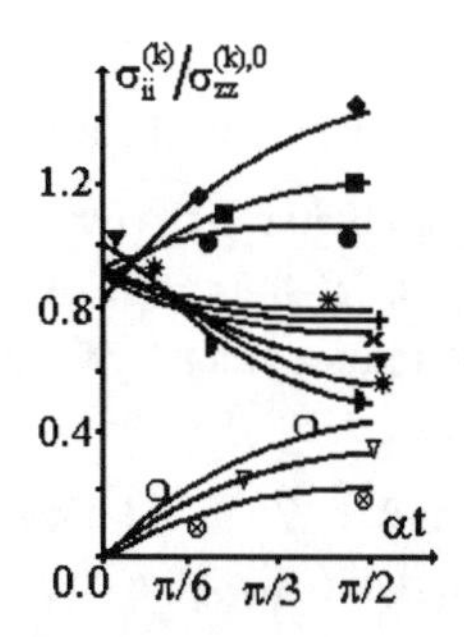

Fig.7.4.4. The distribution of the stresses $\sigma_{\tau\tau}^{(1)}, \sigma_{\tau\tau}^{(2)}, \sigma_{ee}^{(1)}$ with respect to αt in the case where $\theta = \pi/2$, $\varepsilon = 0.015$, κ =50, 100, 200: ⊗,∇,○ -- $\sigma_{ee}^{(1)}/\sigma_{zz}^{(1),0}$, $\chi = 0.5$; ✱,✚,✖ -- $\sigma_{\tau\tau}^{(1)}/\sigma_{zz}^{(1),0}$, $\chi = 0.7$; ●,■,◆ -- $\sigma_{\tau\tau}^{(1)}/\sigma_{zz}^{(1),0}$, $\chi = 0.2$ ▼,✱,◗ -- $\sigma_{\tau\tau}^{(2)}/\sigma_{zz}^{(2),0}$, $\chi = 0.7$.

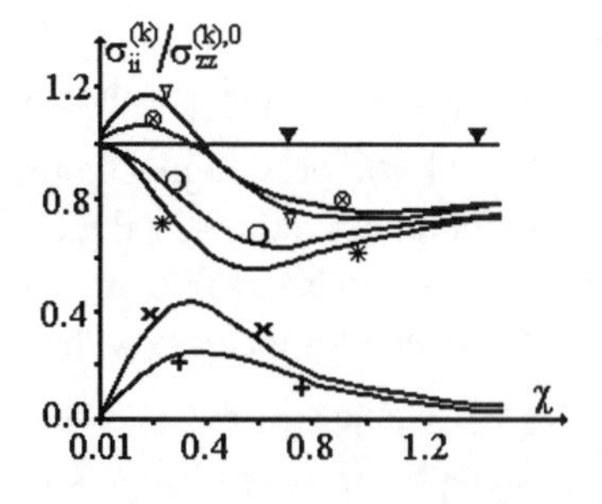

Fig.7.4.5. The graphs of the relations between $\sigma_{\tau\tau}^{(1)}, \sigma_{\tau\tau}^{(2)}, \sigma_{ee}^{(1)}$ and χ in the case where $\varepsilon = 0.015$, $\alpha t = \pi/2$, $\theta = \pi/2$, κ =50, 100: ⊗,∇ -- $\sigma_{\tau\tau}^{(1)}/\sigma_{zz}^{(1),0}$; ○,✱ -- $\sigma_{\tau\tau}^{(2)}/\sigma_{zz}^{(2),0}$; ✚,✖ -- $\sigma_{ee}^{(1)}/\sigma_{zz}^{(1),0}$; ▼ -- $\sigma_{\tau\tau}^{(1)}/\sigma_{zz}^{(1),0} = 1$.

However, the stresses $\sigma_{\tau\tau}^{(2)}$, $\sigma_{ee}^{(1)}$ have unique maxima in relation to χ; the values corresponding to these maxima will be designated by $\chi^{(2)*}$ (for $\sigma_{\tau\tau}^{(2)}$) and $\chi*$ (for $\sigma_{ee}^{(1)}$). The graphs show that when $\chi > \chi^{(2)*}$ ($\chi > \chi*$) and $\chi < \chi^{(2)*}$ ($\chi < \chi*$), the absolute values of the self-balanced part of the stress $\sigma_{\tau\tau}^{(2)}$ ($\sigma_{ee}^{(1)}$) monotonically decrease. For selected values of the elastic moduli ratio, the values of $\chi_{I}^{(1)*}$, $\chi_{II}^{(1)*}$, $\chi^{(2)*}$, $\chi*$ differ appreciably from one another and decrease with an increase in this ratio.

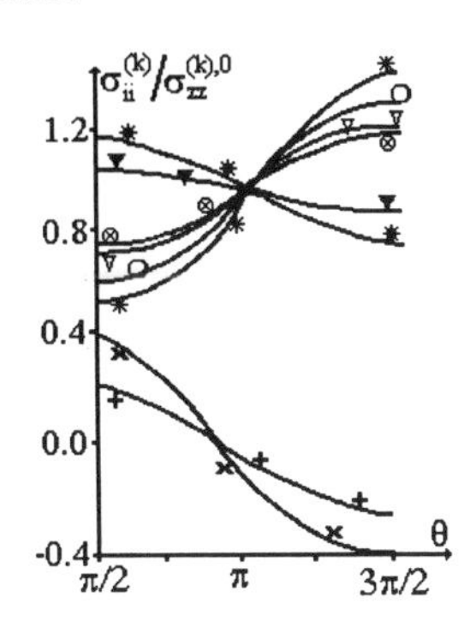

Fig.7.4.6. The graphs of the relations between $\sigma_{\tau\tau}^{(1)}, \sigma_{\tau\tau}^{(2)}, \sigma_{ee}^{(1)}$ and θ in the case where $\varepsilon = 0.015$, $\alpha t = \pi/2$, $\kappa = 50, 100$:
$\otimes, \nabla$ -- $\sigma_{\tau\tau}^{(1)}/\sigma_{zz}^{(1),0}$, $\chi = \chi_{II}^{(1)*} = 1.0$; O, ✱ -- $\sigma_{\tau\tau}^{(2)}/\sigma_{zz}^{(2),0}$, $\chi = \chi^{(2)*} = 0.3$;
▼, ✱ -- $\sigma_{ee}^{(1)}/\sigma_{zz}^{(1),0}$, $\chi = \chi_{I}^{(1)*} = 0.2$;
✚, ✖ -- $\sigma_{ee}^{(1)}/\sigma_{3zz}^{(1),0}$, $\chi = \chi* = 0.3$.

Fig.7.4.6 shows the relations between $\sigma_{\tau\tau}^{(1)}/\sigma_{zz}^{(1),0}$, $\sigma_{\tau\tau}^{(2)}/\sigma_{zz}^{(2),0}$, $\sigma_{ee}^{(1)}/\sigma_{zz}^{(1),0}$ and θ; when $\chi = \chi_{I}^{(1)*}$ the self-balanced part of the stress $\sigma_{\tau\tau}^{(1)}$ is tensile (compressive) for $\theta \in [\pi/2, \pi)$ ($\theta \in (\pi, 3\pi/2]$). The stress $\sigma_{ee}^{(1)}$ is also tensile (compressive) for $\theta \in [\pi/2, \theta*)$ (for $\theta \in (\theta*, 3\pi/2]$). But, for $\chi = \chi_{II}^{(1)*}$ the self-balanced part of the stress $\sigma_{\tau\tau}^{(1)}$ and for $\chi = \chi^{(2)*}$ the self-balanced part of the stress $\sigma_{\tau\tau}^{(2)}$ is compressive (tensile) under $\theta \in [\pi/2, \pi)$ ($\theta \in (\pi, 3\pi/2]$).

Table 7.3. The values of $\sigma_{\tau\tau}^{(1)}/\sigma_{zz}^{(1),0}$ {at $\chi = \chi_{II}^{(1)*} = 1.0$ }, $\sigma_{\tau\tau}^{(2)}/\sigma_{zz}^{(2),0}$ {at $\chi = \chi^{(2)*} = 0.3$}, $\sigma_{ee}^{(1)}/\sigma_{zz}^{(1),0}$ {at $\chi = \chi* = 0.3$ }, under { $\alpha t = \pi/2$, $\theta = \pi/2$ }for various ε and κ.

κ	ε	$\sigma_{\tau\tau}^{(1)}/\sigma_{zz}^{(1),0}$	$\sigma_{\tau\tau}^{(2)}/\sigma_{zz}^{(2),0}$	$\sigma_{ee}^{(1)}/\sigma_{zz}^{(1),0}$
	0.010	0.846	0.755	0.154
50	0.015	0.769	0.633	0.232
	0.020	0.692	0.510	0.311
	0.025	0.615	0.388	0.391
	0.010	0.834	0.694	0.275
	0.015	0.751	0.542	0.415
100	0.020	0.668	0.389	0.556
	0.025	0.585	0.237	0.698
1	2	3	4	5

Table 7.3 (Continuation)

1	2	3	4	5
	0.010	0.825	0.644	0.469
200	0.015	0.738	0.466	0.707
	0.020	0.650	0.288	0.947
	0.025	0.563	0.111	1.188

Table 7.4. The values of $\sigma_{\tau\tau}^{(1)}/\sigma_{zz}^{(1),0}$ {at $\chi=\chi_{II}^{(1)*}=1.0$ }, $\sigma_{\tau\tau}^{(2)}/\sigma_{zz}^{(2),0}$ {at $\chi=\chi^{(2)*}=0.3$}, $\sigma_{ee}^{(1)}/\sigma_{zz}^{(1),0}$ {at $\chi=\chi^*=0.3$ }, under { $\alpha t=\pi/2$, $\theta=\pi/2$ }for some ε and κ in various approximations.

κ	ε	Numb. of appr.	$\sigma_{\tau\tau}^{(1)}/\sigma_{zz}^{(1),0}$	$\sigma_{\tau\tau}^{(2)}/\sigma_{zz}^{(2),0}$	$\sigma_{ee}^{(1)}/\sigma_{zz}^{(1),0}$
		0	1.000	1.000	0.000
	0.010	1	0.846	0.755	0.152
		2	0.846	0.755	0.154
50					
		0	1.000	1.000	0.000
	0.025	1	0.615	0.388	0.381
		2	0.615	0.388	0.391
		0	1.000	1.000	0.000
	0.010	1	0.825	0.644	0.465
200		2	0.825	0.644	0.469
		0	1.000	1.000	0.000
	0.025	1	0.563	0.111	1.164
		2	0.563	0.111	1.188

Table 7.3 shows the values of $\sigma_{\tau\tau}^{(1)}/\sigma_{zz}^{(1),0}$, $\sigma_{\tau\tau}^{(2)}/\sigma_{zz}^{(2),0}$, $\sigma_{ee}^{(1)}/\sigma_{zz}^{(1),0}$ for different ε and κ; self-balanced part of these stresses increase monotonically with parameter ε and ratio κ.

Table 7.4 gives the values of $\sigma_{\tau\tau}^{(1)}/\sigma_{zz}^{(1),0}$, $\sigma_{\tau\tau}^{(2)}/\sigma_{zz}^{(2),0}$, $\sigma_{ee}^{(1)}/\sigma_{zz}^{(1),0}$ obtained in various approximations for some values of problem parameters; convergence of the numerical results is again acceptable.

7.4.2. VISCO-ELASTIC COMPOSITE

Now we assume that the matrix is a visco-elastic with operators (4.6.1)-(4.6.3) and the fiber is elastic. Using the approach proposed in section 4.5, we investigate the influence of the dimensionless rheological parameters ω, α' and dimensionless time t' on the self-balanced normal σ_{nn} and tangent stresses $\sigma_{n\tau}$, σ_{ne} on the fiber-matrix interface S (Fig.7.2.1). Note that the meanings of ω, α' and t' are given in section 4.6. Taking the results discussed in the previous subsection into account we consider the values of the stresses σ_{nn}, $\sigma_{n\tau}$ and σ_{ne} at points $\{\alpha t=\pi/2,\theta=\pi/2\}$, $\{\alpha t=0,\theta=\pi/2\}$ and $\{\alpha t=\pi/2,\theta=\pi\}$ respectively. We assume that $\chi=0.5$, 0.4, 0.3, 0.3, 0.2 for $\kappa_0=E^{(2)}/E_0^{(1)}=20$, 50, 100,150, 200 respectively and $\nu_0^{(1)}=\nu^{(2)}=0.3$.

Fig.7.4.7 shows the variation in the ratios $\sigma_{nn}/\sigma_{zz}^{(1),0}$, $\sigma_{n\tau}/\sigma_{zz}^{(1),0}$, $\sigma_{ne}/\sigma_{zz}^{(1),0}$ in relation to t' ; the process of changing the stresses can be divided into the following three periods: $0 \le t' \le 1$ (period I), $1 \le t' \le 3$ (period II), $t'>3$ (period III). The rate of change of the stresses σ_{nn}, $\sigma_{n\tau}$ and σ_{ne} can with sufficient accuracy be considered constant in periods I and III. However, these rates change greatly in period II. The rate is much greater in the first period than in the third.

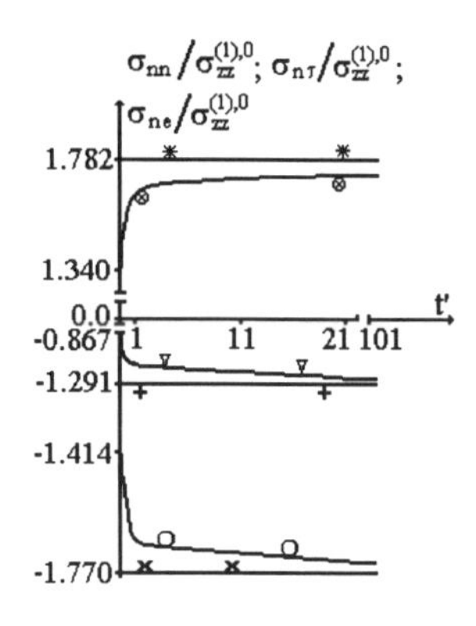

Fig.7.4.7. The relation between $\sigma_{nn}/\sigma_{zz}^{(1),0}$, $\sigma_{n\tau}/\sigma_{zz}^{(1),0}$, $\sigma_{ne}/\sigma_{zz}^{(1),0}$ and t' in the case where $\omega = 2; \alpha' = -0.5$, $\kappa_0 = 100$, $\varepsilon = 0.02$. ⊗ -- $\sigma_{nn}/\sigma_{zz}^{(1),0}$, ✳ -- $\sigma_{nn}/\sigma_{zz}^{(1),0}$ at $t' = \infty$, ∇ -- $\sigma_{n\tau}/\sigma_{zz}^{(1),0}$, ✚ -- $\sigma_{n\tau}/\sigma_{zz}^{(1),0}$ at $t' = \infty$, ○ -- $\sigma_{ne}/\sigma_{zz}^{(1),0}$, ✖ -- $\sigma_{ne}/\sigma_{zz}^{(1),0}$ at $t' = \infty$.

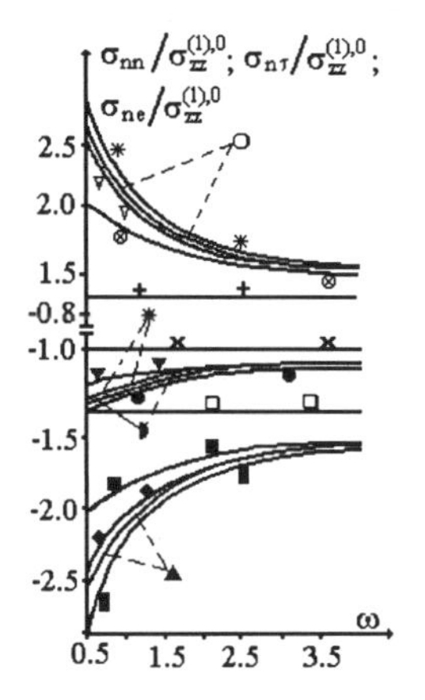

Fig.7.4.8. The relation between $\sigma_{nn}/\sigma_{zz}^{(1),0}$, $\sigma_{n\tau}/\sigma_{zz}^{(1),0}$, $\sigma_{ne}/\sigma_{zz}^{(1),0}$ and ω in the case $\alpha' = -0.5$, κ_0 100, $\varepsilon = 0.02$ for t'=0., 1., 21.,51., ∞ : ✚,⊗,∇,○,✳ -- $\sigma_{nn}/\sigma_{zz}^{(1),0}$, ✖,▼,✱,◗,● -- $\sigma_{n\tau}/\sigma_{zz}^{(1),0}$, ❑,■,◆,▲,▮ -- $\sigma_{ne}/\sigma_{zz}^{(1),0}$.

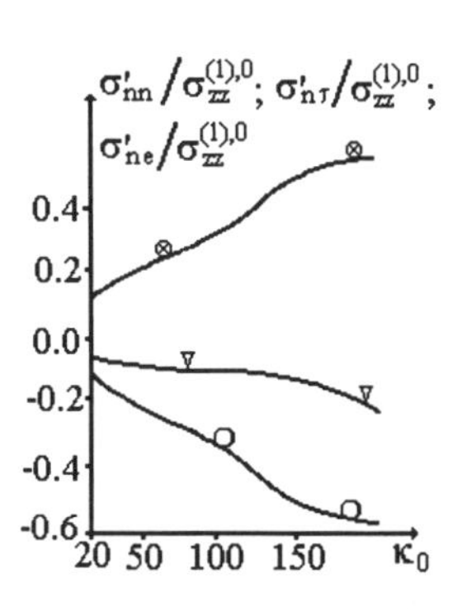

Fig.7.4.9. The graphs show the increase of the viscosity effect with κ_0 in the case where $\varepsilon = 0.02$, $\omega = 2$, $\alpha' = -0.5$. ⊗ -- $\sigma'_{nn}/\sigma_{zz}^{(1),0}$, ∇ -- $\sigma'_{n\tau}/\sigma_{zz}^{(1),0}$, ○ -- $\sigma'_{ne}/\sigma_{zz}^{(1),0}$.

Consider the graphs shown in Fig.7.4.8 which illustrate the relationship between the ratios $\sigma_{nn}/\sigma_{zz}^{(1),0}$, $\sigma_{n\tau}/\sigma_{zz}^{(1),0}$, $\sigma_{ne}/\sigma_{zz}^{(1),0}$ and the parameter ω ; the self-balanced stresses σ_{nn}, $\sigma_{n\tau}$ and σ_{ne} decrease monotonically with an increase in the parameter ω, and approach their corresponding instantaneous values (i.e., the values which are obtained at t'=0). When $0 \le \omega \le 3$, the viscosity of the matrix material has a significant effect on the stresses σ_{nn}, $\sigma_{n\tau}$ and σ_{ne}.

Table 7.5 shows values of $\sigma_{nn}/\sigma_{zz}^{(1),0}$, $\sigma_{n\tau}/\sigma_{zz}^{(1),0}$, $\sigma_{ne}/\sigma_{zz}^{(1),0}$ for different values of the parameters ω and α', which is the exponent of the singularity of the kernel of the operators (4.6.2); the stresses σ_{nn}, $\sigma_{n\tau}$ and σ_{ne} decrease with an increase in the absolute value of α'; a change in α' has a significant effect on these stresses when $\omega = 0.5$, but the effect is negligible when $\omega = 2$ and $\omega = 4$.

Table 7.5. The values of $\sigma_{nn}/\sigma_{zz}^{(1),0}$, $\sigma_{n\tau}/\sigma_{zz}^{(1),0}$ and $\sigma_{ne}/\sigma_{zz}^{(1),0}$ for various ω and α under $\varepsilon = 0.02$, $\kappa_0 = 100$, t'= 51.

Stresses	ω	α'			
		-0.3	-0.5	-0.7	-0.9
$\sigma_{nn}/\sigma_{zz}^{(1),0}$	0.5	2.709	2.603	2.396	2.085
	2.0	1.774	1.763	1.736	1.681
	4.0	1.565	1.562	1.554	1.537
$\sigma_{n\tau}/\sigma_{zz}^{(1),0}$	0.5	-1.269	-1.242	-1.187	-1.102
	2.0	-1.012	-1.008	-1.000	-0.983
	4.0	-0.946	-0.945	-0.942	-0.936
$\sigma_{ne}/\sigma_{zz}^{(1),0}$	0.5	-2.713	-2.607	-2.399	-2.083
	2.0	-1.760	-2.750	-1.722	-1.664
	4.0	-1.540	-1.537	-1.528	-1.510

Table 7.6. The values of $\sigma_{nn}/\sigma_{zz}^{(1),0}$, $\sigma_{n\tau}/\sigma_{zz}^{(1),0}$ and $\sigma_{ne}/\sigma_{zz}^{(1),0}$ for various κ_0 under $\alpha' = -0.5$, $\varepsilon = 0.02$, $\omega = 2$, t'= 51.

Stresses	κ_0				
	20	50	100	150	200
$\sigma_{nn}/\sigma_{zz}^{(1),0}$	0.451	0.988	1.763	2.347	2.804
$\sigma_{n\tau}/\sigma_{zz}^{(1),}$	-0.542	-0.788	-1.008	-1.251	-1.411
$\sigma_{ne}/\sigma_{zz}^{(1),0}$	-0.466	-0.994	-1.750	-2.322	-2.770

Table 7.6 gives the values of $\sigma_{nn}/\sigma_{zz}^{(1),0}$, $\sigma_{n\tau}/\sigma_{zz}^{(1),0}$, $\sigma_{ne}/\sigma_{zz}^{(1),0}$ obtained for various κ_0 are given: the stresses increase monotonically with κ_0. Furthermore, Fig.7.4.9 shows that with an increase in κ_0, the effect of viscosity of the matrix

material is to increase the stresses σ_{nn}, $\sigma_{n\tau}$ and σ_{ne} monotonically. In Fig.7.4.9 the following notation is used:

$$\sigma'_{nn} = \sigma_{nn}\big|_{t'=0} - \sigma_{nn}\big|_{t'=51}, \quad \sigma'_{n\tau} = \sigma_{n\tau}\big|_{t'=0} - \sigma_{n\tau}\big|_{t'=51},$$

$$\sigma'_{ne} = \sigma_{ne}\big|_{t'=0} - \sigma_{ne}\big|_{t'=51}.$$

To illustrate the convergence of the method used here, Table 7.7 shows the values of $\sigma_{nn}/\sigma_{zz}^{(1),0}$, $\sigma_{n\tau}/\sigma_{zz}^{(1),0}$, $\sigma_{ne}/\sigma_{zz}^{(1),0}$ obtained in various approximations for some problem parameters: convergence is acceptable

Table 7.7. The values of $\sigma_{nn}/\sigma_{zz}^{(1),0}$, $\sigma_{n\tau}/\sigma_{zz}^{(1),0}$ and $\sigma_{ne}/\sigma_{zz}^{(1),0}$ obtained in various approximations for $\alpha' = -0.5$, $\varepsilon = 0.025$, $\omega = 2$, $t' = 51$.

κ_0	Numb. of approx.	$\sigma_{nn}/\sigma_{zz}^{(1),0}$	$\sigma_{n\tau}/\sigma_{zz}^{(1),0}$	$\sigma_{ne}/\sigma_{zz}^{(1),0}$
	0	-0.047	0.000	0.000
50	1	1.236	-0.977	-1.242
	2	1.249	-0.977	-1.242
	0	-0.047	0.000	0.000
100	1	2.191	-1.260	-2.187
	2	2.221	-1.260	-2.187

7.5. Screwed Fibers in an Elastic Matrix

Screwed fibers are employed mainly in textile composites; some problems for such composites have been investigated in [149] and elsewhere. The screwing of the fibers arises as a result of the construction requirement of composites, however, this screwing causes self-balanced stresses in the matrix. We employ the method developed in the present chapter to composites with screwed fibers.

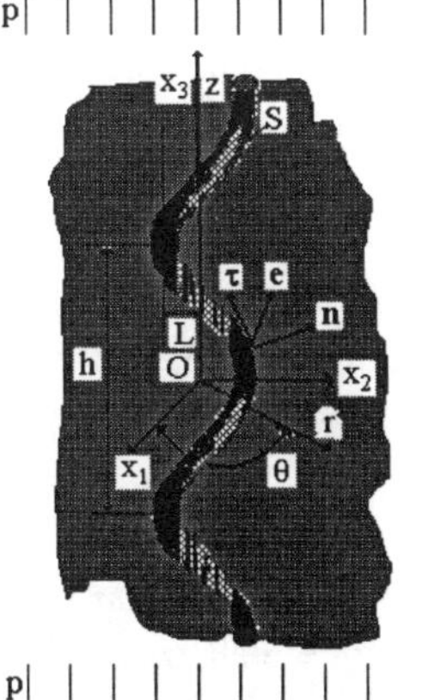

Fig.7.5.1. A screwed fiber in a matrix.

7.5.1. FORMULATION AND METHOD OF SOLUTION

For simplicity we consider an infinite matrix containing a single screwed fiber. The middle line of the fiber we take as a helical line with parameters L and h, where L is a radius and h is a step of the helical line (Fig.7.5.1). With a fiber we associate the Cartesian $Ox_1x_2x_3$ and cylindrical $Or\theta z$ system of coordinates in which the equation of the middle line can be written as follows:

$$z = t, \; \theta = \frac{2\pi}{h} t, \; r = L = \text{const}. \tag{7.5.1}$$

We assume that L<h and introduce the small parameter $\varepsilon = L/h$.

As before we suppose that the cross section of the fiber normal to the middle line of the fiber is a circle with radius R, constant along the entire length of the fiber. We assume that the material of the fiber and matrix are homogeneous and transversely isotropic with symmetry axis Oz. As before we suppose that the body is loaded at infinity by uniformly distributed normal forces with intensity p acting in the direction of the Oz axis (Fig.7.5.1), and that the complete cohesion condition on the fiber-matrix interface S is satisfied. As in section 7.2, using the constancy condition of the fiber cross section and equation (7.5.1) we obtain the following equation for this surface.

$$r = r(\theta, t) = \left[1 + 4\pi^2\varepsilon^2 \sin^2(\alpha t - \theta)\right]^{-1} \left\{\varepsilon h \cos(\alpha t - \theta) + \right.$$

$$\left. R\left[1 + \left(4\pi^2\varepsilon^2 - \frac{h^2\varepsilon^2}{R^2}\left(1 + 4\pi^2\varepsilon^2\right)\right)\sin^2(\alpha t - \theta)\right]^{\frac{1}{2}}\right\},$$

$$z = t + 2\pi\varepsilon r(\theta, t)\sin(\alpha t - \theta), \quad \alpha = 2\pi/h \tag{7.5.2}$$

Going through all the procedures given in sections 7.2 and 7.3 with obvious changes and with equation (7.5.2), we obtain the corresponding system of equations and contact conditions for each approximation (7.2.7). The equations and contact conditions for the zeroth approximation coincide with those obtained in section 7.2, and the solution to this problem is (7.2.29)-(7.2.32). The first and subsequent approximations are also determined as in sections 7.2 and 7.3, by representing the sought values in the $f(r)\cos(\alpha t - \theta)$ or $f(r)\sin(\alpha t - \theta)$ form. The remaining part of the solution procedure for screwed fibers is the same as in sections 7.2 and 7.3; we omit details.

7.5.2. NUMERICAL RESULTS

The fiber and matrix are assumed to be isotropic and homogeneous. We investigate the stress distribution in this material under the action of uniformly distributed normal forces at infinity with intensity p in the direction of the Oz axis. We introduce the parameter $\lambda = \alpha t - \theta$ and taking the screwing periodicity into account, consider only the case where $0 \le \lambda \le \pi$. Moreover, we introduce the parameter $\chi = 2\pi R/h$ and assume that $\nu_1 = \nu_2 = 0.3$. First we investigate the distribution of the normal σ_{nn} and shear $\sigma_{n\tau}$, σ_{ne} stresses on the fiber-matrix interface S.

Fig. 7.5.2 shows the dependencies of $\sigma_{nn}/\sigma_{33}^{(1),0}$, $\sigma_{n\tau}/\sigma_{33}^{(1),0}$, $\sigma_{ne}/\sigma_{33}^{(1),0}$ on λ; the stress σ_{nn} has its maximum at $\lambda = 0;\pi$, and the shear stresses $\sigma_{n\tau}$, σ_{ne} vanish there. These shear stresses have their maxima at the point for which $\sigma_{nn} = 0$ and which corresponds to $\lambda \approx \pi/2$ as suggested by Fig.7.5.2.

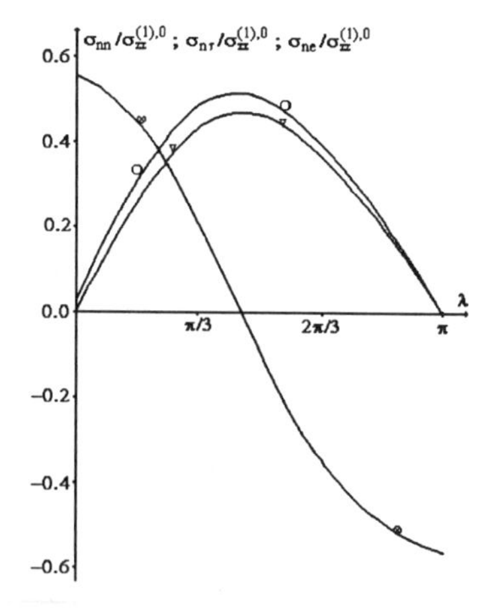

Fig.7.5.2. The relations between $\sigma_{nn}/\sigma_{zz}^{(1),0}$, $\sigma_{n\tau}/\sigma_{zz}^{(1),0}$, $\sigma_{ne}/\sigma_{zz}^{(1),0}$ and λ in the case where $\varepsilon = 0.015$, $\kappa = 50$, $\chi = 0.3$: ⊗ -- $\sigma_{nn}/\sigma_{zz}^{(1),0}$; ∇ -- $\sigma_{n\tau}/\sigma_{zz}^{(1),0}$; O -- $\sigma_{ne}/\sigma_{zz}^{(1),0}$.

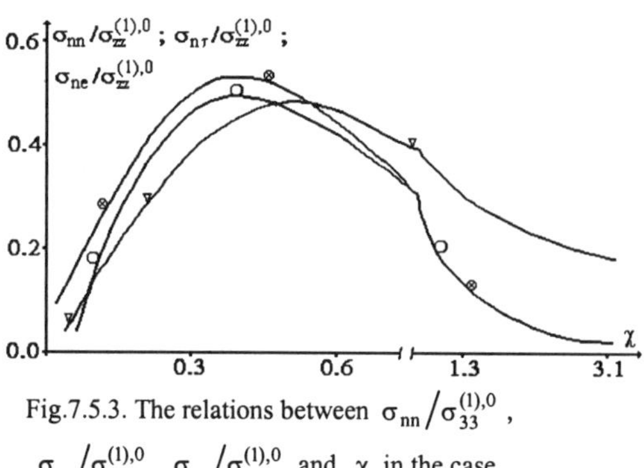

Fig.7.5.3. The relations between $\sigma_{nn}/\sigma_{33}^{(1),0}$, $\sigma_{n\tau}/\sigma_{zz}^{(1),0}$, $\sigma_{ne}/\sigma_{zz}^{(1),0}$ and χ in the case where $\varepsilon = 0.015$, $\kappa = 50$, ⊗ -- $\sigma_{nn}/\sigma_{zz}^{(1),0}$, $\lambda = 0$; ∇ -- $\sigma_{n\tau}/\sigma_{zz}^{(1),0}$, $\lambda = \pi/2$; O -- $\sigma_{ne}/\sigma_{zz}^{(1),0}$, $\lambda = \pi/2$.

Consider the relations between $\sigma_{nn}/\sigma_{zz}^{(1),0}$, $\sigma_{n\tau}/\sigma_{zz}^{(1),0}$, $\sigma_{ne}/\sigma_{zz}^{(1),0}$ and χ. Fig.7.5.3 gives for $\chi \leq 3.1$. This limitation follows from the requirement that different points of the fiber-matrix interface do not coincide. The relations are non-monotonic; there is a value χ^* for which the values of σ_{nn}, $\sigma_{n\tau}$ and σ_{ne} become maximum. These results agree with those for the bending form curving of the fiber discussed in the previous section.

Table 7.8. The values of $\sigma_{nn}/\sigma_{zz}^{(1),0}$ (at $\lambda = 0$), $\sigma_{n\tau}/\sigma_{zz}^{(1),0}$ ($\lambda = \pi/2$), $\sigma_{ne}/\sigma_{zz}^{(1),0}$ ($\lambda = \pi/2$) for various ε, κ under $\chi = 0.3$.

Stresses	κ	ε			
		0.010	0.015	0.020	0.025
$\sigma_{nn}/\sigma_{zz}^{(1),0}$	50	0.355	0.517	0.657	0.769
	100	0.639	0.910	1.120	1.249
	150	0.864	1.209	1.446	1.536
$\sigma_{n\tau}/\sigma_{zz}^{(1),0}$	50	0.280	0.397	0.484	0.532
	100	0.408	0.565	0.664	0.688
	150	0.509	0.691	0.786	0.788
$\sigma_{ne}/\sigma_{zz}^{(1),0}$	50	0.332	0.469	0.569	0.621
	100	0.605	0.832	0.969	0.985
	150	0.821	1.107	1.244	1.266

Table 7.8 shows the influence of κ and ε on the values $\sigma_{nn}/\sigma_{zz}^{(1),0}$, $\sigma_{n\tau}/\sigma_{zz}^{(1),0}$, $\sigma_{ne}/\sigma_{zz}^{(1),0}$: the stresses σ_{nn}, $\sigma_{n\tau}$ and σ_{ne} increase monotonically with κ and ε.

Now we investigate the distribution of the normal stresses $\sigma_{\tau\tau}^{(k)}$, $\sigma_{ee}^{(k)}$ (k=1,2) which act in the directions of the vectors $\boldsymbol{\tau}$ and $\mathbf{e}$ respectively. The values of these stresses are calculated at points of the interface S. As before, in the zeroth approximation when $\nu_1 = \nu_2$ we obtain $\sigma_{ee}^{(1)} = \sigma_{ee}^{(2)} = \sigma_{\theta\theta}^{(1)} = \sigma_{\theta\theta}^{(2)} = 0$ and $\sigma_{\tau\tau}^{(1)} = \sigma_{zz}^{(1),0}$, $\sigma_{\tau\tau}^{(2)} = \sigma_{zz}^{(2),0}$. Therefore the influence of the screwing of the fiber on the distribution of these stresses will be analysed in relation to its self-balanced parts. In this case the numerical results show that the values of $\sigma_{ee}^{(2)}/\sigma_{zz}^{(2),0}$ are obtained as $(\kappa)^{-1}\sigma_{\tau\tau}^{(2)}/\sigma_{zz}^{(2),0}$. Therefore we investigate only the stresses $\sigma_{\tau\tau}^{(k)}$ $(k = 1,2)$ and $\sigma_{ee}^{(1)}$, because $\sigma_{ee}^{(2)} << \sigma_{\tau\tau}^{(2)}$.

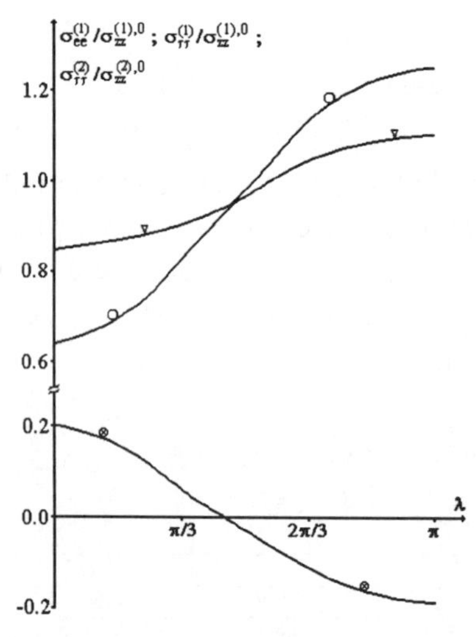

Fig.7.5.4. The relations between $\sigma_{\tau\tau}^{(1)}/\sigma_{zz}^{(1),0}$, $\sigma_{\tau\tau}^{(2)}/\sigma_{zz}^{(2),0}$, $\sigma_{ee}^{(1)}/\sigma_{zz}^{(1),0}$ and λ in the case where $\varepsilon = 0.015$, $\kappa = 50$, $\chi = 0.5$: ⊗ -- $\sigma_{ee}^{(1)}/\sigma_{zz}^{(1),0}$; ∇-- $\sigma_{\tau\tau}^{(1)}/\sigma_{zz}^{(1),0}$; O-- $\sigma_{\tau\tau}^{(2)}/\sigma_{zz}^{(2),0}$.

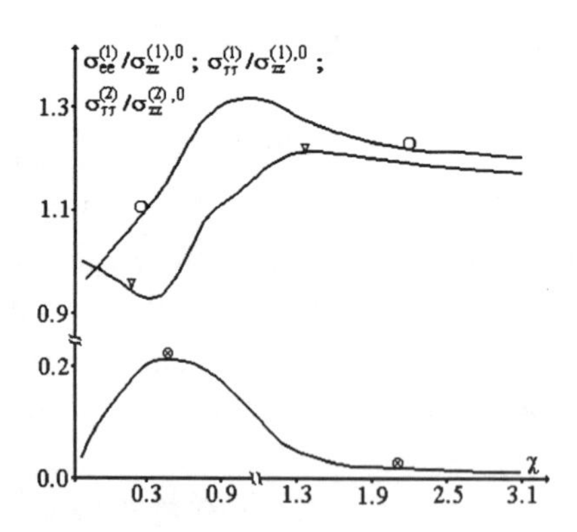

Fig.7.5.5. The relations between $\sigma_{\tau\tau}^{(1)}/\sigma_{zz}^{(1),0}$, $\sigma_{\tau\tau}^{(2)}/\sigma_{zz}^{(2),0}$, $\sigma_{ee}^{(1)}/\sigma_{zz}^{(1),0}$ and χ in the case where $\varepsilon = 0.015$, $\kappa = 50$:
⊗ -- $\sigma_{ee}^{(1)}/\sigma_{zz}^{(1),0}$ at $\lambda = 0$;
∇-- $\sigma_{\tau\tau}^{(1)}/\sigma_{zz}^{(1),0}$ at $\lambda = \pi$;
O-- $\sigma_{\tau\tau}^{(2)}/\sigma_{zz}^{(2),0}$ at $\lambda = \pi$.

Fig. 7.5.4 shows the distribution of the stresses $\sigma_{ee}^{(1)}$, $\sigma_{\tau\tau}^{(2)}$ with respect to λ; the self-balanced part of $\sigma_{\tau\tau}^{(k)}$ and $\sigma_{ee}^{(1)}$ has its maximum at $\lambda = 0$ and $\lambda = \pi$. In this case in the part of the region where $\sigma_{ee}^{(1)}$ is tensile (compressive), the self-balanced parts of the stresses $\sigma_{\tau\tau}^{(1)}$ and $\sigma_{\tau\tau}^{(2)}$ are compressive (tensile); $\sigma_{\tau\tau}^{(2)}$ has a more significant change than the others.

Consider the relation between $\sigma_{\tau\tau}^{(k)}$ (k=1,2), $\sigma_{ee}^{(1)}$ and χ. The graphs of these relations given in Fig. 7.5.5, reveal that they are very similar to those which have been analysed in the previous section.

Table 7.9. The values of $\sigma_{ee}^{(1)}/\sigma_{zz}^{(1),0}$ (at $\chi = 0.5,\ \lambda = 0$), $\sigma_{\tau\tau}^{(1)}/\sigma_{zz}^{(1),0}$ (at $\chi = 0.7,\ \lambda = \pi$), $\sigma_{\tau\tau}^{(2)}/\sigma_{zz}^{(2),0}$ (at $\chi = 0.7, \lambda = \pi$) for various ε, κ under $\chi = 0.3$.

Stresses	κ	ε			
		0.010	0.015	0.020	0.025
$\sigma_{ee}^{(1)}/\sigma_{zz}^{(1),0}$	50	0.143	0.209	0.268	0.319
	100	0.265	0.379	0.472	0.535
	150	0.362	0.509	0.616	0.668
$\sigma_{\tau\tau}^{(1)}/\sigma_{zz}^{(1),0}$	50	1.120	1.169	1.211	1.245
	100	1.135	1.190	1.234	1.270
	150	1.142	1.198	1.244	1.280
$\sigma_{\tau\tau}^{(2)}/\sigma_{zz}^{(2),0}$	50	1.230	1.329	1.414	1.480
	100	1.274	1.391	1.487	1.559
	150	1.294	1.418	1.519	1.593

Table 7.9 shows the values of $\sigma_{nn}/\sigma_{33}^{(1),0}$, $\sigma_{n\tau}/\sigma_{33}^{(1),0}$, $\sigma_{ne}/\sigma_{33}^{(1),0}$ at various ε and κ. These results prove that the self-balanced part of the stresses $\sigma_{\tau\tau}^{(k)}$ (k=1,2) and $\sigma_{ee}^{(1)}$ increase monotonically with ε and κ.

Note that the numerical results analysed here are obtained in the framework of the first four approximations. In this case the difference between the results obtained in the last two approximations is not greater than 3%.

7.6. Bibliographical Notes

The method for investigation of the stress state in the fibrous composites with curved structure was proposed by S.D. Akbarov and A.N. Guz [30]. The development of this method for the cases where interaction between the curved fibers must be taken into account was made by S.D. Akbarov and M.B. Babazade [24]. The numerical results discussed in section 7.4 were obtained by S.D. Akbarov and A.N. Guz [35], and by S.D. Akbarov [4,6]. The problem related to the composite with screwed fiber was explored by S.D.Akbarov [17].

CHAPTER 8

GEOMETRICALLY NON-LINEAR PROBLEMS

In some combinations of geometric and curvature parameters of the filler layers (or fibers) and of the values of the external force intensities it is necessary to investigate of problems using the geometrically non-linear statement. Using the results of these investigations, we can determine the limit of the intensity of the external forces for which the results obtained in the linear statement are acceptable. Furthermore, we can determine the character of the influence of the geometrical non-linearity on the mechanical behaviour of the composites. We will also derive relations which will be very important for investigations of the fracture of unidirectional composites with curved structure in compression.

8.1. Formulation. Governing Relations and Equations

We use the notation of Chapter 5, particularly that in Fig.5.1.1. Thus we associate the corresponding Lagrangian coordinates $O_m^{(k)}x_{1m}^{(k)}x_{2m}^{(k)}x_{3m}^{(k)}$ (k=1,2; m=1,2,...) which in their natural state coincide with Cartesian coordinates and are obtained from $Ox_1x_2x_3$ by parallel transfer along Ox_2 axis, with the middle surface of each layer of the filler and matrix. As in Chapter 5, the reinforcing layers will also be assumed to be located in planes $O_m^{(2)}x_{1m}^{(2)}x_{3m}^{(2)}$ and the thickness of every filler layer will be assumed constant. We suppose that the matrix and filler materials are homogeneous, anisotropic and non-aging viscoelastic.

Now, *in the geometrical non-linear statement*, we investigate the stress distribution under the action of uniformly distributed normal forces of intensity $p_1(p_3)$ in the direction of the $Ox_1(Ox_3)$ axis at infinity. According to [90,101], for each layer we write the equilibrium equations, mechanical and geometrical relations as follows:

$$\frac{\partial}{\partial x_{\underline{jm}}^{(\underline{k})}}\left[\sigma_{jn}^{(\underline{k})\underline{m}}\left(\delta_i^n+\frac{\partial u_i^{(\underline{k})\underline{m}}}{\partial x_{\underline{nm}}^{(\underline{k})}}\right)\right]=0, \tag{8.1.1}$$

$$\sigma_{ij}^{(k)m}=C_{ijrs0}^{(\underline{k})\underline{m}}\varepsilon_{rs}^{(\underline{k})\underline{m}}(t)+\int_0^t C_{ijrs}^{(\underline{k})\underline{m}}(t-\tau)\varepsilon_{rs}^{(\underline{k})\underline{m}}(\tau)d\tau, \tag{8.1.2}$$

$$\varepsilon_{ij}^{(k)m} = \frac{1}{2}\left(\frac{\partial u_i^{(k)\underline{m}}}{\partial x_{j\underline{m}}^{(k)}} + \frac{\partial u_j^{(k)\underline{m}}}{\partial x_{i\underline{m}}^{(k)}} + \frac{\partial u_n^{(k)\underline{m}}}{\partial x_{i\underline{m}}^{(k)}} \frac{\partial u_n^{(k)\underline{m}}}{\partial x_{j\underline{m}}^{(k)}} \right), \quad i,j,r,s = 1,2,3; \quad k = 1,2 . \quad (8.1.3)$$

We assume that, the filler and matrix layers, are complete cohesion; using the notation introduced in Fig.5.1.1 we may write these conditions as follows:

$$\left[\left(\delta_i^n + \frac{\partial u_i^{(1)\underline{m}}}{\partial x_{n\underline{m}}^{(1)}} \right) \sigma_{jn}^{(1)\underline{m}} \right] \Bigg|_{S_m^+} n_j^{\underline{m},+} = \left[\left(\delta_i^n + \frac{\partial u_i^{(2)\underline{m}}}{\partial x_{n\underline{m}}^{(2)}} \right) \sigma_{jn}^{(2)\underline{m}} \right] \Bigg|_{S_m^+} n_j^{\underline{m},+},$$

$$\left[\left(\delta_i^n + \frac{\partial u_i^{(1)\underline{m_0}}}{\partial x_{n\underline{m_0}}^{(1)}} \right) \sigma_{jn}^{(1)\underline{m_0}} \right] \Bigg|_{S_m^-} n_j^{\underline{m},-} = \left[\left(\delta_i^n + \frac{\partial u_i^{(2)\underline{m}}}{\partial x_{n\underline{m}}^{(2)}} \right) \sigma_{jn}^{(2)\underline{m}} \right] \Bigg|_{S_m^-} n_j^{\underline{m},-},$$

$$u_i^{(1)\underline{m}} \Big|_{S_{\underline{m}}^+} = u_i^{(2)\underline{m}} \Big|_{S_{\underline{m}}^+}, \quad u_i^{(1)\underline{m_0}} \Big|_{S_{\underline{m}}^-} = u_i^{(2)\underline{m}} \Big|_{S_{\underline{m}}^-}, \quad (8.1.4)$$

where $m_0 = m - 1$ and $n_j^{m,\pm}$ are the components of the normal to the surfaces $S_m^{\pm}$.

Assume that the middle surface of the m-th filler layer is given by (5.1.3). Performing the operations described in section 5.2 and representing the various quantities in the series form (5.3.1), we obtain a closed system of equations and contact conditions for each approximation. Owing to the linearity of the mechanical relations, each approximation in equation (5.3.1) will be satisfied separately. The remaining relations obtained from equations (8.1.1)-(8.1.4) for the q-th approximation contain the values of all previous approximations. To simplify the exposition, we write the relations for the zeroth and subsequent approximations.

The zeroth approximation. In this case the equations (8.1.1)-(8.1.3) are valid, and from the contact conditions (8.1.4) the following are obtained:

$$\left[\left(\delta_i^n + \frac{\partial u_i^{(1)\underline{m},0}}{\partial x_{n\underline{m}}^{(1)}} \right) \sigma_{jn}^{(1)\underline{m},0} \right] \Bigg|_{\left(t_{1\underline{m}}, -H_{\underline{m}}^{(1)}, t_{3\underline{m}}\right)} = \left[\left(\delta_i^n + \frac{\partial u_i^{(2)\underline{m},0}}{\partial x_{n\underline{m}}^{(2)}} \right) \sigma_{jn}^{(2)\underline{m},0} \right] \Bigg|_{\left(t_{1\underline{m}}, +H_{\underline{m}}^{(2)}, t_{3\underline{m}}\right)},$$

$$\left[\left(\delta_i^n + \frac{\partial u_i^{(1)\underline{m}_0,0}}{\partial x_{n\underline{m}_0}^{(1)}}\right)\sigma_{jn}^{(1)\underline{m}_0,0}\right]\Bigg|_{\left(t_{1\underline{m}_0},+H_{\underline{m}_0}^{(1)},t_{3\underline{m}_0}\right)} = \left[\left(\delta_i^n + \frac{\partial u_i^{(2)\underline{m},0}}{\partial x_{n\underline{m}}^{(2)}}\right)\sigma_{jn}^{(2)\underline{m},0}\right]\Bigg|_{\left(t_{1\underline{m}},-H_{\underline{m}}^{(2)},t_{3\underline{m}}\right)},$$

$$u_i^{(1)\underline{m},0}\Big|_{\left(t_{1\underline{m}},-H_{\underline{m}}^{(1)},t_{3\underline{m}}\right)} = u_i^{(2)\underline{m},0}\Big|_{\left(t_{1\underline{m}},+H_{\underline{m}}^{(2)},t_{3\underline{m}}\right)},$$

$$u_i^{(1)\underline{m}_0,0}\Big|\left(t_{1\underline{m}_0},-H_{\underline{m}_0}^{(1)},t_{3\underline{m}_0}\right) = u_i^{(2)\underline{m},0}\Big|\left(t_{1\underline{m}},-H_{\underline{m}}^{(2)},t_{3\underline{m}}\right). \tag{8.1.5}$$

The first approximation. For this approximation we get the following equilibrium equations and geometrical relations from (8.1.1) and (8.1.3):

$$\frac{\partial}{\partial x_{j\underline{m}}^{(\underline{k})}} T_{ij}^{(\underline{k})\underline{m},1} = 0, \tag{8.1.6}$$

$$\varepsilon_{ij}^{(k)m,1} = \frac{1}{2}\left(\delta_i^n + \frac{\partial u_n^{(\underline{k})\underline{m},0}}{\partial x_{i\underline{m}}^{(\underline{k})}}\right)\frac{\partial u_n^{(\underline{k})\underline{m},1}}{\partial x_{j\underline{m}}^{(\underline{k})}} + \frac{1}{2}\left(\delta_j^n + \frac{\partial u_n^{(\underline{k})\underline{m},0}}{\partial x_{j\underline{m}}^{(\underline{k})}}\right)\frac{\partial u_n^{(\underline{k})\underline{m},1}}{\partial x_{i\underline{m}}^{(\underline{k})}}, \tag{8.1.7}$$

The subsequent approximations. For the q-th approximation we obtain the following equilibrium equations and geometrical relations from (8.1.1) and (8.1.3):

$$\frac{\partial}{\partial x_{j\underline{m}}^{(\underline{k})}} T_{ij}^{(\underline{k})\underline{m},q} = \Psi_{iq}\left(T_{ij}^{(\underline{k})\underline{m},1},\ldots,T_{ij}^{(\underline{k})\underline{m},q-1}\right), \tag{8.1.8}$$

$$\varepsilon_{ij}^{(k)m,q} = \frac{1}{2}\left(\delta_i^n + \frac{\partial u_n^{(\underline{k})\underline{m},0}}{\partial x_{i\underline{m}}^{(\underline{k})}}\right)\frac{\partial u_n^{(\underline{k})\underline{m},q}}{\partial x_{j\underline{m}}^{(\underline{k})}} + \frac{1}{2}\left(\delta_j^n + \frac{\partial u_n^{(\underline{k})\underline{m},0}}{\partial x_{j\underline{m}}^{(\underline{k})}}\right)\frac{\partial u_n^{(\underline{k})\underline{m},q}}{\partial x_{i\underline{m}}^{(\underline{k})}} +$$

$$\sum_{r=1}^{q-1}\frac{\partial u_n^{(\underline{k})\underline{m},q-r}}{\partial x_{i\underline{m}}^{(\underline{k})}}\frac{\partial u_n^{(\underline{k})\underline{m},r}}{\partial x_{i\underline{m}}^{(\underline{k})}}. \tag{8.1.9}$$

The following notation is used in (8.1.6) and (8.1.8):

$$T_{ij}^{(k)\underline{m},q} = \left(\delta_i^n + \frac{\partial u_n^{(k)\underline{m},0}}{\partial x_{n\underline{m}}^{(k)}} \right) \sigma_{jn}^{(k)\underline{m},q} + \sigma_{jn}^{(k)\underline{m},0} \frac{\partial u_n^{(k)\underline{m},q}}{\partial x_{n\underline{m}}^{(k)}} . \tag{8.1.10}$$

The mechanical relations for each approximation can be written as (8.1.2), i.e., as

$$\sigma_{ij}^{(k)\underline{m},q} = C_{ijrs0}^{(k)\underline{m}} \varepsilon_{rs}^{(k)\underline{m},q}(t) + \int_0^t C_{ijrs}^{(k)\underline{m}}(t-\tau) \varepsilon_{rs}^{(k)\underline{m},q}(\tau) d\tau . \tag{8.1.11}$$

Moreover, the contact conditions for the first and subsequent approximations are obtained from relations (8.1.4) as follows:

$$T_{i2}^{(1)\underline{m},q}\Big|_{\left(t_{1\underline{m}},-H_{\underline{m}}^{(1)},t_{3\underline{m}}\right)} - T_{i2}^{(2)\underline{m},q}\Big|_{\left(t_{1\underline{m}},+H_{\underline{m}}^{(2)},t_{3\underline{m}}\right)} = \Phi_{1q}\Big(T_{ij}^{(1)\underline{m},q-1}, T_{ij}^{(2)\underline{m},q-1}, u_i^{(1)\underline{m},q-1},$$

$$u_i^{(2)\underline{m},q-1}, \ldots, \sigma_{ij}^{(1)\underline{m},0}, \sigma_{ij}^{(2)\underline{m},0}, u_i^{(1)\underline{m},0}, u_i^{(2)\underline{m},0}, f_{\underline{m}}\Big),$$

$$T_{i2}^{(1)\underline{m_0},q}\Big|_{\left(t_{1\underline{m_0}},+H_{\underline{m_0}}^{(1)},t_{3\underline{m_0}}\right)} - T_{i2}^{(2)\underline{m},q}\Big|_{\left(t_{1\underline{m}},-H_{\underline{m}}^{(2)},t_{3\underline{m}}\right)} = \Phi_{2q}\Big(T_{ij}^{(1)\underline{m_0},q-1}, T_{ij}^{(2)\underline{m},q-1}, u_i^{(1)\underline{m_0},q-1},$$

$$u_i^{(2)\underline{m},q-1}, \ldots, \sigma_{ij}^{(1)\underline{m_0},0}, \sigma_{ij}^{(2)\underline{m},0}, u_i^{(1)\underline{m_0},0}, u_i^{(2)\underline{m},0}, f_{\underline{m}}\Big),$$

$$u_i^{(1)\underline{m},q}\Big|_{\left(t_{1\underline{m}},-H_{\underline{m}}^{(1)},t_{3\underline{m}}\right)} - u_i^{(2)\underline{m},q}\Big|_{\left(t_{1\underline{m}},+H_{\underline{m}}^{(2)},t_{3\underline{m}}\right)} = \varphi_{1q}\Big(u_i^{(1)\underline{m},q-1}, u_i^{(2)\underline{m},q-1}, \ldots,$$

$$u_i^{(1)\underline{m},0}, u_i^{(2)\underline{m},0}, f_{\underline{m}}\Big),$$

$$u_i^{(1)\underline{m_0},q}\Big|_{\left(t_{1\underline{m_0}},+H_{\underline{m_0}}^{(1)},t_{3\underline{m_0}}\right)} - u_i^{(2)\underline{m},q}\Big|_{\left(t_{1\underline{m}},-H_{\underline{m}}^{(2)},t_{3\underline{m}}\right)} = \varphi_{2q}\Big(u_i^{(1)\underline{m_0},q-1}, u_i^{(2)\underline{m},q-1}, \ldots,$$

$$u_i^{(1)\underline{m_0},0}, u_i^{(2)\underline{m},0}, f_{\underline{m}}\Big). \tag{8.1.12}$$

In equations (8.1.8) and (8.1.12) the Ψ_{iq}, Φ_{1q}, Φ_{2q}, φ_{1q} and φ_{2q} are the known functions defined in the solution procedure; we omit details.

Thus, the investigations reduce to the solution of the series of problems described by equations (8.1.1)-(8.1.3), (8.1.5) - (8.1.12). The zeroth approximation is found from the non-linear equations (8.1.1)-(8.1.3), (8.1.5), and the values of the subsequent approximations are determined from the linear equations (8.1.6)-(8.1.12).

By direct verification we may prove that these linear equations in the first approximation coincide with the equations of the three-dimensional linearized theory of a deformable solid body [57, 59, 60, 90, 101, 139]. *Also, the linear equations obtained for the subsequent approximations are the inhomogeneous equations, the corresponding homogeneous equations of which also coincide with these three-dimensional linearized theory equations.* This situation was observed, first, in [48] and was applied for investigation of the stability loss in a structure composed unidirected composite materials. Composites with curved structures may be used to model numerous fracture problems of composites in compression. Among the application fields of the three dimensional linearized theory there is also stability problems for deformable bodies [53, 56, 60, 94,95, 101]. In some references, for example in [59], this theory is called the *General theory of stability*. Consequently, this theory is also applied for investigation of the stability problems in structures composed of the unidirected composites in compression [54, 92, 102]. However, up to now these applications have been made mainly only for composites fabricated from time independent materials. Layered composites with slightly periodically curved structure can be taken as models for stability investigations in viscoelastic unidirected composites. These and other applications of the equations (8.1.5) - (8.1.12) will be considered in later chapters. Now we consider the determination of each approximation in the geometrical non-linear statement.

8.2. Method of Solution

We consider the two- and three - dimensional problems separately. We use the notation and assumptions introduced in chapters 4 and 5 and all discussions will be made on composites with co-phase periodically curved layers shown in Figs. 4.2.1 and 5.4.1. The materials of the layers will be assumed to be homogeneous and isotropic. In both two- and three-dimensional cases we find that the zeroth approximation corresponds to the stress-deformation state in a laminated composite with an ideal layout of layers and a prescribed form of the external forces. Therefore, in the zeroth order approximation, the non-linear terms in equations (8.1.1)-(8.1.3) and (8.1.5) can be neglected with sufficiently high accuracy. Thus, the zeroth approximation corresponds to geometrical linearity, and this approximation for the plane-strain state is given by (4.2.5)-(4.2.7); for the three-dimensional problem by (5.4.6)-(5.4.9).

8.2.1. PLANE-STRAIN STATE

We consider the determination of the first approximation for which the following equations and relations are obtained from (8.1.6), (8.1.7) and (8.1.12).

The equilibrium equation:

$$c_1^{(k)} \frac{\partial^2 u_1^{(k),1}}{\partial (x_1^{(k)})^2} + c_2^{(k)} \frac{\partial^2 u_2^{(k),1}}{\partial x_1^{(k)} \partial x_2^{(k)}} + c_3^{(k)} \frac{\partial^2 u_1^{(k),1}}{\partial (x_2^{(k)})^2} = 0,$$

$$c_4^{(k)} \frac{\partial^2 u_2^{(k),1}}{\partial (x_1^{(k)})^2} + c_5^{(k)} \frac{\partial^2 u_1^{(k),1}}{\partial x_1^{(k)} \partial x_2^{(k)}} + c_6^{(k)} \frac{\partial^2 u_2^{(k),1}}{\partial (x_2^{(k)})^2} = 0 . \tag{8.2.1}$$

Geometrical relations:

$$\varepsilon_{11}^{(k),1} = \left(1+\Gamma_1^{(k)}\right) \frac{\partial u_1^{(k),1}}{\partial x_1^{(k)}}, \quad \varepsilon_{22}^{(k),1} = \left(1+\Gamma_2^{(k)}\right) \frac{\partial u_2^{(k),1}}{\partial x_2^{(k)}},$$

$$\varepsilon_{12}^{(k),1} = \frac{1}{2}\left(1+\Gamma_1^{(k)}\right) \frac{\partial u_1^{(k),1}}{\partial x_2^{(k)}} + \frac{1}{2}\left(1+\Gamma_2^{(k)}\right) \frac{\partial u_2^{(k),1}}{\partial x_1^{(k)}} . \tag{8.2.2}$$

Contact relations:

$$-f'(t_1)\left(1+\Gamma_1^{(1)}\right)\sigma_{11}^{(1),0} + \left(1+\Gamma_1^{(1)}\right)\sigma_{12}^{(1),1}\Big|_{\left(t_1,\pm H^{(1)}\right)} =$$

$$-f'(t_1)\left(1+\Gamma_1^{(2)}\right)\sigma_{11}^{(2),0} + \left(1+\Gamma_1^{(2)}\right)\sigma_{12}^{(2),1}\Big|_{\left(t_1,\mp H^{(2)}\right)},$$

$$\left(1+\Gamma_2^{(1)}\right)\sigma_{12}^{(1),1}\Big|_{\left(t_1,\pm H^{(1)}\right)} = \left(1+\Gamma_2^{(2)}\right)\sigma_{12}^{(2),1}\Big|_{\left(t_1,\mp H^{(2)}\right)}.$$

$$U_i^{(1),1}\Big|_{\left(t_1,\pm H^{(1)}\right)} = U_i^{(2),1}\Big|_{\left(t_1,\mp H^{(2)}\right)} . \tag{8.2.3}$$

In (8.2.1) - (8.2.3) the following notation is used:

$$\Gamma_1^{(k)} = \frac{\partial u_1^{(k),0}}{\partial x_1^{(k)}}, \quad \Gamma_2^{(k)} = \frac{\partial u_2^{(k),0}}{\partial x_2^{(k)}}, \quad c_1^{(k)} = \left(1+\Lambda^{(k)}\right)\left(1+\Gamma_1^{(k)}\right)^2 + \frac{\sigma_{11}^{(k),0}}{E^{(k)}}\left(1+\nu^{(k)}\right),$$

$$c_2^{(k)} = \Lambda^{(k)}\left(1+\Gamma_1^{(k)}\right)\left(1+\Gamma_2^{(k)}\right) + \frac{1}{2}\left(1+\Gamma_1^{(k)}\right)\left(1+\Gamma_2^{(k)}\right), \quad c_3^{(k)} = \frac{1}{2}\left(1+\Gamma_1^{(k)}\right)^2,$$

$$c_4^{(k)} = \frac{1}{2}\left(1+\Gamma_2^{(k)}\right)^2 + \frac{\sigma_{11}^{(k),0}}{E^{(k)}}\left(1+\nu^{(k)}\right), \quad c_5^{(k)} = \left(1+\Gamma_2^{(k)}\right)\left(1+\Gamma_1^{(k)}\right)\left(\Lambda^{(k)} + \frac{1}{2}\right),$$

$$c_6^{(k)} = \left(1+\Lambda^{(k)}\right)\left(1+\Gamma_2^{(k)}\right)^2, \ \Lambda^{(k)} = \frac{\nu^{(k)}}{1-2\nu^{(k)}}, \quad f'(x_1) = \frac{df}{dx_1}. \tag{8.2.4}$$

The expression for $U_i^{(k),1}$ is given in (4.1.23). When $\sigma_{11}^{(k),0} = 0$, the relations (8.2.1) - (8.2.4) coincide with those obtained in Chapter 4.

For the periodical curving function $f(x_1)$, we take

$$f(x_1) = \ell \sin \alpha x_1, \quad \alpha = 2\pi/\ell. \tag{8.2.5}$$

Noting (8.2.5) and the contact relations (8.2.3) we represent the displacements in the first approximation as follows:

$$u_1^{(k),1} = f_1^{(k),1}\left(x_2^{(k)}\right)\cos \alpha x_1, u_2^{(k),1} = f_2^{(k),1}\left(x_2^{(k)}\right)\sin \alpha x_1. \tag{8.2.6}$$

Substituting (8.2.6) in (8.2.1) and doing some operations we derive the following equation for the function $f_2^{(k),1}$:

$$d_1^{(k)} \frac{d^4 f_2^{(k),1}}{d(x_2^{(k)})^4} + d_2^{(k)} \frac{d^2 f_2^{(k),1}}{d(x_2^{(k)})^2} + d_3^{(k)} f_2^{(k),1} = 0. \tag{8.2.7}$$

where

$$d_1^{(k)} = \frac{c_3^{(k)} c_6^{(k)}}{c_5^{(k)}}, \quad d_2^{(k)} = c_2^{(k)} - \frac{c_1^{(k)} c_6^{(k)}}{c_5^{(k)}} - \frac{c_3^{(k)} c_4^{(k)}}{c_5^{(k)}}, \quad d_3^{(k)} = \frac{c_1^{(k)} c_4^{(k)}}{c_5^{(k)}}. \tag{8.2.8}$$

We select the particular solution to the equation (8.2.7) as

$$f_2^{(k),1} = D_1^{(k)} \exp\left(\lambda^{(k)} \alpha x_2^{(k)}\right). \tag{8.2.9}$$

Substituting (8.2.9) in (8.2.7) we obtain the following characteristic equation :

$$\left(\lambda^{(k)}\right)^4 + a_1^{(k)}\left(\lambda^{(k)}\right)^2 + a_2^{(k)} = 0, \tag{8.2.10}$$

where

$$a_1^{(k)} = \frac{d_2^{(k)}}{d_1^{(k)}}, \quad a_2^{(k)} = \frac{d_3^{(k)}}{d_1^{(k)}}. \tag{8.2.11}$$

We investigate the discriminant $D=\left(a_1^{(k)}\right)^2/4-a_2^{(k)}$ of the equation (8.2.10); taking (8.2.9), (8.2.11) into account and performing some transformations we obtain the following expression:

$$D=\frac{1}{4\left(c_3^{(\underline{k})}c_6^{(\underline{k})}\right)^2}\frac{1+\nu^{(\underline{k})}}{4\left(1-2\nu^{(\underline{k})}\right)^2}\left(\frac{\sigma_{11}^{(\underline{k}),0}}{E^{(\underline{k})}}\right)\left[-8\left(1-\nu^{(k)}\right)^2+\left(1+\nu^{(k)}\right)\right]+O\left(\left(\frac{\sigma_{11}^{(\underline{k}),0}}{E^{(\underline{k})}}\right)\right) \tag{8.2.12}$$

in practice

$$\left|\frac{\sigma_{11}^{(\underline{k}),0}}{E^{(k)}}\right|<<1, \tag{8.2.13}$$

Moreover, we obtain from (8.2.4), (4.2.5)-(4.2.7) that

$$\Gamma_1^{(k)}=\left(1-\left(\nu^{(\underline{k})}\right)^2\right)\frac{\sigma_{11}^{(\underline{k})}}{E^{(\underline{k})}},\ \Gamma_2^{(k)}=-\nu^{(\underline{k})}\left(1+\nu^{(\underline{k})}\right)\frac{\sigma_{11}^{(\underline{k}),0}}{E^{(\underline{k})}}. \tag{8.2.14}$$

Equations (8.2.4), (8.2.8), (8.2.13), (8.2.14) show that $D<0$, from which it follows that the roots of the equation (8.2.10) are complex, and can therefore be represented as follows:

$$\lambda_1^{(k)}=\alpha_1^{(k)}+i\beta_1^{(k)},\ \lambda_2^{(k)}=\alpha_1^{(k)}-i\beta_1^{(k)},\ \lambda_3^{(k)}=-\lambda_1^{(k)},\ \lambda_4^{(k)}=-\lambda_2^{(k)}, \tag{8.2.15}$$

where

$$\alpha_1^{(k)}=\left(\left(\alpha^{(\underline{k})}\right)^2+\left(\beta^{(\underline{k})}\right)^2\right)^{\frac{1}{4}}\left(\frac{1}{2}+\frac{\alpha^{(\underline{k})}}{2}\left(\left(\alpha^{(\underline{k})}\right)^2+\left(\beta^{(\underline{k})}\right)^2\right)^{-\frac{1}{2}}\right)^{\frac{1}{2}},$$

$$\alpha_1^{(k)}=\left(\left(\alpha^{(\underline{k})}\right)^2+\left(\beta^{(\underline{k})}\right)^2\right)^{\frac{1}{4}}\left(\frac{1}{2}-\frac{\alpha^{(\underline{k})}}{2}\left(\left(\alpha^{(\underline{k})}\right)^2+\left(\beta^{(\underline{k})}\right)^2\right)^{-\frac{1}{2}}\right)^{\frac{1}{2}},$$

$$\alpha^{(k)}=-\frac{a_1^{(k)}}{2},\ \beta^{(k)}=\frac{1}{2}\sqrt{4a_2^{(k)}-\left(a_1^{(k)}\right)^2}. \tag{8.2.16}$$

Thus, the function $f_2^{(k),1}$ is determined as follows:

$$f_2^{(k),1}\left(x_2^{(k)}\right) = D_1^{(k)} \cos\left(\beta_1^{(k)} \alpha x_2^{(k)}\right) \cosh\left(\alpha_1^{(k)} \alpha x_2^{(k)}\right) +$$

$$D_2^{(k)} \cos\left(\beta_1^{(k)} \alpha x_2^{(k)}\right) \sinh\left(\alpha_1^{(k)} \alpha x_2^{(k)}\right) + D_3^{(k)} \sin\left(\beta_1^{(k)} \alpha x_2^{(k)}\right) \cosh\left(\alpha_1^{(k)} \alpha x_2^{(k)}\right) +$$

$$D_4^{(k)} \cos\left(\beta_1^{(k)} \alpha x_2^{(k)}\right) \cosh\left(\alpha_1^{(k)} \alpha x_2^{(k)}\right), \quad \alpha = 2\pi/\ell . \tag{8.2.17}$$

Taking the expressions of the contact relations (8.2.3) into account we conclude that the function $f_2^{(k),1}$ must be an even function, i.e., $D_2^{(k)} = D_4^{(k)} = 0$ and

$$f_2^{(k),1}\left(x_2^{(k)}\right) = D_1^{(k)} \cos\left(\beta_1^{(k)} \alpha x_2^{(k)}\right) \cosh\left(\alpha_1^{(k)} \alpha x_2^{(k)}\right) +$$

$$D_4^{(k)} \cos\left(\beta_1^{(k)} \alpha x_2^{(k)}\right) \cosh\left(\alpha_1^{(k)} \alpha x_2^{(k)}\right). \tag{8.2.18}$$

Using the usual procedure we determine the unknown constants $D_1^{(k)}$ and $D_4^{(k)}$ from the contact conditions (8.2.3) and in this way we also find the values of the first approximation. We can determine and subsequent approximations, similarly.

8.2.2. SPATIAL STRESS STATE

The zeroth approximation is determined by the expressions (5.4.6)-(5.4.9). For the first approximation (8.1.6), (8.1.7), (8.1.11) and (8.1.12) we present the following governing equations and relations.

The equilibrium equation:

$$L_{ij}^{(k)} u_j^{(k),1} = 0, \tag{8.2.19}$$

The geometrical relations:

$$\varepsilon_{ij}^{(k),1} = \frac{1}{2}\left(1+\Gamma_j^{(k)}\right)\frac{\partial u_j^{(k),1}}{\partial x_i^{(k)}} + \frac{1}{2}\left(1+\Gamma_i^{(k)}\right)\frac{\partial u_i^{(k),1}}{\partial x_j^{(k)}}, \tag{8.2.20}$$

The contact conditions:

$$\left(1+\Gamma_1^{(2)}\right)\sigma_{12}^{(2),1}\Big|_{\left(t_1, \pm H^{(2)}, t_3\right)} - \left(1+\Gamma_1^{(1)}\right)\sigma_{12}^{(1),1}\Big|_{\left(t_1, \mp H^{(1)}, t_3\right)} =$$

$$-\left.\frac{\partial f}{\partial x_1}\right|_{(t_1,t_3)}\left[\left(1+\Gamma_1^{(1)}\right)\sigma_{11}^{(1),0}-\left(1+\Gamma_1^{(2)}\right)\sigma_{11}^{(2),0}\right],$$

$$\left(1+\Gamma_2^{(2)}\right)\sigma_{22}^{(2),1}\Big|_{\left(t_1,\pm H^{(2)},t_3\right)}=\left(1+\Gamma_2^{(1)}\right)\sigma_{22}^{(1),1}\Big|_{\left(t_1,\mp H^{(1)},t_3\right)},$$

$$\left(1+\Gamma_3^{(2)}\right)\sigma_{23}^{(2),1}\Big|_{\left(t_1,\pm H^{(2)},t_3\right)}-\left(1+\Gamma_3^{(1)}\right)\sigma_{23}^{(1),1}\Big|_{\left(t_1,\mp H^{(1)},t_3\right)}=$$

$$-\left.\frac{\partial f}{\partial x_3}\right|_{(t_1,t_3)}\left[\left(1+\Gamma_3^{(1)}\right)\sigma_{33}^{(1),0}-\left(1+\Gamma_3^{(2)}\right)\sigma_{33}^{(2),0}\right],$$

$$U_i^{(1),1}\Big|_{\left(t_1,\mp H^{(1)},t_3\right)}=U_i^{(2),1}\Big|_{\left(t_1,\pm H^{(1)},t_3\right)}. \tag{8.2.21}$$

In equations (8.2.19), the operators $L_{ij}^{(k)}$ are defined by the following expressions:

$$L_{11}^{(k)}=\sum_{i=1}^{3}C_{1i}^{(\underline{k})}\frac{\partial^2}{\partial(x_i^{(\underline{k})})^2},\quad L_{12}^{(k)}=C_{14}^{(\underline{k})}\frac{\partial^2}{\partial x_1^{(\underline{k})}\partial x_2^{(\underline{k})}},\ L_{13}^{(k)}=C_{15}^{(\underline{k})}\frac{\partial^2}{\partial x_1^{(\underline{k})}\partial x_3^{(\underline{k})}},$$

$$L_{21}^{(k)}=C_{24}^{(\underline{k})}\frac{\partial^2}{\partial x_1^{(\underline{k})}\partial x_2^{(\underline{k})}},\quad L_{22}^{(k)}=\sum_{i=1}^{3}C_{2i}^{(\underline{k})}\frac{\partial^2}{\partial\left(x_i^{(\underline{k})}\right)^2},$$

$$L_{23}^{(k)}=C_{25}^{(\underline{k})}\frac{\partial^2}{\partial x_2^{(\underline{k})}\partial x_3^{(\underline{k})}},\quad L_{31}^{(k)}=C_{34}^{(\underline{k})}\frac{\partial^2}{\partial x_1^{(\underline{k})}\partial x_3^{(\underline{k})}},$$

$$L_{32}^{(k)}=C_{35}^{(\underline{k})}\frac{\partial^2}{\partial x_2^{(\underline{k})}\partial x_3^{(\underline{k})}},\quad L_{33}^{(k)}=\sum_{i=1}^{3}C_{3i}^{(\underline{k})}\frac{\partial^2}{\partial(x_i^{(\underline{k})})^2}. \tag{8.2.22}$$

The following notation is used in (8.2.22):

$$C_{11}^{(k)}=\left(1+\Gamma_1^{(\underline{k})}\right)^2\left(1+\Lambda^{(k)}\right)+\frac{\sigma_{11}^{(\underline{k}),0}}{E^{(\underline{k})}}\left(1+\nu^{(\underline{k})}\right),\ C_{12}^{(\underline{k})}=\frac{1}{2}\left(1+\Gamma_1^{(k)}\right)^2,$$

$$C_{13}^{(k)}=\frac{1}{2}\left(1+\Gamma_1^{(k)}\right)^2+\frac{\sigma_{33}^{(\underline{k})}}{E^{(\underline{k})}}\left(1+\nu^{(\underline{k})}\right),\ C_{14}^{(k)}=\left(\frac{1}{2}+\Lambda^{(\underline{k})}\right)\left(1+\Gamma_1^{(\underline{k})}\right)\left(1+\Gamma_2^{(\underline{k})}\right),$$

$$C_{15}^{(k)} = \left(\frac{1}{2} + \Lambda^{(\underline{k})}\right)\left(1+\Gamma_1^{(\underline{k})}\right)\left(1+\Gamma_3^{(\underline{k})}\right), \; C_{21}^{(k)} = \frac{1}{2}\left(1+\Gamma_2^{(\underline{k})}\right)^2 + \frac{\sigma_{11}^{(\underline{k}),0}}{E^{(\underline{k})}}\left(1+\nu^{(\underline{k})}\right),$$

$$C_{22}^{(k)} = \left(1+\Lambda^{(\underline{k})}\right)\left(1+\Gamma_2^{(\underline{k})}\right)^2, \; C_{23}^{(k)} = \frac{1}{2}\left(1+\Gamma_2^{(\underline{k})}\right)^2 + \frac{\sigma_{33}^{(\underline{k}),0}}{E^{(\underline{k})}}\left(1+\nu^{(\underline{k})}\right),$$

$$C_{24}^{(k)} = C_{14}^{(k)}, \; C_{25}^{(k)} = C_{15}^{(k)}, \; C_{31}^{(k)} = \frac{1}{2}\left(1+\Gamma_3^{(k)}\right)^2 + \frac{\sigma_{11}^{(\underline{k}),0}}{E^{(\underline{k})}}\left(1+\nu^{(\underline{k})}\right),$$

$$C_{33}^{(k)} = \left(1+\Gamma_3^{(\underline{k})}\right)^2\left(1+\Lambda^{(k)}\right) + \frac{\sigma_{33}^{(\underline{k}),0}}{E^{(\underline{k})}}\left(1+\nu^{(\underline{k})}\right), \; C_{34}^{(k)} = C_{15}^{(k)}, \; C_{35}^{(k)} = C_{25}^{(k)}, \quad (8.2.23)$$

where

$$\Gamma_i^{(k)} = \frac{\partial u_i^{(\underline{k}),0}}{\partial x_i^{(\underline{k})}}, \quad \Lambda^{(k)} = \frac{\nu^{(\underline{k})}}{1-2\nu^{(\underline{k})}}. \quad (8.2.24)$$

and the expressions for $U_i^{(k),1}$ are given by (5.3.9).

Now we assume that the curving of the reinforcing layer is determined by the function (5.4.1) and introduce the parameters (5.4.2) and (5.4.3). Further, for simplicity we suppose that the composite shown in Fig 5.4.1 is loaded only in the direction of the Ox_1 axis and $\nu^{(1)} = \nu^{(2)} = \nu$. In this case it is follows from (5.4.7)-(5.4.9) that $\sigma_{33}^{(1),0} = \sigma_{33}^{(2),0} = 0$.

Thus, it follows from the contact conditions (8.2.21) and from the expression of the function $f(x_1, x_3)$ (5.4.1) that the displacements in the first approximation can be represented as

$$u_1^{(k),1} = \varphi_1^{(\underline{k})}\left(x_2^{(\underline{k})}\right)\cos\alpha x_1 \cos\beta x_3, \; u_2^{(k),1} = \varphi_2^{(\underline{k})}\left(x_2^{(\underline{k})}\right)\sin\alpha x_1 \cos\beta x_3,$$

$$u_3^{(k),1} = \varphi_3^{(\underline{k})}\left(x_2^{(\underline{k})}\right)\sin\alpha x_1 \sin\beta x_3 . \quad (8.2.25)$$

Substituting equations (8.2.25) into (8.2.19), we obtain the following system of ordinary differential equations to find of the unknown functions $\varphi_i^{(\underline{k})}\left(x_2^{(\underline{k})}\right)$, $(i = 1,2,3)$:

$$\frac{d^2\varphi_1^{(\underline{k})}}{d(x_2^{(\underline{k})})^2} + \alpha_{11}^{(\underline{k})}\varphi_1^{(\underline{k})} + \alpha_{12}^{(\underline{k})}\frac{d\varphi_2^{(\underline{k})}}{dx_2^{(\underline{k})}} + \alpha_{13}^{(\underline{k})}\varphi_3^{(\underline{k})} = 0,$$

$$\frac{d^2\varphi_2^{(k)}}{d(x_2^{(k)})^2}+\alpha_{11}^{(k)}\varphi_2^{(k)}+\alpha_{22}^{(k)}\frac{d\varphi_2^{(k)}}{dx_2^{(k)}}+\alpha_{23}^{(k)}\varphi_3^{(k)}=0,$$

$$\frac{d^2\varphi_3^{(k)}}{d(x_2^{(k)})^2}+\alpha_{11}^{(k)}\varphi_3^{(k)}+\alpha_{33}^{(k)}\frac{d\varphi_2^{(k)}}{dx_2^{(k)}}+\alpha_{32}^{(k)}\varphi_1^{(k)}=0, \tag{8.2.26}$$

where

$$\alpha_{11}^{(k)}=-\frac{C_{11}^{(k)}+C_{13}^{(k)}\gamma^2}{C_{12}^{(k)}},\ \alpha_{12}^{(k)}=\frac{C_{14}^{(k)}}{C_{12}^{(k)}},\ \alpha_{13}^{(k)}=\frac{C_{15}^{(k)}}{C_{12}^{(k)}}\gamma,$$

$$\alpha_{21}^{(k)}=-\frac{C_{21}^{(k)}+C_{23}^{(k)}\gamma^2}{C_{22}^{(k)}},\ \alpha_{22}^{(k)}=-\frac{C_{24}^{(k)}}{C_{22}^{(k)}},\ \alpha_{23}^{(k)}=\frac{C_{25}^{(k)}}{C_{22}^{(k)}}\gamma,$$

$$\alpha_{31}^{(k)}=-\frac{C_{31}^{(k)}+C_{33}^{(k)}\gamma^2}{C_{32}^{(k)}},\ \alpha_{32}^{(k)}=\frac{C_{34}^{(k)}}{C_{32}^{(k)}},\ \alpha_{33}^{(k)}=-\frac{C_{31}^{(k)}}{C_{32}^{(k)}}\gamma. \tag{8.2.27}$$

Following some transformations, the representations for the functions $\varphi_i^{(k)}$ $(k=1,2;i=1,2,3)$ are obtained from (8.26), (8.27) as follows:

$$\varphi_1^{(k)}\left(x_2^{(k)}\right)=\left[\frac{d^2}{d(x_2^{(k)})^2}-\gamma^2-\frac{1-2\nu}{1-\nu}\frac{\kappa(1+\nu)}{(1-\nu\kappa)^2}-\frac{1-2\nu}{2(1-\nu)}\right]X^{(k)}$$

$$\varphi_2^{(k)}(x_2^{(k)})=-\alpha_{22}^{(k)}\frac{d}{dx_2^{(k)}}X^{(k)}+\left[\left(\frac{\alpha_{13}^{(k)}\alpha_{33}^{(k)}}{\alpha_{12}^{(k)}}+\alpha_{11}^{(k)}\right)+\right.$$

$$\left.\left(\frac{\alpha_{13}^{(k)}\alpha_{23}^{(k)}}{\alpha_{12}^{(k)}}-\alpha_{31}^{(k)}\right)\left(\alpha_{11}^{(k)}\alpha_{31}^{(k)}-\alpha_{13}^{(k)}\alpha_{32}^{(k)}\right)\right]\Psi^{(k)},$$

$$\varphi_3^{(k)}\left(x_2^{(k)}\right)=\left(\alpha_{22}^{(k)}\alpha_{33}^{(k)}-\left(\alpha_{32}^{(k)}\right)^2\right)X^{(k)}-\left[\alpha_{33}^{(k)}\left(\frac{\alpha_{13}^{(k)}\alpha_{33}^{(k)}}{\alpha_{12}^{(k)}}-\alpha_{31}^{(k)}\right)+\right.$$

$$+\alpha_{11}^{(k)}\alpha_{33}^{(k)}-\alpha_{12}^{(k)}\alpha_{32}^{(k)}\Big]\frac{d}{dx_2^{(k)}}\Psi^{(k)}, \qquad (8.2.28)$$

where

$$\kappa=\frac{\sigma_{11}^{(1),0}}{E^{(1)}}=\frac{\sigma_{11}^{(2),0}}{E^{(2)}}=\varepsilon_{11}^{(1),0}=\varepsilon_{11}^{(2),0}, \qquad (8.2.29)$$

and $X^{(k)}$, $\Psi^{(k)}$ are the solutions to the following equations:

$$\left(\frac{d^2}{d(x_2^{(k)})^2}-b_0^{(k)}\right)\Psi^{(k)}=0, \qquad (8.2.30)$$

$$\left(\frac{d^4}{d(x_2^{(k)})^4}+b_1^{(k)}\frac{d^2}{d(x_2^{(k)})^2}+b_2^{(k)}\right)X^{(k)}=0. \qquad (8.2.31)$$

In equations (8.2.30) and (8.2.31), $b_0^{(k)}$, $b_1^{(k)}$ and $b_2^{(k)}$ are

$$b_0^{(k)}=\frac{\alpha_{13}^{(k)}\alpha_{33}^{(k)}}{\alpha_{12}^{(k)}}-\alpha_{31}^{(k)}, \quad b_1^{(k)}=-\alpha_{12}^{(k)}\alpha_{22}^{(k)}+\alpha_{11}^{(k)}+\alpha_{31}^{(k)}-\frac{\alpha_{23}^{(k)}\alpha_{12}^{(k)}\alpha_{31}^{(k)}}{\alpha_{13}^{(k)}},$$

$$b_2^{(k)}=-\alpha_{13}^{(k)}\alpha_{32}^{(k)}+\alpha_{31}^{(k)}\alpha_{11}^{(k)}+\alpha_{23}^{(k)}\alpha_{12}^{(k)}\alpha_{32}^{(k)}-\frac{\alpha_{23}^{(k)}\alpha_{12}^{(k)}\alpha_{31}^{(k)}\alpha_{11}^{(k)}}{\alpha_{13}^{(k)}}. \qquad (8.2.32)$$

Taking into account that $\kappa \ll 1$ in practice, we prove by direct verification that the roots of characteristic equations corresponding to the differential equations (8.2.30) and (8.2.31) are respectively real and complex numbers. Furthermore, taking into account that according to the contact conditions (8.1.21) the stresses $\sigma_{12}^{(k),1}$ and $\sigma_{23}^{(k),1}$ must be even functions with respect to $x_2^{(k)}$, the solutions of equations (8.2.30), (8.2.31) are selected in the following form:

$$\Psi^{(k)}=A_1^{(k)}\cosh\left(\sqrt{a_0^{(k)}}\alpha x_2^{(k)}\right),$$

$$X^{(k)}=B_1^{(k)}\cosh\left(\kappa_1^{(k)}\alpha x_2^{(k)}\right)\sin\left(\kappa_2^{(k)}\alpha x_2^{(k)}\right)+$$

$$B_2^{(k)} \sinh\left(\kappa_1^{(k)} \alpha x_2^{(k)}\right) \cos\left(\kappa_2^{(k)} \alpha x_2^{(k)}\right), \tag{8.2.33}$$

where

$$\kappa_{1,2}^{(k)} = \left[\left(\kappa_{11}^{(k)}\right)^2 + \left(\kappa_{21}^{(k)}\right)^2\right]^{\frac{1}{4}} \left[\frac{1}{2} \pm \frac{\kappa_{11}^{(k)}}{2}\left(\left(\kappa_{11}^{(k)}\right)^2 + \left(\kappa_{21}^{(k)}\right)^2\right)^{-1/2}\right]^{1/2},$$

$$\kappa_{11}^{(k)} = -\frac{b_1^{(k)}}{2}, \quad \kappa_{21}^{(k)} = \frac{1}{2}\left(4b_2^{(k)} - \left(b_1^{(k)}\right)^2\right)^{1/2}. \tag{8.2.34}$$

Using these relations we obtain algebraic equations for unknown constants $A_1^{(k)}$, $B_1^{(k)}$ and $B_2^{(k)}$, which enter the expressions (8.2.33), from contact conditions (8.2.21). In this way we get the values of the first and subsequent approximations.

The anti-phase case can be analysed in a similar way.

8.3. Numerical Results

Up to now a numerical investigation of the geometrically non linear statement has been made only for the composites shown in Figs. 4.2.1 and 4.2.3, i.e., with co-phase and anti-phase periodically plane-curved layers, under the action of uniformly distributed normal forces with intensity p at infinity. Here we analyse these numerical results and assume that the corresponding notation and restriction accepted in Chapter 4 holds. We introduce a parameter $e = p/E^{(1)}$ and consider the influence of this parameter on the values of $\sigma_{nn}^{+}/\sigma_{11}^{(1),0}$, $\sigma_{nn}^{-}/\sigma_{11}^{(1),0}$ and $\sigma_{n\tau}^{+}/\sigma_{11}^{(1),0}$ which will be calculated at points C, D and A (Fig.4.2.1) respectively, for co-phase curving case and at points A, C and B (Fig.4.2.3) respectively, for the anti-phase curving case. We assume that $E^{(2)}/E^{(1)} = 50$, $\varepsilon = 0.015$, $\nu^{(1)} = \nu^{(2)} = 0.3$. Moreover according to data given in [73,74] and elsewhere we assume that $10^{-3} \le |e| \le 10^{-1}$. In both co- and anti-phase curving cases we consider tension ($e > 0$) and compression ($e < 0$) separately.

8.3.1. CO-PHASE CURVING

Uniaxial tension(e>0). Table 8.1 shows the values of $\sigma_{nn}^{+}/\sigma_{11}^{(1),0}$, $\sigma_{nn}^{-}/\sigma_{11}^{(1),0}$ and $\sigma_{n\tau}^{+}/\sigma_{11}^{(1),0}$ for various e and $\eta^{(2)}$ for $\chi = 0.3$; the values of σ_{nn}^{+}, σ_{nn}^{-} and $\sigma_{n\tau}^{+}$ increase monotonically with e. In this case, when $10^{-3} \le e \le 5*10^{-3}$, these values coincide (with very high accuracy) with those obtained in the linear statement. When e>0 the effect of the geometrical non-linearity on the stresses σ_{nn}^{+}, σ_{nn}^{-} and $\sigma_{n\tau}^{+}$ decreases with increasing $\eta^{(2)}$.

Table 8.2 gives the values of ϑ_1 for various $\eta^{(2)}$ and χ where

$$\vartheta_1 = \frac{q_2 - q_1}{q_1} * 100\%, \tag{8.3.1}$$

where

$$q_2 = \sigma_{n\tau}^{+} / \sigma_{11}^{(1),0} \Big|_{|e| = 9\times10^{-3}}, \quad q_1 = \sigma_{n\tau}^{+} / \sigma_{11}^{(1),0} \Big|_{|e| = 10^{-3}}. \tag{8.3.2}$$

These results prove that for all selected $\eta^{(2)}$ with increasing $|e|$ the effect of the geometrical non-linearity decreases. Note that similar results not given here, are obtained for $\sigma_{nn}^{+} / \sigma_{11}^{(1),0}$ and $\sigma_{nn}^{-} / \sigma_{11}^{(1),0}$.

Uniaxial compression (e<0). The numerical results related to this case and obtained for various e and $\eta^{(2)}$ under $\chi = 0.3$ are given in Table 8.3; under compression the values of $\sigma_{nn}^{+} / \sigma_{11}^{(1),0}$, $\sigma_{nn}^{-} / \sigma_{11}^{(1),0}$ and $\sigma_{n\tau}^{+} / \sigma_{11}^{(1),0}$ increase with increasing $|e|$. In this case the results obtained for $|e| = 10^{-3}$ coincide with very high accuracy with the corresponding ones obtained in tension and given in Table 8.1. Moreover these results show that an increase in the values of $\eta^{(2)}$ causes a decrease in the values of σ_{nn}^{+}, σ_{nn}^{-} and $\sigma_{n\tau}^{+}$. Table 8.4 shows the values of ϑ_1 calculated for compression under various χ and $\eta^{(2)}$; the effect of the geometrical non-linearity decreases monotonically with χ.

Table 8.1. The values of $\sigma_{nn}^{+} / \sigma_{11}^{(1),0}$, $\sigma_{nn}^{-} / \sigma_{11}^{(1),0}$ and $\sigma_{n\tau}^{+} / \sigma_{11}^{(1),0}$ in tension for various e and $\eta^{(2)}$ (co-phase curving).

Stresses	$e \times 10^3$	$\eta^{(2)}$			
		0.02	0.1	0.2	0.5
$\sigma_{nn}^{+} / \sigma_{11}^{(1),0}$	1	0.662	0.653	0.545	0.225
	5	0.645	0.648	0.532	0.224
	50	0.513	0.592	0.513	0.212
	90	0.445	0.551	0.490	0.203
$\sigma_{nn}^{-} / \sigma_{11}^{(1),0}$	1	-0.571	-0.671	-0.592	-0.425
	5	-0.656	-0.666	-0.589	-0.426
	50	-0.517	-0.611	-0.560	-0.419
	90	-0.428	-0.570	-0.537	-0.414
1	2	3	4	5	6

Table 8.1 (Continuation)

1	2	3	4	5	6
$\sigma_{n\tau}^{+}/\sigma_{11}^{(1),0}$	1	-0.465	-0.479	-0.621	-1.137
	5	-0.455	-0.476	-0.618	-1.137
	50	-0.245	-0.444	-0.550	-1.107
	90	-0.006	-0.419	-0.468	-1.087

Table 8.2. The values of ϑ_1 (8.3.1) in tension for various χ and $\eta^{(2)}$.

$\eta^{(2)}$	χ			
	0.1	0.3	0.5	0.7
0.1	12.5	12.0	7.0	4.8
0.2	11.4	8.4	4.0	2.4
0.5	8.3	4.3	2.2	1.3

Table 8.3. The values of $\sigma_{nn}^{+}/\sigma_{11}^{(1),0}$, $\sigma_{nn}^{-}/\sigma_{11}^{(1),0}$ and $\sigma_{n\tau}^{+}/\sigma_{11}^{(1),0}$ in compression for various e and $\eta^{(2)}$ (co-phase curving).

Stresses	$\lvert e\rvert \times 10^3$	$\eta^{(2)}$			
		0.02	0.1	0.2	0.5
$\sigma_{nn}^{+}/\sigma_{11}^{(1),0}$	1	0.671	0.656	0.547	0.225
	5	0.689	0.661	0.549	0.226
	50	0.995	0.733	0.574	0.239
	90	1.655	0.812	0.618	0.252
$\sigma_{nn}^{-}/\sigma_{11}^{(1),0}$	1	-0.679	-0.674	-0.593	-0.425
	5	-0.696	-0.679	-0.596	-0.425
	50	-0.963	-0.746	-0.630	-0.431
	90	-1.540	-0.826	-0.664	-0.435
$\sigma_{n\tau}^{+}/\sigma_{11}^{(1),0}$	1	-0.469	-0.480	-0.622	-1.137
	5	-0.476	-0.483	-0.625	-1.147
	50	-0.332	-0.521	-0.657	-1.177
	90	-0.287	-0.561	-0.690	-1.200

Table 8.4. The values of ϑ_1 (8.3.1) in compression for various χ and $\eta^{(2)}$.

$\eta^{(2)}$	χ			
	0.1	0.3	0.5	0.7
0.1	18.8	16.7	8.4	3.8
0.2	16.6	10.8	5.2	2.7
0.5	10.1	5.5	2.3	1.3

Note that all the numerical results analysed here are obtained in the framework of the first three approximations of the series (4.1.15): the differences between the results of the last two approximations were not greater than 2%.

8.3.2. ANTI-PHASE CURVING

Uniaxial tension (e>0). Table 8.5 shows the values of $\sigma_{nn}^{+}/\sigma_{11}^{(1),0}$, $\sigma_{nn}^{-}/\sigma_{11}^{(1),0}$ and $\sigma_{n\tau}^{+}/\sigma_{11}^{(1),0}$ obtained for various e and $\eta^{(2)}$ under $\chi = 0.3$; the absolute values of $\sigma_{nn}^{+}/\sigma_{11}^{(1),0}$, $\sigma_{nn}^{-}/\sigma_{11}^{(1),0}$ and $\sigma_{n\tau}^{+}/\sigma_{11}^{(1),0}$ decrease with e and $\eta^{(2)}$.

Table 8.5. The values of $\sigma_{nn}^{+}/\sigma_{11}^{(1),0}$, $\sigma_{nn}^{-}/\sigma_{11}^{(1),0}$ and $\sigma_{n\tau}^{+}/\sigma_{11}^{(1),0}$ in tension for various e and $\eta^{(2)}$ (anti - phase curving).

Stresses	$e\times10^3$	$\eta^{(2)}$			
		0.02	0.1	0.2	0.5
$\sigma_{nn}^{+}/\sigma_{11}^{(1),0}$	1	0.662	0.716	0.868	1.390
	5	0.645	0.711	0.864	1.390
	50	0.513	0.654	0.829	1.380
	90	0.445	0.611	0.801	1.380
$\sigma_{nn}^{-}/\sigma_{11}^{(1),0}$	1	-0.671	-0.687	-0.793	-0.971
	5	-0.656	-0.682	-0.790	-0.970
	50	-0.517	-0.626	-0.757	-0.964
	90	-0.428	-0.584	-0.731	-0.958
$\sigma_{n\tau}^{+}/\sigma_{11}^{(1),0}$	1	-0.465	-0.471	-0.281	-0.008
	5	-0.455	-0.468	-0.280	-0.008
	50	-0.245	-0.442	-0.276	-0.008
	90	-0.006	-0.423	-0.273	-0.008

In Table 8.6 the values of

$$\vartheta_2 = \frac{r_2 - r_1}{r_1} * 100\%, \tag{8.3.3}$$

where

$$r_2 = \sigma_{nn}^{+}/\sigma_{11}^{(1),0}\Big|_{|e|=9\times10^{-3}}, \quad r_1 = \sigma_{nn}^{+}/\sigma_{11}^{(1),0}\Big|_{|e|=10^{-3}}. \tag{8.3.4}$$

are tabulated for various $\eta^{(2)}$ and χ; the effect of the geometrical non-linearity decreases with $\eta^{(2)}$. In the anti-phase curving case for $\eta^{(2)} \geq 0.2$ this effect is less than that in the co-phase curving case. Moreover, in the anti-phase curving case the dependence between ϑ_2 and χ is non-monotonic.

Table 8.6. The values of ϑ_2 (8.3.3) in tension for various χ and $\eta^{(2)}$.

$\eta^{(2)}$	χ			
	0.1	0.3	0.5	0.7
0.1	5.0	14.5	11.4	8.4
0.2	0.9	7.7	6.6	4.8
0.5	0.1	0.7	1.7	2.3

Table 8.7. The values of $\sigma_{nn}^{+}/\sigma_{11}^{(1),0}$, $\sigma_{nn}^{-}/\sigma_{11}^{(1),0}$ and $\sigma_{n\tau}^{+}/\sigma_{11}^{(1),0}$ in compression for various e and $\eta^{(2)}$ (anti - phase curving).

Stresses	$\lvert e \rvert \times 10^3$	$\eta^{(2)}$			
		0.02	0.1	0.2	0.5
$\sigma_{nn}^{+}/\sigma_{11}^{(1),0}$	1	0.671	0.719	0.809	1.390
	5	0.689	0.725	0.872	1.390
	50	0.995	0.798	0.911	1.400
	90	1.655	0.877	0.949	1.400
$\sigma_{nn}^{-}/\sigma_{11}^{(1),0}$	1	-0.679	-0.690	-0.795	-0.971
	5	-0.696	-0.696	-0.798	-0.972
	50	-0.963	-0.766	-0.834	-0.978
	90	-1.540	-0.844	-0.869	-0.984
$\sigma_{n\tau}^{+}/\sigma_{11}^{(1),0}$	1	-0.469	-0.472	-0.281	-0.070
	5	-0.476	-0.475	-0.281	-0.070
	50	-0.332	-0.509	-0.286	-0.070
	90	-0.387	-0.546	-0.291	-0.006

Table 8.8. The values of ϑ_2 (8.3.3) in compression for various χ and $\eta^{(2)}$.

$\eta^{(2)}$	χ			
	0.1	0.3	0.5	0.7
0.1	5.6	21.9	15.5	10.6
0.2	0.7	9.2	7.8	5.4
0.5	0.0	0.7	2.6	2.3

Uniaxial compression (e<0). The numerical results related to this case are given in Tables 8.7 and 8.8; the absolute values of $\sigma_{nn}^{+}/\sigma_{11}^{(1),0}$, $\sigma_{nn}^{-}/\sigma_{11}^{(1),0}$ increase with $\lvert e \rvert$. However, the dependence between $\sigma_{n\tau}^{+}/\sigma_{11}^{(1),0}$ and $\lvert e \rvert$ is more complicated.

8.4. Bibliographical Notes

The fact that the equations and relations of the first and subsequent approximations obtained under the geometrical non-linear statement are those of the three-dimensional linearized theory of deformable solid body, was established firstly by S.D.Akbarov, T.Sisman and N.Yahnioglu [48]. The method for determination of the values of each approximation was developed by S.D. Akbarov [20] (for three-dimensional problems) and by S.D. Akbarov and Z.R. Djamalov [26] (for plane strain state). The numerical results were discussed in the paper by S.D. Akbarov, A.N. Guz, Z.R. Djamalov and E.A. Movsumov [42].

CHAPTER 9

NORMALIZED MODULUS OF ELASTICITY

9.1. Basic Equations

Many researchers have investigated the influence of the reinforcing layers and fibers curvature on the values of the normalized moduli of elasticity, for example [56, 62, 64, 82, 112, 127, 128, 130, 140, 149]. These investigations used various approximate theories and different hypotheses. We will use the analyses developed in Chapters 3-8 and apply it to various particular problems.

We consider the composite material with co-phase (Fig.9.1.1) and anti-phase (Fig.9.1.2) periodically spatially-curved layers and use the notation of Chapter 5.

Denote by $\widetilde{D}_m^{(k)}$ the region occupied by the m-th layer. For each region $\widetilde{D}_m^{(k)}$ we choose a strip $D_m^{(k)}$, the thickness $\left(2H_m^{(k)}\right)$ of which is equal to that of the layer occupied by the region $\widetilde{D}_m^{(k)}$. The coordinates of the points of the region $\widetilde{D}_m^{(k)}$ and strip $D_m^{(k)}$ in the system of coordinates $O_m^{(k)}x_{1m}^{(k)}x_{2m}^{(k)}x_{3m}^{(k)}$ we denote by $\widetilde{x}_{im}^{(k)}$ and $x_{im}^{(k)}$ respectively, where $\widetilde{x}_{1m}^{(k)}$, $\widetilde{x}_{3m}^{(k)}$, $x_{1m}^{(k)}$, $x_{3m}^{(k)} \in (-\infty, +\infty)$, $-H_m^{(k)} \le x_{2m}^{(k)} \le H_m^{(k)}$. It is required that between the coordinates $\widetilde{x}_{im}^{(k)}$ and $x_{im}^{(k)}$ the following relations should be satisfied:

For matrix layer in the co-phase curving case (Fig.9.1.1),

$$\widetilde{x}_1^{(1)} = x_1^{(1)} + \varepsilon \frac{x_2^{(1)}}{H^{(1)}} H^{(2)} \frac{\partial f}{\partial x_1^{(1)}} - \varepsilon^3 \frac{x_2^{(1)}}{H^{(1)}} \frac{H^{(2)}}{2} \frac{\partial f}{\partial x_1^{(1)}} \Phi\left(x_1^{(1)}, x_3^{(1)}\right) +$$

$$\varepsilon^5 \frac{3}{8} \frac{x_2^{(1)}}{H^{(1)}} H^{(2)} \frac{\partial f}{\partial x_1^{(1)}} \left[\Phi\left(x_1^{(1)}, x_3^{(1)}\right)\right]^2 + \ldots,$$

$$\widetilde{x}_2^{(1)} = x_2^{(1)} + \varepsilon f\left(x_1^{(1)}, x_3^{(1)}\right) + \varepsilon^2 \frac{x_2^{(1)}}{2} \frac{H^{(2)}}{H^{(1)}} \Phi\left(x_1^{(1)}, x_3^{(1)}\right) -$$

$$\varepsilon^4 \frac{3}{8} x_2^{(1)} \frac{H^{(2)}}{H^{(1)}} \left[\Phi\left(x_1^{(1)}, x_3^{(1)}\right)\right]^2 + \ldots,$$

$$\tilde{x}_3^{(1)} = x_3^{(1)} + \varepsilon \frac{x_2^{(1)}}{H^{(1)}} H^{(2)} \frac{\partial f}{\partial x_3^{(1)}} - \varepsilon^3 \frac{x_2^{(1)}}{H^{(1)}} \frac{H^{(2)}}{2} \frac{\partial f}{\partial x_3^{(1)}} \Phi\left(x_1^{(1)}, x_3^{(1)}\right) +$$

$$\varepsilon^5 \frac{3}{8} \frac{x_2^{(1)}}{H^{(1)}} H^{(2)} \frac{\partial f}{\partial x_3^{(1)}} \left[\Phi\left(x_1^{(1)}, x_3^{(1)}\right)\right]^2 + \ldots \quad . \tag{9.1.1}$$

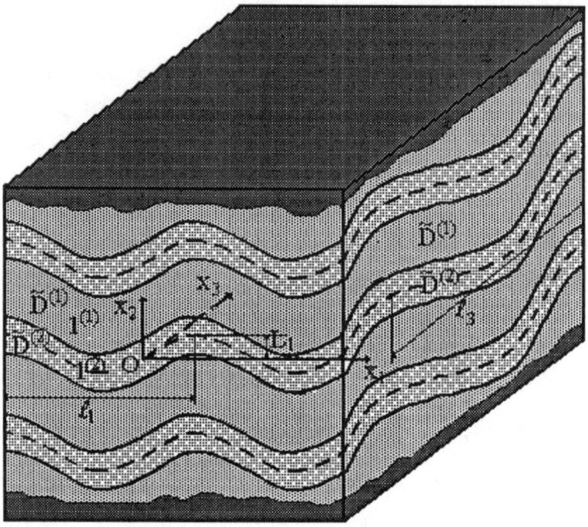

Fig.9.1.1. A part of composite with periodically spatially co-phase curved layers.

For the reinforcing layer in both co-phase (Fig.9.1.1) and anti- phase (Fig.9.1.2) curving cases,

$$\tilde{x}_1^{(2)} = x_1^{(2)} - \varepsilon x_2^{(2)} \frac{\partial f}{\partial x_1^{(2)}} + \varepsilon^2 \frac{1}{2} x_2^{(2)} \frac{\partial f}{\partial x_1^{(2)}} \Phi\left(x_1^{(2)}, x_3^{(2)}\right) -$$

$$\varepsilon^5 \frac{3}{8} x_2^{(2)} \frac{\partial f}{\partial x_1^{(2)}} \left[\Phi\left(x_1^{(2)}, x_3^{(2)}\right)\right]^2 + \ldots,$$

$$\tilde{x}_2^{(2)} = x_2^{(2)} + \varepsilon f\left(x_1^{(2)}, x_3^{(2)}\right) - \varepsilon^2 \frac{1}{2} x_2^{(2)} \Phi\left(x_1^{(2)}, x_3^{(2)}\right) + \varepsilon^4 \frac{3}{8} x_2^{(2)} \left[\Phi\left(x_1^{(2)}, x_3^{(2)}\right)\right]^2 + \ldots,$$

$$\tilde{x}_3^{(2)} = x_3^{(2)} - \varepsilon x_2^{(2)} \frac{\partial f}{\partial x_3^{(2)}} + \varepsilon^2 \frac{1}{2} x_2^{(2)} \frac{\partial f}{\partial x_3^{(2)}} \Phi\left(x_1^{(2)}, x_3^{(2)}\right) -$$

$$\varepsilon^5 \frac{3}{8} x_2^{(2)} \frac{\partial f}{\partial x_3^{(2)}} \left[\Phi\left(x_1^{(2)}, x_3^{(2)}\right)\right]^2 + \ldots . \tag{9.1.2}$$

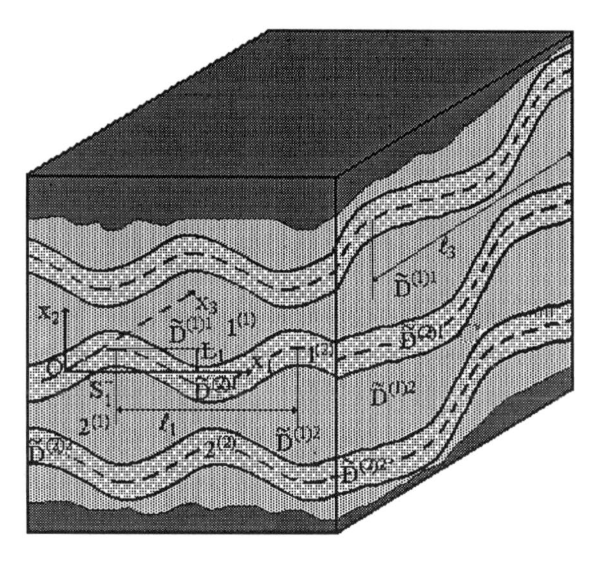

Fig.9.1.2. A part of composite with periodically spatially anti-phase curved layers.

For the matrix layer in the anti-phase curving case (Fig.9.1.2),

$$\tilde{x}_1^{(1)} = x_1^{(1)} - \varepsilon \frac{\left(x_2^{(1)}\right)^2}{\left(H^{(1)}\right)^2} H^{(2)} \frac{\partial f}{\partial x_1^{(1)}} + \varepsilon^3 \frac{\left(x_2^{(1)}\right)^4}{\left(H^{(1)}\right)^4} \frac{H^{(2)}}{2} \frac{\partial f}{\partial x_1^{(1)}} \Phi\left(x_1^{(1)}, x_3^{(1)}\right) -$$

$$\varepsilon^5 \frac{3}{8} \frac{\left(x_2^{(1)}\right)^6}{\left(H^{(1)}\right)^6} H^{(2)} \frac{\partial f}{\partial x_1^{(1)}} \left[\Phi\left(x_1^{(1)}, x_3^{(1)}\right)\right]^2 + \ldots ,$$

$$\tilde{x}_2^{(1)} = x_2^{(1)} - \varepsilon \frac{x_2^{(1)}}{H^{(1)}} f\left(x_1^{(1)}, x_3^{(1)}\right) + \varepsilon^2 \frac{\left(x_2^{(1)}\right)^3}{2} \frac{H^{(2)}}{\left(H^{(1)}\right)^3} \Phi\left(x_1^{(1)}, x_3^{(1)}\right) -$$

$$\varepsilon^4 \frac{3}{8} \left(x_2^{(1)}\right)^5 \frac{H^{(2)}}{\left(H^{(1)}\right)^5} \left[\Phi\left(x_1^{(1)}, x_3^{(1)}\right)\right]^2 + \ldots ,$$

$$\tilde{x}_3^{(1)} = x_3^{(1)} - \varepsilon \frac{\left(x_2^{(1)}\right)^2}{\left(H^{(1)}\right)^2} H^{(2)} \frac{\partial f}{\partial x_3^{(1)}} + \varepsilon^3 \frac{\left(x_2^{(1)}\right)^4}{\left(H^{(1)}\right)^4} \frac{H^{(2)}}{2} \frac{\partial f}{\partial x_3^{(1)}} \Phi\left(x_1^{(1)}, x_3^{(1)}\right) -$$

$$\varepsilon^5 \frac{3}{8} \frac{\left(x_2^{(1)}\right)^6}{\left(H^{(1)}\right)^6} H^{(2)} \frac{\partial f}{\partial x_3^{(1)}} \left[\Phi\left(x_1^{(1)}, x_3^{(1)}\right)\right]^2 + \ldots, \tag{9.1.3}$$

where

$$\Phi(x_1, x_3) = \left(\frac{\partial f}{\partial x_1}\right)^2 + \left(\frac{\partial f}{\partial x_3}\right)^2, \quad f = \ell_1 \sin \alpha x_1 \cos \beta x_3, \quad \alpha = 2\pi/\ell_1, \quad \beta = 2\pi/\ell_3. \tag{9.1.4}$$

In (9.1.1)-(9.1.4) the index m is omitted and it is assumed that $x_{1m}^{(k)} = x_1^{(k)} = x_1$, $x_{3m}^{(k)} = x_3^{(k)} = x_3$. Thus, in this way the points of the strip $D_m^{(k)}$ are uniquely mapped onto points of the region (layer) $\widetilde{D}_m^{(k)}$. The formulae (9.1.1)-(9.1.3) are obtained by direct verification.

The essence of the method described in Chapter 5 may be formulated as follows. *The solutions to the equations (5.1.1) in the regions* $\widetilde{D}_m^{(k)}$ *satisfying the contact conditions (5.1.2) and the corresponding boundary conditions are reduced to the solutions of the successive boundary-value problems in the regions (strips)* $D_m^{(k)}$. *The functions describing the stress-deformation state in layers (regions)* $\widetilde{D}_m^{(k)}$ *are determined by the following formulae with functions which are the solutions of this series boundary-value problems for regions* $D_m^{(k)}$:

$$\widetilde{\sigma}_{ij}^{(k)}\left(\widetilde{x}_1^{(k)}, \widetilde{x}_2^{(k)}, \widetilde{x}_3^{(k)}\right) = \sum_{q=1}^{\infty} \varepsilon^q p_{ij}^{(k),q}\left(x_1^{(k)}, x_2^{(k)}, x_3^{(k)}\right),$$

$$\widetilde{\varepsilon}_{ij}^{(k)}\left(\widetilde{x}_1^{(k)}, \widetilde{x}_2^{(k)}, \widetilde{x}_3^{(k)}\right) = \sum_{q=1}^{\infty} \varepsilon^q e_{ij}^{(k),q}\left(x_1^{(k)}, x_2^{(k)}, x_3^{(k)}\right). \tag{9.1.5}$$

We write the expressions of $p_{ij}^{(k),q}$. Note that these expressions for $q = 0,1$ are the following ones

$$p_{ij}^{(k),0} = \sigma_{ij}^{(k),0}, \quad p_{ij}^{(k),1}\left(x_1^{(k)}, x_2^{(k)}, x_3^{(k)}\right) = \sigma_{ij}^{(k),1}\left(x_1^{(k)}, x_2^{(k)}, x_3^{(k)}\right). \tag{9.1.6}$$

But for $q \geq 2$ they are different for matrix and filler layers. Here we write these expressions for $q = 2$.

For matrix layer in the co-phase curving case(Fig.9.1.1):

$$P_{ij}^{(1),2}\left(x_1^{(1)},x_2^{(1)},x_3^{(1)}\right)=\sigma_{ij}^{(1),2}\left(x_1^{(1)},x_2^{(1)},x_3^{(1)}\right)+\frac{x_2^{(1)}}{H^{(1)}}H^{(2)}\frac{\partial f}{\partial x_1^{(1)}}\frac{\partial \sigma_{ij}^{(1),1}}{\partial x_1^{(1)}}+$$

$$f\left(x_1^{(1)},x_3^{(1)}\right)\frac{\partial \sigma_{ij}^{(1),1}}{\partial x_2^{(1)}}+\frac{x_2^{(1)}}{H^{(1)}}H^{(2)}\frac{\partial f}{\partial x_3^{(1)}}\frac{\partial \sigma_{ij}^{(1),1}}{\partial x_3^{(1)}}. \quad (9.1.7)$$

For the reinforcing layer:

$$P_{ij}^{(2),2}\left(x_1^{(2)},x_2^{(2)},x_3^{(2)}\right)=\sigma_{ij}^{(2),2}\left(x_1^{(2)},x_2^{(2)},x_3^{(2)}\right)-x_2^{(2)}\frac{\partial f}{\partial x_1^{(2)}}\frac{\partial \sigma_{ij}^{(2),1}}{\partial x_1^{(2)}}+$$

$$f\left(x_1^{(2)},x_3^{(2)}\right)\frac{\partial \sigma_{ij}^{(2),1}}{\partial x_2^{(2)}}-x_2^{(2)}\frac{\partial f}{\partial x_3^{(2)}}\frac{\partial \sigma_{ij}^{(2),1}}{\partial x_3^{(2)}}. \quad (9.1.8)$$

For the matrix layer in the anti-phase curving case (Fig.9.1.2):

$$P_{ij}^{(1),2}\left(x_1^{(1)},x_2^{(1)},x_3^{(1)}\right)=\sigma_{ij}^{(1),2}\left(x_1^{(1)},x_2^{(1)},x_3^{(1)}\right)-\frac{\left(x_2^{(1)}\right)^2}{\left(H^{(1)}\right)^2}H^{(2)}\frac{\partial f}{\partial x_1^{(1)}}\frac{\partial \sigma_{ij}^{(1),1}}{\partial x_1^{(1)}}-$$

$$\frac{x_2^{(1)}}{H^{(1)}}f\left(x_1^{(1)},x_3^{(1)}\right)\frac{\partial \sigma_{ij}^{(1),1}}{\partial x_2^{(1)}}-\frac{\left(x_2^{(1)}\right)^2}{\left(H^{(1)}\right)^2}H^{(2)}\frac{\partial f}{\partial x_3^{(1)}}\frac{\partial \sigma_{ij}^{(1),1}}{\partial x_3^{(1)}}. \quad (9.1.9)$$

In (9.1.7)-(9.1.9) replacing $\sigma_{ij}^{(k),q}$ with $\varepsilon_{ij}^{(k),q}$ we obtain the expressions for $e_{ij}^{(k),q}$.

Note that the expressions for $P_{ij}^{(k),q}\left(x_1^{(k)},x_2^{(k)},x_3^{(k)}\right)$ are obtained from expressions (5.3.8)-(5.3.12) by applying the following replacements:

For the matrix layer in the co-phase curving case (Fig.9.1.1)

$$\pm H^{(2)},\ \left(t_1,\pm H^{(1)},t_3\right) \quad \text{with} \quad \left(-\frac{x_2^{(1)}}{H^{(1)}}\right)H^{(2)},\ \left(x_1^{(1)},x_2^{(1)},x_3^{(1)}\right) \text{ respectively.} \quad (9.1.10)$$

For the reinforcing layer:

$$\pm H^{(2)},\ \left(t_1,\pm H^{(2)},t_3\right) \text{ with } x_2^{(2)},\ \left(x_1^{(2)},x_2^{(2)},x_3^{(2)}\right) \text{ respectively.} \quad (9.1.11)$$

For the matrix layer in the anti-phase curving case (Fig.9.1.2):

$$\left(t_1, \pm H^{(1)}, t_3\right),\ \ f\left(t_1, t_3\right),\ \ \pm H^{(2)} \text{ with } \left(x_1^{(1)}, x_2^{(1)}, x_3^{(1)}\right),\ -x_2^{(1)} f\left(x_1^{(1)}, x_3^{(1)}\right)\Big/ H^{(1)},$$

$$-x_2^{(1)} H^{(2)} \big/ H^{(1)}, \text{ respectively.} \qquad (9.1.12)$$

Note that by writing $\left(\tilde{x}_1^{(k)}, \tilde{x}_2^{(k)}, \tilde{x}_3^{(k)}\right)$, as represented by (9.1.1)-(9.1.3), instead of $\left(x_1^{(k)}, x_2^{(k)}, x_3^{(k)}\right)$ in (5.3.1) and expanding the values of each approximation in the series form in the vicinity of $\left(x_1^{(k)}, x_2^{(k)}, x_3^{(k)}\right)$ and grouping by equal powers of ε, we can obtain the expressions (9.1.6) - (9.1.9) and the expressions for $p_{ij}^{(k),q}\left(x_1^{(k)}, x_2^{(k)}, x_3^{(k)}\right)$ and $e_{ij}^{(k),q}\left(x_1^{(k)}, x_2^{(k)}, x_3^{(k)}\right)$ in (9.1.5) under $q > 2$.

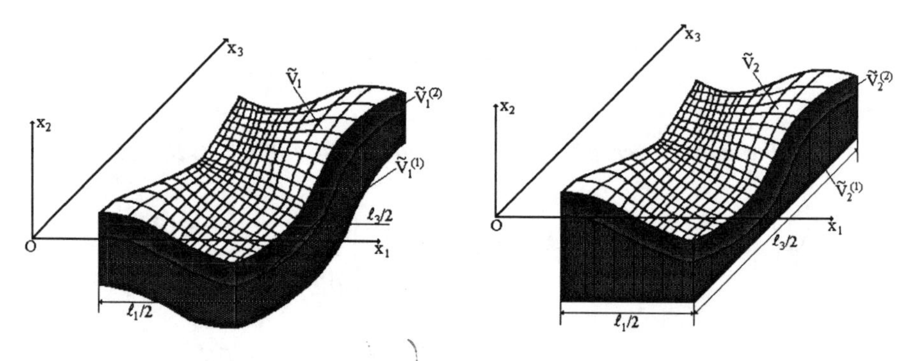

Fig.9.2.1. A representative element of the composite shown in Fig.9.1.1.

Fig.9.2.2. A representative element of the composite shown in Fig.9.1.2.

9.2. Normalized Moduli

Now we attempt to determine the normalised moduli of elasticity in the direction of Ox_1 axis of the composites shown in Figs. 9.1.1 and 9.1.2. First we select the representative elements as shown in Figs.9.2.1 and 9.2.2.

The volumes of these elements are denoted by $\tilde{V}_1$ (Fig.9.2.1) and $\tilde{V}_2$ (Fig.9.2.2), and represented in the form $\tilde{V}_k = \tilde{V}_k^{(1)} + \tilde{V}_k^{(2)}$, $(k = 1,2)$, where $\tilde{V}_k^{(1)} \left(\tilde{V}_k^{(2)}\right)$ is the part of $\tilde{V}_k$ filled by the matrix (reinforcing) layer. The coordinates of points of $\tilde{V}_k$ are determined from equations (9.1.1)-(9.1.4) where

$$\frac{\ell_1}{4} \le x_1^{(k)} \le \frac{3\ell_1}{4},\ \ 0 \le x_3^{(k)} \le \frac{\ell_3}{2},\ -H^{(1)} \le x_2^{(1)} \le 0,\ \ 0 \le x_2^{(2)} \le H^{(2)}. \quad (9.2.1)$$

After separations of these representative elements, we determine the average values of the stress and strains over their volume $\widetilde{V}_k$, i.e.,

$$\langle \sigma_{ij} \rangle = \eta^{(1)} \langle \widetilde{\sigma}_{ij}^{(1)} \rangle + \eta^{(2)} \langle \widetilde{\sigma}_{ij}^{(2)} \rangle, \quad \langle \varepsilon_{ij} \rangle = \eta^{(1)} \langle \widetilde{\varepsilon}_{ij}^{(1)} \rangle + \eta^{(2)} \langle \widetilde{\varepsilon}_{ij}^{(2)} \rangle,$$

$$\left\{ \begin{matrix} \langle \widetilde{\sigma}_{ij}^{(k)} \rangle \\ \langle \widetilde{\varepsilon}_{ij}^{(k)} \rangle \end{matrix} \right\} = \frac{1}{\widetilde{V}_n^{(k)}} \iiint_{\widetilde{V}_n^{(k)}} \left\{ \begin{matrix} \widetilde{\sigma}_{ij}^{(k)} \\ \widetilde{\varepsilon}_{ij}^{(k)} \end{matrix} \right\} \left(\widetilde{x}_1^{(k)}, \widetilde{x}_2^{(k)}, \widetilde{x}_3^{(k)} \right) d\widetilde{V}_n^{(k)}, \qquad k = 1,2; \; n = 1,2. \tag{9.2.2}$$

Taking into account that the change of the coordinates $\widetilde{x}_i^{(k)}$ in the volume $\widetilde{V}_n^{(k)}$ corresponds to the change of the coordinates $x_i^{(k)}$ in the region determined by (9.2.1), we obtain from (9.1.5) that

$$\left\{ \begin{matrix} \langle \widetilde{\sigma}_{ij}^{(k)} \rangle \\ \langle \widetilde{\varepsilon}_{ij}^{(k)} \rangle \end{matrix} \right\} = \sum_{q=0}^{\infty} \varepsilon^q \left\{ \begin{matrix} \langle P_{ij}^{(k),q} \rangle \\ \langle e_{ij}^{(k),q} \rangle \end{matrix} \right\}, \tag{9.2.3}$$

where

$$\left\{ \begin{matrix} \langle P_{ij}^{(1)} \rangle \\ \langle e_{ij}^{(1)} \rangle \end{matrix} \right\} = \frac{4}{H^{(1)} \ell_1 \ell_3} \int_{\ell_1/4}^{3\ell_1/4} \left(\int_{-H^{(1)}}^{0} \left(\int_{0}^{\ell_3/2} \left\{ \begin{matrix} P_{ij}^{(1),q} \\ e_{ij}^{(1),q} \end{matrix} \right\} \left(x_1^{(1)}, x_2^{(1)}, x_3^{(1)} \right) dx_3^{(1)} \right) dx_2^{(1)} \right) dx_1^{(1)},$$

$$\left\{ \begin{matrix} \langle P_{ij}^{(2)} \rangle \\ \langle e_{ij}^{(2)} \rangle \end{matrix} \right\} = \frac{4}{H^{(2)} \ell_1 \ell_3} \int_{\ell_1/4}^{3\ell_1/4} \left(\int_{0}^{H^{(2)}} \left(\int_{0}^{\ell_3/2} \left\{ \begin{matrix} P_{ij}^{(2),q} \\ e_{ij}^{(2),q} \end{matrix} \right\} \left(x_1^{(2)}, x_2^{(2)}, x_3^{(2)} \right) dx_3^{(2)} \right) dx_2^{(2)} \right) dx_1^{(2)}. \tag{9.2.4}$$

Writing the expressions for $P_{ij}^{(k),q}$, $e_{ij}^{(k),q}$ through $\sigma_{ij}^{(k),q}$, $\varepsilon_{ij}^{(k),q}$ and calculating the integrals (9.2.4), we determine the values of $\langle \sigma_{ij} \rangle$, $\langle \varepsilon_{ij} \rangle$ from (9.2.2), (9.2.3). We reduce the calculation of the integrals (9.2.2) to that of the integrals (9.2.4). After these operations we define the deformation energy U accumulated in a selected representative element. For the composites shown in Figs. 9.1.1 and 9.1.2 under the action uniformly distributed forces at infinity with intensity p_1 in the direction of the Ox_1 axis, this energy can be represented as follows:

$$U = \frac{1}{2} E_1 \left(1 + \varepsilon^2 \alpha_2 + \varepsilon^4 \alpha_4 + \ldots \right) \left(\varepsilon_{11}^{(1),0} \right)^2, \tag{9.2.5}$$

where

$$E_1 = \frac{\eta^{(2)}E^{(1)} + \eta^{(2)}E^{(2)}}{1-\eta^{(2)}\left(\nu^{(2)} - \nu^{(1)}\right)P_{33}}, \quad P_{33} = -\frac{\nu^{(1)}}{e_1} + \frac{e_2}{e_1}\left(e_1 - \nu^{(1)}e_2\right)\left((e_1)^2 - (e_2)^2\right)^{-1},$$

$$\alpha_k = \alpha_k\left(\eta^{(2)}, \chi, E^{(2)}/E^{(1)}, \nu^{(1)}, \nu^{(2)}\right) \quad k = 2,4,.... \tag{9.2.6}$$

In (9.2.6) the following notation is used.

$$e_1 = \eta^{(1)}\frac{E^{(1)}}{E^{(2)}} + \eta^{(2)}, \quad e_2 = \nu^{(1)}\eta^{(2)} + \nu^{(2)}\eta^{(1)}\frac{E^{(1)}}{E^{(2)}}. \tag{9.2.7}$$

As an example, we write the expression for α_2.

$$\alpha_2 = \left\langle \overline{P}_{11}^{;2} \right\rangle + \frac{E_1}{E^{(2)}}\left\langle \overline{\varepsilon}_{11}^{;2} \right\rangle + \frac{E_1}{E^{(2)}}\overline{\varepsilon}_{33}^{(2),0}\left\langle \overline{P}_{33}^{;2} \right\rangle + \frac{2E_1}{E^{(2)}}\left\langle \overline{P}_{13}^{;1} \right\rangle\left\langle \overline{\varepsilon}_{13}^{;1} \right\rangle, \tag{9.2.8}$$

where

$$\left\langle \overline{P}_{ij}^{;q} \right\rangle = \left(\varepsilon_{11}^{(1),0}\right)^{-1}\left\langle P_{ij}^{;q} \right\rangle, \quad \left\langle \overline{\varepsilon}_{ij}^{;q} \right\rangle = \left(\varepsilon_{11}^{(1),0}\right)^{-1}\left\langle \varepsilon_{11}^{;q} \right\rangle,$$

$$\left\langle P_{ij}^{;q} \right\rangle = \eta^{(1)}\left\langle P_{ij}^{(1),q} \right\rangle + \eta^{(2)}\left\langle P_{ij}^{(2),q} \right\rangle, \quad \left\langle \varepsilon_{ij}^{;q} \right\rangle = \eta^{(1)}\left\langle \varepsilon_{ij}^{(1),q} \right\rangle + \eta^{(2)}\left\langle \varepsilon_{ij}^{(2),q} \right\rangle. \tag{9.2.9}$$

According to the well known procedure, by differentiating the expression U (9.2.5) with respect to $\varepsilon_{11}^{(1),0}$, we define

$$\left\langle \sigma_{11} \right\rangle = \frac{\partial U}{\partial \varepsilon_{11}^{(1),0}} = \widetilde{E}_1 \varepsilon_{11}^{(1),0}. \tag{9.2.10}$$

We derive from (9.2.3), (9.2.10) that

$$\widetilde{E}_1 = E_1\left(1 + \varepsilon^2\alpha_2 + \varepsilon^4\alpha_4 + ...\right). \tag{9.2.11}$$

Thus, we obtain the expression for the normalized moduli of elasticity $\widetilde{E}_1$ of the composite materials shown in Figs.9.1.1 and 9.1.2 . Note that in (9.2.5), (9.2.6) and (9.2.11) there is a modulus of elasticity for these composites in the case where their layers are uncurved. In other words, it follows from (9.2.11) that in the case where

$\varepsilon = 0$ we obtain that $\tilde{E}_1 = E_1$. We can determine other normalized mechanical properties similarly.

9.3. Numerical Results

Consider some numerical results related to the modulus of elasticity $\tilde{E}_1$. We analyse the results obtained for the composites with periodically plane-curved (Figs.4.2.1 and 4.2.3) and spatially curved (Fig.9.1.1) structures separately. For plane curved structures the expression of E_1 is

$$E_1 = \frac{\eta^{(1)}E^{(1)}}{1-\left(\nu^{(1)}\right)^2} + \frac{\eta^{(2)}E^{(2)}}{1-\left(\nu^{(2)}\right)^2}. \tag{9.3.1}$$

All numerical results, have been obtained in the framework of the following approximation:

$$\tilde{E}_1 \approx E_1\left(1 + \varepsilon^2\alpha_2\right). \tag{9.3.2}$$

9.3.1. PLANE-CURVED STRUCTURES

Table 9.1 shows the values of $\tilde{E}_1/E_1$ for the composite with co-phase curved structure (Fig.4.2.1) with various $E^{(2)}/E^{(1)}$, χ, $\eta^{(2)}$ and $\varepsilon = 0.02$, $\nu^{(1)} = \nu^{(2)} = 0.3$. The results for anti-phase curvature (Fig.4.2.3) are given in Table 9.2, for the same values of the problem parameters.

Table 9.1. The values of $\tilde{E}_1/E_1$ for the composite with co-phase periodically plane-curved structure (Fig.4.2.1).

$\frac{E^{(2)}}{E^{(1)}}$	χ	$\eta^{(2)}$			
		0.2	0.3	0.4	0.5
20	0.02	0.939	0.917	0.904	0.899
	0.1	0.941	0.921	0.909	0.904
	0.3	0.954	0.943	0.937	0.934
	0.5	0.973	0.967	0.964	0.962
50	0.02	0.832	0.777	0.743	0.731
	0.1	0.843	0.797	0.770	0.760
	0.3	0.902	0.889	0.881	0.878
	0.5	0.952	0.947	0.944	0.943
100	0.02	0.653	0.543	0.476	0.453
	0.1	0.696	0.618	0.573	0.557
	0.3	0.857	0.844	0.837	0.834
	0.5	0.940	0.935	0.933	0.933

These results show that for co-phase curving the values of $\tilde{E}_1/E_1$ decrease with increasing $\eta^{(2)}$ and $E^{(2)}/E^{(1)}$, but increase with χ. In the anti-phase curving case the influence of the parameters on the ratio $\tilde{E}_1/E_1$ is more complicated. An increase in $E^{(2)}/E^{(1)}$, basically, causes a decrease in $\tilde{E}_1/E_1$. However, the influence of $\eta^{(2)}$ depends on the values of χ, and the dependence on χ and $E^{(2)}/E^{(1)}$ is not strictly monotonic.

9.3.2. SPATIALLY CURVED STRUCTURES

Table 9.2. The values of $\tilde{E}_1/E_1$ for the composite with anti-phase periodically plane-curved structure (Fig.4.2.3).

$E^{(2)}/E^{(1)}$	χ	$\eta^{(2)}$			
		0.2	0.3	0.4	0.5
20	0.02	0.980	0.977	0.975	0.972
	0.1	0.970	0.971	0.971	0.970
	0.3	0.944	0.942	0.948	0.954
	0.5	0.967	0.956	0.948	0.948
50	0.02	0.977	0.975	0.972	0.970
	0.1	0.949	0.958	0.962	0.963
	0.3	0.887	0.888	0.904	0.922
	0.5	0.945	0.930	0.918	0.917
100	0.02	0.974	0.973	0.971	0.969
	0.1	0.918	0.939	0.950	0.955
	0.3	0.829	0.825	0.849	0.878
	0.5	0.932	0.914	0.897	0.891

Table 9.3. The values of $\tilde{E}_1/E_1$ for the composite with co-phase periodically spatially-curved structure (Fig.9.1.1) for $\eta^{(2)} = 0.2$.

$E^{(2)}/E^{(1)}$	χ	γ			
		0.1	0.5	1.0	1.5
20	0.02	0.971	0.975	0.981	0.985
	0.1	0.971	0.976	0.982	0.986
	0.3	0.977	0.982	0.988	0.992
	0.5	0.985	0.989	0.994	0.996
50	0.02	0.922	0.935	0.956	0.969
	0.1	0.927	0.941	0.961	0.974
	0.3	0.954	0.967	0.982	0.989
	0.5	0.976	0.983	0.991	0.995
100	0.02	0.842	0.870	0.915	0.944
	0.1	0.861	0.891	0.923	0.956
	0.3	0.934	0.955	0.978	0.988
	0.5	0.971	0.981	0.990	0.995

Tables 9.3 and 9.4 show the values of $\widetilde{E}_1/E_1$ for the composite given in Fig.9.1.1 for $\eta^{(2)} = 0.2$ and 0.5 respectively, with various χ, $E^{(2)}/E^{(1)}$ and $\gamma = \ell_1/\ell_3$. We take $\nu^{(1)} = \nu^{(2)} = 0.3$ and $\varepsilon = 0.02$ from which follows that the values of $\widetilde{E}_1/E_1$ decrease with increasing $\eta^{(2)}$ and $E^{(2)}/E^{(1)}$, and increase with increasing χ and γ.

These results can be improved by increasing the approximation order in (9.3.3), and can be used to test the accuracy of other approaches.

Table 9.4. The values of $\widetilde{E}_1/E_1$ for the composite with co-phase periodically spatially-curved structure (Fig.9.1.1) for $\eta^{(2)} = 0.5$.

$\frac{E^{(2)}}{E^{(1)}}$	χ	γ			
		0.1	0.5	1.0	1.5
	0.02	0.954	0.961	0.972	0.980
20	0.1	0.956	0.963	0.975	0.982
	0.3	0.970	0.977	0.986	0.991
	0.5	0.982	0.987	0.993	0.997
	0.02	0.878	0.900	0.934	0.956
50	0.1	0.891	0.914	0.948	0.968
	0.3	0.945	0.961	0.980	0.989
	0.5	0.974	0.983	0.992	0.996
	0.02	0.753	0.799	0.872	0.918
100	0.1	0.800	0.848	0.915	0.953
	0.3	0.925	0.951	0.977	0.988
	0.5	0.969	0.980	0.991	0.996

9.4. Bibliographical Notes

The calculation of the normalized moduli of elasticity of composites with periodically curved structures using the piecewise homogeneous body model was proposed by S.D. Akbarov, M.S. Guliev and E.A. Movsumov [28]. The determination of the normalized non-linear mechanical properties of these composites was made by S.D. Akbarov [20].

CHAPTER 10

FRACTURE PROBLEMS

10.1. Fiber Separation

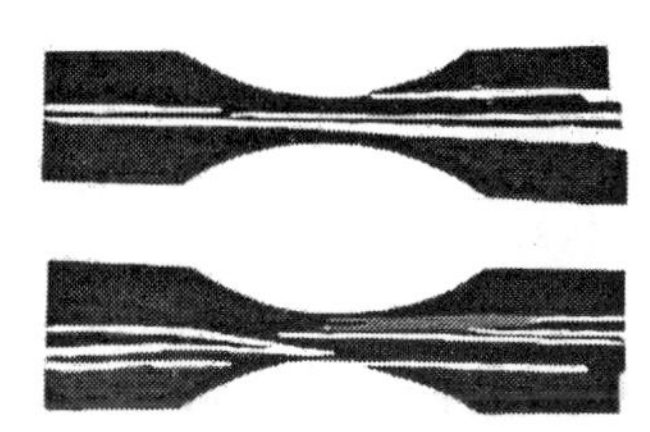

Fig.10.1.1. The character of fracture of the unidirected glass-fiber reinforced plastic in compression along reinforcements. This picture is taken from [129].

The fracture of composite materials has different peculiarities depending on the structure of the composite, the type of loading and other factors. The following is one of these peculiarities.

Experimental studies of the fracture of composite materials under uniaxial loading along reinforcing elements has shown that a peculiar fracture manifesting itself as a partition of material into separate parts along the line of action of the external loading, as shown in Fig.10.1.1. Such experiments are described in detail in [58,102, 129, 141,142] and elsewhere. This phenomenon is not observed for homogeneous materials and is characteristic only of composite materials; the full fracture arises over planes and surfaces along the reinforcing elements. This phenomenon, called *Fiber separation,* is attributed to the action of forces directed perpendicularly to the reinforcing elements; but in uniaxial loading along the reinforcing elements the *external* loading is applied only *along* the reinforcing elements.

To explain the mechanism we must ask why there are normal and tangential stresses acting on places parallel to the direction of the external loading. One reason may be a slight curving of reinforcing elements. Many researchers have considered similar views [68, 75, 114, 141, 142] and elsewhere. However, these researchers had insufficient quantitative information on the distribution of stresses due to curving of reinforcing elements. To explain the effect we must determine the values of stresses arising from the curving of the reinforcing elements; to do this we use the results obtained in Chapters 1-8. As a criterion of material fracture, we can use a macroscopic criterion of fracture by maximum normal and tangential stresses [70, 73, 142].

We consider a composite consisting of layers perpendicular to the Ox_2 axis loaded along the Ox_1 axis with the uniformly distributed normal forces of intensity p_1. We introduce the following notation for this composite without curvature: $\Pi_1^{\pm}$ are the ultimate strengths, + and - representing respectively tension and compression along the Ox_1 axis, Π_2^{+} is the ultimate strength in tension in the direction of the Ox_2 axis, Π_{12} the ultimate strength in shear in the Ox_1x_2 plane. Since the values of Π_2^{+} and Π_{12} are determined mainly by the strength of the matrix material we have

$$\Pi_2^{\pm}/\Pi_1^{+} \ll 1, \qquad (10.1.1)$$

$$\Pi_1^{\pm}/\Pi_{12} \gg 1 \qquad (10.1.2)$$

Table 10.1 shows the values of the ratios $\Pi_2^{\pm}/\Pi_1^{+}$, $\Pi_1^{\pm}/\Pi_{12}$ for some composites, taken from reference [142].

Table 10.1. The mechanical and failure characteristic of some plastics.

Plastics	E_1/G_{12}	$\Pi_1^{\pm}/\Pi_{12}$	E_1/E_2	Π_1^{+}/Π_2^{+}
	Glass-fiber reinforced			
Unidirectional (1:0)	25 - 50	30 - 40	5 - 8	10 -18
Layered (1:1)	15 - 20	15 -10	≤ 5	≤ 5
	Boron-fiber reinforced			
Unidirectional (1:0)	≤ 100	---	10 - 15	7 - 10

According to the mechanical consideration the fracture of the composite occurs when one of the following criteria is satisfied:

$$\sigma_{11} = \Pi_1^{\pm} \qquad (10.1.3)$$

$$\sigma_{22} = \Pi_2^{\pm}, \qquad (10.1.4)$$

$$\sigma_{12} = \Pi_{12}. \qquad (10.1.5)$$

Under loading along the Ox_1 axis the stresses σ_{22} and σ_{12} arise as a result of the curving. Consequently, if the curvature parameters provide the satisfaction of the criterion (10.1.4) or (10.1.5) before that of the criterion (10.1.3), then the *fiber separation* effect occurs.

10.1.1. CONTINUUM APPROACH

Periodically plane-curved structure.

It follows from the expression (1.9.8) that the fiber separation effect takes place when the following relations are satisfied:

$$|\sigma_{11}| = |p_1| < \Pi_1^{\pm}, \quad |\sigma_{22}| = \varepsilon^2 1.75 |p_1| = \Pi_2^{\pm}, \qquad (10.1.6)$$

From (10.1.6) we determine the values of ε (1.3.23) for which this effect occurs:

$$\varepsilon > \sqrt{\Pi_2^{\pm}\left(1.75\Pi_1^{\pm}\right)^{-1}}. \qquad (10.1.7)$$

It follows from (1.3.23) and (10.1.7) that

$$H > \frac{\pi}{8}\Lambda\sqrt{\Pi_2^{\pm}\left(1.75\Pi_1^{\pm}\right)^{-1}}. \quad (10.1.8)$$

As a numerical example, we consider a glass-fiber reinforced plastic and assume that $\Pi_1^{\pm}/\Pi_2^{\pm} \approx 18$; then we obtain from (10.1.8) that

$$H > 0.07\Lambda. \quad (10.1.9)$$

The relation (10.1.9) corresponds to very real cases where the fiber separation can be caused by curving.

Periodically spatially-curved structure.
We use the results given by expressions (2.9.2) and assume that $\gamma \geq 1$ ($\gamma = \Lambda_1/\Lambda_3$, where Λ_1 (Λ_3) is a curvature half-wave length in the direction of the Ox_1 (Ox_3) axis). Equation (2.9.2) shows that the values of the shear stresses σ_{12} and σ_{23} are of order ε^1 (2.7.3), but, σ_{22} are of order ε^2. Moreover equation (2.9.2) shows that among the stresses which arise as a result of the curvature in the structure of the composite material, the shear stress σ_{12} dominates; we take the fiber separation as follows:

$$|\sigma_{11}| = |p_1| < \Pi_1^{\pm}, \quad |\sigma_{12}| = \varepsilon|p_1|\frac{\gamma^2}{1+\gamma^2} = \Pi_{12}. \quad (10.1.10)$$

We obtain from (10.1.10) that

$$\varepsilon\frac{\gamma^2}{1+\gamma^2} > \frac{\Pi_{12}}{\Pi_1^{\pm}}. \quad (10.1.11)$$

Thus, from (2.7.3) and (10.1.11) we derive the following relation between the curvature and failure parameters for fiber separation:

$$H > \frac{\Lambda_1\left(1+\gamma^2\right)\Pi_{12}}{\pi\gamma^2\Pi_1^{\pm}}. \quad (10.1.12)$$

Consider glass-fiber reinforced plastic and, according to Table 10.1, assume that $\Pi_1^{\pm}/\Pi_{12} \approx 40$; (10.1.12) shows that

$$H > 0.0079\Lambda_1\frac{1+\gamma^2}{\gamma^2}. \quad (10.1.13)$$

If $\Lambda_1 = \Lambda_3$ we obtain from (10.1.13) that

$$H > 0.0158\Lambda_1 . \qquad (10.1.14)$$

The inequalities (10.1.9) and (10.1.14) show that under spatial curving the fiber separation can occur for smaller curvature (i.e., ε) than that under plane curving.

10.1.2. PIECE-WISE HOMOGENEOUS MODEL

We use the numerical results given in Chapter 4 for composites with alternating co-phase (Fig.4.2.1) and anti-phase (Fig.4.2.3) periodically plane-curved layers. We introduce the following notation: $\Pi_M^{(1)+}$ and $\Pi_M^{(12)}$ are respectively the ultimate strengths in tension and in shear of the matrix material. Taking into account that the values of Π_2^+ and Π_{12} are determined mainly by matrix material ultimate strength in tension and in shear respectively we assume that

$$\Pi_M^{(1)+} \approx \Pi_2^+ , \; \Pi_M^{(12)} \approx \Pi_{12} , \qquad (10.1.15)$$

According to the assumption given in the previous subsection fiber separation can arise if

$$\sigma_{nn} / \sigma_{11}^{(1),0} > 1 , \; \sigma_{11}^{(1),0} < \Pi_M^{(1)+} , \; \sigma_{nn} = \Pi_M^{(1)+} , \qquad (10.1.16)$$

or

$$\sigma_{n\tau} / \sigma_{11}^{(1),0} > \gamma_p , \; \sigma_{n\tau} = \Pi_M^{(12)} , \qquad (10.1.17)$$

where $\gamma_p = \Pi_M^{(12)} / \Pi_M^{(1)+}$. According to Table 10.1 and (10.1.15) for glass-fiber reinforced plastics γ_p ranges between 0.25 and 0.60.

Equations (10.1.15) - (10.1.17) and the numerical results obtained in section 4.3 show that under co-phase curving of the layers fiber separation occurs when

$$\chi \le 0.2 , \; 0.2 \le \eta^{(2)} \le 0.5 , \; E^{(2)}/E^{(1)} \ge 50 , \; \varepsilon \ge 0.015 , \qquad (10.1.18)$$

but under anti-phase the effect takes place when

$$0.2 \le \chi \le 0.4 , \; 0.2 \le \eta^{(2)} \le 0.5 , \; E^{(2)}/E^{(1)} \ge 50 , \; \varepsilon \ge 0.015 . \qquad (10.1.19)$$

The notation in (10.1.18), (10.1.19) is defined in sections 4.2 and 4.3. The relations (10.1.18), (10.1.19) show that the explanation of the fiber separation in the framework of the piece-wise homogeneous body model is very real, and gives the possibility for optimal selection of the material and curvature parameters for prevention of this type of fracture.

10.1.3. LOCAL FIBER SEPARATION

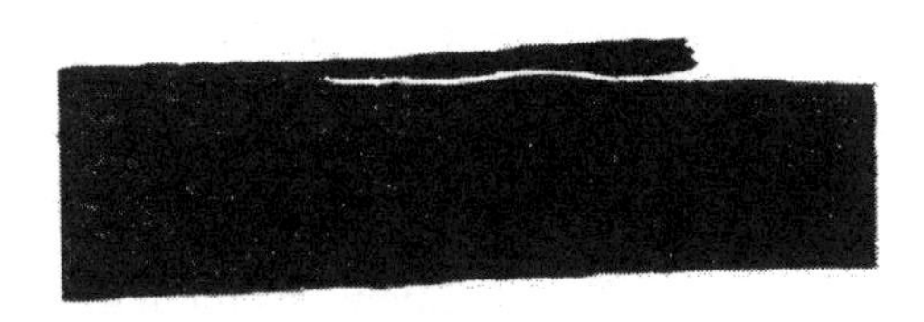

Fig.10.1.2. Local "Fiber separation" effect under uniaxial tension of the unidirected glass-fiber reinforced plastic[142].

Many experiments show that the fiber separation effect is sometimes observed in certain local parts of the region occupied by a unidirected composite under uniaxial loading along the reinforcing fibers or layers; there is a local curving of the reinforcing fibers in these parts. As an example Fig.10.1.2 shows this failure effect for unidirected glass-fiber reinforced plastic under uniaxial tension along the fibers; failure occurs because of the local curving of the fibers in the failure region. Such effects can be predicted by using the approach developed in Chapter 6. To illustrate this, we use the numerical results obtained for alternating co-phase local curving layers. We use the local curving form given by (6.2.6), and the failure criteria (10.1.16), (10.1.17). Analysis of the numerical results, given in Chapter 6, demonstrates that if

$$1 \le \delta \le 5\,,\ m \ge 1\,,\ 0.2 \le \eta^{(2)} \le 0.5\,,\ 0.04 \le \varepsilon \le 0.1\,,$$

$$0.01 \le H^{(2)}/L \le 0.3\,,\quad E^{(2)}/E^{(1)} \ge 20\,, \tag{10.1.20}$$

then, according to the criteria (10.1.16), (10.1.17), local failure takes place in the characteristic parts of the interface surfaces S_1^+ and S_1^-. Predictions and explanations can also be made by using the continuum approach developed in Chapter 2.

10.2. Crack Problems

A macroscopic fracture criterion can give only a rough estimate of fracture and cannot describe a precise fracture mechanism; we need to solve crack problems. We use the piece-wise homogeneous model based on the exact equations of theory of elasticity. According to the *Desai-Mac-Harry* hypothesis [75], the cracks are assumed to be in the most dangerous parts of matrix layers, and their edges parallel to the direction of external forces.

10.2.1. FORMULATION

For simplicity we shall consider only two-dimensional problems and investigate plane deformation in composites with curved layers. Consider an infinite elastic body reinforced by of non-intersecting curved layers of a filler. The values related to the matrix will be denoted by upper indices (1), and the values related to the filler by (2). The rectangular Cartesian system of coordinates $O_m^{(k)}x_{1m}^{(k)}x_{2m}^{(k)}x_{3m}^{(k)}$ $(k = 1,2;\ m = 1,2,3,...)$ obtained from the coordinate system $Ox_1x_2x_3$ by parallel transfer along the Ox_2 axis and related to middle surfaces of the corresponding layers will refer to each layer. The curving of these layers is assumed to be independent of x_3.

We consider that the reinforcing layers are located in planes $O_m^{(2)}x_{1m}^{(2)}x_{3m}^{(2)}$ and the thickness of each filler layer is constant. The matrix and filler layer materials are assumed to be homogeneous, isotropic and linearly elastic.

We investigate plane deformation under loading at infinity by uniformly distributed normal forces of intensity p in the direction of the Ox_1 axis. For each layer the equilibrium equations, Hooke's law and Cauchy relations may be written as

$$\frac{\partial \sigma_{ij}^{(\underline{k})\underline{m}}}{\partial x_{j\underline{m}}^{(\underline{k})}} = 0, \qquad \sigma_{ij}^{(k)m} = \lambda^{(\underline{k})\underline{m}} \theta^{(\underline{k})\underline{m}} \delta_i^j + \mu^{(\underline{k})\underline{m}} \varepsilon_{ij}^{(\underline{k})\underline{m}},$$

$$\varepsilon_{ij}^{(k)m} = \frac{1}{2}\left(\frac{\partial u_i^{(\underline{k})\underline{m}}}{\partial x_{j\underline{m}}^{(\underline{k})}} + \frac{\partial u_j^{(\underline{k})\underline{m}}}{\partial x_{i\underline{m}}^{(\underline{k})}} \right), \quad \theta^{(k)m} = \frac{\partial u_i^{(\underline{k})\underline{m}}}{\partial x_{i\underline{m}}^{(\underline{k})}}, \quad i; j; k = 1,2. \tag{10.2.1}$$

We use the notation introduced in Chapter 4.

Suppose that there is complete cohesion at the interfaces between matrix and filler material. Denoting the upper surfaces of the $m^{(2)}$-th layer by S_m^+, the lower surfaces through S_m^-, and introducing the notation $m_1 = m-1$, we can write these conditions in the following form:

$$\sigma_{ij}^{(1)\underline{m}}\Big|_{S_{\underline{m}}^+} n_j^{\underline{m},+} = \sigma_{ij}^{(2)\underline{m}}\Big|_{S_{\underline{m}}^+} n_j^{\underline{m},+}; \quad u_i^{(1)\underline{m}}\Big|_{S_{\underline{m}}^+} = u_i^{(2)\underline{m}}\Big|_{S_{\underline{m}}^+};$$

$$\sigma_{ij}^{(1)m_1}\Big|_{S_{\underline{m}}^-} n_j^{\underline{m},-} = \sigma_{ij}^{(2)\underline{m}}\Big|_{S_{\underline{m}}^-} n_j^{\underline{m},-}; \quad u_i^{(1)m_1}\Big|_{S_{\underline{m}}^-} = u_i^{(2)\underline{m}}\Big|_{S_{\underline{m}}^-}, \tag{10.2.2}$$

where $n_j^{m,\pm}$ are the orthonormal components to the surface $S_m^\pm$.

The equation of the middle surface of the $m^{(2)}$-th layer of the filler is written as

$$x_{2m}^{(2)} = F_{\underline{m}}\left(x_{1\underline{m}}^{(2)}\right) = \varepsilon f_{\underline{m}}\left(x_{1\underline{m}}^{(2)}\right), \tag{10.2.3}$$

where, ε is a dimensionless small parameter and $\varepsilon \in [0,1)$.

Assume that there are free cracks of finite length $2\ell_{mn}$ located in the planes $x_{2m}^{(1)} = C_{mn}^{(1)}$ (where $C_{mn}^{(1)} = \text{const}_n$, $n = 1,2,\ldots$) in matrix layers, and we add the following conditions on the edges of these cracks to the relations (10.2.1) - (10.2.3):

$$\sigma_{i2}^{(1)\underline{m}}\left(x_{1\underline{m}}^{(1)}, C_{\underline{mn}}^{(1)} + 0\right) = \sigma_{i2}^{(1)\underline{m}}\left(x_{1\underline{m}}^{(1)}, C_{\underline{mn}}^{(1)} - 0\right) = 0 \text{ at } x_{1m}^{(1)} \in \left(a_{mn}, b_{mn}\right), \qquad (10.2.4)$$

where $b_{mn} - a_{mn} = 2\ell_{mn}$; a_{mn}, b_{mn} are the abscissas of the left and right tips of cracks, respectively.

This formulates the crack problems; we consider particular cases.

Case I. Consider an infinite elastic body reinforced by two neighbouring anti-phase periodically curved layers (Fig.10.2.1). We associate the corresponding Cartesian coordinate system with the middle surface of each layer of the filler and the matrix $1^{(1)}$. Let the equilibrium equations be satisfied within the layered $0^{(2)}$, $1^{(1)}$, $1^{(2)}$, and also in the half-spaces $0^{(1)}$ and $2^{(1)}$. The contact conditions (10.2.2) can be written as

$$\sigma_{ij}^{(2)1}\Big|_{S_1^+} n_j^{1,+} = \sigma_{ij}^{(1)2}\Big|_{S_1^+} n_j^{1,+}; \quad u_i^{(2)1}\Big|_{S_1^+} = u_i^{(1)2}\Big|_{S_1^+};$$

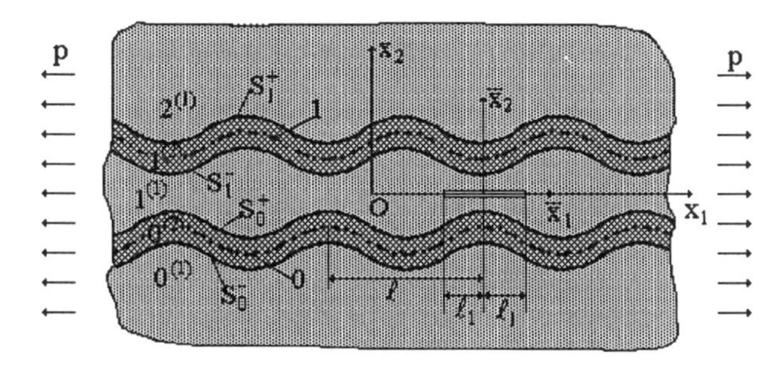

Fig.10.2.1. The structure of infinite body reinforced by two anti-phase periodically curved layers of the filler and with one crack of the finite length in matrix.

$$\sigma_{ij}^{(2)1}\Big|_{S_1^-} n_j^{1,-} = \sigma_{ij}^{(1)1}\Big|_{S_1^-} n_j^{1,-}; \quad u_i^{(2)1}\Big|_{S_1^-} = u_i^{(1)1}\Big|_{S_1^-};$$

$$\sigma_{ij}^{(2)0}\Big|_{S_0^+} n_j^{0,+} = \sigma_{ij}^{(1)1}\Big|_{S_0^+} n_j^{0,-}; \quad u_i^{(2)0}\Big|_{S_0^+} = u_i^{(1)1}\Big|_{S_0^+};$$

$$\sigma_{ij}^{(2)0}\Big|_{S_0^-} n_j^{0,-} = \sigma_{ij}^{(1)1}\Big|_{S_0^-} n_j^{0,-}; \quad u_i^{(2)0}\Big|_{S_0^-} = u_i^{(1)0}\Big|_{S_0^-}. \qquad (10.2.5)$$

The following decay conditions should also be added to (10.2.5).

$$\sigma_{ij}^{(1)2} \to p\delta_i^1\delta_j^1 \text{ at } x_{21}^{(2)} \to +\infty \ , \ \sigma_{ij}^{(1)0} \to p\delta_i^1\delta_j^1 \text{ at } x_{20}^{(2)} \to -\infty . \tag{10.2.6}$$

We write the equation of the middle surface of the layer $1^{(2)}$ as

$$x_{21}^{(2)} = L\sin\alpha x_{11}^{(2)}, \tag{10.2.7}$$

and the equation of the middle surface of the layer $0^{(2)}$ as

$$x_{20}^{(2)} = -L\sin\alpha x_{10}^{(2)}, \tag{10.2.8}$$

where L is the amplitude of curving, $\alpha = 2\pi/\ell$ and ℓ is the wavelength of the middle surface. Assuming that $L < \ell$ and introducing the small parameter $\varepsilon = L/\ell$, we rewrite the equations (10.2.7), (10.2.8) in the following form:

$$x_{21}^{(2)} = \varepsilon\ell\sin\alpha x_{11}^{(2)}, \quad x_{20}^{(2)} = -\varepsilon\ell\sin\alpha x_{10}^{(2)}. \tag{10.2.9}$$

We formulate the following problems.

Problem I. To define the stress intensity factor (SIF) at the tips of the crack in the matrix layer $1^{(1)}$ (Fig.10.2.1) at $x_{21}^{(1)} = 0$, $3\ell/4 - \ell_1 \le x_{11}^{(1)} \le 3\ell/4 + \ell_1$, where $2\ell_1$ is the length of the crack. The condition (10.2.4) in this case will be of the following form:

$$\sigma_{i2}^{(1)1}\left(x_{11}^{(1)},+0\right) = \sigma_{i2}^{(1)1}\left(x_{11}^{(1)},-0\right) = 0 \text{ at } x_{11}^{(1)} \in \left(3\ell/4 - \ell_1, 3\ell/4 + \ell_1\right). \tag{10.2.10}$$

Problem II. To determine the SIF at the tips of two collinear cracks (taking into account their mutual influence) in the matrix layer $1^{(1)}$ at $x_{21}^{(1)} = 0$, $3\ell/4 - \ell_1 \le x_{11}^{(1)} \le 3\ell/4 + \ell_1$ and $-\ell/4 - \ell_1 \le x_{11}^{(1)} \le -\ell/4 + \ell_1$. In this case the conditions (10.2.10) will also be satisfied at $x_{11}^{(1)} \in \left(-\ell/4 - \ell_1, -\ell/4 + \ell_1\right)$ (Fig.10.2.2).

Problem III. To determine the SIF at the periodically located row cracks (taking into account their interaction) in matrix layer $1^{(1)}$ at $x_{21}^{(1)} = 0$, $n\ell - \ell_1 \le x_1 \le \ell_1 + n\ell$, $n = 0, \pm 1, \pm 2, \ldots$. In this case the conditions will be satisfied at $x_{11}^{(1)} \in \left(n\ell/4 - \ell_1, n\ell/4 + \ell_1\right)$ (Fig.10.2.3).

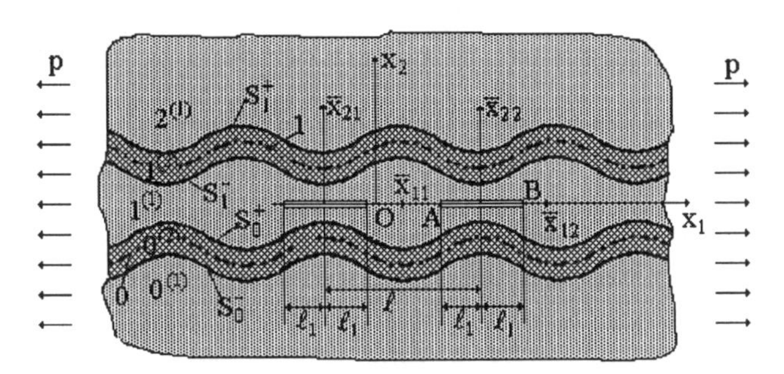

Fig.10.2.2. The structure of infinite body reinforced by two anti-phase periodically curved layers of the filler and with two collinear cracks in matrix.

We will define the SIF in the next subsection. Now, consider another case.

Case II. Consider the composite with alternating anti-phase periodically (Fig.10.2.4) and locally (Fig.10.2.5) curved layers. Associate the corresponding Cartesian systems of coordinates $O_m^{(k)} x_{1m}^{(k)} x_{2m}^{(k)}$ obtained from Ox_1x_2 (Figs.10.2.4 and 10.2.5) by parallel transfer along the Ox_2 axis with middle surface of each layer of the filler and matrix. Taking into account the periodicity of the composite structure shown in Figs.10.2.4 and 10.2.5 in the direction of the Ox_2 axis with the period $4\left(H^{(2)} + H^{(1)}\right)$ (where $2H^{(1)}$ is the mean thickness of the matrix layer, $2H^{(2)}$ is the thickness of the filler layer), among the layers considered we single out four of them, i.e., $1^{(1)}$, $1^{(2)}$,

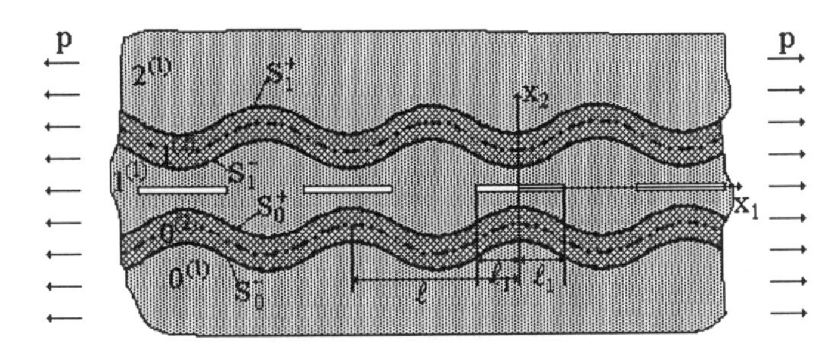

Fig.10.2.3. An infinite body reinforced by two anti-phase periodically curved layers with periodically located row cracks in the matrix

$2^{(1)}$, $2^{(2)}$, and discuss them below. In the notation in Figs. 10.2.4 and 10.2.5, the contact conditions (10.2.2) can be written in the following form:

$$\sigma_{ij}^{(2)1}\Big|_{S_1^+} n_j^{1,+} = \sigma_{ij}^{(1)1}\Big|_{S_1^+} n_j^{1,+}, \quad u_i^{(2)1}\Big|_{S_1^+} = u_i^{(1)1}\Big|_{S_1^+} ,$$

$$\sigma_{ij}^{(2)1}\Big|_{S_1^-} n_j^{1,-} = \sigma_{ij}^{(1)2}\Big|_{S_1^-} n_j^{1,-}, \quad u_i^{(2)1}\Big|_{S_1^-} = u_i^{(1)2}\Big|_{S_1^-},$$

$$\sigma_{ij}^{(2)2}\Big|_{S_2^+} n_j^{2,+} = \sigma_{ij}^{(1)2}\Big|_{S_2^+} n_j^{2,+}, \quad u_i^{(2)2}\Big|_{S_2^+} = u_i^{(1)2}\Big|_{S_2^+},$$

$$\sigma_{ij}^{(2)2}\Big|_{S_2^-} n_j^{2,-} = \sigma_{ij}^{(1)1}\Big|_{S_2^-} n_j^{2,-}, \quad u_i^{(2)2}\Big|_{S_2^-} = u_i^{(1)1}\Big|_{S_2^-}. \tag{10.2.11}$$

The equations of the middle surfaces of filler layers $1^{(2)}$ and $2^{(2)}$ in Fig.10.2.4 may be presented by

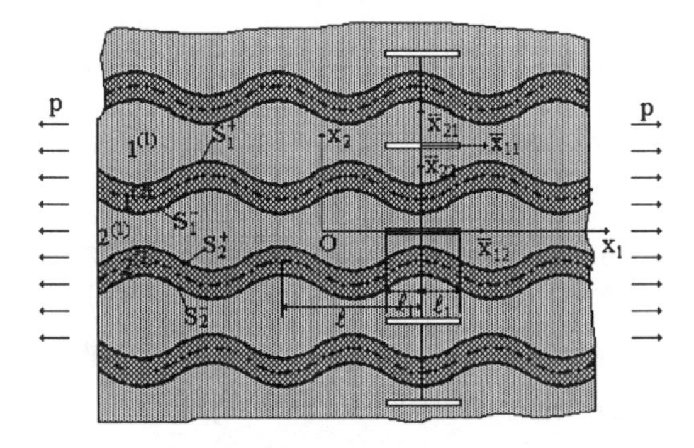

Fig.10.2.4. Alternating anti-phase periodically curved layers with one crack of finite length in each matrix layer.

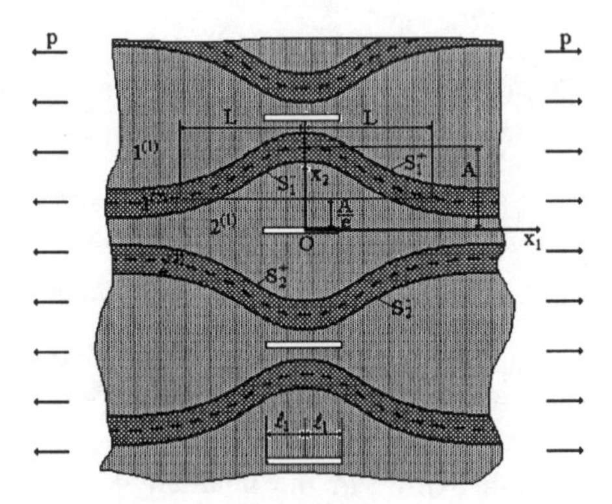

Fig.10.2.5. Alternating anti-phase locally curved layers with one crack of finite length in each matrix layer

$$x_{21}^{(2)} = L\sin\alpha x_{11}^{(2)} = \varepsilon\ell\sin\alpha x_{11}^{(2)}, \quad x_{22}^{(2)} = -L\sin\alpha x_{12}^{(2)} = -\varepsilon\ell\sin\alpha x_{12}^{(2)}. \tag{10.2.12}$$

In (10.2.12) L, ℓ and ε represent the same quantities as before. For the local curving case (Fig.10.2.5), the equations are taken as

$$x_{21}^{(2)} = A\exp\left(-\left(x_{11}^{(2)}/L\right)^2\right), \quad x_{22}^{(2)} = -A\exp\left(-\left(x_{12}^{(2)}/L\right)^2\right). \tag{10.2.13}$$

where A and L are geometrical parameters shown in Fig.10.2.5. Assuming that A<L we introduce a small parameter $\varepsilon = A/L$.

As before formulate the following problems.

Problem IV. To determine the SIF in the crack tips of each matrix layer at $x_{2m}^{(1)} = 0$, $3\ell/4 - \ell_1 \le x_{1m} \le 3\ell/4 + \ell_1$ of the composite shown in Fig.10.2.4. In this case the conditions (10.2.4) have the form

$$\sigma_{i2}^{(1)m}\left(x_{1m}^{(1)}, +0\right) = \sigma_{i2}^{(1)m}\left(x_{1m}^{(1)}, -0\right) = 0 \quad \text{at} \quad x_{1m}^{(1)} \in \left(3\ell/4 - \ell_1, 3\ell/4 + \ell_1\right). \tag{10.2.14}$$

Problem V. To determine the SIF in the crack tips of each matrix layer at $x_{2m}^{(1)} = 0$, $-\ell_1 \le x_{1m} \le +\ell_1$ of the composite shown in Fig.10.2.5. In this case the conditions (10.2.14) are satisfied at $x_{1m}^{(1)} \in \left(-\ell_1, \ell_1\right)$.

Problem VI. To determine the SIF in the crack tips of two collinear cracks of each matrix layer at $x_{2m}^{(1)} = 0$, $-\ell/4 - \ell_1 \le x_{1m} \le -\ell/4 + \ell_1$ and $3\ell/4 - \ell_1 \le x_{1m} \le 3\ell/4 + \ell_1$ of the composite shown in Fig.10.2.6. In this case the conditions (10.2.14) will be satisfied at $x_{1m}^{(1)} \in \left(-\ell/4 - \ell_1, -\ell/4 + \ell_1\right)$ and $x_{1m}^{(1)} \in \left(3\ell/4 - \ell_1, 3\ell/4 + \ell_1\right)$.

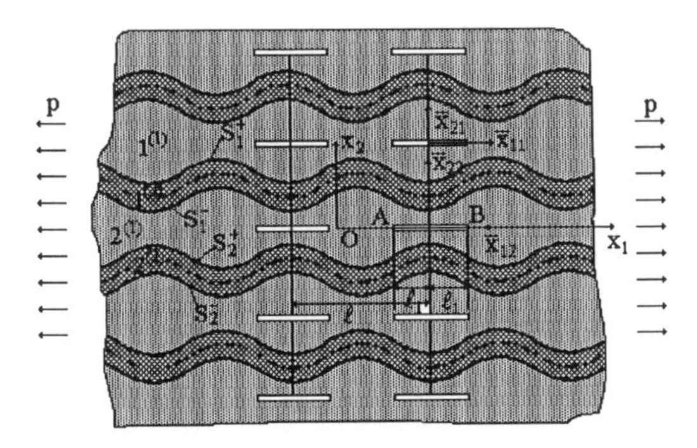

Fig. 10.2.6. Alternating anti-phase periodically curved layers with two collinear cracks in each matrix layer.

10.2.2. METHOD OF SOLUTION

As usual, we reduce the solutions of the problems to the solution of two successive problems: the stress-deformed state in the material without cracks; the stressed state in the materials with cracks on which the stresses determined in stage I act. The first stage was considered in Chapters 4 -7; we consider stage II; we determine the stress state when the external force is absent, and the matrix layers have the cracks with stresses on their edges determined in stage I. The following conditions are satisfied at the crack edges:

$$\sigma_{i2}^{(1)m}\left(x_{1m}^{(1)},+0\right)=\sigma_{i2}^{(1)m}\left(x_{1m}^{(1)},-0\right)=\sigma_{i}^{(1)m}\left(x_{1m}^{(1)}\right). \tag{10.2.15}$$

Taking into account the symmetries relative to planes over which the cracks are located, we can write the conditions (10.2.15) in the following form:

$$\sigma_{12}^{(1)m}\left(x_{1m}^{(1)},+0\right)=\sigma_{12}^{(1)m}\left(x_{1m}^{(1)},-0\right)=0\,,$$

$$\sigma_{22}^{(1)m}\left(x_{1m}^{(1)},+0\right)=\sigma_{22}^{(1)m}\left(x_{1m}^{(1)},-0\right)=\sigma^{(1)m}\left(x_{1m}^{(1)}\right). \tag{10.2.16}$$

We use the assumptions and conditions in Chapters 4 - 6, and assume ideal contact between matrix and filler.

At stage I the function $\sigma^{(1)m}\left(x_{1m}^{(1)}\right)$ has the following representation:

$$\sigma^{(1)m}(x_{1m}^{(1)})=\sum_{q=1}^{\infty}\varepsilon^{q}\sigma^{(1)m,q}\left(x_{1m}^{(1)}\right). \tag{10.2.17}$$

where the values of $\sigma^{(1)m,q}(x_{1m}^{(q)})$ are defined by the solution of boundary value problems corresponding to the q-th approximation.

This means that the values characterizing the stress-deformed state at stage II in each layer are power series of the small parameter ε, as in (4.1.15) . Now we consider the determination of the values of each approximation of the series (4.1.15).

At stage II the values of the zeroth approximation are identically equal to zero; we consider the first and subsequent approximations. We follow this determination procedure in *Problem III* (Fig.10.2.3) and *Problem IV* (Fig.10.2.4).

Problem III. The first and subsequent approximations reduce to the investigation of crack problems in an infinite body with two neighbouring parallel reinforcing layers as shown in Fig.10.2.7 . Within the layers $0^{(2)}$, $1^{(1)}$, $1^{(2)}$, and half-planes $0^{(1)}$ and $2^{(1)}$, the governing equations (10.2.1) are satisfied separately.

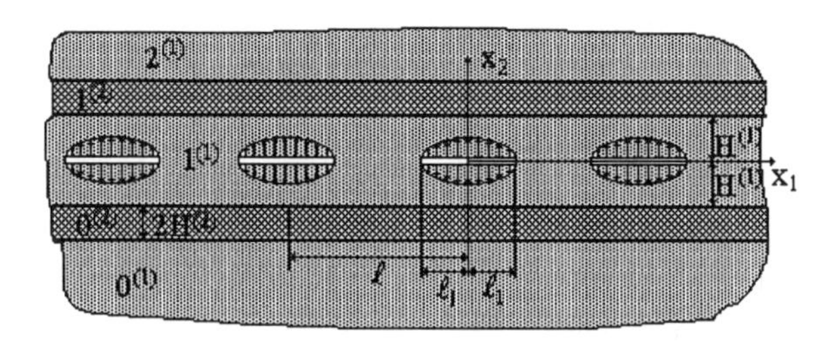

Fig.10.2.7. The structure of a body corresponding to Problem III at stage II.

Moreover, between these layers and half-planes the complete cohesion conditions are satisfied. We consider the determination of the values of the first approximation. With the middle plane of each layer we associate the corresponding Cartesian system of coordinates $O_m^{(k)}x_{1m}^{(k)}x_{2m}^{(k)}$ $(m = 0,1;\ k = 1,2)$, $x_{10}^{(2)} = x_{11}^{(1)} = x_{11}^{(2)} = x_1$, which are obtained from the Cartesian coordinate system Ox_1x_2 (Fig.10.2.7) by parallel transfer along the Ox_2 axis. Using these system of coordinates and the symmetry of the material structure with respect to $x_2 = 0$ plane, we write the complete cohesion conditions only between the layers $1^{(1)}$, $1^{(2)}$ and half plane $2^{(1)}$ and follow all the solution procedure on these. Thus, for the first approximation the complete cohesion conditions can be written as follows:

$$\sigma_{i2}^{(1)2,1}\left(x_1, H^{(2)}\right) = \sigma_{i2}^{(2)1,1}\left(x_1, H^{(2)}\right), \quad \sigma_{i2}^{(2)1,1}\left(x_1, -H^{(2)}\right) = \sigma_{i2}^{(1)1,1}\left(x_1, H^{(2)}\right),$$

$$u_i^{(1)2,1}\left(x_1, H^{(2)}\right) = u_i^{(2)1,1}\left(x_1, H^{(2)}\right), \quad u_i^{(2)1,1}\left(x_1, -H^{(2)}\right) = u_i^{(1)1,1}\left(x_1, H^{(2)}\right). \tag{10.2.18}$$

The following conditions must be satisfied:

$$\sigma_{12}^{(1)1,1}(x_1, +0) = \sigma_{12}^{(1)1,1}(x_1, -0) = 0 \quad \text{at } -\infty < x_1 < +\infty,$$

$$\sigma_{22}^{(1)1,1}(x_1, +0) = \sigma_{22}^{(1)1,1}(x_1, -0) \quad \text{at } -\infty < x_1 < +\infty,$$

$$\sigma_{22}^{(1)1,1}(x_1, +0) = \sigma_{22}^{(1)1,1}(x_1, -0) = \sigma^{(1)}(x_1) \text{ at } x_1 \in \bigcup_{n=-\infty}^{+\infty}(n\ell - \ell_1, n\ell + \ell_1),$$

$$u_1^{(1)1,1}(x_1, +0) = u_1^{(1)1,1}(x_1, -0) \quad \text{at } -\infty < x_1 < +\infty,$$

$$u_2^{(1)1,1}(x_1, +0) = u_2^{(1)1,1}(x_1, -0) \text{ at } x_1 \notin \bigcup_{n=-\infty}^{+\infty}(n\ell - \ell_1, n\ell + \ell_1). \tag{10.2.19}$$

We must also add to the conditions (10.2.18), (10.2.19) the following decay conditions

$$\sigma_{ij}^{(1)2,1} \to 0 \text{ at } x_{21}^{(2)} \to +\infty . \tag{10.2.20}$$

Thus, finding the first approximation is reduced to the solution of the system of equations (10.2.1) within the framework of the conditions (10.2.18) - (10.2.19). We use the Papkovich -Neuber representations (4.2.9), and apply the exponential Fourier transformation with respect to x_1 after which the unknown harmonic functions $\Phi_0^{(k)m,1}$, $\Phi_2^{(k)m,1}$ are taken as follows:

$$\Phi_0^{(1)2,1} = \int_{-\infty}^{+\infty} D_0(s) e^{-|s|\left(x_{21}^{(2)} - H^{(2)}\right)} e^{-isx_1} ds, \quad \Phi_2^{(1)2,1} = \int_{-\infty}^{+\infty} D_2(s) e^{-|s|\left(x_{21}^{(2)} - H^{(2)}\right)} e^{-isx_1} ds,$$

$$\Phi_0^{(2)1,1} = \int_{-\infty}^{+\infty} \left[D_{01}(s)\cosh\left(sx_{21}^{(2)}\right) + D_{02}(s)\sinh\left(sx_{21}^{(2)}\right)\right] e^{-isx_1} ds,$$

$$\Phi_2^{(2)1,1} = \int_{-\infty}^{+\infty} \left[D_{21}(s)\cosh\left(sx_{21}^{(2)}\right) + D_{22}(s)\sinh\left(sx_{21}^{(2)}\right)\right] e^{-isx_1} ds,$$

$$\Phi_0^{(1)1,1} = \int_{-\infty}^{+\infty} A_{01}(s)\cosh\left(sx_{21}^{(1)}\right) e^{-isx_1} + \begin{cases} \int_{-\infty}^{+\infty} A_0(s) e^{-|s|x_{21}^{(1)}} e^{-isx_1} ds & \text{at } x_{21}^{(1)} > 0, \\ \int_{-\infty}^{+\infty} B_0(s) e^{|s|x_{21}^{(1)}} e^{-isx_1} ds & \text{at } x_{21}^{(1)} < 0, \end{cases}$$

$$\Phi_2^{(1)1,1} = \int_{-\infty}^{+\infty} A_{01}(s)\cosh\left(sx_{21}^{(1)}\right) e^{-isx_1} + \begin{cases} \int_{-\infty}^{+\infty} A_2(s) e^{-|s|x_{21}^{(1)}} e^{-isx_1} ds & \text{at } x_{21}^{(1)} > 0, \\ \int_{-\infty}^{+\infty} B_2(s) e^{|s|x_{21}^{(1)}} e^{-isx_1} ds & \text{at } x_{21}^{(1)} < 0. \end{cases} \tag{10.2.21}$$

Expressing the first approximation values by functions (10.2.21) and performing some transformations we obtain from the first condition (10.2.19)

$$A_2(s) = \frac{2|s|}{1-\kappa} A_0(s), \quad B_2(s) = -\frac{2|s|}{1-\kappa} B_0(s), \quad \kappa = 3 - 4\nu^{(1)} \tag{10.2.22}$$

To satisfy of the fourth condition of (10.2.19) one has to define

$$B_0(s) = A_0(s), \quad B_2(s) = A_2(s). \tag{10.2.23}$$

Within the framework of (10.2.22), (10.2.23), the second condition of (10.2.19) is satisfied automatically.

Introducing the function

$$Q_1(x_1) = \frac{\partial u_2^{(1)1,1}(x_1,+0)}{\partial x_1} - \frac{\partial u_2^{(1)1,1}(x_1,-0)}{\partial x_1}, \tag{10.2.24}$$

applying the exponential Fourier transformation to the function (10.2.24) and using the last condition of (10.2.19) we obtain

$$\overline{Q}_1(s) = \frac{1}{2\pi}\int_{-\ell_1}^{\ell_1} Q_1(\eta_1)\left(\sum_{n=-\infty}^{+\infty} e^{isn\ell}\right) e^{is\eta_1} d\eta_1 . \tag{10.2.25}$$

Using (10.2.25) and doing some transformations we deduce from (10.2.24) that

$$A_0(s) = i\frac{1-\kappa}{(1+\kappa)2\pi}\int_{-\ell_1}^{+\ell_1} Q(\eta_1)\frac{1}{s|s|}\left(\sum_{n=-\infty}^{+\infty} e^{isn\ell}\right) e^{is\eta_1} d\eta_1, \tag{10.2.26}$$

where

$$Q(\eta_1) = G^{(1)} Q_1(\eta_1). \tag{10.2.27}$$

In (10.2.27) $G^{(1)}$ is a shear modulus of the matrix material.

Using (10.2.26), we define the expression of $\sigma_{22}^{(1)1,1}$ from Hooke's law (10.2.1) as

$$\sigma_{22}^{(1)1,1}(x_1,+0) = L_1(x_1) + L_2(x_1), \tag{10.2.28}$$

where

$$L_1(x_1) = \int_{-\infty}^{+\infty}\left[-s^2 A_0(s) + |s|(1-\kappa)A_2(s)\left(\frac{\lambda^{(1)}}{2G^{(1)}+1}\right)\right] e^{-isx_1} ds,$$

$$L_2(x_1) = \int_{-\infty}^{+\infty}\left[-s^2 A_{01}(s) + s(\kappa-1)A_{22}(s)\left(\frac{\lambda^{(1)}}{2G^{(1)}+1}\right)\right] e^{-isx_1} ds. \tag{10.2.29}$$

In (10.2.29) $\lambda^{(1)}$ is a Lamé constant of the matrix material.

Taking into account (10.2.26) and the first correlation (10.2.22) we may transform the expression of $L_1(x_1)$ in (10.2.29) into the following form:

$$L_1(x_1) = \frac{(2\Lambda-1)(\kappa-1)}{\pi(1+\kappa)} \int_{-\ell}^{+\ell} Q(\eta_1) \left[\int_0^{+\infty} \sum_{n=0}^{+\infty} \frac{\gamma_n}{2} \left[\sin(sn\ell + s(\eta_1 - x_1)) - \right. \right.$$

$$\left. \left. \sin(sn\ell - s(\eta_1 - x_1)) \right] ds \right] d\eta_1 ,$$

(10.2.30)

where $\Lambda = \nu^{(1)} / (2G^{(1)}) + 1, \gamma_0 = 1, \ \gamma_n = 2 \ \text{if} \ n \neq 0$.

Replacing the order of integration in (10.2.30) with respect to s and using the relation

$$\int_0^{+\infty} \sin(sn\ell \pm s(\eta_1 - x_1)) ds = \frac{1}{n\ell + (\eta_1 - x_1)} \tag{10.2.31}$$

we may rewrite the expression (10.2.30) as follows

$$L_1(x_1) = \frac{(2\Lambda-1)(\kappa-1)}{\pi(1+\kappa)} \int_{-\ell_1}^{+\ell_1} Q(\eta_1) \left\{ \frac{\pi}{\ell} \frac{1}{\frac{\pi}{\ell}(\eta_1 - x_1)} + \right.$$

$$\left. \frac{\pi}{\ell} \sum_{n=1}^{+\infty} \left(\frac{1}{\frac{\pi}{\ell}(\eta_1 - x_1) + n\pi} - \frac{1}{n\pi - \frac{\pi}{\ell}(\eta_1 - x_1)} \right) \right\} d\eta_1 . \tag{10.2.32}$$

Using the expansion

$$\cot(z) = \frac{1}{z} + \sum_{n=1}^{\infty} \frac{2z}{z^2 - n^2\pi^2} , \tag{10.2.33}$$

and introducing the new variables $x = x_1/\ell_1$, $\eta = \eta_1/\ell_1$ we may write the integral (10.2.32) as

$$L_1(x) = \frac{(2\Lambda-1)(\kappa-1)}{\pi(1+\kappa)} \frac{\pi\ell_1}{\ell} \int_{-1}^{+1} Q(\eta) \cot\left(\frac{\pi\ell_1}{\ell}(\eta - x) \right) d\eta . \tag{10.2.34}$$

Consider the second integral of (10.2.29). Note that this integral contains $A_{01}(s)$ and $A_{02}(s)$ which are defined from the contact conditions (10.2.18) by $A_0(s)$; we write

$$A_{01}(s) = R_{01}(\ell_1 s)A_0(s),\ A_{22}(s) = R_{22}(\ell_1 s)A_0(s). \tag{10.2.35}$$

The expressions for $R_{01}(\ell_1 s)$ and $R_{22}(\ell_1 s)$ are obtained from the contact conditions (10.2.18). Equations (10.2.29), (10.2.35) show that $A_{01}(s)$ is an even function , but $A_{22}(s)$ is an odd with respect to s. Taking this situation into account and introducing the notation

$$R(\ell_1 s) = -R_{01}(\ell_1 s) + (\kappa - 1)\Lambda R_{22}(\ell_1 s)/s \tag{10.2.36}$$

we may represent the integral $L_2(x_1)$ as

$$L_2(x_1) = \int_{-\infty}^{+\infty} R(\ell_1 s)s^2 A_0(s)\, e^{-isx_1} ds. \tag{10.2.37}$$

Substituting for $A_0(s)$ its expression (10.2.26) into (10.2.37) and performing some transformations we obtain

$$L_2(x_1) = \frac{(\kappa-1)}{\pi(\kappa+1)} \int_{-\ell_1}^{+\ell_1} Q(\eta_1)\left[\int_0^{+\infty} R(s\ell_1)\left(\sum_{n=0}^{+\infty} \gamma_n \cos(ns\ell)\right)\sin(s(\eta_1 - x_1))ds\right]d\eta_1. \tag{10.2.38}$$

As for convergence of the generalized functions, the series in (10.2.38) converges and it may be written as

$$\sum_{n=0}^{+\infty} \gamma_n \cos(ns\ell) = 2\pi \sum_{k=-\infty}^{+\infty} \delta(s\ell - 2\pi k), \tag{10.2.39}$$

where $\delta(s\ell - 2\pi k)$ is a generalized function.

Using (10.2.39) and introducing the new variables $\eta = \eta_1/\ell_1$, $x = x_1/\ell_1$ we obtain from (10.2.38) that

$$L_2(x) = \frac{2(\kappa-1)}{1+\kappa}\frac{\ell_1}{\ell} \int_{-1}^{+1} Q(\eta)\left[\sum_{k=-\infty}^{+\infty} R\left(2k\pi\frac{\ell_1}{\ell}\right)\sin\left(2k\pi\frac{\ell_1}{\ell}(\eta - x)\right)\right]d\eta. \tag{10.2.40}$$

Thus, substituting for $L_1(x)$, $L_2(x)$ their expression (10.2.34), (10.2.40) into (10.2.28) and satisfying the third condition of (10.2.19) we obtain the following singular integral equation, with Hilbert's kernel with respect to unknown function $Q(\eta)$:

$$\frac{(2\Lambda-1)(\kappa-1)}{\kappa+1}\frac{\ell_1}{\ell}\int_{-1}^{+1}Q(\eta)\cot\left(\pi\frac{\ell_1}{\ell}(\eta-x)\right)d\eta+$$

$$\frac{2(k-1)}{1+\kappa}\frac{\ell_1}{\ell}\int_{-1}^{+1}Q(\eta)\left[\sum_{k=-\infty}^{+\infty}R\left(2\pi k\frac{\ell_1}{\ell}\right)\sin\left(2\pi k\frac{\ell_1}{\ell}(\eta-x)\right)\right]d\eta=-\sigma'^{1}(x). \quad (10.2.41)$$

As $\ell_1/\ell \to 0$ in the equation (10.2.41) the Hilbert kernel changes to a Cauchy kernel and the series in (10.2.41) converges to the Fredholm's kernel (i.e. to a Riemann sum). In other words, for $\ell_1/\ell \to 0$ the equation (10.2.41) transforms to the following:

$$\frac{(2\Lambda-1)(\kappa-1)}{\kappa+1}\int_{-1}^{+1}\frac{Q(\eta)}{\eta-x}dx+\frac{2(\kappa-1)}{\kappa+1}\int_{-1}^{+1}Q(\eta)R^*(\eta,x)d\eta=\sigma'^{1}(x), \quad (10.2.42)$$

where

$$R^*(\eta,x)=\lim_{\ell_1/\ell\to 0}\left[\sum_{k=-\infty}^{+\infty}R(2\pi k\frac{\ell_1}{\ell})\sin\left(2\pi k\frac{\ell_1}{\ell}(\eta-x)\right)\right]d\eta=-\sigma'^{1}(x). \quad (10.2.43)$$

The equation (10.2.42) corresponds to the case in which the matrix layer $1^{(1)}$ (Fig.10.2.7) has a single isolated crack, in other words, the equation (10.2.42) corresponds to the *problem I* (Fig. 10.2.1).

Note that using the expansion

$$x\cot(x)=1-\sum_{n=1}^{\infty}\frac{2^{2n}B_n}{(2n)!}x^{2n} \quad \text{at} \quad |x|<\pi, \quad (10.2.44)$$

where B_n are Bernoulli's numbers, we can reduce the equation (10.2.41) to the following :

$$\int_{-1}^{+1}\frac{Q(\eta)}{\eta-x}d\eta+\int_{-1}^{+1}Q(\eta)R_1(x,\eta)d\eta=-\sigma_1'^{1}(x), \quad |x|\le 1, \quad (10.2.45)$$

where

$$R_1(x,\eta)=\frac{2}{2\Lambda-1}\sum_{k=-\infty}^{+\infty}R\left(2\pi k\frac{\ell_1}{\ell}\right)\sin\left(2k\pi\frac{\ell_1}{\ell}(\eta-x)\right)-$$

$$\sum_{k=1}^{\infty} \frac{2^{2k} B_k}{(2k)!} (\eta - x)^{2k-1}, \quad \sigma_1^{;1}(x) = \sigma^{,1}(x) \frac{\ell}{\ell_1} \frac{(1+\kappa)}{(\kappa - 1)(2\Lambda - 1)}. \qquad (10.2.46)$$

Now we add to these equations the uniqueness condition of the displacement of the crack edge point:

$$\int_{-1}^{+1} Q(t)dt = 0 . \qquad (10.2.47)$$

Thus, the first approximation at stage II is reduced to solution of the singular integral equations (10.2.45) - (10.2.47) ((10.2.42), (10.2.43), (10.2.47)) for *problem III* (*problem I*).

We can obtain the second and subsequent approximations for Problems III and I similarly.

Problem IV. Consider the first approximation at stage II.

We consider a composite with ideal layout of (uncurved) layers of the same material and with the same thickness for the composite shown in Fig.10.2.4. The matrix layers have cracks (Fig.10.2.8) with normal stresses $\sigma^{(1)m}(x_{1m}^{(1)})$ on their edges in the direction of the Ox_2 axis.

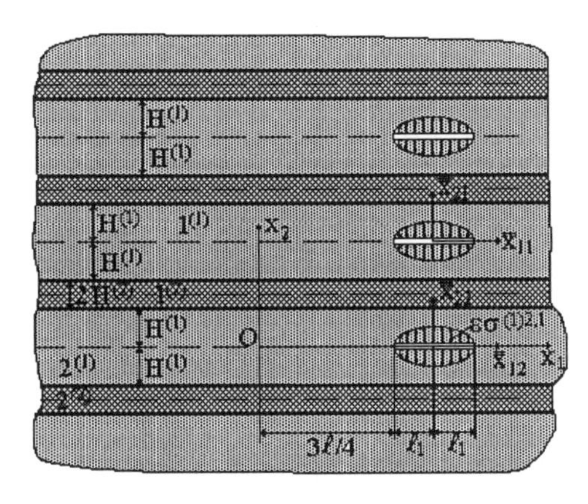

Fig.10.2.8. The structure of a composite corresponding to the Problem IV at stage II.

We associate the corresponding Cartesian system of coordinates $O_m^{(k)} x_{1m}^{(k)} x_{2m}^{(k)}$ obtained from Ox_1x_2 coordinate system by parallel transfer along the Ox_2 axis with the middle plane of every $m^{(k)}$ th strip. Furthermore, we associate a local Cartesian system of coordinates $\overline{O}_m^{(1)} \overline{x}_{1m}^{(1)} \overline{x}_{2m}^{(1)}$, which are obtained from one another by parallel transport along the Ox_2 axis (Fig.10.2.8), i.e. $\overline{x}_{11}^{(1)} = \overline{x}_{12}^{(1)} = ... = \overline{x}_1$, with the centre of each crack. From the formulation of the Problem IV it follows that

$$\sigma^{(1)m}(\bar{x}_1) = -\sigma^{(1)m_1}(\bar{x}_1), \quad m_1 = m-1. \tag{10.2.48}$$

Below, for the matrix strip we shall use the coordinate system $\overline{O}_m^{(1)}\bar{x}_{1m}^{(1)}\bar{x}_{2m}^{(1)}$ associated with the cracks, and omit the over-bars. Moreover, instead of $x_{1m}^{(2)}$ we will use $x_1^{(1)}$, assuming that $x_{1m}^{(2)} = x_1^{(1)} + 3\ell/4$.

Taking (10.2.48) and the periodicity of the structure shown in Fig.10.2.4 into account, we single out and deal only with four strips, i.e. $1^{(1)}$, $1^{(2)}$, $2^{(1)}$, $2^{(2)}$ (Fig.10.2.4). The complete cohesion conditions between these strips can be written as follows:

$$\sigma_{i2}^{(1)2,1}\left(x_1,-H^{(1)}\right) = \sigma_{i2}^{(2)2,1}\left(x_1,H^{(2)}\right), \quad u_i^{(1)2,1}\left(x_1,-H^{(1)}\right) = u_i^{(2)2,1}\left(x_1,H^{(2)}\right),$$

$$\sigma_{i2}^{(2)1,1}\left(x_1,-H^{(2)}\right) = \sigma_{i2}^{(1)2,1}\left(x_1,+H^{(1)}\right), \quad u_i^{(2)1,1}\left(x_1,-H^{(2)}\right) = u_i^{(1)2,1}\left(x_1,+H^{(1)}\right),$$

$$\sigma_{i2}^{(2)1,1}\left(x_1,+H^{(2)}\right) = \sigma_{i2}^{(1)1,1}\left(x_1,-H^{(1)}\right), \quad u_i^{(2)1,1}\left(x_1,+H^{(2)}\right) = u_i^{(1)1,1}\left(x_1,-H^{(1)}\right),$$

$$\sigma_{i2}^{(1)1,1}\left(x_1,+H^{(1)}\right) = \sigma_{i2}^{(2)2,1}\left(x_1,-H^{(2)}\right), \quad u_i^{(1)1,1}\left(x_1,+H^{(1)}\right) = u_i^{(2)2,1}\left(x_1,-H^{(2)}\right). \tag{10.2.49}$$

Moreover the following will be added to these conditions:

$$u_1^{(1)j,1}(x_1,+0) = u_1^{(1)j,1}(x_1,-0) \quad \text{at} \quad -\infty < x_1 < +\infty,$$

$$u_2^{(1)j,1}(x_1,+0) = u_2^{(1)j,1}(x_1,-0) \quad \text{at} \quad |x_1| > \ell_1,$$

$$\sigma_{12}^{(1)j,1}(x_1,+0) = \sigma_{12}^{(1)j,1}(x_1,-0) = 0 \quad \text{at} \quad -\infty < x_1 < +\infty,$$

$$\sigma_{22}^{(1)j,1}(x_1,+0) = \sigma_{22}^{(1)j,1}(x_1,-0) \quad \text{at} \quad -\infty < x_1 < +\infty,$$

$$\sigma_{22}^{(1)j,1}(x_1,+0) = \sigma_{22}^{(1)j,1}(x_1,-0) = \sigma^{(1)j,1}(x_1) \quad \text{at} \quad |x_1| < \ell_1, \quad j=1,2. \tag{10.2.50}$$

Thus, the first approximation is reduced to the solution to the system of equations (10.2.1) within the framework of the conditions (10.2.49), (10.2.50). As before, we use the Papkovich-Neuber representations (4.2.9) and apply the exponential Fourier transformation with respect to x_1, after which the unknown harmonic functions $\Phi_0^{(k)m,1}$, $\Phi_2^{(k)m,1}$ are taken as follows:

$$\Phi_n^{(2)m,1} = \int_{-\infty}^{+\infty} \left[C_{n1}^{(2)m,1}(s)\cosh(sx_{2m}^{(2)}) + C_{n2}^{(2)m,1}(s)\sinh(sx_{2m}^{(2)}) \right] e^{-isx_1} ds ,$$

$$\Phi_n^{(1)m,1} = \overline{\Phi}_n^{(1)m,1} + \overline{\overline{\Phi}}_n^{(1)m,1}, \qquad n = 0,2 ,$$

$$\overline{\Phi}_n^{(1)m,1} = \int_{-\infty}^{+\infty} \left[C_{n1}^{(1)m,1}(s)\cosh(sx_{2m}^{(1)}) + C_{n2}^{(1)m,1}(s)\sinh(sx_{2m}^{(1)}) \right] e^{-isx_1} ds ,$$

$$\overline{\overline{\Phi}}_n^{(1)m,1} = \begin{cases} \int_{-\infty}^{+\infty} D_{n1}^{(1)m,1}(s) e^{-|s|x_{2m}^{(1)}} e^{-isx_1} ds & \text{at} \quad x_{2m}^{(1)} > 0 \\ \int_{-\infty}^{+\infty} D_{n1}^{(1)m,1}(s) e^{|s|x_{2m}^{(1)}} e^{-isx_1} ds & \text{at} \quad x_{2m}^{(1)} < 0 \end{cases} \qquad (10.2.51)$$

Introducing the functions

$$Q^{(1)m}(x_1) = \frac{\partial}{\partial x_1} \left[u_2^{(1)m,1}(x_1,+0) - u_2^{(1)m,1}(x_1,-0) \right], \qquad (10.2.52)$$

proceeding as in *problem III* , using the contact conditions (10.2.49) and conditions (10.2.50), we express the unknowns entered into (10.2.51) through $Q^{(1)1}(x_1)$. Then, taking into account that

$$Q^{(1)1}(x_1) = -Q^{(1)2}(x_1) = Q(x_1), \quad \sigma^{(1)1,1}(x_1) = -\sigma^{(1)2,1}(x_1) = \sigma(x_1), \qquad (10.2.53)$$

and applying certain transformations, we obtain a singular integral equation:

$$\int_{-1}^{+1} \frac{Q(t)}{t-x} dt + \int_{-1}^{+1} Q(t) r(x,t) dt = 2\pi\sigma(x), \quad x = x_1/\ell_1 , \ |x \le 1| . \qquad (10.2.54)$$

We do not show the expression for $r(x,t)$ explicitly. As before, we add the uniqueness condition (10.2.47) for the displacement of the crack-edges points.

We can obtain subsequent approximation similarly. This approach may be used for *problem V* and other problems also.

10.2.3. NUMERICAL RESULTS

The integral equations are solved numerically with the use of the algorithm proposed in [78, 87, 111]. According to [78, 87, 111] $Q(x)$ is sought in the form

$$Q(x) = \frac{g(x)}{\sqrt{1-x^2}}, \tag{10.2.55}$$

where $g(x)$ is a bounded function at [-1,+1].

If we substitute $x = \cos\theta$ $(t = \cos\tau)$ and the expression (10.2.55) into equation (10.2.54) where the argument θ changes within limits $0 \le \theta \le \pi$, then equations (10.2.54) take the form:

$$\int_0^\pi \frac{g(\tau)d\tau}{\cos\tau - \cos\theta} + \int_0^\pi r(\cos\theta, \cos\tau)g(\tau)d\tau = 2\pi\sigma(\theta). \tag{10.2.56}$$

Substituting the function $g(\tau)$ by Lagrange's interpolation polynomial on the Chebyshev nodes, we obtain

$$x_m = \cos\theta_m, \ \theta_m = \frac{2m-1}{2n}\pi, \quad m = 1,2,\dots,n,$$

$$L_n[g,x] = \frac{1}{n}\sum_{k=1}^{n}(-1)^{k+1} g_k \frac{\cos n\theta \sin\theta_k}{\cos\theta - \cos\theta_k}, \quad x = \cos\theta, \quad g_k = g(x_k). \tag{10.2.57}$$

Equation (10.2.56) is approximated by

$$\sum_{k=1}^{n} a_{mk} g_k = 2\pi\sigma_m, \ m = 1,2,\dots,n, \tag{10.2.58}$$

where

$$a_{mk} = \frac{1}{2n}\left[\frac{1}{\sin\theta_m}\cot\frac{\theta_m + (-1)^{|m-k|}\theta_k}{2} + r(\cos\theta_m, \cos\theta_k)\right], \quad \sigma_m = \sigma(x_m). \tag{10.2.59}$$

Note the following conditions. In solving the equations (10.2.58), it is necessary to calculate the $r(\cos\theta_m, \cos\theta_k)$ values through sine and cosine of Fourier transforms. The numerical algorithm is based on that proposed in [122].

Now consider numerical results obtained from the first-order approximation. Assume that $E^{(2)}/E^{(1)} = 50$, $\nu^{(1)} = \nu^{(2)} = 0.3$.

Define

$$\overline{K}_I = 2K_I \Big/ \left(\sigma_{11}^{(1),0}\sqrt{\pi\ell_1}\right), \tag{10.2.60}$$

Table 10.2. The values of $\overline{K}_{IA}$ (Fig.10.2.2) (Problem II).

$2\pi H^{(1)}/\ell$	$2\pi H^{(2)}/\ell$	$2\pi\ell_1/\ell$			
		$\pi/8$	$\pi/6$	$\pi/4$	$\pi/2$
	0.1	109.06	105.56	97.32	55.21
0.1	0.2	177.97	169.32	159.81	75.47
	0.3	203.41	189.60	175.60	65.30
	0.1	108.47	10.59	98.21	56.62
0.2	0.2	180.93	172.92	160.77	82.03
	0.3	210.39	194.71	173.70	70.23
	0.1	73.48	73.49	72.20	55.87
1.0	0.2	115.58	114.54	110.02	75.90
	0.3	117.56	114.52	106.33	64.55

Table 10.3. The values of $\overline{K}_{IB}$ (Fig.10.2.2) (Problem II).

$2\pi H^{(1)}/\ell$	$2\pi H^{(2)}/\ell$	$2\pi\ell_1/\ell$			
		$\pi/8$	$\pi/6$	$\pi/4$	$\pi/2$
	0.1	108.68	104.49	95.14	55.24
0.1	0.2	173.28	157.23	138.02	92.72
	0.3	192.26	163.25	136.09	99.24
	0.1	108.52	106.05	98.32	56.62
0.2	0.2	178.17	165.77	147.43	94.05
	0.3	204.27	180.19	150.74	97.87
	0.1	73.58	73.77	72.72	52.43
1.0	0.2	115.90	114.54	110.04	76.51
	0.3	117.53	114.42	106.31	67.48

Table 10.4. The values of $\overline{K}_I$ (Fig.10.2.3) (Problem III).

$2\pi H^{(1)}/\ell$	$2\pi H^{(2)}/\ell$	$2\pi\ell_1/\ell$				
		$\pi/8$	$\pi/6$	$\pi/4$	$\pi/2\ \times 0.95$	$\pi/2 \times 0.951$
	0.1	81.9	86.37	88.28	0.06	-0.003
0.1	0.2	90.7	91.21	89.90	0.03	-0.013
	1.0	71.9	71.20	68.10	0.03	-0.012
	0.1	115.0	119.1	124.0	0.01	-0.041
0.2	0.2	133.1	131.1	130.1	0.03	-0.056
	1.0	112.6	109.4	101.4	0.02	-0.015

where εK_I is the SIF. The stresses σ_{22} near the crack tips are

$$\sigma_{22}^{(1)2}(x_1,0) \approx \frac{\varepsilon K_{IA}}{\sqrt{2(x_1-\ell_1)}} \quad \text{at} \quad 0 < x_1 - \ell_1 << 1,$$

$$\sigma_{22}^{(1)2}(x_1,0) \approx \frac{\varepsilon K_{IA}}{\sqrt{2(\ell_1 - x_1)}} \quad \text{at } 0 < \ell_1 - x_1 << 1 \tag{10.2.61}$$

These are tabulated in Tables 10.2-10.8 for different values of the problem parameters.

Table 10.5. The values of $\overline{K}_I$ (Fig.10.2.4) (Problem IV).

$\frac{2\pi H^{(2)}}{\ell}$	$\eta^{(2)}$	$2\pi\ell_1/\ell$			
		$\pi/8$	$\pi/6$	$\pi/4$	$\pi/2$
0.1	0.1	68.17	74.91	95.45	105.91
	0.2	89.82	114.58	232.83	---
	0.4	84.98	107.56	173.50	---
0.2	0.1	58.65	58.95	59.67	62.18
	0.2	114.95	121.16	139.40	298.30
	0.4	117.21	125.12	148.12	245.60
0.3	0.1	26.72	26.35	25.25	19.53
	0.2	93.27	94.03	96.15	109.90
	0.4	141.33	141.08	146.04	166.90

Table 10.6. The values of $\overline{K}_I$ (Fig.10.2.5) (Problem V).

$\frac{H^{(2)}}{L}$	$\eta^{(2)}$	ℓ_1/L			
		0.1	0.3	0.5	0.8
0.1	0.1	154.07	158.24	166.15	187.23
	0.2	338.94	383.94	480.58	914.95
	0.4	520.84	570.71	695.10	915.32
	0.5	513.44	549.81	647.61	736.63
0.2	0.1	39.26	39.08	38.74	38.04
	0.2	101.54	102.05	103.30	107.34
	0.4	236.87	217.07	207.64	196.65
	0.5	295.22	249.82	238.34	202.49

Table 10.7. The values of $\overline{K}_{IA}$ (Fig.10.2.6) (Problem VI).

$\frac{2\pi H^{(2)}}{\ell}$	$\eta^{(2)}$	$2\pi\ell_1/\ell$			
		$\pi/8$	$\pi/6$	$\pi/4$	$\pi/2$
0.1	0.1	66.18	70.64	82.35	355.80
	0.2	73.92	80.00	91.82	392.60
	0.4	70.79	75.18	82.45	176.79
0.2	0.1	58.41	58.46	58.29	51.79
	0.2	114.46	120.05	135.28	302.76
	0.4	126.50	132.08	146.03	257.16
0.3	0.1	26.60	26.13	24.76	17.83
	0.2	93.10	93.63	94.71	89.91
	0.4	149.61	151.10	157.26	162.04

It follows from solution operation that

$$\begin{Bmatrix} K_{IA} \\ K_{IB} \end{Bmatrix} = \begin{Bmatrix} K_{IA} \\ K_{IB} \end{Bmatrix} \left(H^{(1)}/\ell, H^{(2)}/\ell, \ell_1/\ell, \eta^{(2)}, E^{(2)}/E^{(1)} \right) \tag{10.2.62}$$

This is confirmed by the data shown in Tables 10.2 - 10.8, which indicate the change of the $\overline{K}_I$ (10.2.60) versus all the parameters of (10.2.62) (except for $E^{(2)}/E^{(1)}$). Note that in these tables, dashes indicate the $\overline{K}_I$ value is negative

Table 10.8. The values of $\overline{K}_{IB}$ (Fig.10.2.6) (Problem VI).

$2\pi H^{(2)}/\ell$	$\eta^{(2)}$	$2\pi\ell_1/\ell$			
		$\pi/8$	$\pi/6$	$\pi/4$	$\pi/2$
0.1	0.1	66.55	71.64	84.83	274.80
	0.2	75.02	83.33	100.62	232.06
	0.4	71.71	77.93	89.31	121.37
0.2	0.1	58.48	58.62	58.86	56.23
	0.2	114.81	120.55	136.18	272.83
	0.4	126.88	133.04	147.85	209.56
0.3	0.1	26.61	26.13	24.78	17.67
	0.2	93.21	93.91	95.72	100.50
	0.4	148.97	149.38	153.40	173.68

The numerical results show that the relationship between $\overline{K}_I$ and the problem parameters is a complicated. In spite of this fact, these results allow us to draw a conclusion about the definition of the cases where the mutual influence between cracks can be neglected. Moreover, these results show the influence of a crack length to the values of $\overline{K}_I$. In particular, Table 10.6 shows that in some cases of increase of crack length, i.e. increase of ℓ_1/L, the values of $\overline{K}_I$ grow monotonously. It is evident that this situation can cause the instantaneous fracture of the composite material.

10.3. Fracture in Compression

One of the major mechanisms of the fracture of unidirectional composites under uniaxial compression along the reinforcing elements is structural instability. This fracture mechanism is suggested in [77, 132], is acknowledged in almost all monographs on the fracture of the composite materials, for example in [67, 75, 81] and confirmed by experimental data in [77, 88, 89] and others; the instability occurs at a critical external force. A review of previous investigations is given in [54, 105, 106]; see also [102]. Currently in the piece-wise homogeneous body model two approaches are used: approximate theories governing the interaction between components [for example, as in [132, 133, 135]; and three-dimensional linearized theory (TDLT) discussed in Chapter 8. The TDLT investigations use the Euler approach; stability problems reduce to eigenvalue problems.

Euler's approach to stability loss problems is valid for static loading and under certain restrictions on the external forces. Euler's approach can be applied to cases where the reinforcing fibres are rectilinear and parallel, but not when there is some initial curvature. In this section we solve concrete problems for a unidirectional composite with periodical curved structure in compression. The fracture (stability loss) criterion is the compressive force for which the initial curvature of the layer starts to increase and grows indefinitely.

10.3.1. FORMULATION AND METHOD OF SOLUTION

We again consider an infinite elastic body, reinforced by an arbitrary number of non-intersecting slightly curved filler layers. Using the assumptions and notation employed in Chapter 8 we associate the corresponding Lagrangian coordinates $O_m^{(k)}x_{1m}^{(k)}x_{2m}^{(k)}x_{3m}^{(k)}$, $(k = 1,2; m = 1,2,...)$ which in their natural state coincide with Cartesian coordinates and are obtained from $Ox_1x_2x_3$ by parallel transfer along Ox_2 axis, with the middle surface of each layer of the filler and matrix (Fig.5.1.1). As in section 8.1 we assume that the reinforcing layers are located in planes $O_m^{(2)}x_{1m}^{(2)}x_{3m}^{(2)}$, the thickness of each filler layer is constant, and that the matrix and filler materials are homogeneous, anisotropic and non-aging linearly viscoelastic. We investigate the stress under compression at infinity by uniformly distributed normal forces of intensity p_1 (p_3) in the direction of the Ox_1 (Ox_3) axis.

We assume that the equilibrium equations (8.1.1), mechanical relations (8.1.2) and geometrical relations (8.1.3) hold for each layer, and that the complete cohesion conditions (8.1.4) are satisfied. Supposing that the middle surface of the m-th filler layer is given by (5.1.3), performing operations described in section 5.2 and representing the sought values in the series form (5.3.1) we obtain a closed system of equations and contact conditions for each approximation. In this case owing to the linearity of the mechanical relations, (5.3.1) will be satisfied for each approximation separately. The remaining relations for the q-th approximation contain the values of all previous approximations, and these relations are given by formulae (8.1.5)-(8.1.12). In section 8.1 we proved that the relations related to the first and subsequent approximations are those of TDLT. *This means that we can investigate the stability loss (fracture) in a unidirected composite material with curved layers or fibers in compression.* First we find the stress state in the composite in compression by using the approach described in Chapter 8, then we select the failure criteria. According to [77, 92, 93, 102, 132, 134, 135], we assume that stability loss corresponds to the limit of the load-carrying capacity; displacements increase significantly or become infinite for this finite limiting external compressive load. We restrict our attention to the zeroth and first approximations.

10.3.2. CO-PHASE PERIODICALLY CURVED LAYERS

We consider the problem investigated in subsection 8.2.1, i.e., the determination of the stress in composite material with co-phase periodically curved layers (Fig.4.2.1) in the case where $p < 0$. We investigate the development of the initial imperfection curving of the filler layers with increasing $|p|$. As in subsection 8.2.1, we obtain a system of non-homogeneous algebraic equations for the unknown constants $D_1^{(k)}$ and $D_2^{(k)}$ $(k = 1,2)$, in formulae (8.2.18). For the composite shown in Fig.10.3.1,

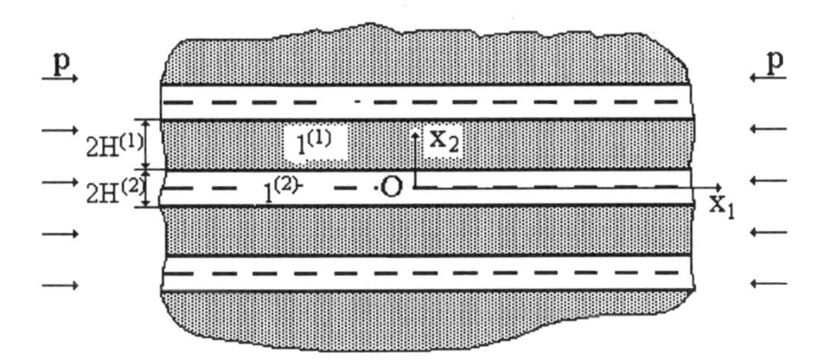

Fig.10.3.1. Ideal layout of the filler layers in the structure of the composite material under compression

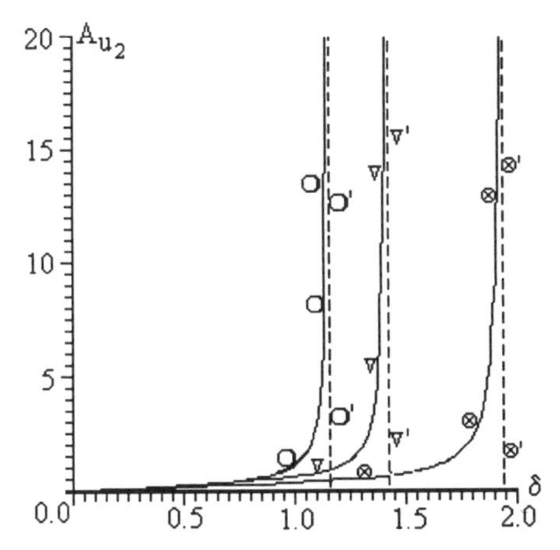

Fig.10.3.2. The relation between A_{u_2} and δ in the case where $\chi = 0.1$, $\eta^{(2)} = 0.5$, $\nu^{(1)} = \nu^{(2)} = 0.3$ with various $E^{(2)}/E^{(1)}$: ⊗ - $E^{(2)}/E^{(1)}$ =100, ∇ -- $E^{(2)}/E^{(1)}$ =150, ○ - $E^{(2)}/E^{(1)}$ =200; ⊗', ○', ∇' the critical values of δ obtained in the framework of Euler approach for $E^{(2)}/E^{(1)}$ =100,150 and 200 respectively.

in the framework of the TDLT of stability [101] under employing Euler approach, the unknown perturbations of the displacements are also presented as (8.2.6); i. e. the function $f_2^{(k),1}(x_2^{(k)})$ is selected as in equation (8.2.18). Furthermore, for the unknown constants $D_1^{(k)}$ and $D_2^{(k)}$ $(k = 1,2)$ the homogeneous algebraic equations are obtained from the contact conditions which coincide with (8.2.3) under $f = 0$, $df/dx_1 = 0$. Thus, a characteristic equation (determinant) is obtained for the determination of the critical value of the compressive force p for the composite in Fig. 10.3.1.

However, in the framework of the approach proposed in the previous subsection the equations obtained for the unknown constants $D_1^{(k)}$ and $D_2^{(k)}$ $(k = 1,2)$ are inhomogeneous equations.

Moreover, the determinant consisting of the coefficients of the unknown constants $D_1^{(k)}$, $D_2^{(k)}$ coincide with that obtained for the composite in Fig.10.3.1. Consequently, if the fracture criterion is

$$\max_{x_1 \in (-\infty,+\infty)} \left| u_2^{(2)}(x_1,0) \right| \to +\infty \quad \text{under } p \to p^* \tag{10.3.1}$$

and the values of p^* are taken as the values of the external forces for which fracture in compression occurs, then p^* must coincide with that obtained for the composite in Fig.10.3.1in the framework of Euler approach. To explain this situation, consider some numerical results.

Introduce the parameter

$$\delta = \frac{p \times 10^2 \left(1-\left(\nu^{(1)}\right)^2\right)}{E^{(1)}\left(1-\left(\nu^{(2)}\right)^2\right)\left(\eta^{(1)}+\eta^{(2)}E^{(2)}\left(1-\left(\nu^{(1)}\right)^2\right)\left(1-\left(\nu^{(2)}\right)^2\right)^{-1}\right)}$$

$$= 10^2 \times \varepsilon_{11}^{(1),0} = 10^2 \times \varepsilon_{11}^{(2),0} \tag{10.3.2}$$

and investigate the dependence between

$$A_{u_2} = \max_{0 \le x_1 \le \ell} \left| u_2(x_1,0) \right| / \ell \tag{10.3.3}$$

and the parameter δ. Fig. 10.3.2 shows this dependence for fixed problem parameters and various $E^{(2)}/E^{(1)}$; the critical values of external forces obtained from the equation (10.3.1) and the zeroth and first approximations coincide with those obtained from the Euler approach. Other investigations (not given here) show that the second and subsequent approximations do not change the critical values; they merely improve only the accuracy of the stress-strain distributions in the precritical state. Other results on the critical values of δ (denote it as $\delta_{cr.}$) are given in Table 10. 9.

These results show that compressive force p corresponding to fracture of the composite shown in Fig.4.2.1 depend significantly on the parameters $E^{(2)}/E^{(1)}$, $\eta^{(2)}$ and $\chi = 2\pi H^{(2)} / \ell$ (where ℓ is the length of the period of the curving form); only χ characterizes the curvature of the reinforcing layers. Such a parameter is used under investigation of stability loss of ideal structure under compression for which ℓ in expression of χ is the length of the period of the stability loss form of the layers and the value of χ is determined from the relation between p and χ. This relation is constructed from the solution to the corresponding characteristic equation. If this relation is non-monotonic and has a minimum for $\chi \neq 0$, then it is assumed that the fracture (internal or structural stability loss) occurs, and the values of p and χ corresponding to this minimum are taken as their critical values. However, for curved structures the value of the parameter χ is given in advance as a structural parameter of composites. This raises the possibility for *optimal* selection of the parameter χ.

Table 10. 9. The values of $\delta_{cr.}$ obtained for the composite shown in Fig.4.2.1 under $\nu^{(1)} = \nu^{(2)} = 0.3$.

$E^{(2)}/E^{(1)}$	χ	$\eta^{(2)}$			
		0.02	0.10	0.20	0.50
100	0.05	9.620	3.678	2.205	1.468
	0.10	5.845	3.698	2.410	1.703
	0.20	4.276	3.965	3.208	2.606
	0.30	4.849	4.807	4.457	3.990
150	0.05	6.694	2.521	1.509	1.008
	0.10	4.064	2.609	1.727	1.247
	0.20	3.291	3.085	2.572	2.165
	0.30	4.149	4.119	3.885	3.570
200	0.05	5.136	1.928	1.156	0.778
	0.10	3.151	2.052	1.381	1.019
	0.20	2.791	2.638	2.251	1.943
	0.30	3.794	3.773	3.597	3.359

10.3.3. A SINGLE PERIODICALLY CURVED LAYER

Consider a composite consisting of alternating isotropic, homogeneous layers of two materials, and assume that only one filler layer has initial periodical curving (Fig.10.3.3). All other layers, are ideal. Suppose that the curving of the filler layer $0^{(2)}$ depends only on x_1, and the uniformly distributed normal compressive forces are applied at infinity with the intensity p in the direction of the Ox_1 axis. We investigate fracture under the criterion (10.3.1).

Using the geometrical *non-linear* statement, the fracture criterion (10.3.1), and the zeroth and first approximations we investigate the plane-strain state. Find the critical value of compressive force p under which the fracture occurs. The corresponding linear problem was investigated in subsection 4.4.3.

Because of the symmetry of the problem, the zeroth approximation may be written as follows:

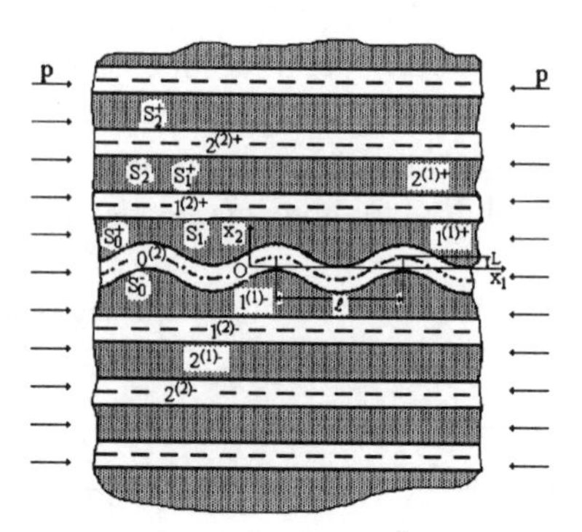

Fig.10.3.3. Structure of the composite material with a single periodically curved layer.

$$\sigma_{ij}^{(k)m,0} = \sigma_{ij}^{(k),0}, \quad u_1^{(k)m,0} = u_1^{(k),0},$$

$$u_2^{(k)m,0} = -\frac{\nu^{(\underline{k})}\left(1+\nu^{(\underline{k})}\right)\sigma_{11}^{(\underline{k}),0}}{E^{(\underline{k})}} x_{2\underline{m}}^{(\underline{k})} + C^{(\underline{k})\underline{m}}, \quad C^{(k)m} = \text{const}, \quad (10.3.4)$$

where $\sigma_{ij}^{(k),0}$, $u_i^{(k),0}$ are given by formulae (4.2.5)-(4.2.7).

For the first approximation we obtain the following contact conditions on the surface S_0^+ (Fig.10.3.3):

$$\left(1+\Gamma_i^{(1)}\right)\sigma_{i2}^{(1)1,1}\Big|_{\left(t_1,-H^{(1)}\right)} - \left(1+\Gamma_i^{(2)}\right)\sigma_{i2}^{(2)0,1}\Big|_{\left(t_1,H^{(2)}\right)} =$$

$$2\pi\cos(\alpha t_1)\left(\left(1+\Gamma_1^{(1)}\right)\sigma_{11}^{(1),0} - \left(1+\Gamma_1^{(2)}\right)\sigma_{11}^{(2),0}\right)\delta_i^1,$$

$$u_i^{(1)1,1}\Big|_{\left(t_1,-H^{(1)}\right)} - u_i^{(2)0,1}\Big|_{\left(t_1,+H^{(2)}\right)} = \ell\sin(\alpha t_1)\left(\Gamma_2^{(1)} - \Gamma_2^{(2)}\right)\delta_i^2, \quad (10.3.5)$$

where $\Gamma_i^{(k)}$ $(i=1,2;\ k=1,2)$ are determined by (8.2.4), $\alpha = 2\pi/\ell$, δ_i^j are Kronecker symbols. The contact conditions between the uncurved filler and matrix layers are obtained as

$$\left(1+\Gamma_i^{(1)}\right)\sigma_{i2}^{(1)m,1}\Big|_{\left(x_{2m}^{(1)}=+H^{(1)}\right)} = \left(1+\Gamma_i^{(2)}\right)\sigma_{i2}^{(2)m,1}\Big|_{\left(x_{2m}^{(2)}=-H^{(2)}\right)},$$

$$U_i^{(1)m,1}\Big|_{\left(x_{2m}^{(1)}=+H^{(1)}\right)} = U_i^{(2)m,1}\Big|_{\left(x_{2m}^{(2)}=-H^{(2)}\right)},$$

$$\left(1+\Gamma_i^{(2)}\right)\sigma_{i2}^{(2)m,1}\Big|_{\left(x_{2m}^{(2)}=+H^{(2)}\right)}=\left(1+\Gamma_i^{(1)}\right)\sigma_{i2}^{(1)m,1}\Big|_{\left(x_{2m}^{(1)}=-H^{(1)}\right)},$$

$$U_i^{(2)m,1}\Big|_{\left(x_{2m}^{(2)}=+H^{(2)}\right)}=U_i^{(1)m,1}\Big|_{\left(x_{2m}^{(1)}=-H^{(1)}\right)}, \tag{10.3.6}$$

where $U_i^{(k)m,1}$ is determined by expressions (4.1.23).

The following limiting conditions hold:

$$\tilde{\sigma}_{ij}^{(k)m,1}=\left(\sigma_{ij}^{(k)m,1}-\sigma_{ij}^{(k)m,0}\right)\to 0,\ \text{under } m\to\infty . \tag{10.3.7}$$

As in subsection 4.4.3, using (10.3.7), we assume that after a certain layer with the number m^* all components of the tensor $\tilde{\sigma}_{ij}^{(k)m,1}$ are zero; an infinite number of layers is transformed into a finite number and equation (10.3.7) is changed to

$$\tilde{\sigma}_{ij}^{(k)m,1}=\left(\sigma_{ij}^{(\underline{k})\underline{m},1}-\sigma_{ij}^{(\underline{k})\underline{m},0}\right)\Big|_{x_{2\underline{m}^*}^{(\underline{k})}=H^{(\underline{k})}}=0 . \tag{10.3.8}$$

According to (10.3.5), (10.3.6) the solution to the equations (8.2.1), (8.2.2) we select as follows:

$$u_1^{(k)m,1}=\varphi_1^{(k)m}\cos(\alpha x_1),\ u_2^{(k)m,1}=\varphi_2^{(k)m}\sin(\alpha x_1). \tag{10.3.9}$$

Substituting (10.3.9) into (8.2.1), (8.2.2) and performing some operations we obtain the following expressions for the functions $\varphi_1^{(k)m}$, $\varphi_2^{(k)m}$:

$$\varphi_2^{(2)0}\left(x_{20}^{(2)}\right)=A_1^{(2)0}\sinh\theta_{10}^{(2)}\sin\theta_{20}^{(2)}+A_2^{(2)0}\cosh\theta_{10}^{(2)}\cos\theta_{20}^{(2)},$$

$$\varphi_1^{(2)0}\left(x_{20}^{(2)}\right)=A_1^{(2)0}\left(\kappa_{11}^{(2)}I_{ss}^{(2)}\left(x_{20}^{(2)}\right)+\kappa_{12}^{(2)}I_{cc}^{(2)}\left(x_{20}^{(2)}\right)\right)+A_2^{(2)0}\left(\kappa_{11}^{(2)}I_{cc}^{(2)}\left(x_{20}^{(2)}\right)-\kappa_{12}^{(2)}I_{ss}^{(2)}\left(x_{20}^{(2)}\right)\right)$$

$$\varphi_2^{(\underline{k})\underline{m}}\left(x_{2\underline{m}}^{(\underline{k})}\right)=A_1^{(\underline{k})\underline{m}}\sinh\theta_{1\underline{m}}^{(\underline{k})}\sin\theta_{2\underline{m}}^{(\underline{k})}+A_2^{(\underline{k})\underline{m}}\cosh\theta_{1\underline{m}}^{(\underline{k})}\cos\theta_{2\underline{m}}^{(\underline{k})}+$$

$$A_3^{(\underline{k})\underline{m}}\cosh\theta_{1\underline{m}}^{(\underline{k})}\sin\theta_{2\underline{m}}^{(\underline{k})}+A_2^{(\underline{k})\underline{m}}\sinh\theta_{1\underline{m}}^{(\underline{k})}\cos\theta_{2\underline{m}}^{(\underline{k})},$$

$$\varphi_1^{(\underline{k})\underline{m}}\left(x_{2\underline{m}}^{(\underline{k})}\right)=A_1^{(\underline{k})\underline{m}}\left(\kappa_{11}^{(\underline{k})}I_{ss}^{(\underline{k})}\left(x_{2\underline{m}}^{(\underline{k})}\right)+\kappa_{12}^{(\underline{k})}I_{cc}^{(\underline{k})}\left(x_{2\underline{m}}^{(\underline{k})}\right)\right)+A_2^{(\underline{k})\underline{m}}\left(\kappa_{11}^{(\underline{k})}I_{cc}^{(\underline{k})}\left(x_{2\underline{m}}^{(\underline{k})}\right)-\kappa_{12}^{(\underline{k})}I_{ss}^{(\underline{k})}\left(x_{2\underline{m}}^{(\underline{k})}\right)\right)+$$

$$A_3^{(\underline{k})m}\left(\kappa_{11}^{(\underline{k})}I_{sc}^{(\underline{k})}\left(x_{2\underline{m}}^{(\underline{k})}\right)-\kappa_{12}^{(\underline{k})}I_{cs}^{(\underline{k})}\left(x_{2\underline{m}}^{(\underline{k})}\right)\right)+A_4^{(\underline{k})m}\left(\kappa_{11}^{(\underline{k})}I_{cs}^{(\underline{k})}\left(x_{2\underline{m}}^{(\underline{k})}\right)-\kappa_{12}^{(\underline{k})}I_{sc}^{(\underline{k})}\left(x_{2\underline{m}}^{(\underline{k})}\right)\right), \tag{10.3.10}$$

where

$$I_{cs}^{(\underline{k})}\left(x_{2\underline{m}}^{(\underline{k})}\right)=-\kappa_1^{(k)}\cosh\theta_{1\underline{m}}^{(\underline{k})}\cos\theta_{2\underline{m}}^{(\underline{k})}+\kappa_2^{(\underline{k})}\sinh\theta_{1\underline{m}}^{(\underline{k})}\sin\theta_{2\underline{m}}^{(\underline{k})},$$

$$I_{sc}^{(\underline{k})}\left(x_{2\underline{m}}^{(\underline{k})}\right)=\kappa_2^{(\underline{k})}\cosh\theta_{1\underline{m}}^{(\underline{k})}\cos\theta_{2\underline{m}}^{(\underline{k})}+\kappa_1^{(k)}\sinh\theta_{1\underline{m}}^{(\underline{k})}\sin\theta_{2\underline{m}}^{(\underline{k})},$$

$$I_{ss}^{(\underline{k})}\left(x_{2\underline{m}}^{(\underline{k})}\right)=\kappa_2^{(\underline{k})}\cosh\theta_{1\underline{m}}^{(\underline{k})}\sin\theta_{2\underline{m}}^{(\underline{k})}-\kappa_1^{(k)}\sinh\theta_{1\underline{m}}^{(\underline{k})}\cos\theta_{2\underline{m}}^{(\underline{k})},$$

$$I_{cc}^{(\underline{k})}\left(x_{2\underline{m}}^{(\underline{k})}\right)=\kappa_1^{(k)}\cosh\theta_{1\underline{m}}^{(\underline{k})}\sin\theta_{2\underline{m}}^{(\underline{k})}+\kappa_2^{(\underline{k})}\sinh\theta_{1\underline{m}}^{(\underline{k})}\cos\theta_{2\underline{m}}^{(\underline{k})}. \tag{10.3.11}$$

In (10.3.10) and (10.3.11) the following notation is used:

$$\theta_{1m}^{(k)}=2\pi\alpha_1^{(k)}x_{2m}^{(\underline{k})}/\ell,\quad \theta_{2m}^{(k)}=2\pi\beta_1^{(k)}x_{2m}^{(\underline{k})}/\ell,\quad \kappa_1^{(k)}=\beta_1^{(\underline{k})}\Big/\left(\left(\alpha_1^{(\underline{k})}\right)^2+\left(\beta_1^{(\underline{k})}\right)^2\right),$$

$$\kappa_2^{(k)}=\alpha_1^{(\underline{k})}\Big/\left(\left(\alpha_1^{(\underline{k})}\right)^2+\left(\beta_1^{(\underline{k})}\right)^2\right),\quad \kappa_{11}^{(k)}=C_4^{(\underline{k})}\left(\left(\alpha_1^{(\underline{k})}\right)^2-\left(\beta_1^{(\underline{k})}\right)^2\right)-C_3^{(\underline{k})},$$

$$\kappa_{12}^{(\underline{k})}=2C_4^{(\underline{k})}\alpha_1^{(\underline{k})}\beta_1^{(\underline{k})},\ C_1^{(k)}=c_1^{(\underline{k})}/c_2^{(\underline{k})},\ C_2^{(k)}=c_3^{(\underline{k})}/c_2^{(\underline{k})},$$

$$C_3^{(k)}=c_4^{(\underline{k})}/c_5^{(\underline{k})},\ C_4^{(k)}=c_6^{(\underline{k})}/c_5^{(\underline{k})}. \tag{10.3.12}$$

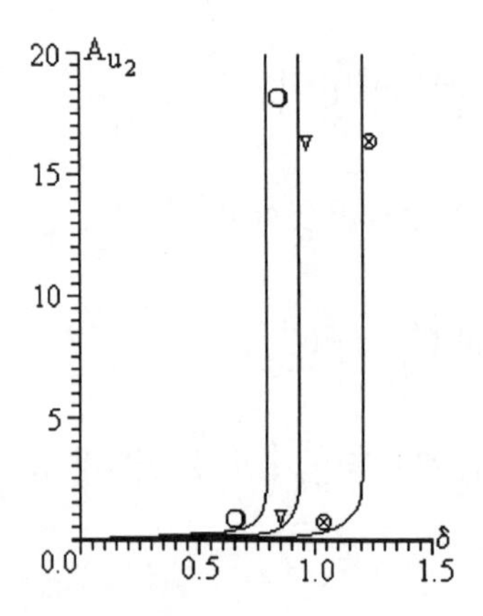

Fig.10.3.4. The relation between A_{u_2} and δ in the case where $\chi=0.1$, $\eta^{(2)}=0.5$, $\nu^{(1)}=\nu^{(2)}=0.3$ with various $E^{(2)}/E^{(1)}$: ⊗ - $E^{(2)}/E^{(1)}$ =100, ∇ -- $E^{(2)}/E^{(1)}$ =150 , ❍ - $E^{(2)}/E^{(1)}$ =200;

Note that the expressions of the constants $c_i^{(k)}$ $(i = 1,2,...,6;\ k = 1,2)$, $\alpha_1^{(k)}$ and $\beta_1^{(k)}$ are given by expressions (8.2.4), (8.2.16).

Thus, using (10.3.9) - (10.3.12) and we obtain algebraic equations for unknown constants $A_i^{(k)m}$ $(i = 1,2,3,4)$ from the contact conditions (10.3.6) and (10.3.8), determine first approximation, and investigate the relation between the parameter δ (10.3.2) and

Table 10. 10. The values of $\delta_{cr.}$ obtained for the composite shown in Fig.10.3.3 for $\nu^{(1)} = \nu^{(2)} = 0.3$.

$E^{(2)}/E^{(1)}$	χ	$\eta^{(2)}$			
		0.02	0.10	0.20	0.50
	0.05	9.619	2.504	1.443	0.774
100	0.10	5.802	2.671	1.749	1.189
	0.20	4.201	3.005	2.668	2.180
	0.30	4.846	4.742	3.981	3.621
	0.05	6.690	1.716	1.004	0.578
150	0.10	4.002	1.888	1.299	0.916
	0.20	3.201	2.461	2.216	1.880
	0.30	4.145	4.100	3.575	3.326
	0.05	5.111	1.306	0.780	0.479
200	0.10	3.146	1.480	1.048	0.769
	0.20	2.780	2.133	1.985	1.729
	0.30	3.791	3.788	3.345	3.178

Table 10. 11. The values of $\delta_{cr.}$ obtained for various m*, and $E^{(2)}/E^{(1)} = 200$.

$\eta^{(2)}$	Number of equations (layers)	χ		
		0.05	0.10	0.20
	10 (3)	5.111	3.106	2.740
0.02	14 (4)	5.111	3.106	2.740
	54 (13)	5.111	3.146	2.780
	10 (3)	1.262	1.393	2.002
0.10	14 (4)	1.306	1.480	2.133
	54 (13)	1.306	1.480	2.133
	10 (3)	0.758	1.003	1.895
0.20	14 (4)	0.780	1.048	1.985
	54 (13)	0.780	1.048	1.985
	10 (3)	0.470	0.769	1.711
0.50	14 (4)	0.479	0.778	1.729
	54 (13)	0.479	0.769	1.729

$$A_{u_2} = \max_{0 \le x_1 \le \ell} \left| u_2^{(2)0,1}(x_1,0) \right| / \ell . \tag{10.3.13}$$

The fracture criteria (10.3.1) determines the external compressive force p for fracture of the composite shown in Fig.10.3.3.

Fig.10.3.4 shows the dependence of A_{u_2} on δ. The results show that a single periodically curved layer causes the local stability loss. The critical values are less than those obtained for the composite shown in Fig. 4.2.1; the composite shown in Fig.10.3.3 fails under uniaxial compression before that shown in Fig.4.2.1.

Table 10.10 shows the critical values of δ for various $E^{(2)}/E^{(1)}$, $\eta^{(2)}$ and χ; with a decrease of filler concentration $\eta^{(2)}$, they approach to those given in Table 10.9.

Table 10.11 shows the critical values of δ for various m*, and confirms the effectiveness of the solution method.

10.4. Bibliographical Notes

Fiber separation in the continuum theory of Chapter 1 was first explained by A.N. Guz [99]. The explanation using the piece-wise homogeneous body model was given by S.D. Akbarov and A.N. Guz [36]. The *local* fiber separation effect was detailed by S.D. Akbarov [16]; crack problems was investigated by S.D. Akbarov, F.G. Maksudov, P.G. Panakhov and A.I. Seyfullayev [46]; particular crack problems were investigated by S.D. Akbarov [18,19]; fracture problems in compression were investigated by S.D. Akbarov, T. Sisman and N. Yahnioglu [48]

SUPPLEMENT 1

VISCOELASTIC UNIDIRECTIONAL COMPOSITES IN COMPRESSION

S.1.1. Fracture of Unidirectional Viscoelastic Composites in Compression

Up to now some attempts, which are cited in [54, 102, 105], had been made for investigations of fracture of viscoelastic unidirected composites in compression with the use of TDLT of stability. However, in these investigations the critical time is found by employing of the critical deformation method [86]. In the present supplement the composite with initial insignificant imperfection is used as a model for investigation of these problems and the approach proposed in section 10.3 is developed for viscoelastic composites.

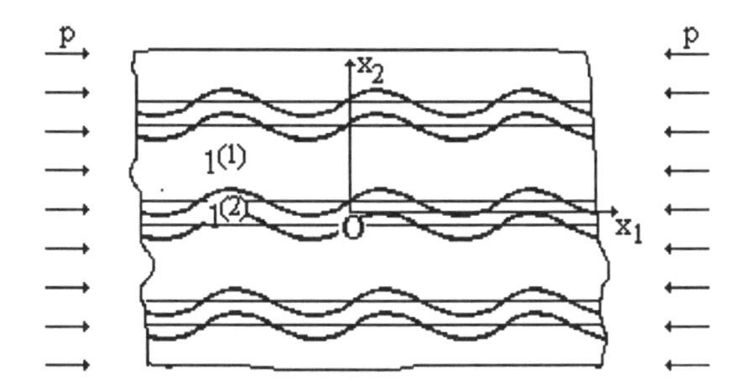

Fig.S.1.1.1. The initial imperfection with the periodically curving form in the structure of the unidirected composite.

For simplicity we take a composite material with a small periodical curving in the structure, as shown in Fig.S1.1.1, under compression along the reinforcing layers at infinity by uniformly distributed normal forces of intensity p. The reinforcing layers will be assumed to be located in planes which are parallel to the plane Ox_1x_3 and the thickness of filler layer will be assumed constant. Values related to the matrix will be denoted by upper indices (1); and those related to the filler, by (2). The material of layers of filler is assumed to be elastic with mechanical characteristics $E^{(2)}$ (Young's modulus) and $\nu^{(2)}$ (Poisson's ratio), and the material of the layers of the matrix is assumed to be linearly viscoelastic with operators

$$E^{(1)} = E_0^{(1)}\left[1 - \omega_0 \Pi_{\alpha'}^{*}(-\omega_0 - \omega_\infty)\right], \quad \nu^{(1)} = \nu_0^{(1)}\left[1 + \frac{1-2\nu_0^{(1)}}{2\nu_0^{(1)}}\omega_0 \Pi_{\alpha'}^{*}(-\omega_0 - \omega_\infty)\right], \tag{S.1.1.1}$$

where $E_0^{(1)}$ and $\nu_0^{(1)}$ are reference values of $E^{(1)}$ and $\nu^{(1)}$; ω_0, ω_∞ and α' are the rheological parameters of the matrix material; Π_α^* is the fractional-exponential operator of Rabotnov [131]:

$$\Pi_{\alpha'}^{*} = \int_0^t \Pi_{\alpha'}(\beta, t-\tau)d\tau, \tag{S.1.1.2}$$

where

$$\Pi_{\alpha'} = t^{\alpha'} \sum_{n=0}^{\infty} \frac{\beta^n t^{n(1+\alpha')}}{\Gamma[(n+1)(1+\alpha')]}. \tag{S.1.1.3}$$

In (S.1.1.3) $\Gamma(x)$ is the Gamma function.

Taking into account the periodicity of the composite material structure shown in Fig.S.1.1.1 in the direction of the Ox_2 axis with period $2(H^{(2)} + H^{(1)})$, (where $2H^{(2)}$ is the thickness of the filler, $2H^{(1)}$ the thickness of the matrix layer), among the layers we single out two of them, i.e. $1^{(1)}, 1^{(2)}$ and discuss them below. We associate the corresponding Lagrangian coordinates $O^{(k)}x_1^{(k)}x_2^{(k)}x_3^{(k)}$ (k=1,2; $x_1^{(1)} = x_1^{(2)} = x_1$; $x_3^{(1)} = x_3^{(2)} = x_3$) which in their natural state coincide with Cartesian coordinates and are obtained from $Ox_1x_2x_3$ by parallel transfer along the Ox_2 axis, with the middle surface of each layer of the filler and matrix. We assume that the small periodical curving of the reinforcing layers exists only in the direction of the Ox_1 axis and we investigate the plane deformation state in this composite in the framework of the piece-wise homogeneous body model with the use of the exact equations of the geometrical non-linear theory of viscoelasticity. For each layer, we write the equilibrium equations, mechanical relations as follows:

$$\frac{\partial}{\partial x_j^{(k)}}\left[\sigma_{jn}^{(k)}\left(\delta_i^n + \frac{\partial u_i^{(k)}}{\partial x_n^{(k)}}\right)\right] = 0, \quad i,j,n = 1,2,$$

$$\sigma_{ij}^{(k)} = \lambda^{(\underline{k})}\theta^{(\underline{k})}\delta_i^j + 2\mu^{(\underline{k})}\varepsilon_{ij}^{(\underline{k})}, \quad \theta^{(k)} = \varepsilon_{11}^{(k)} + \varepsilon_{22}^{(k)},$$

$$\lambda^{(k)} = \frac{E^{(\underline{k})}\nu^{(k)}}{(1+\nu^{(\underline{k})})(1-2\nu^{(\underline{k})})}, \quad \mu^{(k)} = \frac{E^{(\underline{k})}}{2(1+\nu^{(\underline{k})})},$$

$$\varepsilon_{ij}^{(k)} = \frac{1}{2}\left(\frac{\partial u_i^{(\underline{k})}}{\partial x_j^{(\underline{k})}} + \frac{\partial u_j^{(\underline{k})}}{\partial x_i^{(\underline{k})}} + \frac{\partial u_n^{(\underline{k})}}{\partial x_j^{(\underline{k})}}\frac{\partial u_n^{(\underline{k})}}{\partial x_j^{(\underline{k})}}\right). \quad \text{(S.1.1.4)}$$

In (S.1.1.4) $E^{(1)}$, $\nu^{(1)}$ are the operators (S.1.1.1)-(S.1.1.3) and the other notation is conventional.

We assume that, between the filler and matrix layers there is complete cohesion. The upper surface of the $1^{(2)}$ th layer is denoted by S^+, the lower surface by S^-.

$$\left[\left(\delta_i^n + \frac{\partial u_i^{(1)}}{\partial x_n^{(1)}}\right)\sigma_{jn}^{(1)}\right]\Bigg|_{S^\pm} n_j^{\pm} = \left[\left(\delta_i^n + \frac{\partial u_i^{(2)}}{\partial x_n^{(2)}}\right)\sigma_{jn}^{(2)}\right]\Bigg|_{S^\pm} n_j^{\pm}, \quad u_i^{(1)}\Big|_{S^\pm} = u_i^{(2)}\Big|_{S^\pm},$$

(S.1.1.5)

where $n_j^{\pm}$ is the orthonormal component to the surfaces $S^{\pm}$.

The initial small periodical curving of the filler layers is given through the equation of the middle surface of $1^{(2)}$ th filler layer:

$$x_2^{(2)} = A\sin\alpha x_1 = \varepsilon\ell\sin\alpha x_1, \quad \varepsilon = A/\ell, \quad \alpha = 2\pi/\ell. \quad \text{(S.1.1.6)}$$

We suppose that $L << \ell$ and consequently $\varepsilon << 1$.

For the solution of the problem (S.1.1.1) - (S.1.1.6) we use the method developed in chapters 4 and 8. Taking into account the condition of constant thickness of the filler and the equation of the middle surface of layer $1^{(2)}$, we derive the equations for the medium interfaces $S^{\pm}$ of the matrix and filler and the orthonormal components to these surfaces as follows:

$$x_i^{(2)\pm} = x_i^{(2)\pm}\left(t_1, H^{(2)}, \varepsilon f(t_1)\right), \quad n_i^{\pm} = n_i^{\pm}\left(t_1, H^{(2)}, \varepsilon f(t_1)\right), \quad \text{(S.1.1.7)}$$

where $f(t_1) = \ell\sin\alpha t_1$ and t_1 is a parameter, $t_1 \in (-\infty, +\infty)$. The explicit form of equations (S.1.1.7) are given by expressions (4.1.13), (4.1.14).

We seek the stress-deformation state of arbitrary $1^{(k)}$ th layer in series form in the parameter ε

$$\left\{\sigma_{ij}^{(k)};\varepsilon_{ij}^{(k)};u_i^{(k)}\right\}=\sum_{q=0}^{+\infty}\varepsilon^q\left\{\sigma_{ij}^{(k),q};\varepsilon_{ij}^{(k),q};u_i^{(k),q}\right\}. \qquad (S.1.1.8)$$

Considering the expression (S.1.1.7), we expand the values of each approximation (S.1.1.8) in series in the vicinity of $\left(t_1,\pm H^{(k)}\right)$; after some operations we obtain from (S.1.1.5) the corresponding contact relations; the q-th approximation contains the values of all the previous approximations. Substituting (S.1.1.8) into (S.1.1.4) and comparing equal powers of ε to describe each approximation, we obtain the corresponding closed system of equations. Owing to the linearity of the mechanical relations considered, they will be satisfied for each approximation (S.1.1.8) separately.

In Chapter 8 we showed that the values of the zeroth approximation are determined from the non-linear equation (S.1.1.4), the values of subsequent approximations are determined from the linear equations, and by direct verification we proved that these linear equations are the equations of three-dimensional linearized theory of stability (TDLTS) of the deformable solid body.

For the fracture of the composite shown in Fig.S.1.1.1 in compression, we need to select the failure criteria. Following [77,132,134,135] we assume that the exhaustion of the load-carrying capacity of the composite occurs as the result of stability loss. However the structure becomes unstable when the initially infinitesimal periodical curving amplitude grows significantly or even becomes infinite. This stability loss criterion is the analogue of the well-known stability loss criterion of beams, plates and shells fabricated from the time-dependent material.

Thus, in the framework of these considerations we determine the zeroth and first approximations. As in Chapter 8, for the zeroth approximation, we can neglect the non-linear terms in (S.1.1.4) and apply the principle of correspondence:

$$\overline{\varphi}(s)=\int_0^{+\infty}\varphi(t)e^{-st}dt \qquad (S.1.1.9)$$

with real parameter s>0, we obtain

$$\overline{\sigma}_{11}^{(1),0}=p\left(\eta^{(1)}+\eta^{(2)}\frac{1-\left(\overline{\nu}^{(1)}\right)^2}{1-\left(\nu^{(2)}\right)^2}\frac{E^{(2)}}{\overline{E}^{(1)}}\right)^{-1},\quad \overline{\sigma}_{12}^{(k),0}=\overline{\sigma}_{22}^{(k),0}=0,$$

$$\overline{u}_1^{(k),0}=\frac{1-\left(\overline{\nu}_1^{(k)}\right)^2}{\overline{E}^{(\underline{k})}}\overline{\sigma}_{11}^{(\underline{k}),0}x_1,\quad \overline{\sigma}_{11}^{(2),0}=\frac{E^{(2)}}{\overline{E}^{(1)}}\frac{1-\left(\overline{\nu}^{(1)}\right)^2}{1-\left(\nu^{(2)}\right)^2}\overline{\sigma}_{11}^{(1),0}, \qquad (S.1.1.10)$$

$$\overline{u}_2^{(\underline{k}),0}=-\frac{\overline{\nu}^{(\underline{k})}\left(1+\overline{\nu}^{(\underline{k})}\right)\overline{\sigma}_{11}^{(\underline{k}),0}}{\overline{E}^{(\underline{k})}}x_2^{(\underline{k})}+C^{(\underline{k})},\ C^{(k)}=\text{const},\ \eta^{(k)}=\frac{H^{(k)}}{H^{(1)}+H^{(2)}}.$$

For the first order approximation we obtain the following linearized equations from (S.1.1.4) and (S.1.1.5):

the equilibrium equations

$$\left(1+\Gamma_1^{(\underline{k})}\right)\frac{\partial\sigma_{11}^{(\underline{k})}}{\partial x_1}+\sigma_{11}^{(\underline{k}),0}\frac{\partial^2 u_1^{(\underline{k}),1}}{\partial x_1^2}+\left(1+\Gamma_1^{(\underline{k})}\right)\frac{\partial\sigma_{12}^{(\underline{k}),1}}{\partial x_2^{(\underline{k})}}=0;$$

$$\left(1+\Gamma_2^{(\underline{k})}\right)\frac{\partial\sigma_{12}^{(\underline{k})}}{\partial x_1}+\sigma_{11}^{(\underline{k}),0}\frac{\partial^2 u_2^{(\underline{k}),1}}{\partial x_1^2}+\left(1+\Gamma_2^{(\underline{k})}\right)\frac{\partial\sigma_{22}^{(\underline{k}),1}}{\partial x_2^{(\underline{k})}}=0. \qquad (S.1.1.11)$$

the mechanical relations

$$\sigma_{11}^{(k),1}=\frac{E^{(\underline{k})}}{(1+\nu^{(\underline{k})})(1-2\nu^{(\underline{k})})}\left[\left(1+\Gamma_1^{(\underline{k})}\right)\left(1-\nu^{(\underline{k})}\right)\frac{\partial u_1^{(k)}}{\partial x_1}+\left(1+\Gamma_2^{(\underline{k})}\right)\nu^{(\underline{k})}\frac{\partial u_2^{(k)}}{\partial x_2^{(\underline{k})}}\right],$$

$$\sigma_{22}^{(k),1}=\frac{E^{(\underline{k})}}{(1+\nu^{(\underline{k})})(1-2\nu^{(\underline{k})})}\left[\left(1+\Gamma_1^{(\underline{k})}\right)\nu^{(\underline{k})}\frac{\partial u_1^{(k)}}{\partial x_1}+\left(1+\Gamma_2^{(\underline{k})}\right)\left(1-\nu^{(\underline{k})}\right)\frac{\partial u_2^{(k)}}{\partial x_2^{(\underline{k})}}\right],$$

$$\sigma_{11}^{(k),1}=\frac{E^{(\underline{k})}}{2(1+\nu^{(\underline{k})})}\left[\left(1+\Gamma_1^{(\underline{k})}\right)\frac{\partial u_1^{(k)}}{\partial x_2^{(\underline{k})}}+\left(1+\Gamma_2^{(\underline{k})}\right)\frac{\partial u_2^{(k)}}{\partial x_1}\right]. \qquad (S.1.1.12)$$

the contact conditions

$$\left(1+\Gamma_i^{(1)}\right)\sigma_{i2}^{(1),1}\Big|_{(t_1,\pm H^{(1)})}-\left(1+\Gamma_i^{(2)}\right)\sigma_{i2}^{(2),1}\Big|_{(t_1,\mp H^{(2)})}=$$

$$\left.\frac{df}{dx_1}\right|_{x_1=t_1}\left(\left(1+\Gamma_1^{(1)}\right)\sigma_{11}^{(1),0}-\left(1+\Gamma_1^{(2)}\right)\sigma_{11}^{(2),0}\right)\delta_i^1,$$

$$u_i^{(1),1}\Big|_{(t_1,\pm H^{(1)})}-u_i^{(2),1}\Big|_{(t_1,\mp H^{(2)})}=f\big|_{x_1=t_1}\left(\Gamma_2^{(1)}-\Gamma_2^{(2)}\right)\delta_i^2,\quad i=1,2. \qquad (S.1.1.13)$$

In (S.1.1.11)-(S.1.1.13) $E^{(1)}$, $\nu^{(1)}$ are the operators determined by (S.1.1.1)-(S.1.1.3), and $\Gamma_i^{(k)}=\partial u_{\underline{i}}^{(\underline{k}),0}/\partial x_{\underline{i}}^{(\underline{k})}$.

Thus the determination of the first approximation is reduced to the solution to the quasistatic problem, given by equations (S.1.1.11)-(S.1.1.13). For simplicity we introduce the dimensionless rheological parameter $\omega = \omega_\infty / \omega_0$ and dimensionless time $t' = t\,\omega_0^{1/(1+\alpha')}$. Note that in the equations (S.1.1.11)-(S.1.1.13) the coefficients of the sought values are functions of time t'; causes difficulties: we cannot employ the principle of correspondence. We propose the following approach.

At first, we investigate the behaviour of the functions

$$\varphi_1(t') = \frac{\sigma_{11}^{(1),0}}{E_0^{(1)}}, \quad \varphi_2(t') = \frac{\sigma_{11}^{(2),0}}{E^{(2)}}, \quad \varphi_3^{(k)}(t') = 1 + \Gamma_1^{(k)}, \quad \varphi_4^{(k)}(t') = 1 + \Gamma_2^{(k)}, \quad \text{(S.1.1.14)}$$

which are the coefficients of equations (S.1.1.11)-(S.1.1.11). As an example, consider the case where $\omega = 0.5$, $\alpha' = -0.7$, $E^{(2)}/E_0^{(1)} = 100$, $p/E_0^{(1)} = 0.3$. Figs.S.1.1.2 , S.1.1.3 show typical graphs of equation (S.1.1.14) with $\eta^{(2)} = 0.2, 0.5$.

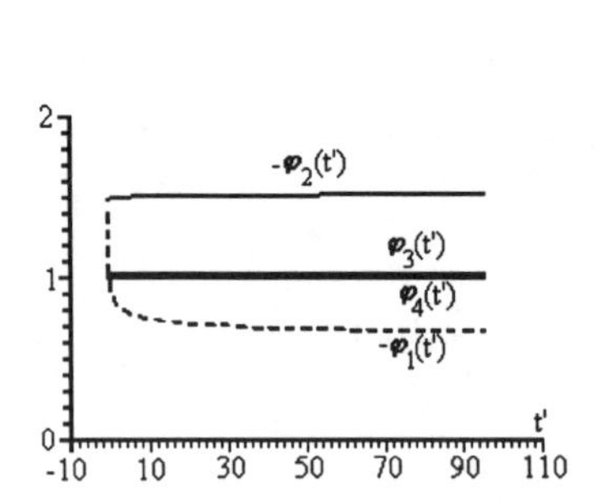

Fig.S.1.1.2. Graphs of the relation between t he functions $\varphi_i^{(k)}(t')$ and t' under $\eta^{(2)} = 0.2$.

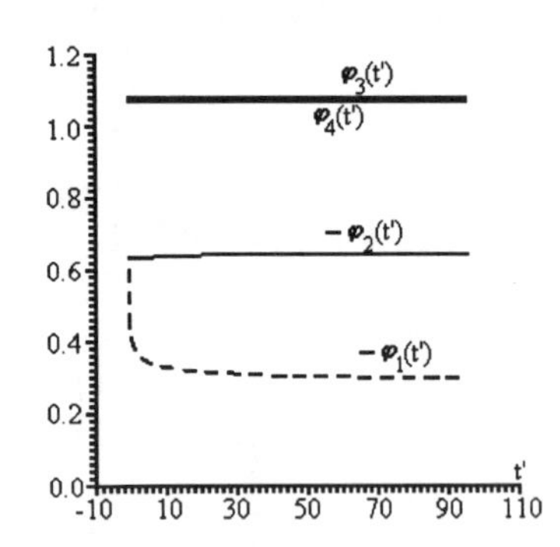

Fig.S.1.1.3. Graphs of the relation between the functions $\varphi_i^{(k)}(t')$ and t' under $\eta^{(2)} = 0.5$.

These graphs show that the functions $\varphi_2(t')$, $\varphi_3^{(k)}(t')$, $\varphi_4^{(k)}(t')$ may be taken to be constants with respect to t'. However, the function $\varphi_1(t')$ depends significantly on t' at an early stage of the compression. Therefore, in equations (S.1.1.11)-(S.1.1.13) we replace the functions $\varphi_2(t')$, $\varphi_3^{(k)}(t')$, $\varphi_4^{(k)}(t')$ by their values at $t' = 0$ or $t' = \infty$. Now we apply the Laplace transform (S.1.1.9) to equations (S.1.1.11)-(S.1.1.13). The Laplace transforms of the values related to the filler layer are obtained as equations (S.1.1.11)-(S.1.1.13) where k=2 and $\sigma_{ij}^{(2),1}$, $u_i^{(2),1}$ have been replaced by their Laplace transforms $\bar{\sigma}_{ij}^{(2),1}$, $\bar{u}_i^{(2),1}$. The convolution theorem for the Laplace transform gives the following equations from equations (S.1.1.11), (S.1.1.12) with $k = 1$:

$$\left(1+\Gamma_1^{(1)}\right)\frac{\partial\overline{\sigma}_{11}^{(1),1}}{\partial x_1}+E_0^{(1)}\int_0^{+\infty}\varphi_1(t')\frac{\partial^2 u_1^{(1),1}}{\partial x_1^2}e^{-st'}dt'+\left(1+\Gamma_1^{(1)}\right)\frac{\partial\overline{\sigma}_{12}^{(1),1}}{\partial x_2^{(1)}}=0,$$

$$\left(1+\Gamma_2^{(1)}\right)\frac{\partial\overline{\sigma}_{12}^{(1),1}}{\partial x_1}+E_0^{(1)}\int_0^{+\infty}\varphi_1(t')\frac{\partial^2 u_2^{(1),1}}{\partial x_1^2}e^{-st'}dt'+\left(1+\Gamma_2^{(1)}\right)\frac{\partial\overline{\sigma}_{22}^{(1),1}}{\partial x_2^{(1)}}=0,$$

$$\overline{\sigma}_{11}^{(1),1}=\frac{\overline{E}^{(1)}}{(1+\overline{\nu}^{(1)})(1-2\overline{\nu}^{(1)})}\left[\left(1+\Gamma_1^{(1)}\right)\left(1-\overline{\nu}^{(1)}\right)\frac{\partial\overline{u}_1^{(1)}}{\partial x_1}+\left(1+\Gamma_2^{(1)}\right)\overline{\nu}^{(1)}\frac{\partial\overline{u}_2^{(1)}}{\partial x_2^{(1)}}\right],$$

$$\overline{\sigma}_{22}^{(1),1}=\frac{\overline{E}^{(1)}}{(1+\overline{\nu}^{(1)})(1-2\overline{\nu}^{(1)})}\left[\left(1+\Gamma_1^{(1)}\right)\overline{\nu}^{(1)}\frac{\partial\overline{u}_1^{(1)}}{\partial x_1}+\left(1+\Gamma_2^{(1)}\right)\left(1-\overline{\nu}^{(1)}\right)\frac{\partial\overline{u}_2^{(1)}}{\partial x_2^{(1)}}\right],$$

$$\overline{\sigma}_{11}^{(1),1}=\frac{\overline{E}^{(1)}}{2(1+\overline{\nu}^{(1)})}\left[\left(1+\Gamma_1^{(1)}\right)\frac{\partial\overline{u}_1^{(1)}}{\partial x_2^{(1)}}+\left(1+\Gamma_2^{(1)}\right)\frac{\partial\overline{u}_2^{(1)}}{\partial x_1}\right]. \quad \text{(S.1.1.15)}$$

Now consider the integral terms in the first two equations of (S.1.1.15), i.e. consider the integrals

$$\int_0^{+\infty}\varphi_1(t')\frac{\partial^2 u_j^{(1),1}}{dx_1^2}e^{-st'}dt'\,;\; j=1,2. \quad \text{(S.1.1.16)}$$

Since

$$\varphi_1(t')\le 0 \text{ and } |\varphi_1(\infty)|\le|\varphi_1(t')|\le|\varphi_1(0)| \text{ for each } t'\in(0,+\infty), \quad \text{(S.1.1.17)}$$

we can write the following relations for each fixed s>0:

$$\left|\varphi_1(\infty)\int_0^{+\infty}\frac{\partial^2 u_i^{(1),1}}{\partial x_1^2}e^{-st'}dt'\right|\le\left|\int_0^{+\infty}\varphi_1(t')\frac{\partial^2 u_i^{(1),1}}{\partial x_1^2}e^{-st'}dt'\right|\le\left|\varphi_1(0)\int_0^{+\infty}\frac{\partial^2 u_i^{(1),1}}{\partial x_1^2}e^{-st'}dt'\right|. \quad \text{(S.1.1.18)}$$

So that, if $s>0$,

$$\int_0^{+\infty}\varphi_1(t')\frac{\partial^2 u_i^{(1),1}}{\partial x_1^2}e^{-st'}dt'\cong\varphi_1(t'_*)\int_0^{+\infty}\frac{\partial^2 u_i^{(1),1}}{\partial x_1^2}e^{-st'}dt'=\varphi_1(t'_*)\frac{\partial^2\overline{u}_i^{(1),1}}{\partial x_1^2}, \quad \text{(S.1.1.19)}$$

where t'_* is some fixed value of t'.

Consider the selection of the value of t'_* in equation (S.1.1.19). Although it is difficult to find t'_* exactly. Physical intuition and equation (S.1.1.18) show that critical t' lies between those obtained for $t'_* = 0$ and $t'_* = \infty$, i.e. $t'_{cr.\infty} \leq t'_{cr.} \leq t'_{cr.0}$: $t'_{cr.0}$, $t'_{cr.\infty}$ are the critical times for $t'_* = 0$, $t'_* = \infty$ respectively, and find $t'_{cr.0}$ and $t'_{cr.\infty}$.

We consider some numerical results. The inverse transform of the investigated functions are found using Schapery's method [136,137]. Higher approximations may be determined similarly, but have not an effect on the critical values. The higher approximations do increase the accuracy of the precritical stress-strain distribution.

Thus, we analyse the change of

$$A_{u_2} = \max_{0 \leq x_1 \leq \ell} \left| u_2^{(2)}(x_1, 0, t') \right| / \ell \qquad \text{(S.1.1.20)}$$

with respect to t' under

$$\chi = 2\pi H^{(2)}/\ell = 0.1, \quad \omega = 0.5, \quad \alpha' = -0.7, \quad E^{(2)}/E_0^{(1)} = 100. \qquad \text{(S.1.1.21)}$$

Consider the graphs given in Figs.S.1.1.4 - S.1.1.7. They show that the values of A_{u_2} obtained for various t'_* are bounded by those for $t'_* = \infty, t'_* = 0$. With an increase in filler concentration, and decrease in $p/E_0^{(1)}$, the values of $t'_{cr.}$ grow.

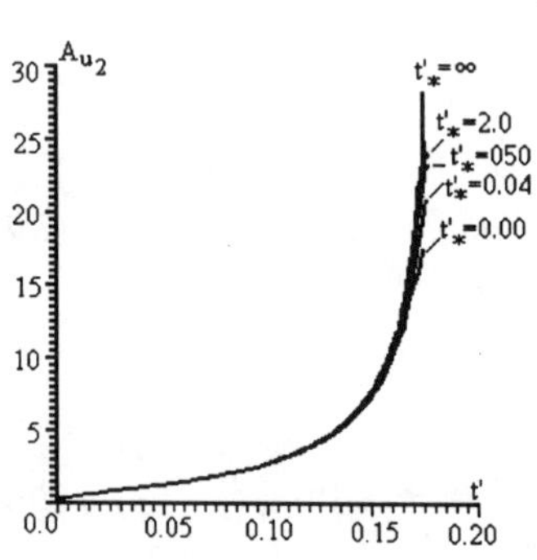

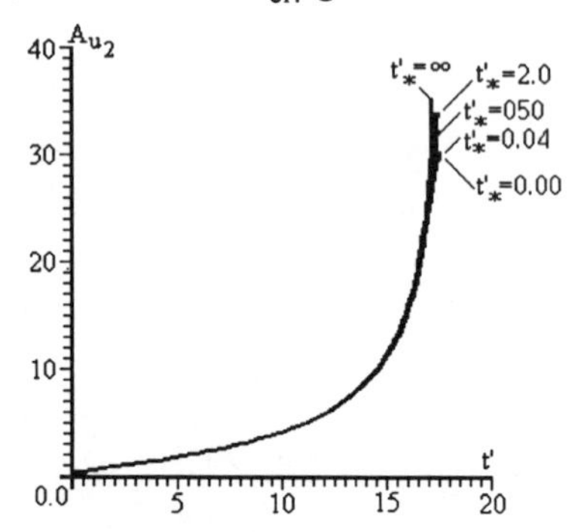

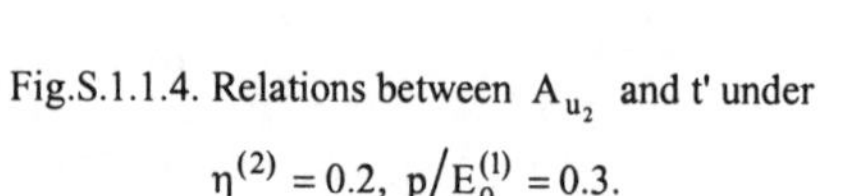

Fig.S.1.1.4. Relations between A_{u_2} and t' under $\eta^{(2)} = 0.2$, $p/E_0^{(1)} = 0.3$.

Fig.S.1.1.5. Relations between A_{u_2} and t' under $\eta^{(2)} = 0.2$, $p/E_0^{(1)} = 0.2$

We now compare results with those obtained by the critical deformation method [86]. Assume that there are uniformly distributed normal forces of intensity p acting at infinity in the direction of the Ox_1 axis, filler layer is purely elastic, and matrix layer viscoelastic with operators given in equation (S.1.1.1)-(S.1.1.3). The critical deformation method gives $p^{e}_{cr.}$, the elastic critical value from the relation

$$\varepsilon^{(1),0}_{11.cr} = \frac{p^{e}_{cr.}}{E^{(1)}_{0}}\left(1-\left(\nu^{(1)}_{0}\right)^{2}\right)\left(\eta^{(1)}+\eta^{(2)}\frac{\left(1-\left(\nu^{(1)}_{0}\right)^{2}\right)E^{(2)}}{\left(1-\left(\nu^{(2)}\right)^{2}\right)E^{(1)}_{0}}\right)^{-1}. \qquad \text{(S.1.1.22)}$$

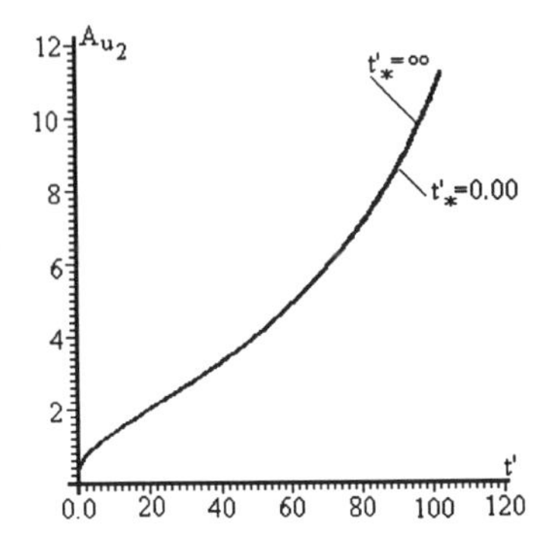

Fig.S.1.1.6. Relations between A_{u_2} and t' under $\eta^{(2)}=0.5$, $p/E^{(1)}_{0}=0.3$

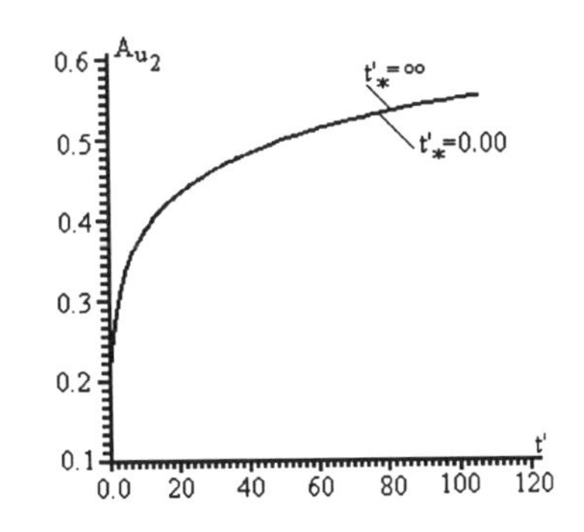

Fig.S.1.1.7. Relations between A_{u_2} and t' under $\eta^{(2)}=0.5$, $p/E^{(1)}_{0}=0.4$

The corresponding viscoelastic strain is

$$\varepsilon^{(1),0}_{11.v.e.} = \frac{p_{v.e.}}{E^{(1)}}\left(1-\left(\nu^{(1)}\right)^{2}\right)\left(\eta^{(1)}+\eta^{(2)}\frac{\left(1-\left(\nu^{(1)}\right)^{2}\right)E^{(2)}}{\left(1-\left(\nu^{(2)}\right)^{2}\right)E^{(1)}}\right)^{-1}, \qquad \text{(S.1.1.23)}$$

where $p_{v.e.}$ is the external normal compressive force in the viscoelastic case. We obtain the following equation for the determination of the critical time by equating equations (S.1.1.22) and (S.1.1.23):

$$\frac{p_{v.e}}{p^{e}_{cr.}} = \frac{\left(1-\left(\nu^{(1)}\right)^{2}\right)E^{(1)}\left(E^{(1)}_{0}\right)^{-1}}{\left(1-\left(\nu^{(1)}_{0}\right)^{2}\right)}\times$$

$$\frac{\left(\eta^{(1)}+\eta^{(2)}\left(1-\left(\nu_0^{(1)}\right)^2\right)\left(1-\left(\nu^{(2)}\right)^2\right)^{-1}E^{(2)}\left(E_0^{(1)}\right)^{-1}\right)}{\left(\eta^{(1)}+\eta^{(2)}\left(1-\left(\nu^{(1)}\right)^2\right)\left(1-\left(\nu^{(2)}\right)^2\right)^{-1}E^{(2)}\left(E^{(1)}\right)^{-1}\right)}, \qquad (S.1.1.24)$$

where $E^{(1)}$ and $\nu^{(1)}$ given by (S.1.1.1)-(S.1.1.3).

We compare numerical results obtained from equation (S.1.1.24) for the critical time with those obtained in our approach. Assume that $E^{(2)}/E_0^{(1)}=100$, $\nu^{(2)}=\nu_0^{(1)}=0.3$, $\eta^{(2)}=0.2$, $2\pi H^{(2)}/\ell=0.1$, $\alpha=-0.7$, $\omega=0.5$. We find $p_{cr.}^{e}=0.550E_0^{(1)}$ from equation (10.3.2) and from the numerical results given in Table 10.9. Consider the following two cases: $p_{v.e.}=0.3E_0^{(1)}$ (case 1); $p_{v.e.}=0.2E_0^{(1)}$ (case 2). Equation (S.1.24) gives the critical time: $t'_{cr.}=0.06$ (for case 1); $t'_{cr.}=1.0$ (for case 2). Fig.S.1.1.4 (case 1) gives $t'_{cr.}=0.17$, while Fig.S.1.1.5 (case 2) gives $t'_{cr.}=17$. The critical deformation method [86] underestimates the critical time.

S.1.2. Compressive Strength in Compression of Viscoelastic Unidirectional Composites

A review of the theoretical investigations of fracture is given in [54, 105]; these papers use the piecewise homogeneous model with TDLTS. Continuum TDLTS investigations were first carried out in [60, 92]; further development were made in [S2-S6]. We call the problems involving piecewise homogeneous models, the *first group*, and those investigated with the continuum approach, the *second group*.

The *first group* investigations in [54, 105] use of the bifurcation criterion for TDLTS; it is assumed that the reinforcing layers or fibers lose stability in a periodic form; the wave-generation parameter $\chi=2\pi H^{(2)}/\ell$ is introduced for layered materials; and $\chi=\pi R/\ell$ for fibrous composite materials. Here $2H^{(2)}$ is the thickness of the reinforcing layer, R is a radius of the fiber cross-section, ℓ is the half-wavelength of the instability form. The value of failure force is determined by investigating the dependence between χ and the critical values of external compressive force.

For the *second group*, composite materials are modelled as a structurally homogeneous orthotropic material with normalized mechanical characteristics. According to [60, 92] and [S3 - S6] the TDLTS equations are analysed for the action of uniformly distributed compressive forces at infinity. Fracture occurs when the TDLTS equations lose their ellipticity. The critical external forces corresponding to loss of ellipticity give the *theoretical strength limit in compression* (TSLC). References [60, 92] and [S2-S6] show that TSLC gives the load carrying capacity in the macro-volume of the composite material; it has considerable theoretical and practical significance.

The results related to the first group in [54, 105] were obtained for brittle fracture (polymer matrix) and plastic failure (metal matrix). For the viscous matrix, the results [54, 105] were based on an approximate criterion [86].

In section S.1.1, the fracture (stability loss) criterion was the value of the compressive force for which the initial infinitesimal periodic curvature starts to increase and keeps growing indefinitely. These investigations used the piece-wise homogeneous model and the exact three-dimensional geometrical non-linear equations. This method was used for the first group problems for viscoelastic composites. Now we use the analyses of S.1.1 to the determine the TSLC a layered viscoelastic composite. We assume that there are initial infinitesimal local imperfections (curvings) of the reinforcing layers and take the fracture criterion as the period of time for which this infinitesimal local curvature starts to increase and keeps growing indefinitely. Moreover we take into account the following considerations.

According to [60, S3-S6] the values of TSLC determined from the continuum approach, must depend only on the normalized mechanical properties of the composite material or only on the micro-mechanical parameters through which these normalized mechanical properties are determined, not on the initial local imperfection form or the parameter χ. By direct verification we will prove that for the elastic composites the TSLC so determined coincides with that in [60].

Consider a composite consisting of alternating layers of two materials with local curving, as shown in Fig.S.2.1.

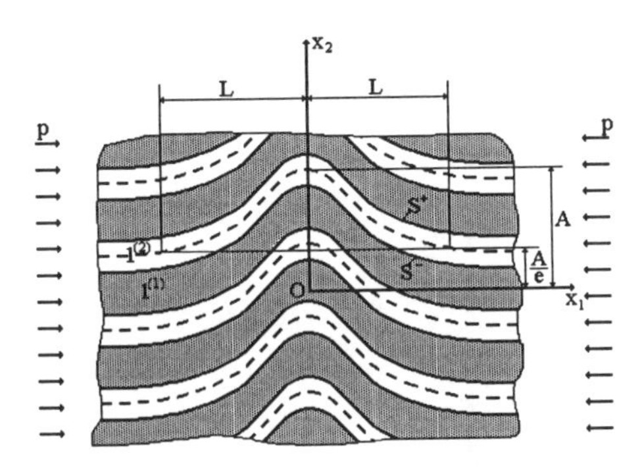

Fig.S.1.2.1. The initial imperfection with local curving form in the structure of the composite material

As usual we single out two layers, i.e. $1^{(1)}, 1^{(2)}$; we assume that the layer $1^{(2)}$ is elastic, the layer $1^{(1)}$ is linearly viscoelastic with operators (S.1.1.1) - (S.1.1.3). Within each layer we write the governing equations (S.1.1.4) and contact conditions (S.1.1.5).

The initial slight local curving form we select as follows:

$$x_2^{(2)} = A\exp\left\{-\left[\left(\frac{x_1}{L}\right)^2\right]^{\gamma}\right\}\cos\left(n\frac{x_1}{L}\right) = \varepsilon L\exp\left\{-\left[\left(\frac{x_1}{L}\right)^2\right]^{\gamma}\right\}\cos\left(n\frac{x_1}{L}\right). \quad \text{(S.1.2.1)}$$

The parameters γ, n and L are defined in section 6.2. We assume that A<<L, and that $\varepsilon = A/L$.

Applying the usual procedure we find the critical time $t_{cr.0}$ and $t_{cr.\infty}$ from the criterion

$$u_2^{(2),1}(0,0)/L \to \infty \text{ under } t \to t_{cr.0} \text{ at } t_* = 0 \text{ or under } t \to t_{cr.\infty} \text{ at } t_* = \infty . \quad \text{S.1.2.2)}$$

To determine $u_2^{(2),1}(0,0)$ we use the algorithm in section 6.2, and Shapery's method [136, 137].

Since $t_{cr.\infty} \le t_{cr.} \le t_{cr.0}$, our analysis is reduced to the determination of $t_{cr.0}$ and $t_{cr.\infty}$. Again we have found that it is sufficient to use the zeroth and the first approximations.

Thus, introducing the dimensionless rheological parameter $\omega = \omega_\infty/\omega_0$ and the dimensionless time $t' = t\omega_0^{1/(1+\alpha')}$ we consider some numerical results; we assume that the external force p does not depend on time t, and $\nu_0^{(1)} = \nu^{(2)} = 0.3$.

Figs.S.1.2.2 and S.1.2.3 show the dependence of $u_2^{(2),1}(0,0)/L$ on t' for fixed problem parameters $t'_* = 0$ (Fig.S.1.2.2) and $t'_* = \infty$ (Fig.S.1.2.3). We find that the graphs are independent of γ, and that $t'_{cr.0}$, $t'_{cr.\infty}$ do not depend on χ.

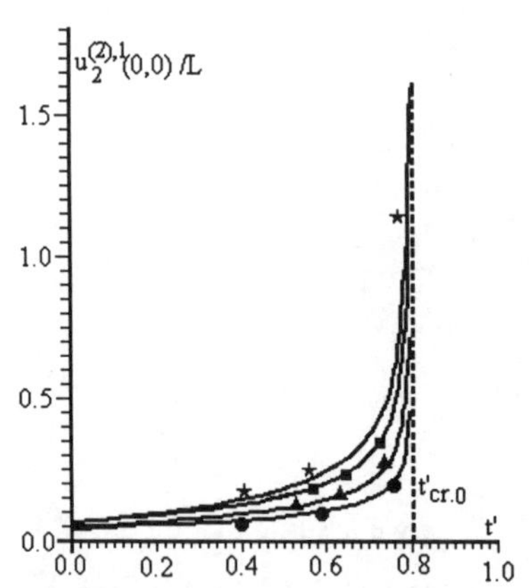

Fig.S.1.2.2. The graphs of the dependence of $u_2^{(2),1}(0,0)/L$ on t', for $t'_* = 0$, $E^{(2)}/E_0^{(1)} = 100$, $p/E_0^{(1)} = 0.25$, $\omega = 0.5$, $\alpha' = -0.5$; ★-- $\chi = 0.1$, n=0, $\gamma = 1,2,3$; ■ -- $\chi = 0.1$, $\gamma = 1$, n=1; ▲ -- $\chi = 0.2$, $\gamma = 1$, n=0; ● -- $\chi = 0.3$, $\gamma = 1$, n=0.

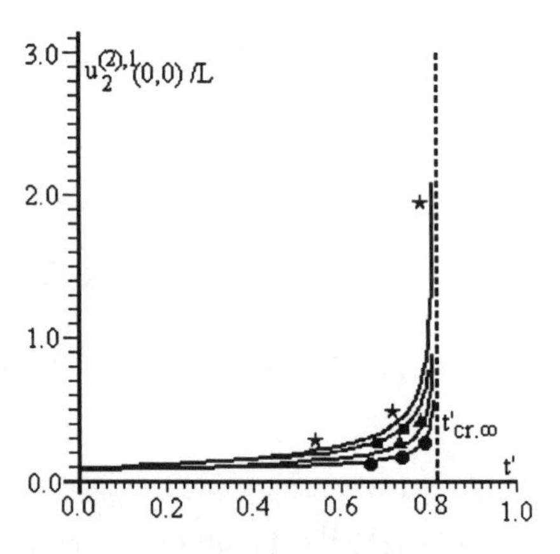

Fig.S.1.2.3. The graphs of the dependence of $u_2^{(2),1}(0,0)/L$ on t' for $t'_* = \infty$, $E^{(2)}/E_0^{(1)} = 100$, $p/E_0^{(1)} = 0.25$, $\omega = 0.5$, $\alpha' = -0.5$; ★-- $\chi = 0.1$, n=0, $\gamma = 1,2,3$; ■ -- $\chi = 0.1$, $\gamma = 1$, n=1; ▲ -- $\chi = 0.2$, $\gamma = 1$, n=0; ● -- $\chi = 0.3$, $\gamma = 1$, n=0.

Table S.1.1 shows $t'_{cr.0}$ and $t'_{cr.\infty}$ obtained for various $p/E_0^{(1)}$; increasing $p/E_0^{(1)}$, the values of $t'_{cr.0}$ and $t'_{cr.\infty}$ monotonically decrease, and $p/E_0^{(1)}$ approaches the $G_{12.0}/E_0^{(1)}$ which is the TSLC for the corresponding pure elastic composite material [60]. Here $G_{12.0}$ is the normalized shear modulus of the composite material in the case where this material has an ideal (uncurved) structure. It is known that

$$G_{12.0} = \frac{G_0^{(1)}G^{(2)}}{G_0^{(1)}\eta^{(2)} + G^{(2)}\eta^{(1)}}, \quad G_0^{(1)} = \frac{E_0^{(1)}}{2(1+\nu_0^{(1)})}, \quad G^{(2)} = \frac{E^{(2)}}{2(1+\nu^{(2)})}. \quad \text{(S.1.2.3)}$$

Table S.1.1. The values of $t'_{cr.0}$ and $t'_{cr.\infty}$ at various $p/E_0^{(1)}$ under $\eta^{(2)} = 0.2$, $E^{(2)}/E_0^{(1)} = 100$, $\omega = 0.5$, $\alpha' = -0.3$.

$p/E_0^{(1)}$	$t'_{cr.0}$	$t'_{cr.\infty}$
0.36	0.1150	0.1150
0.38	0.0839	0.0839
0.40	0.0590	0.0599
0.42	0.0350	0.0350
0.44	0.0209	0.0205
0.47	0.0054	0.0054

Mechanical considerations and numerical results (not shown here) show that for long time fracture the external compressive force p must satisfy $G_{12.\infty} \le p \le G_{12.0}$, where $G_{12.\infty}$ is determined by (S.1.2.3) by replacing $G_0^{(1)}$ and $\nu_0^{(1)}$ by $G_\infty^{(1)}$ and $\nu_\infty^{(1)}$ respectively. Here $G_\infty^{(1)}$ and $\nu_\infty^{(1)}$ are the values of $G^{(1)}$ and $\nu^{(1)}$ at $t' = \infty$.

Consider the influence of $\eta^{(2)}$, $E^{(2)}/E_0^{(1)}$ and of the rheological parameters ω and α on the values of $t'_{cr.0}$ and $t'_{cr.\infty}$. Table 2 shows $t'_{cr.0}$ and $t'_{cr.\infty}$ for various

Table S.1.2. The values of $t'_{cr.0}$ and $t'_{cr.\infty}$ at various $E^{(2)}/E_0^{(1)}$ under $\eta^{(2)} = 0.2$, $p/E_0^{(1)} = 0.28$, $\omega = 0.5$, $\alpha' = -0.3$.

$E^{(2)}/E_0^{(1)}$	$t'_{cr.0}$	$t'_{cr.\infty}$
50	0.4900	0.4799
100	0.4699	0.4650
150	0.4499	0.4499
200	0.4000	0.4000

$E^{(2)}/E_0^{(1)}$; for increasing $E^{(2)}/E_0^{(1)}$ the TSLC (i.e. $t'_{cr.0}$ and $t'_{cr.\infty}$) decreases. The influence the rheological parameter ω on $t'_{cr.0}$ and $t'_{cr.\infty}$ is given in Table S.1.3; It follows from these numerical results as ω increases, (i.e. $\left(E_0^{(1)} - E_\infty^{(1)}\right)$ decreases, where

$E_0^{(1)}\left(E_\infty^{(1)}\right)$ is the value of $E^{(1)}$ at $t'=0(t'=\infty)$) the values of $t'_{cr.0}$ and $t'_{cr.\infty}$ increase monotonically.

Table S.1.3. The values of $t'_{cr.0}$ and $t'_{cr.\infty}$ at various ω under $\eta^{(2)}=0.2$, $E^{(2)}/E_0^{(1)}=100$, $p/E_0^{(1)}=0.42$, $\alpha'=-0.3$.

ω	$t'_{cr.0}$	$t'_{cr.\infty}$
0.5	0.0350	0.0350
1.0	0.0390	0.0390
2.0	0.0544	0.0549
3.0	0.0649	0.0649
5.0	0.1890	0.1890
6.0	0.6490	0.6490

Table S.1.4 shows $t'_{cr.0}$ and $t'_{cr.\infty}$, for various α' and $\eta^{(2)}$; as $|\alpha|$ and $\eta^{(2)}$ increase $t'_{cr.0}$ and $t'_{cr.\infty}$ increase monotonically.

Table S.1.4. The values of $t'_{cr.0}$ and $t'_{cr.\infty}$ for various α' and $\eta^{(2)}$ and $E^{(2)}/E_0^{(1)}=100$, $p/E_0^{(1)}=0.32$, $\omega=0.5$.

α'	$\eta^{(2)}$	$t'_{cr.0}$	$t'_{cr.\infty}$
	0.1	0.100	0.100
-0.3	0.2	0.209	0.209
	0.5	2.500	2.500
	0.1	0.075	0.075
-0.5	0.2	0.170	0.170
	0.5	4.900	4.900
	0.1	0.023	0.022
-0.7	0.2	0.088	0.086
	0.5	22.000	22.000

The numerical results given in Tables S.1.1- S.1.4 show that $t'_{cr.0}$ and $t'_{cr.\infty}$ close; either may be used for $t'_{cr.}$.

S.1.3. Bibliographical Notes

The results are based on the papers by S.D. Akbarov , T. Sisman and N. Yahnioglu [48], and S.D. Akbarov, A. Cilli and A.N. Guz [25].

SUPPLEMENT 2

GEOMETRICAL NON-LINEAR AND STABILITY PROBLEMS

S.2.1. Geometrical Non-linear Bending of the Strip

We will consider some plane deformation problems for the continuum theory presented in Chapter 2. It will be assumed that the plane deformation takes place in the Ox_1x_2 plane for the rectangular region shown in Fig.3.1.1 where Ox_1x_2 are the Lagrangian coordinates which in their natural state coincide with Cartesian coordinates. We will only consider the rigid edge conditions for this region-strip and use the geometrical non-linear exact equations of the theory of elasticity. All results will be obtained numerically by employing FEM.

S.2.1.1. FORMULATION

Assume that the curvature is unidirectional, in the Ox_1 direction, and this strip occupies the region

$$\Omega = \{0 < x_1 < \ell; 0 < x_2 < h\} . \tag{S.2.1.1}$$

In this region the following equations are satisfied:

The equilibrium equation.

$$\frac{\partial}{\partial x_j}\left[\sigma_{jn}\left(\delta_i^n + \frac{\partial u_i}{\partial x_n}\right)\right] = 0, \quad i, j, n = 1,2, \tag{S.2.1.2}$$

The mechanical relations.

$$\sigma_{11} = A_{11}\varepsilon_{11} + A_{12}\varepsilon_{22} + 2A_{16}\varepsilon_{12}, \quad \sigma_{22} = A_{12}\varepsilon_{11} + A_{22}\varepsilon_{22} + 2A_{26}\varepsilon_{12},$$

$$\sigma_{12} = A_{16}\varepsilon_{11} + A_{26}\varepsilon_{22} + 2A_{66}\varepsilon_{12}. \tag{S.2.1.3}$$

The geometrical relations.

$$\varepsilon_{ij} = \frac{1}{2}\left(\frac{\partial u_i}{\partial x_j} + \frac{\partial u_j}{\partial x_i} + \frac{\partial u_n}{\partial x_j}\frac{\partial u_n}{\partial x_i}\right). \tag{S.2.1.4}$$

In (S.2.1.2) $A_{11}, A_{12},..., A_{66}$ are functions of x_1 and, according to the continuum theory of Chapter 2, are determined through the expressions (3.1.1) and (3.1.2). We assume that at the edges of this strip are the rigidly supported:

$$u_1\big|_{x_1=0,\ell} = u_2\big|_{x_1=0,\ell} = 0, \tag{S.2.1.5}$$

and that on the upper and lower plane of this strip the following conditions are given:

$$\sigma_{2n}\left(\delta_2^n + \frac{\partial u_2}{\partial x_n}\right)\Bigg|_{x_2=0} = 0, \quad \sigma_{2n}\left(\delta_2^n + \frac{\partial u_2}{\partial x_n}\right)\Bigg|_{x_2=h} = -p,$$

$$\sigma_{2n}\left(\delta_1^n + \frac{\partial u_1}{\partial x_n}\right)\Bigg|_{x_2=0,h} = 0. \tag{S.2.1.6}$$

Thus, geometrical non-linear bending problem of the rigidly supported strip is reduced to the solution of the equations (S.2.1.2) - (S.2.1.6). We suppose that the curving is periodic, given by the function

$$x_2 = \varepsilon f(x_1) = H\sin\left(\pi\frac{x_1}{\Lambda_1} + \delta\right) = \frac{H}{\Lambda_1}\Lambda_1 \sin\left(\pi\frac{x_1}{\Lambda_1} + \delta\right) =$$

$$\varepsilon\Lambda_1 \sin\left(\pi\frac{x_1}{\Lambda_1} + \delta\right), \qquad \varepsilon = \frac{H}{\Lambda_1}. \tag{S.2.1.7}$$

Here H is the rise of the curve; Λ_1 is the half wavelength of curvature; δ characterizes the distance by which the origin of the system of coordinates Ox_1x_2 (Fig.3.1.1) shifts (along the Ox_1 axis) relative to the first "nodal" point of the curving form. We assume that $H < \Lambda_1$; the parameter $\varepsilon = H/\Lambda_1$ estimates the degree of curving; if $\varepsilon = 0$, then there is no curving.

S.2.1.2. FEM MODELLING

Consider the following functional

$$\Pi = \frac{1}{2}\iint_{\Omega}\sigma_{ij}\varepsilon_{ij}d\Omega - \int_{S_1}P_i u_i dS, \tag{S.2.1.8}$$

which is the total potential energy of the strip. In (S.2.1.8) S_1 is the part of the surface of this strip on which the external forces P_i $(i = 1,2)$ are given. We assume that the equations (S.2.1.3), (S.2.1.4) and edge conditions (S.2.1.5) are satisfied for the virtual displacements δu_i. According to the principle of virtual work, the equilibrium state is obtained from

$$\delta\Pi = \iint_{\Omega} \sigma_{ij}\delta\varepsilon_{ij}d\Omega - \int_{S_1} P_i\delta u_i dS = 0\,, \tag{S.2.1.9}$$

where

$$\delta\varepsilon_{ij} = \frac{1}{2}\left(\left(\delta_i^n + \frac{\partial u_n}{\partial x_i}\right)\frac{\partial \delta u_n}{\partial x_j} + \left(\delta_j^n + \frac{\partial u_n}{\partial x_j}\right)\frac{\partial \delta u_n}{\partial x_i}\right). \tag{S.2.1.10}$$

Integration by parts applied to equation (S.2.1.9) yields the equation (S.2.1.2) and boundary conditions (S.2.1.6).

Dividing the region Ω (S.2.1.1) into a finite number of rectangular elements Ω_k (Fig.3.1.2) within which the displacements are prescribed through a finite number of nodal displacements $\mathbf{a}$ and Shape Functions $\mathbf{N}$ (3.1.14) and following the solution procedure of subsection 3.1.2 we obtain a non-linear set of equations

$$\boldsymbol{\Psi}(\mathbf{a}) = \mathbf{0}\,. \tag{S.2.1.11}$$

In (S.2.1.11) $\boldsymbol{\Psi}$ represents the sum of external and internal generalized forces.

To solve equation (S.2.1.11) we use the Newton-Raphson [NR] method [156] and Modified Newton-Raphson [MNR] method [156] are used. According to Newton-Raphson method the non-linear equation (S.2.1.11) is linearized as follows

$$\boldsymbol{\Psi}(\mathbf{a} + \boldsymbol{\delta}\mathbf{a}) = \boldsymbol{\Psi}(\mathbf{a}) + \frac{\partial \boldsymbol{\Psi}}{\partial \mathbf{a}}\boldsymbol{\delta}\mathbf{a}\,. \tag{S.2.1.11}$$

In the present investigation as the zeroth iteration we take the solution to the corresponding linear problem which has been investigated in Chapter 3. Denoting the solution to this linear problem as $\mathbf{a}_0$ we obtain from (S.2.1.11), (S.2.1.12) the following equation for the first iteration

$$\mathbf{K}_T(\mathbf{a}_0)\delta\mathbf{a}_1 = -\boldsymbol{\Psi}(\mathbf{a}_0) \tag{S.2.1.12}$$

where

$$\mathbf{K}_T = \left.\frac{\partial \boldsymbol{\Psi}}{\partial \mathbf{a}}\right|_{\mathbf{a}=\mathbf{a}_0} \tag{S.2.1.13}$$

is the Tangential Stiffness Matrix for the first iteration. Note that the equation (S.2.1.12) is linear with respect to the correction $\delta\mathbf{a}_1$; we apply the solution technique of Chapter 3 to find $\delta\mathbf{a}_1$. After the first iteration, the nodal displacements are determined as

$$\mathbf{a}_1 = \mathbf{a}_0 + \delta\mathbf{a}_1 . \tag{S.2.1.14}$$

For the n-th iteration, we have

$$\mathbf{a}_n = \mathbf{a}_{n-1} + \delta\mathbf{a}_n \tag{S.2.1.15}$$

where $\delta\mathbf{a}_n$ is the solution to the linearized equation

$$\mathbf{K}_T(\mathbf{a}_{n-1})\delta\mathbf{a}_n = -\mathbf{\Psi}(\mathbf{a}_{n-1}), \tag{S.2.1.16}$$

and $\mathbf{K}_T(\mathbf{a}_{n-1})$ is a tangential Stiffness Matrix for the n-th iteration.

There are various modifications of this basic procedure; see [156]. In our investigation under MNR method we will understand the replacement of (S.2.1.16) with

$$\mathbf{K}\delta\mathbf{a}_n = -\mathbf{\Psi}(\mathbf{a}_{n-1}), \tag{S.2.1.17}$$

where $\mathbf{K}$ is the Stiffness Matrix of the corresponding linear problem.

S.2.1.3. NUMERICAL RESULTS

Assume that the strip material is fabricated from alternating isotropic layers of two materials, and these layers are perpendicular to the Ox_2 axis in the absence of the curving. For these layers we introduce the following notation: E_1, E_2 are the Young's moduli; ν_1, ν_2 are the Poisson's ratios; η_1, η_2 are the concentrations of the constituents. We will use the expression for the normalized elasticity coefficients A_{11}^0, A_{12}^0, A_{22}^0 and $A_{66}^0 = G_{12}^0$ which are given by the formulas (1.9.1), (1.9.3). Using these coefficients and the expression (S.2.1.7) from (3.1.1), (3.1.2) we determine the functions $A_{11}(x_1)$, $A_{12}(x_1)$, $A_{22}(x_1)$, $A_{16}(x_1)$, $A_{26}(x_1)$ and $A_{66}(x_1)$ which enter (S.2.1.3). We take $\eta_1 = \eta_2 = 0.5$, $\nu_1 = \nu_2 = 0.3$, $\ell/\Lambda_1 = 16$, $\delta = \pi/2$. Using the problem symmetry with respect to $x_1 = \ell/2$ we consider the sub-domain $\{0 \le x_1 \le \ell/2; 0 \le x_2 \le h\}$. In this case 4 and 20 rectangular elements (Fig.3.1.2) are taken in the directions of Ox_2 and Ox_1 axes respectively.

First, we investigate the convergence of the numerical results obtained for various iterations. We find the upper limit of p/E_1, denote it by p^*/E_1, for which the iterations converge. As a convergence criterion we take

$$\frac{\max\limits_{x_1,x_2\in\Omega}\left|\sigma_{22}^{(n)}-\sigma_{22}^{(n-1)}\right|}{|p|}\leq\frac{\max\limits_{x_1,x_2\in\Omega}\left|\sigma_{22}^{(n-1)}-\sigma_{22}^{(n-2)}\right|}{|p|}\leq\alpha_\sigma\approx O\left(10^{-2}\right), \qquad (S.2.1.18)$$

where n is a number of the iteration.

Table S.2.1 shows the values of p^*/E_1 and corresponding iteration number obtained for various E_2/E_1. These results show that the values of p^*/E_1 for the MNR method are significantly greater than those for NR, but MNR needs more iterations than NR.

Table S.2.1. The values of p^*/E_1 for various E_2/E_1 and $h/\ell=0.1$, $\varepsilon=0.1$.

E_2/E_1	MNR method (iter. number)	NR method (iter. number)
5	0.002 (13)	0.0011 (13)
20	0.006 (34)	0.0016 (5)
50	0.007 (46)	0.0030 (4)

Table S.2.2. The values of σ_{22}/p obtained under $p/E_1=0.001$ for $E_2/E_1=5$, $h/\ell=0.1$, $x_2/h=0.5$, $\varepsilon=0.1$.

x_1/ℓ	MNR method. 5-th iteration	NR method. 5-th iteration
0.0625	-3.078	-3.079
0.1250	0.873	0.877
0.1875	-1.043	-1.047
0.2600	-1.012	-1.012
0.3125	0.422	0.414
0.3750	-2.061	-2.067
0.4375	1.052	1.065
0.5000	-2.448	-2.456

Table S.2.3. The values of σ_{22}/p obtained under $p/E_1=0.003$ for $E_2/E_1=50$, $h/\ell=0.1$, $x_2/h=0.5$, $\varepsilon=0.1$.

x_1/ℓ	MNR method. 4-th iteration	NR method. 2-nd iteration
0.0625	-6.357	-6.334
0.1250	3.206	3.184
0.1875	-1.837	-1.817
0.2500	-1.535	-1.546
0.3125	1.746	1.753
0.3750	-4.202	-4.175
0.4375	3.452	3.404
0.5000	-5.153	-5.090

Table S.2.4. The values of σ_{22}/p for various p/E_1, ε, x_1/ℓ under $x_2/h = 0.5$, $E_2/E_1 = 5$, $h/\ell = 0.1$, $\ell/\Lambda_1 = 16$.

ε	x_1/ℓ	p/E_1: Linear state.	0.0005	0.001	0.002
		Iteration number			
		-------	4	5	13
	0.0625	-0.492	-0.565	-0.637	-0.767
	0.1250	-0.504	-0.527	-0.550	-0.591
	0.1875	-0.500	-0.504	-0.507	-0.512
	0.2500	-0.500	-0.506	-0.513	-0.526
0.00	0.3125	-0.500	-0.521	-0.542	-0.580
	0.3750	-0.500	-0.537	-0.574	-0.641
	0.4375	-0.500	-0.551	-0.602	-0.693
	0.5000	-0.500	-0.555	-0.609	-0.707
	0.0625	-2.901	-2.995	-3.078	-3.177
	0.1250	0.972	0.937	0.873	0.740
	0.1875	-1.053	-1.039	-1.043	-1.020
0.10	0.2500	-0.980	-0.986	-1.012	-1.036
	0.3125	0.509	0.478	0.422	0.312
	0.3750	-1.942	-2.003	-2.061	-2.127
	0.4375	1.204	1.137	1.052	0.864
	0.5000	-2.297	-2.380	-2.448	-2.527

Table S.2.5. The values of σ_{11}/p for various p/E_1, ε, x_1/ℓ under $x_2/h = 1.0$, $E_2/E_1 = 5$, $h/\ell = 0.1$, $\ell/\Lambda_1 = 16$.

ε	x_1/ℓ	p/E_1: Linear state.	0.0005	0.001	0.002
		Iteration number			
		-------	2	3	9
	0.0625	32.326	32.891	33.291	33.533
	0.1250	17.509	18.126	18.549	19.227
	0.1875	7.365	7.980	8.457	9.355
	0.2500	-8.516	-7.934	-7.350	-6.154
0.00	0.3125	-15.528	-14.973	-14.336	-13.015
	0.3750	-21.153	-20.621	-19.942	-18.503
	0.4375	-24.622	-24.110	-23.413	-21.897
	0.5000	-25.840	-25.331	-24.624	-23.069
	0.0625	38.911	39.576	39.919	40.006
	0.1250	15.073	15.724	16.158	16.687
	0.1875	7.470	8.106	8.566	9.237
0.10	0.2500	-7.959	-7.351	-6.749	-5.587
	0.3125	-18.290	-17.528	-17.003	-15.504
	0.3750	-18.042	-17.503	-16.817	-15.302
	0.4375	-29.140	-28.607	-27.815	-25.913
	0.5000	-21.983	-21.479	-20.770	-19.117

Note that if p/E_1 is smaller than the p^*/E_1 for the NR method, the numerical results for both methods coincide. The values of σ_{22} given in these tables are calculated at points

$$x_1/\ell = n\Lambda_1/\ell, \quad n = 1,2,...,8, \tag{S.2.1.19}$$

which correspond to the vertex "above" or vertex "below" of the curvature in the strip material structure; the maximal effect of the geometrical non-linearity on the stresses σ_{22} and σ_{11} is observed there.

First, we note that the character of the distribution of the stresses with respect to x_1 or x_2 is similar to that obtained in the corresponding linear statement of Chapter 3. Therefore we do not show these distributions and consider only the influence of the geometrical non-linearity on the values of the stresses σ_{22} and σ_{11} obtained at the points (S.2.1.19) for various p/E_1 , as shown in Tables S.2.4. and S.2.5.

Table S.2.4 shows that as a result of the geometrical non-linearity the absolute values of σ_{22}/p at points for which $\sigma_{22}/p<0$ $(\sigma_{22}/p>0)$ increase (decrease) monotonically with increasing p/E_1. Table S.2.5 shows that at points for which $\sigma_{11}/p>0$ $(\sigma_{11}/p<0)$ the absolute values of σ_{11}/p increase (decrease) monotonically with p/E_1. This effect is observed for other values of the problem parameters E_2/E_1, h/ℓ, ε, ℓ/Λ_1 also. The corresponding results show that the effect of the geometrical non-linearity on the foregoing stress distribution decays with increasing E_2/E_1 and h/ℓ.

S.2.2. Stability Loss of the Strip

We again consider the strip occupying the region (S.2.1.1) and fabricated from the composite with periodically curved structure, and use three-dimensional linearized theory of stability (TDLTS) [101] to investigate the stability loss under compression along the Ox_1 axis (Fig.3.1.1). We study the influence of the curvature parameters entering (S.2.1.7) on the critical values of the compression force under simply supported and clamped conditions at the edges of this strip. We assume that the external forces are the "dead " forces and apply the Euler (bifurcation) approach. The precritical state in the strip, denoted by upper index 0, will be determined from the linear theory of elasticity.

S.2.2.1. FORMULATION. FEM MODELLING

We assume that in the precritical state within the strip the following equations are satisfied.

The equilibrium equation.

$$\frac{\partial \sigma_{ij}^{0}}{\partial x_j} = 0, \quad i; j = 1,2. \tag{S.2.2.1}$$

The mechanical relations.

$$\sigma_{11}^{0} = A_{11}\varepsilon_{11}^{0} + A_{12}\varepsilon_{22}^{0} + 2A_{16}\varepsilon_{12}^{0}, \quad \sigma_{22}^{0} = A_{12}\varepsilon_{11}^{0} + A_{22}\varepsilon_{22}^{0} + 2A_{26}\varepsilon_{12}^{0},$$

$$\sigma_{12}^{0} = A_{16}\varepsilon_{11}^{0} + A_{26}\varepsilon_{22}^{0} + 2A_{66}\varepsilon_{12}^{0}. \tag{S.2.2.2}$$

The geometrical relations.

$$\varepsilon_{ij}^{0} = \frac{1}{2}\left(\frac{\partial u_i^0}{\partial x_j} + \frac{\partial u_j^0}{\partial x_i}\right). \tag{S.2.2.3}$$

The boundary conditions on the upper and lower plane of the strip.

$$\sigma_{12}^{0}\Big|_{x_2=0;h} = \sigma_{22}^{0}\Big|_{x_2=0;h} = 0. \tag{S.2.2.4}$$

The edge conditions for simply supported strip.

$$u_2^0\Big|_{x_1=0;\ell} = 0, \quad \sigma_{11}^{0}\Big|_{x_1=0;\ell} = p. \tag{S.2.2.5}$$

The edge conditions for clamped strip.

$$u_2^0\Big|_{x_2=0;\ell} = 0, \quad u_1^0\Big|_{x_1=0} = -u_1^0\Big|_{x_1=\ell} = U = \text{const.} \tag{S.2.2.6}$$

For the clamped case we assume that clamp compresses the strip by normal forces with intensity p along the Ox_1 axis. In this case the relation between U (S.2.2.6) and p is determined after the solution procedure related to the precritical state from the equation

$$p = \frac{1}{h\Lambda_1}\int_{7\Lambda_1}^{8\Lambda_1}\left(\int_0^h \sigma_{11}^{0} dx_2\right)dx_1. \tag{S.2.2.7}$$

In equation (S.2.2.7) it is assumed that $\ell/\Lambda_1 = 16$, $\delta = \pi/2$ in the expression (S.2.1.7). Moreover note that in (S.2.2.2) $A_{11}, A_{12}, \ldots, A_{66}$ are functions on x_1 and are determined as before.

Thus, the precritical state is determined from equations (S.2.2.1)-(S.2.2.7). We obtain the following eigenvalue problem to find the critical values of the external compressive force p.

The governing equations of TDLTS.

$$\frac{\partial \sigma_{ij}}{\partial x_j} + \frac{\partial}{\partial x_j}\left(\sigma_{jn}^0 \frac{\partial u_i}{\partial x_n}\right) = 0, \quad i; j; n = 1,2,$$

$$\sigma_{11} = A_{11}\varepsilon_{11} + A_{12}\varepsilon_{22} + 2A_{16}\varepsilon_{12}, \quad \sigma_{22} = A_{12}\varepsilon_{11} + A_{22}\varepsilon_{22} + 2A_{26}\varepsilon_{12},$$

$$\sigma_{22} = A_{16}\varepsilon_{11} + A_{26}\varepsilon_{22} + 2A_{66}\varepsilon_{12}, \quad \varepsilon_{ij} = \frac{1}{2}\left(\frac{\partial u_j}{\partial x_i} + \frac{\partial u_i}{\partial x_j}\right). \tag{S.2.2.8}$$

The condition for the upper and lower plane of the strip.

$$\sigma_{12}\big|_{x_2=0;h} = \sigma_{22}\big|_{x_2=0;h} = 0. \tag{S.2.2.9}$$

The edge conditions for the simply supported case.

$$u_2\big|_{x_1=0;\ell} = 0, \quad \sigma_{11}\big|_{x_1=0;\ell} = 0. \tag{S.2.2.10}$$

The edge conditions for the clamped case.

$$u_1\big|_{x_1=0;\ell} = u_2\big|_{x_1=0;\ell} = 0. \tag{S.2.2.11}$$

The boundary-value problems (S.2.2.1)-(S.2.2.6) are solved by employing FEM, as in Chapter 3. But, for solution of the eigenvalue problems (S.2.2.8)-(S.2.2.11) we use the FEM modelling described in section S.2.1 and obtain

$$\mathbf{K}_T(\mathbf{a}_0)\delta\mathbf{a}_1 = \mathbf{0}, \tag{S.2.2.12}$$

where $\mathbf{a}_0$ is the nodal displacement of the precritical state governed by equations (S.2.2.1)-(S.2.2.6), $\delta\mathbf{a}_1$ is the nodal values of displacements u_i which enter the equations (S.2.2.8)-(S.2.2.11). The requirement that equation (S.2.2.12) has a non-trivial solution gives the critical values of compression force p.

S.2.2.2. NUMERICAL RESULTS

As before, we assume that the strip consists of alternating layers of two homogeneous isotropic materials. We take $\ell/\Lambda_1 = 16$, $\nu_1 = \nu_2 = 0.3$. Table S.2.6 shows the values of $p_{cr.}/E_1$ for various values of the problem parameters for the simply supported case. If $\varepsilon = 0$ we have $\sigma_{11}^0 = p$, $\sigma_{22}^0 = \sigma_{12}^0 = 0$ and corresponding eigenvalue problem (S.2.2.8)-(S.2.2.10) can be solved analytically. The values of $p_{cr.}/E_1$ determined from the condition (S.2.2.9) using these analytical expressions coincide (with accuracy 10^{-4}) with those determined by FEM modelling, confirming the trustworthiness of the FEM approach.

Table S.2.6. The values $p_{cr.}/E_1$ for simply supported strip

h/ℓ	E_2/E_1	ε = 0.00	0.05	0.10	0.15	0.20
0.10	5	0.0256	0.0239	0.0220	0.0199	0.0182
	10	0.0455	0.0411	0.0356	0.0301	0.0258
	20	0.0817	0.0708	0.0574	0.0453	0.0364
	50	0.1682	0.1350	0.1034	0.0804	0.0620
0.15	5	0.0541	0.0469	0.0433	0.0392	0.0356
	10	0.0929	0.0782	0.0670	0.0564	0.0463
	20	0.1571	0.1261	0.0991	0.0750	0.0579
	50	0.2842	0.2075	0.1501	0.1115	0.0819
0.20	5	0.0885	0.0679	0.0628	0.0569	0.0517
	10	0.1462	0.1099	0.0939	0.0775	0.0649
	20	0.2325	0.1677	0.1303	0.0989	0.0769
	50	0.3758	0.2479	0.1748	0.1277	0.0944

Table S.2.7. The values $p_{cr.}/E_1$ for clamped strip

h/ℓ	E_2/E_1	ε = 0.00	0.05	0.10	0.15	0.20
0.10	5	0.0886	0.0875	0.0830	0.0771	0.0719
	10	0.1453	0.1439	0.1320	0.1179	0.1054
	20	0.2310	0.2265	0.2031	0.1746	0.1503
	50	0.3700	0.3635	0.3227	0.2727	0.2251
0.15	5	0.1616	0.1657	0.1623	0.1552	0.1464
	10	0.2460	0.2553	0.2461	0.2268	0.2047
	20	0.3513	0.3632	0.3459	0.3090	0.2671
	50	0.4823	0.4998	0.4695	0.4199	0.3609
0.20	5	0.2255	0.2351	0.2363	0.2303	0.2203
	10	0.3215	0.3431	0.3431	0.3247	0.2990
	20	0.4232	0.4558	0.4486	0.4168	0.3700
	50	0.5385	0.5721	0.5493	0.5099	0.4496

Table S.2.6 shows that in the simply supported case the curving causes a decrease in the values $p_{cr.}/E_1$, i.e. the values of $p_{cr.}/E_1$ decrease monotonically with ε. However, in the clamped case the relation between $p_{cr.}/E_1$ and ε is more complicated. The data given in Table S.2.7 show that up to some value ε^* of ε the $p_{cr.}/E_1$ increases with ε and than decreases as shown in Tables S.2.8 and S.2.9.

Table S.2.8. The values of $p_{cr.}/E_1$ for the rigidly supported case under relatively small values of ε with various h/ℓ in the case where $E_2/E_1 = 5$.

h/ℓ	ε						
	0.00	0.02	0.03	0.05	0.08	0.09	0.10
0.10	0.0886	0.0888	0.0884	0.0875	0.0850	0.0843	0.0830
0.15	0.1616	0.1642	0.1646	0.1657	0.1641	0.1635	0.1623
0.20	0.2255	0.2304	0.2325	0.2351	0.2372	0.2369	0..2363

Table S.2.9. The values of $p_{cr.}/E_1$ for the rigidly supported case under relatively small values of ε with various E_2/E_1 in the case where $h/\ell = 0.2$

E_2/E_1	ε						
	0.00	0.02	0.03	0.05	0.08	0.09	0.10
5	0.2255	0.2304	0.2325	0.2351	0.2372	0.2369	0.2363
10	0.3215	0.3323	0.3373	0.3431	0.3448	0.3448	0.3431
20	0.4232	0.4422	0.4467	0.4558	0.4557	0.4527	0.4486

S.2.3. Bibliographical Notes

The results given in this supplement are based on the papers by N.Yahnioglu and S.Selim [S7], and S.D. Akbarov and S.Selim [S1].

REFERENCES

1. Akbarov, S.D.: Loss of stability in two fibers in an elastic matrix with large elastic strains, *Soviet Applied Mechanics* January (1982), 626-630.
2. Akbarov, S.D.: On the stress state in an infinite body, reinforced by a single curved layer of the filler, in *Proceedings of the 10-th Scientific Conference of Young Scientists of the Institute of Mechanics of the Academy of Sciences of the Ukrainian SSR, Kiev,* June 12-14, (1984), part I, pp 2-7 (deposited in VINITI 30.07.1984, No 5535-84) (in Russian).
3. Akbarov, S.D.: A method of solving problems in the mechanics of composite materials with curved viscoelastic layers, *Soviet Applied Mechanics* September (1985), 221-226.
4. Akbarov, S.D.: Normal stresses in a fiber composite with curved structures having a low concentration of filler, *Soviet Applied Mechanics* May (1986), 1065-1069.
5. Akbarov, S.D.: On the distribution of normal stresses in multilayered composite material with curved structures, *Izv Akad Nauk Azerb SSR Ser Phys Tech Nath Sci* **3**, (1986), 158-163 (in Russian).
6. Akbarov, S.D.: Stress state in a viscoelastic fibrous composite with curved structures and a low fiber concentration, *Soviet Applied Mechanics* December (1986), 506-513.
7. Akbarov, S.D.: The influence of rheological parameters of matrix material on the distribution of self-equilibrated stresses in multilayered composite with curved structures, *Mech Comp Mater,* 4 (1986), 610-617.
8. Akbarov, S.D.: The distribution of stresses in multilayered composite material with curved structure in shear (the model of the piecewise-homogeneous body), *Preprint, Akad Nauk Azer SSR, Inst Phys,* Baku, No 162 (1986), 35 pp (in Russian).
9. Akbarov, S.D.: On mechanics of composite materials with local curvings in the structure, *Prikl Mech* **23**, No 1(1987), 119-122 (in Russian).
10. Akbarov, S.D.: Stress distribution in a multilayered composite with small-scale antiphase curvatures in the structure, *Soviet Applied Mechanics* August (1987), 107-112.
11. Akbarov, S.D.: On the distribution of self-equilibrated stresses in multilayered composite material with curved structures, in *Mathematical methods and physical mechanics of the field,* series 26 (1987), 83-89 (in Russian).
12. Akbarov, S.D.: Stress state in a laminated composite with local curvatures in the structure, *Soviet Applied Mechanics,* November (1988), 452-461.
13. Akbarov, S.D.: Distributions of self-balanced stresses in a laminated composite material with antiphase locally distorted structures, *Soviet Applied Mechanics* December (1988), 560-565.
14. Akbarov, S.D.: Effect of the modes of small-scale antiphase local curvature in the structure of laminated composites on the distribution of self-balanced stresses, *Soviet Applied Mechanics* January (1989), 658-663.
15. Akbarov, S.D.: Solution of problems of the stress-strain state of composite materials with curvilinearly anisotropic layers, *Soviet Applied Mechanics* July (1989), 12-21.
16. Akbarov, S.D.: On fracture mechanics of composite materials, *Dokl Akad Nauk Azer SSR* **45,** No 11 (1989), 18-21(in Russian).
17. Akbarov, S.D.: The distribution of self-equilibrated stresses in fibrous composite materials with twisted fibers, *Mech Comp Mater,5*(1990),803-812.

18. Akbarov, S.D.: On the crack problems in composite materials with locally curved layers, *Mech Comp Mater,* 6 (1994), 750-759.
19. Akbarov, S.D.: On the row cracks problem in composite materials with periodically curved layers, *Bull. Tech. Univers. Istanbul.* **47**, No3(1994), 577-593.
20. Akbarov, S.D.: On the determination of normalized non-linear mechanical properties of composite materials with periodically curved layers, *Int. J. Solids and Structures,* **32**, No21(1995), 3229-3143.
21. Akbarov, S.D. and Aliev, S.A.: The stress state in laminated composite materials with partial curving in the structure, *Preprint, Akad Nauk Azer SSR, Inst Phys,* Baku, No 252 (1987), 78 pp (in Russian).
22. Akbarov, S.D. and Aliev, S.A.: On the influence of the curving of a single load-carrying layer on the distribution of stresses in the multilayered plate, *Izv Akad Nauk Azerb SSR Ser Phys Tech Nath Sci* **3,** (1988), 54-59 (in Russian).
23. Akbarov, S.D. and Aliev, S.A.: Stress state in laminar composite material with partial distortion in structure, *Soviet Applied Mechanics* June (1991), 1127-1132.
24. Akbarov, S.D. and Babazade, M.B.: On the method of problem solving in mechanics of fibrous composite materials with curved fibers, deposited in *VINITI, CI, ONT,* No 4993-B87 (1987), 67pp. (in Russian).
25. Akbarov, S.D., Cilli, A. and Guz, A.N.: The theoretical strength limit in compression of viscoelastic layered composite materials, *Composites, Part B,* Discussion Paper No.B28.
26. Akbarov, S.D. and Djamalov, Z.R. Influence of geometric non-linearity calculation on stress distribution in laminar composites with curved structures, *Mech Comp Mater,* 6 (1992), 799-812.
27. Akbarov, S.D. and Guliev, G.M.: Quasihomogeneous states in composite materials with small-scale spatially periodically curving in the structure, deposited in *VINITI, CI, ONT,* No 511-B91 (1991), 90pp (in Russian).
28. Akbarov, S.D., Guliev, M.S. and Movsumov, E.A.: Determination of corrected mechanical properties for composites with periodically curved layers, *Int Appl Mech* January (1994), 508-515.
29. Akbarov, S.D. and Guz, A.N.: Method of solving problems in mechanics of composite materials with bent layers, *Soviet Applied Mechanics* October (1984), 299-304.
30. Akbarov, S.D. and Guz, A.N.: Method off solving problems in the mechanics of fiber composites with curved structures, *Soviet Applied Mechanics* March (1985), 777-785.
31. Akbarov, S.D. and Guz A.N.: On the stress state in composite material with curved layers and low filler concentration, *Mech Comp Mater,* 6 (1984), 990-996.
32. Akbarov, S.D. and Guz, A.N.: Stability of two fibers in an elastic matrix with small strains, *Soviet Applied Mechanics* July (1985), 1-7.
33. Akbarov, S.D. and Guz, A.N.: Model of a piecewise-homogeneous body in the mechanics of laminar composites with fine-scale curvatures, *Soviet Applied Mechanics* October (1985), 313-319.
34. Akbarov, S.D. and Guz, A.N.: On mechanics of composite materials with curved structures, *Dokl Akad Nauk USSR* **281**, No1, (1985), 37- 41(in Russian).

35. Akbarov, S.D. and Guz, A.N.: Stress state of a fiber composite with curved structures with a low fiber concentration, *Soviet Applied Mechanics* December (1985), 560-565.
36. Akbarov, S.D. and Guz, A.N.: On one effect in mechanics of composite materials fracture, *Dokl Akad Nauk USSR* **290,** No1, (1986), 23-26 (in Russian).
37. Akbarov, S.D. and Guz, A.N.: Distribution of stresses in multilayered composite material with curved structures (the model of the piecewise-homogeneous body), *Mech Comp Mater,* 4 (1987), 592-599.
38. Akbarov, S.D. and Guz, A.N.: Continuum theory in the mechanics of composite materials with small-scale structural distortion, *Soviet Applied Mechanics* August (1991), 107-117.
39. Akbarov, S.D. and Guz, A.N.: Mechanics of composite materials with distorted structure (Review). Continuum theory, fiber composites, *Soviet Applied Mechanics* November (1991), 429-443.
40. Akbarov, S.D. and Guz, A.N.: Mechanics of composite materials with curved structures (survey). Composite laminates, *Soviet Applied Mechanics* December (1991), 535-550.
41. Akbarov, S.D. and Guz, A.N.: Statics of laminated and fibrous composites with curved structures, *Appl Mech Rev* (published by the American Society of Mechanical Engineers), **45**, No2 (1992), 17-35.
42. Akbarov, S.D., Guz, A.N., Dzhamalov, Z.R. and Movsumov, E.A.: Solution of problems involving the stress of composite materials with curved layers in the geometrically non-linear statement, *Int Appl Mech* December (1992), 343-347.
43. Akbarov, S.D., Guz, A.N. and Mustafaev S.M.: Mechanics of composite materials with anisotropic distorted layers, *Soviet Applied Mechanics* December (1987), 528-534.
44. Akbarov, S.D., Guz, A.N. and Yahnioglu, N.: Mechanics of composite materials with curved structures and elements of constructions (review), *Int Appl Mech,* May (1999), 1067-1079.
45. Akbarov, S.D., Guz, A.N. and Zamanov, A.D.: Natural vibrations of composite materials having structures with small-scale curvatures, *Int Appl Mech* June (1993), 794-800.
46. Akbarov, S.D., Maksudov, F.G., Panakhov P.G. and Seyfullayev, A.I.: On the crack problems in composite materials with curved layers, *Int J Eng Sci* , **32**, No 6 (1994), 1003-1016.
47. Akbarov, S.D. and Mustafaev, S.M.: Distribution of self-balanced stresses in composite materials with curved curvilinearly anisotropic layers, *Soviet Applied Mechanics* June (1992), 1225-1228.
48. Akbarov, S.D., Sisman, T. and Yahnioglu, N.: On the fracture of the unidirectional composites in compression, *Int J Eng Sci* , **35**, No12/13 (1997), 1115-1135.
49. Akbarov, S.D., Verdiev, M.D. and Guz, A.N.: Stress and deformation in a layered composite material with distorted layers, *Soviet Applied Mechanics* June (1989), 1146-1153.
50. Akbarov, S.D. and Yahnioglu, N.: Stress distribution in a strip fabricated from a composite material with small-scale curved structure, *Int Appl Mech* March, (1997), 684-690.

51. Akbarov, S.D. and Yahnioglu, N.: On the finite element analysis of the influence of the local structural damage of the multilayered thick plate material to the stress distribution, in *"Damage and Fracture Mechanics"* Eds: C.A. Birebbia and A. Carpinteri, Computational Mech. Pub., Southampton, UK., Boston USA, 1998, pp:187-196
52. Akbarov, S.D. and Yahnioglu, N.: The influence of the local structural damage of the multilayered thick plate material to the Stress distribution., *Int App Mech* . Mach (1999), 873-878.
53. Babich, I. Yu. and Guz, A.N.: Stability of rods, plates, and shells of composite materials (three-dimensional formulation): survey, *Soviet Applied Mechanics,* April,(1984), 835-849.
54. Babich, I.Yu. and Guz, A.N.: Stability of fibrous composites, *Appl Mech Rev* (published by the American Society of Mechanical Engineers), **45**, No2 (1992), 61-80.
55. Bachvalov, N.S. and Panasenko, G.P.: *Averaging of the processes in periodical medium,* Nauka, Moscow, 1984 (in Russian).
56. Bazhant, Z.P.: The influence of the curvature of reinforced fibres on the elasticity modulus and strength of composite materials, *Mechanica Polimerov,* No.2 (1968), 314-321 (in Russian).
57. Bazhant, Z.P.: Correlation study of formulation of incremental deformation and stability of continuous bodies, *Trans Amer Soc Mech Eng Ser E* No4 (1971), 344-358.
58. Belyankin, F.P., Yachenko, V.F. and Margolin, G.G.: *Strength and deformations of glass-reinforced plastics under two-axial compression,* Nauk Dumka, Kiev, 1971 (in Russian)
59. Biezeno, C.B. and Hencky, H.: On the general theory of elastic stability, *K.Akad Wet Amsterdam Proc* ,32, (1930), 444-456.
60. Biot, M.A.: *Mechanics of incremental deformations,* Willey, New York, 1965.
61. Bolotin, V.V.: Vibration of layered elastic plates, *Proceeding Vibration Problems,* Vol.**4**, No.4(1963),331-346.
62. Bolotin, V.V.: The theory of reinforced layered material with random distortions, *Mechanica Polimerov,* No.1 (1966), 11-19 (in Russian).
63. Bolotin, V.V.: Layered elastic and viscoelastic materials with initial random distortions, *Eng J Mech Tverd Tela,* No 3 (1966), 59-65 (in Russian).
64. Bolotin, V.V. and Novichkov, Yu.N.: *Mechanics of multilayered structures,* Mashinostroyeniye, Moscow,1980 (in Russian).
65. Byun, J.-H., Lee, S.-K. and Um, M.-K.: Analytic characterisation of 3-D angle-interlock woven composites, in *proceedings book of ECCM-8* , Vol.**4**, 1998, pp.603-610.
66. Cherepanov, G.P.: *Mechanics of brittle fracture,* Mc Graw-Hill, New York, 1979.
67. Cherepanov, G.P.: *Fracture mechanics of composite materials,* Nauka, Moscow, 1983. (in Russian).
68. Chernin, I.M. and Gul, V.E.: On the mechanism of tensional failure of cloth-based glass-reinforced plastics, *Plastic Mass,* 3, (1964), 45-48 (in Russian).
69. Cheung, Y.K.: *Finite strip method in structural analysis*, Pergamon Press, Oxford, 1976.
70. Christensen, R.M.: *Mechanics of composite materials.* Willey, New York, 1979.

71. Christensen, R.M. and Lo, K.H.: Solutions for effective shear properties in three phase sphere and cylinder models, *J Mech and Phys Solids,* **27**, No 4, (1979), .
72. Chou, T.W., Mc. Cullough R.L. and Pipes R.B.: Composites, *J Scientific American,* No 10, (1986), 193-203.
73. *Composite materials,* Vol. **1-7**, Eds. By Broutman L.J. and Krock R.H., Academic Press, New York and London, 1974.
74. *Composite materials handbook,* Mel. M. Schwartz, Editor in Chief, Second Edition, Mc Graw-Hill, Inc. New York, 1992.
75. Corten, H.T.: Fracture of reinforcing plastics, in *Modern composite materials,* Eds. By Broutman and Krock, R.H., Addison-Wesley, Reading, Massachusetts, 1967, 27-100.
76. Dimitrenco, Yu.: Inorganic matrix composite materials: peculiarities, modelling, testing, in *proceedings book of ECCM-8* , Vol.**4**, 1998, pp.201-207.
77. Dow, N.F. and Grunfest, I.J.: Determination of most needed potentially possible improvements in materials for ballistic and space vehicles, *General Electric Co, Space Sci Lab,* TISR 60 SD 389, June, (1960).
78. Erdogan, F. and Gupta, G.D.: On the numerical solution of singular integral equations, *Quart Appl Math,* **29**, (1972), 525-534.
79. Eringen, A.C. and Suhubi, E.S.: *Elastodinamics, Vol 1. Finite motions,* Acad Press, New York, 1975.
80. Eringen, A.C. and Suhubi, E.S.: *Elastodinamics, Vol 2. Linear theory,* Acad Press, New York, 1975.
81. *Fracture,* an advanced treatise, Vol.**1-7**, Ed. By Liebowitz, H., Academic Press, New York and London, 1968-1972.
82. Feng, Z-N., Allen, H.G. and Moy, S.S.: Micromechanical analysis of a woven composite, in *proceedings book of ECCM-8* , Vol.**4**, 1998, pp.619-625.
83. Ganesh, V.K. and Naik, N.K.: Failure behaviour of plane weave fabric laminates under on-axis uniaxial tensile loading: 1-laminate geometry, *J Composite Materials,* Vol. **30,** (1996), 1748-1778.
84. Ganesh, V.K. and Naik, N.K.: Failure behaviour of plane weave fabric laminates under on-axis uniaxial tensile loading: 1-analytical predictions, *J Composite Materials,* Vol. **30,** (1996), 1779-1882.
85. Ganesh, V.K. and Naik, N.K.: Failure behaviour of plane weave fabric laminates under on-axis uniaxial tensile loading: III- effect of fabric geometry, *J Composite Materials,* Vol. **30,** (1996), 1823-1856.
86. Gerard, F and Gilbert, A.: Methods of critical deformation for buckling of plates and shells at creep, *Mekhanika,* No. 2, (1959), 113-125 (in Russian, translated from English).
87. Gladwell, G.M.L.: *Contact Problems in the classical theory of elasticity*, Noordhoff, 1980.
88. Greszczuk, L.B.: Fracture of composite reinforced by circular fibers from loss of stability of fibers, *J. AIAA* **13**, No.10, (1975), 67-75.
89. Greszczuk, L.B.: On types of fracture of unidirectional composites in compression, in *Strength and fracture of composite materials,* Zinatne, Riga, (1983), 304-312 (in Russian).
90. Green, A.E. and Adkins, J.E.: *Large elastic deformations and non-linear continuum mechanics,* Oxford at the Clarendon Press, 1960.

91. Guz, A.N.: On the approximate method of stress concentration determination near curvilinear holes in shells, *Prikl Mekh* **8**, No 6, (1962), 605-612 (in Ukrainian).
92. Guz, A.N.: Determination of the theoretical compressive strength of reinforced materials, *Dokl Akad Nauk Ukr SSR, Ser A Phys-Math Tech Sci* **3**, (1969), 236-238 (in Russian).
93. Guz, A.N.: Construction of a theory of stability of unidirectional fiber composites, *Soviet Applied Mechanics,* February, (1969).
94. Guz, A.N. *Stability of three-dimensional bodies,* Naukova Dumka, Kiev, 1971 (in Russian).
95. Guz, A.N.: *Stability of elastic bodies with finite strains,* Naukova Dumka, Kiev, 1973 (in Russian).
96. Guz, A.N.: On the continual theory of composite materials with small scale curvings in the structure, *Dokl Akad Nauk USSR* **268**, No 2, (1983), 307-313 (in Russian).
97. Guz, A.N.: On the vibration theory of the composite materials with small-scale curved structures, *Dokl Akad Nauk USSR* **270**, No 5, (1983), 1078-1081 (in Russian).
98. Guz, A.N.: Mechanics of composite materials wit a small-scale structural flexure, *Soviet Applied Mechanics* November (1983), 383-393.
99. Guz, A.N.: Quasi-uniform states in composites with small-scale curvatures in the structure, *Soviet Applied Mechanics* December (1983), 479-490.
100. Guz, A.N.: *Mechanics of the brittle fracture of materials with initial stresses,* Naukova Dumka, Kiev, 1983 (in Russian).
101. Guz, A.N.: *Fundamentals of the three-Dimensional Theory of stability of deformable bodies,* Springer-Verlag Berlin, 1999.
102. Guz, A.N.: *Mechanics of fracture of composite materials in compression,* Naukova Dumka, Kiev, 1990 (in Russian).
103. Guz, A.N. and Akbarov, S.D.: Problems of mechanics of composite materials with curved structures (model of the piecewise-homogeneous medium), *Mech Comp Mater,* 5 (1989), 788-798.
104. Guz, A.N. and Guz', G.V.: On the mechanics of composite materials with large-scale curving of the reinforcing elements, *Mech Comp Mater,* 4 (1982), 634-641.
105. Guz, A.N. and Chekhov, Vik.N.: Stability of laminated composites, *Appl Mech Rev* (published by the American Society of Mechanical Engineers), **45**, No2 (1992), 81-101.
106. Guz, A.N. and Lapusta, Yu.N.: On the method of investigation of stability of the fiber in the elastic semi-infinite matrix near the free surface, *Prikl Mathem and Mech ,* **53**, No. 4, (1989), 693-697 (in Russian).
107. Guz, A.N. and Nemish Yu.N.: Perturbation of boundary shape in continuum mechanics (review), *Soviet Applied Mechanics* September (1987), 799-822.
108. Guz, A.N., Tomashevski, V.T., Shulga, N.A. and Yakovlev, V.S.: *Technological stresses and strains in composite materials,* Vitsaya skola, Kiev, 1988 (in Russian).
109. Hinton, E. and Campell, J.: Local and global smoothing of discontinuous finite element function using a least squares method, *Int J Num Meth Eng* No 8, (1979), 461-480.
110. Hyer, M.W., Mass, L.C. and Puch, H.P.: The influence of layer waveness on the stress state in hydrostatically loaded cylinders, *J Reinforced Plastics Composites,* No 7, (1988), 601-613.

111.Kalandia, A.I.: *Mathematical methods of two-dimensional elasticity theory,* Nauka, Moscow, 1973 (in Russian).
112.Kalmkarov, A.L., Kudryavcev B.A. and Parton, V.Z.: The problem on the curved layer from the composite materials with periodical waveness surfaces in the structures, *Prikl Mathem and Mech.* **51**, No 1, (1987),68-75. (in Russian).
113.Kantarovich, L.V. and Krilov, V.I.: *Approximate methods in advance calculus,* Fizmatgiz, Moscow, 1962 (in Russian).
114.Karpenko, L.I., Terletskii, V.A. and Lyashenko, B.A. A mechanism of the failure of oriented plastics, *Problemi Prochnosti,* **1**, (1972), 50-55 (in Russian).
115.Kashtalian, M.: On 3-D analysis of laminated composite plates with perturbed interface shape, in *proceedings book of ECCM-8* , Vol.**1**, 1998, pp.165-172.
116.Kelly, A.: Composite materials: impediments to wider use and some suggestions to overcome these, in *proceedings book of ECCM-8* , Vol.**1**, 1998, pp.15-18.
117.Kevorkian, J. and Cole, J.D.: *Perturbation method in applied mathematics,* Springer-Verlag, New York, 1981.
118.Kregers, A.F.: Determination of the properties of composite materials reinforced with the spatially-curvlinear filler, *Mech Polim,* No 5,(1969), 790-793 (in Russian).
119.Kregers, A.F. and Teters G.A.: Structural model of deformation of anisotrop spatially reinforced composites, *Mech Comp Mater,* No1 (1982), 14-22.
120.Kromm, A.: Verallgeneinerte theorie der plattenstatik, *Ingenieur-Archiv,* No 21, (1953), 266-286.
121.Kromm, A.: Über die randguer kraftebei gestutzten platten, *ZAMM,* No 35 (1955), 231-242.
122.Krylov, V.I. and Shulgina, L.T.: *Handbook on numerical integration,* Nauka, Moscow, 1966 (in Russian).
123.Kutug, Z.: Natural vibration of the beam-strip fabricated from a composite material with small-scale curvings in the structure, *Mech Comp Mater,* Vol.32., No.4, (1996), 502-512.
124.Kutuk, Z.: Natural vibration of composite plates having periodical curvings in the plate material structure, in *proceedings book of ECCM-8* , Vol.**2**, 1998, pp.251-258.
125.Lekhnitskii, S.G.: *Theory of elasticity of an anisotropic body,* Holden Day, San Francisco, CA, 1963.
126.Love, A.E.H.: *Mathematical theory of elasticity,* 4th Ed., Cambridge University Press (reprinted by Dover Publications, New York, 1944).
127.Makarov B.P. and Nikolayev, V.P.: The influence of the curving of the reinforcing elements to the mechanical and heat-physical properties of composite materials, *Mech Polim,* No 6,(1971), 1036-1039 (in Russian).
128.Mansfied E.H. and Purslow D.: The influence of fibre waviness on the moduli of unidirectional fibre reinforced composites, *Aeronautical Research Council Current Paper,* No 1339, (1974), 30pp.
129.*Methods of statical experiments of reinforcing plastics.* Zinatne, Riga, 1972.(in Russian).
130.Nosarev, A.V.: The influence of the fibers curving to the elastic properties of the unidirected reinforced plastics, *Mech Polim,* No 5,(1967), 858- (in Russian)
131.Rabotnov, Yu.N.: *Elements of hereditary mechanics of solid bodies,* Nauka, Moscow, 1977 (in Russian).

132.Rosen, B.W.: Mechanics of reinforcement of composites, in *Fiber composite materials*, Amer Soc for Metals, Metals Park, Ohio, 1965, pp.37-78.
133.Sadovsky, M.A., Pu, S. L. and Hussain M.A.: Buckling of microfibers, *Jbid,* **34**, No. 4, (1967), 295-302.
134.Schuerch, H.: *Boron filament composite materials for space structures.* Prt 1: Compressive Strength of boron metal composite, *Astro Research Corp. Santa Barbara* (Ca),1964, Rep. N ARC-R-168.
135.Schuerch, H.: Prediction of compressive strength in uniaxial boron fibermetal matrix composite materials, *AIAA Journal*, **4**, No.1, (1966), 102-106.
136.Schapery, R.A.: Approximate methods of transform inversion for viscoelastic stress analysis, *Proc US Natl Cong:Appl ASME,* No.4, (1966), 1075-1085.
137.Schapery, R.A.: A viscoelastic behaviour of composite materials, in *Composite materials,* Vol. **1-7**, Eds. By Broutman L.J. and Krock R.H., Mir, Moscow, 1978, (translated from English), Vol. **2**.: *Mechanics of composite materials*, pp.102-195 (in Russian, translated from English).
138.Scida, D, Aboura, Z., Benzeggagh, M.L. and Bocherens, E.: Analysis of satin and twill weave composite damage under tensile load: analytical modelling of failure behaviour, in *proceedings book of ECCM-8* , Vol.**4**, 1998, pp.611-618.
139.Southwell, R.V.: On the general theory of elastic stability, *Philos Soc London Ser A* 213, (1913), 187-244.
140.Swift, D.G.: Elastic moduli of fibrous composites containing misalined fibres, *J. Phys. D: Appl. Phys.*, Vol. **8**, (1975), 223-240.
141.Tamuzs, V.P. and Kuksenko, V.A.: *Micromechanics of fracture of polymer materials,* Zinatne, Riga, 1978 (in Russian).
142.Tarnopolsky, Yu.M. and Rose, A.V.: *Special features of design of parts fabricated from reinforced plastics,* Zinatne, Riga, 1969 (in Russian).
143.Tarnopolsky, Yu.M., Jigun, I.G. and Polyakov, V.A.: *Spatially-reinforced composite materials: Handbook,* Mashinosroyenia, Moscow, 1987 (in Russian).
144.Timoshenko, S. and Goodier, J.N.: *Theory of elasticity*, Int Student Edit, New York, 1951.
145.Vandeurzen, Ph., Ivens, J. and Verpoest, I.: A three-dimensional micromechanical analysis of woven fabric composites: I. Geometric analysis, *Comp Sci Technol,* Vol. **56**, (1996), 1303-1315.
146.Vandeurzen, Ph., Ivens, J. and Verpoest, I..: A three-dimensional micromechanical analysis of woven fabric composites: II. elastic analysis, *Comp Sci Technol,* Vol. **56**, (1996), 1317-1327.
147.Watson, G.N.: *Theory of Bessel functions*, *Part 1*,Izd Inostr. Liter, Moscow, 1949 (in Russian, translated from English).
148.Wen-Yi Chung and Testa, R.B.: The elastic stability of fibers in a compressive plate, *J Composite Materials,* Jan. (1969), 149-157.
149.Whitney, J.M., Geometrical effects of filament twist on the modulus and strength of graphite fiber, *Reinforced Composites, Textile Res.J.,* Vol.36, (1966), 765-770.
150.Whitney, T.J. and Chou, T-W.: Modelling of 3-D angle-interlock textile structural composites, *J Composite Materials,* Vol. **23,** (1989), 890-911.
151.Yahnioglu, N.: FEM analyses of the boundary-value problems corresponding of the static of elements of constructions fabricated from the composite materials with curved structures. *Doktoral Diss.* Yildiz Technical University, Istanbul, 1996.

152.Yahnioglu, N.: Three-dimensional analysis of stress fields in the plate fabricated from composite materials with small-scale structural curving, *Mech Comp Mater,* **33,** No3 (1997), 340-348.

153.Yang, J.M., Ma, C.L. and Chou, T-W.: Fiber inclination model of three-dimensional textile structural composites, *J Composite Materials,* Vol. **20,** (1986), 472-484.

154.Zamanov, A.D.: On the stress distribution in the thick plate fabricated from the composite material with curved structures under forced vibration. *Mech Comp Mater,* 4 (1999), 447-454.

155.Zienkiewicz, O.C. and Taylor, R.L.: *The finite element method. 4th Ed. Vol. 1, Basic formulation and linear problems*, Mc Grow-Hill Book Company, London,1989

156.Zienkiewicz, O.C. and Taylor, R.L.: *The finite element method. 4th Ed. Vol. 2, Solid and fluid mechanics dynamics and non-linearity,* Mc Grow-Hill Book Company, London,1991.

References Supplement

S1. Akbarov, S.D. and Selim, S.: Stability loss of the strip fabricated from the composite with curved structure. *Int Appl. Mech.* (to be published).

S2. Guz, A.N.: Mechanics of composite material failure under axial compression (Brittle Failure), *Soviet Applied Mechanics* April (1982), 863-873.

S3. Guz, A.N.: Mechanics of composite material failure under axial compression (Plastic Failure), *Soviet Applied Mechanics* May (1983), 970-877.

S4. Guz, A.N.: Continuum theory of fracture in the compression of composite materials with a metallic matrix, *Soviet Applied Mechanics* June (1983), 1045-1053.

S5. Guz, A.N.: Three-dimensional stability theory of deformed bodies. Internal instability, *Soviet Applied Mechanics* May (1986), 1023-1035.

S6. Guz, A.N.: Continuous theory of failure of composite materials under compression in the case of a complex stressed state, *Soviet Applied Mechanics* October (1986), 301-316.

S7. Yahnioglu, N. and Selim, S. Geometrical non-linear bending problem of the strip fabricated from the composite with curved structure, *Mech Comp Mater* (to be published)

INDEX

E

F

G

H

I

K

L

M

N

O

P

Q

R

S

Mechanics

SOLID MECHANICS AND ITS APPLICATIONS

Series Editor: G.M.L. Gladwell

Aims and Scope of the Series

The fundamental questions arising in mechanics are: *Why?*, *How?*, and *How much?* The aim of this series is to provide lucid accounts written by authoritative researchers giving vision and insight in answering these questions on the subject of mechanics as it relates to solids. The scope of the series covers the entire spectrum of solid mechanics. Thus it includes the foundation of mechanics; variational formulations; computational mechanics; statics, kinematics and dynamics of rigid and elastic bodies; vibrations of solids and structures; dynamical systems and chaos; the theories of elasticity, plasticity and viscoelasticity; composite materials; rods, beams, shells and membranes; structural control and stability; soils, rocks and geomechanics; fracture; tribology; experimental mechanics; biomechanics and machine design.

1. R.T. Haftka, Z. Gürdal and M.P. Kamat: *Elements of Structural Optimization.* 2nd rev.ed., 1990 ISBN 0-7923-0608-2
2. J.J. Kalker: *Three-Dimensional Elastic Bodies in Rolling Contact.* 1990 ISBN 0-7923-0712-7
3. P. Karasudhi: *Foundations of Solid Mechanics.* 1991 ISBN 0-7923-0772-0
4. *Not published*
5. *Not published.*
6. J.F. Doyle: *Static and Dynamic Analysis of Structures.* With an Emphasis on Mechanics and Computer Matrix Methods. 1991 ISBN 0-7923-1124-8; Pb 0-7923-1208-2
7. O.O. Ochoa and J.N. Reddy: *Finite Element Analysis of Composite Laminates.* ISBN 0-7923-1125-6
8. M.H. Aliabadi and D.P. Rooke: *Numerical Fracture Mechanics.* ISBN 0-7923-1175-2
9. J. Angeles and C.S. López-Cajún: *Optimization of Cam Mechanisms.* 1991 ISBN 0-7923-1355-0
10. D.E. Grierson, A. Franchi and P. Riva (eds.): *Progress in Structural Engineering.* 1991 ISBN 0-7923-1396-8
11. R.T. Haftka and Z. Gürdal: *Elements of Structural Optimization.* 3rd rev. and exp. ed. 1992 ISBN 0-7923-1504-9; Pb 0-7923-1505-7
12. J.R. Barber: *Elasticity.* 1992 ISBN 0-7923-1609-6; Pb 0-7923-1610-X
13. H.S. Tzou and G.L. Anderson (eds.): *Intelligent Structural Systems.* 1992 ISBN 0-7923-1920-6
14. E.E. Gdoutos: *Fracture Mechanics.* An Introduction. 1993 ISBN 0-7923-1932-X
15. J.P. Ward: *Solid Mechanics.* An Introduction. 1992 ISBN 0-7923-1949-4
16. M. Farshad: *Design and Analysis of Shell Structures.* 1992 ISBN 0-7923-1950-8
17. H.S. Tzou and T. Fukuda (eds.): *Precision Sensors, Actuators and Systems.* 1992 ISBN 0-7923-2015-8
18. J.R. Vinson: *The Behavior of Shells Composed of Isotropic and Composite Materials.* 1993 ISBN 0-7923-2113-8
19. H.S. Tzou: *Piezoelectric Shells.* Distributed Sensing and Control of Continua. 1993 ISBN 0-7923-2186-3
20. W. Schiehlen (ed.): *Advanced Multibody System Dynamics.* Simulation and Software Tools. 1993 ISBN 0-7923-2192-8
21. C.-W. Lee: *Vibration Analysis of Rotors.* 1993 ISBN 0-7923-2300-9
22. D.R. Smith: *An Introduction to Continuum Mechanics.* 1993 ISBN 0-7923-2454-4
23. G.M.L. Gladwell: *Inverse Problems in Scattering.* An Introduction. 1993 ISBN 0-7923-2478-1

Mechanics

SOLID MECHANICS AND ITS APPLICATIONS

Series Editor: G.M.L. Gladwell

24. G. Prathap: *The Finite Element Method in Structural Mechanics*. 1993 ISBN 0-7923-2492-7
25. J. Herskovits (ed.): *Advances in Structural Optimization*. 1995 ISBN 0-7923-2510-9
26. M.A. González-Palacios and J. Angeles: *Cam Synthesis*. 1993 ISBN 0-7923-2536-2
27. W.S. Hall: *The Boundary Element Method*. 1993 ISBN 0-7923-2580-X
28. J. Angeles, G. Hommel and P. Kovács (eds.): *Computational Kinematics*. 1993 ISBN 0-7923-2585-0
29. A. Curnier: *Computational Methods in Solid Mechanics*. 1994 ISBN 0-7923-2761-6
30. D.A. Hills and D. Nowell: *Mechanics of Fretting Fatigue*. 1994 ISBN 0-7923-2866-3
31. B. Tabarrok and F.P.J. Rimrott: *Variational Methods and Complementary Formulations in Dynamics*. 1994 ISBN 0-7923-2923-6
32. E.H. Dowell (ed.), E.F. Crawley, H.C. Curtiss Jr., D.A. Peters, R. H. Scanlan and F. Sisto: *A Modern Course in Aeroelasticity*. Third Revised and Enlarged Edition. 1995 ISBN 0-7923-2788-8; Pb: 0-7923-2789-6
33. A. Preumont: *Random Vibration and Spectral Analysis*. 1994 ISBN 0-7923-3036-6
34. J.N. Reddy (ed.): *Mechanics of Composite Materials*. Selected works of Nicholas J. Pagano. 1994 ISBN 0-7923-3041-2
35. A.P.S. Selvadurai (ed.): *Mechanics of Poroelastic Media*. 1996 ISBN 0-7923-3329-2
36. Z. Mróz, D. Weichert, S. Dorosz (eds.): *Inelastic Behaviour of Structures under Variable Loads*. 1995 ISBN 0-7923-3397-7
37. R. Pyrz (ed.): *IUTAM Symposium on Microstructure-Property Interactions in Composite Materials*. Proceedings of the IUTAM Symposium held in Aalborg, Denmark. 1995 ISBN 0-7923-3427-2
38. M.I. Friswell and J.E. Mottershead: *Finite Element Model Updating in Structural Dynamics*. 1995 ISBN 0-7923-3431-0
39. D.F. Parker and A.H. England (eds.): *IUTAM Symposium on Anisotropy, Inhomogeneity and Nonlinearity in Solid Mechanics*. Proceedings of the IUTAM Symposium held in Nottingham, U.K. 1995 ISBN 0-7923-3594-5
40. J.-P. Merlet and B. Ravani (eds.): *Computational Kinematics '95*. 1995 ISBN 0-7923-3673-9
41. L.P. Lebedev, I.I. Vorovich and G.M.L. Gladwell: *Functional Analysis*. Applications in Mechanics and Inverse Problems. 1996 ISBN 0-7923-3849-9
42. J. Menčik: *Mechanics of Components with Treated or Coated Surfaces*. 1996 ISBN 0-7923-3700-X
43. D. Bestle and W. Schiehlen (eds.): *IUTAM Symposium on Optimization of Mechanical Systems*. Proceedings of the IUTAM Symposium held in Stuttgart, Germany. 1996 ISBN 0-7923-3830-8
44. D.A. Hills, P.A. Kelly, D.N. Dai and A.M. Korsunsky: *Solution of Crack Problems*. The Distributed Dislocation Technique. 1996 ISBN 0-7923-3848-0
45. V.A. Squire, R.J. Hosking, A.D. Kerr and P.J. Langhorne: *Moving Loads on Ice Plates*. 1996 ISBN 0-7923-3953-3
46. A. Pineau and A. Zaoui (eds.): *IUTAM Symposium on Micromechanics of Plasticity and Damage of Multiphase Materials*. Proceedings of the IUTAM Symposium held in Sèvres, Paris, France. 1996 ISBN 0-7923-4188-0
47. A. Naess and S. Krenk (eds.): *IUTAM Symposium on Advances in Nonlinear Stochastic Mechanics*. Proceedings of the IUTAM Symposium held in Trondheim, Norway. 1996 ISBN 0-7923-4193-7
48. D. Ieşan and A. Scalia: *Thermoelastic Deformations*. 1996 ISBN 0-7923-4230-5

Mechanics

SOLID MECHANICS AND ITS APPLICATIONS

Series Editor: G.M.L. Gladwell

49. J.R. Willis (ed.): *IUTAM Symposium on Nonlinear Analysis of Fracture*. Proceedings of the IUTAM Symposium held in Cambridge, U.K. 1997 ISBN 0-7923-4378-6
50. A. Preumont: *Vibration Control of Active Structures*. An Introduction. 1997 ISBN 0-7923-4392-1
51. G.P. Cherepanov: *Methods of Fracture Mechanics: Solid Matter Physics*. 1997 ISBN 0-7923-4408-1
52. D.H. van Campen (ed.): *IUTAM Symposium on Interaction between Dynamics and Control in Advanced Mechanical Systems*. Proceedings of the IUTAM Symposium held in Eindhoven, The Netherlands. 1997 ISBN 0-7923-4429-4
53. N.A. Fleck and A.C.F. Cocks (eds.): *IUTAM Symposium on Mechanics of Granular and Porous Materials*. Proceedings of the IUTAM Symposium held in Cambridge, U.K. 1997 ISBN 0-7923-4553-3
54. J. Roorda and N.K. Srivastava (eds.): *Trends in Structural Mechanics*. Theory, Practice, Education. 1997 ISBN 0-7923-4603-3
55. Yu.A. Mitropolskii and N. Van Dao: *Applied Asymptotic Methods in Nonlinear Oscillations*. 1997 ISBN 0-7923-4605-X
56. C. Guedes Soares (ed.): *Probabilistic Methods for Structural Design*. 1997 ISBN 0-7923-4670-X
57. D. François, A. Pineau and A. Zaoui: *Mechanical Behaviour of Materials*. Volume I: Elasticity and Plasticity. 1998 ISBN 0-7923-4894-X
58. D. François, A. Pineau and A. Zaoui: *Mechanical Behaviour of Materials*. Volume II: Viscoplasticity, Damage, Fracture and Contact Mechanics. 1998 ISBN 0-7923-4895-8
59. L.T. Tenek and J. Argyris: *Finite Element Analysis for Composite Structures*. 1998 ISBN 0-7923-4899-0
60. Y.A. Bahei-El-Din and G.J. Dvorak (eds.): *IUTAM Symposium on Transformation Problems in Composite and Active Materials*. Proceedings of the IUTAM Symposium held in Cairo, Egypt. 1998 ISBN 0-7923-5122-3
61. I.G. Goryacheva: *Contact Mechanics in Tribology*. 1998 ISBN 0-7923-5257-2
62. O.T. Bruhns and E. Stein (eds.): *IUTAM Symposium on Micro- and Macrostructural Aspects of Thermoplasticity*. Proceedings of the IUTAM Symposium held in Bochum, Germany. 1999 ISBN 0-7923-5265-3
63. F.C. Moon: *IUTAM Symposium on New Applications of Nonlinear and Chaotic Dynamics in Mechanics*. Proceedings of the IUTAM Symposium held in Ithaca, NY, USA. 1998 ISBN 0-7923-5276-9
64. R. Wang: *IUTAM Symposium on Rheology of Bodies with Defects*. Proceedings of the IUTAM Symposium held in Beijing, China. 1999 ISBN 0-7923-5297-1
65. Yu.I. Dimitrienko: *Thermomechanics of Composites under High Temperatures*. 1999 ISBN 0-7923-4899-0
66. P. Argoul, M. Frémond and Q.S. Nguyen (eds.): *IUTAM Symposium on Variations of Domains and Free-Boundary Problems in Solid Mechanics*. Proceedings of the IUTAM Symposium held in Paris, France. 1999 ISBN 0-7923-5450-8
67. F.J. Fahy and W.G. Price (eds.): *IUTAM Symposium on Statistical Energy Analysis*. Proceedings of the IUTAM Symposium held in Southampton, U.K. 1999 ISBN 0-7923-5457-5
68. H.A. Mang and F.G. Rammerstorfer (eds.): *IUTAM Symposium on Discretization Methods in Structural Mechanics*. Proceedings of the IUTAM Symposium held in Vienna, Austria. 1999 ISBN 0-7923-5591-1

Mechanics

SOLID MECHANICS AND ITS APPLICATIONS

Series Editor: G.M.L. Gladwell

69. P. Pedersen and M.P. Bendsøe (eds.): *IUTAM Symposium on Synthesis in Bio Solid Mechanics*. Proceedings of the IUTAM Symposium held in Copenhagen, Denmark. 1999 ISBN 0-7923-5615-2
70. S.K. Agrawal and B.C. Fabien: *Optimization of Dynamic Systems*. 1999 ISBN 0-7923-5681-0
71. A. Carpinteri: *Nonlinear Crack Models for Nonmetallic Materials*. 1999 ISBN 0-7923-5750-7
72. F. Pfeifer (ed.): *IUTAM Symposium on Unilateral Multibody Contacts*. Proceedings of the IUTAM Symposium held in Munich, Germany. 1999 ISBN 0-7923-6030-3
73. E. Lavendelis and M. Zakrzhevsky (eds.): *IUTAM/IFToMM Symposium on Synthesis of Nonlinear Dynamical Systems*. Proceedings of the IUTAM/IFToMM Symposium held in Riga, Latvia. 2000 ISBN 0-7923-6106-7
74. J.-P. Merlet: *Parallel Robots*. 2000 ISBN 0-7923-6308-6
75. J.T. Pindera: *Techniques of Tomographic Isodyne Stress Analysis*. 2000 ISBN 0-7923-6388-4
76. G.A. Maugin, R. Drovot and F. Sidoroff (eds.): *Continuum Thermomechanics*. The Art and Science of Modelling Material Behaviour. 2000 ISBN 0-7923-6407-4
77. N. Van Dao and E.J. Kreuzer (eds.): *IUTAM Symposium on Recent Developments in Non-linear Oscillations of Mechanical Systems*. 2000 ISBN 0-7923-6470-8
78. S.D. Akbarov and A.N. Guz: *Mechanics of Curved Composites*. 2000 ISBN 0-7923-6477-5
79. M.B. Rubin: *Cosserat Theories: Shells, Rods and Points*. 2000 ISBN 0-7923-6489-9

Kluwer Academic Publishers – Dordrecht / Boston / London

ICASE/LaRC Interdisciplinary Series in Science and Engineering

1. J. Buckmaster, T.L. Jackson and A. Kumar (eds.): *Combustion in High-Speed Flows.* 1994 ISBN 0-7923-2086-X
2. M.Y. Hussaini, T.B. Gatski and T.L. Jackson (eds.): *Transition, Turbulence and Combustion.* Volume I: Transition. 1994 ISBN 0-7923-3084-6; set 0-7923-3086-2
3. M.Y. Hussaini, T.B. Gatski and T.L. Jackson (eds.): *Transition, Turbulence and Combustion.* Volume II: Turbulence and Combustion. 1994 ISBN 0-7923-3085-4; set 0-7923-3086-2
4. D.E. Keyes, A. Sameh and V. Venkatakrishnan (eds): *Parallel Numerical Algorithms.* 1997 ISBN 0-7923-4282-8
5. T.G. Campbell, R.A. Nicolaides and M.D. Salas (eds.): *Computational Electromagnetics and Its Applications.* 1997 ISBN 0-7923-4733-1
6. V. Venkatakrishnan, M.D. Salas and S.R. Chakravarthy (eds.): *Barriers and Challenges in Computational Fluid Dynamics.* 1998 ISBN 0-7923-4855-9

KLUWER ACADEMIC PUBLISHERS – DORDRECHT / BOSTON / LONDON

ERCOFTAC SERIES

1. A. Gyr and F.-S. Rys (eds.): *Diffusion and Transport of Pollutants in Atmospheric Mesoscale Flow Fields*. 1995 ISBN 0-7923-3260-1
2. M. Hallbäck, D.S. Henningson, A.V. Johansson and P.H. Alfredsson (eds.): *Turbulence and Transition Modelling*. Lecture Notes from the ERCOFTAC/IUTAM Summerschool held in Stockholm. 1996 ISBN 0-7923-4060-4
3. P. Wesseling (ed.): *High Performance Computing in Fluid Dynamics*. Proceedings of the Summerschool held in Delft, The Netherlands. 1996 ISBN 0-7923-4063-9
4. Th. Dracos (ed.): *Three-Dimensional Velocity and Vorticity Measuring and Image Analysis Techniques*. Lecture Notes from the Short Course held in Zürich, Switzerland. 1996 ISBN 0-7923-4256-9
5. J.-P. Chollet, P.R. Voke and L. Kleiser (eds.): *Direct and Large-Eddy Simulation II*. Proceedings of the ERCOFTAC Workshop held in Grenoble, France. 1997 ISBN 0-7923-4687-4

KLUWER ACADEMIC PUBLISHERS – DORDRECHT / BOSTON / LONDON